Communications in Computer and Information Science 1824

Editorial Board Members

Joaquim Filipe, *Polytechnic Institute of Setúbal, Setúbal, Portugal*
Ashish Ghosh, *Indian Statistical Institute, Kolkata, India*
Raquel Oliveira Prates, *Federal University of Minas Gerais (UFMG), Belo Horizonte, Brazil*
Lizhu Zhou, *Tsinghua University, Beijing, China*

Rationale

The CCIS series is devoted to the publication of proceedings of computer science conferences. Its aim is to efficiently disseminate original research results in informatics in printed and electronic form. While the focus is on publication of peer-reviewed full papers presenting mature work, inclusion of reviewed short papers reporting on work in progress is welcome, too. Besides globally relevant meetings with internationally representative program committees guaranteeing a strict peer-reviewing and paper selection process, conferences run by societies or of high regional or national relevance are also considered for publication.

Topics

The topical scope of CCIS spans the entire spectrum of informatics ranging from foundational topics in the theory of computing to information and communications science and technology and a broad variety of interdisciplinary application fields.

Information for Volume Editors and Authors

Publication in CCIS is free of charge. No royalties are paid, however, we offer registered conference participants temporary free access to the online version of the conference proceedings on SpringerLink (http://link.springer.com) by means of an http referrer from the conference website and/or a number of complimentary printed copies, as specified in the official acceptance email of the event.

CCIS proceedings can be published in time for distribution at conferences or as post-proceedings, and delivered in the form of printed books and/or electronically as USBs and/or e-content licenses for accessing proceedings at SpringerLink. Furthermore, CCIS proceedings are included in the CCIS electronic book series hosted in the SpringerLink digital library at http://link.springer.com/bookseries/7899. Conferences publishing in CCIS are allowed to use Online Conference Service (OCS) for managing the whole proceedings lifecycle (from submission and reviewing to preparing for publication) free of charge.

Publication process

The language of publication is exclusively English. Authors publishing in CCIS have to sign the Springer CCIS copyright transfer form, however, they are free to use their material published in CCIS for substantially changed, more elaborate subsequent publications elsewhere. For the preparation of the camera-ready papers/files, authors have to strictly adhere to the Springer CCIS Authors' Instructions and are strongly encouraged to use the CCIS LaTeX style files or templates.

Abstracting/Indexing

CCIS is abstracted/indexed in DBLP, Google Scholar, EI-Compendex, Mathematical Reviews, SCImago, Scopus. CCIS volumes are also submitted for the inclusion in ISI Proceedings.

How to start

To start the evaluation of your proposal for inclusion in the CCIS series, please send an e-mail to ccis@springer.com.

Bernabé Dorronsoro · Francisco Chicano ·
Gregoire Danoy · El-Ghazali Talbi
Editors

Optimization and Learning

6th International Conference, OLA 2023
Malaga, Spain, May 3–5, 2023
Proceedings

Springer

Editors
Bernabé Dorronsoro
University of Cadiz
Cadiz, Spain

Francisco Chicano
University of Malaga
Malaga, Spain

Gregoire Danoy
University of Luxembourg
Esch-sur-Alzette, Luxembourg

El-Ghazali Talbi
University of Lille
Lille, France

ISSN 1865-0929 ISSN 1865-0937 (electronic)
Communications in Computer and Information Science
ISBN 978-3-031-34019-2 ISBN 978-3-031-34020-8 (eBook)
https://doi.org/10.1007/978-3-031-34020-8

This Springer imprint is published by the registered company Springer Nature Switzerland AG
The registered company address is: Gewerbestrasse 11, 6330 Cham, Switzerland

Preface

This book compiles the best papers submitted to the Sixth International Conference on Optimization and Learning (OLA 2023). The choice was made according to the score of all accepted long papers in the blind review process of the conference. OLA 2023 took place in Malaga, Spain, May 3 to May 5. The main objective of OLA 2023 was to bring together influential researchers from all over the world in the fields of complex problems optimization, machine and deep learning, to benefit from synergies between the two research fields, and to promote their application to real-world problems. The conference offered a nice atmosphere where relevant researchers presented their innovative work.

Three categories of papers were considered in OLA 2023, namely ongoing research work, high-impact journal publications (both of them in the shape of an extended abstract), or regular papers with novel contents and important contributions. A selection of the best papers in this latter category is published in this book.

Sixty papers were presented at OLA 2023, arranged in eleven sessions, covering topics such as Deep Learning, the synergies between optimization and learning techniques, their application to problems with uncertainty, reinforcement learning, logistics, advanced optimization techniques, or parallelism. Also, five special sessions were organized:

1. Reinforcement Learning and (multi-objective) optimization. Organizers: Ann Nowé (VUB Brussels, Belgium) and Grégoire Danoy (University of Luxembourg, Luxembourg).
2. Optimisation and Learning in Energy Demand Site Management. Organizers: Gülgün Kayakutlu, M. Özgür Kayalica, and Üner Çolak (Istanbul Technical University Energy Institute, Turkey).
3. Computational Intelligence for Smart Cities. Organizers: Jamal Toutouh and Christian Cintrano (Universidad de Málaga, Spain), Sergio Nesmachnow, and Renzo Massobrio (Universidad de la República, Uruguay).
4. Advanced Methods for Anomalies Forecasting and Detection. Organizers: M. Pavone, F. Zito, C. Cavallaro, and V. Cutello (University of Catania, Italy).
5. Artificial Intelligence for Sustainability. Organizers: Bernabé Dorronsoro, Juan Carlos de la Torre, Jose Miguel Aragón, and Javier Jareño (University of Cádiz, Spain).

The conference received a total of 78 papers, from which 32 compose this book, making 41% of all submitted papers.

May 2023

Bernabé Dorronsoro
Francisco Chicano
Gregoire Danoy
El-Ghazali Talbi

Organization

Conference Chair

Francisco Chicano	University of Malaga, Spain

Conference Program Chair

Grégoire Danoy	University of Luxembourg, Luxembourg

Workshops and Invited Sessions Chair

Mario Pavone	University of Catania, Italy

Conference Steering Committee Chair

El-Ghazali Talbi	University of Lille & Inria, France

Finance Chairs

Amir Nakib	Paris-East Créteil University, France
Rachid Ellaia	EMI, Mohammed V University, Morocco

Proceedings Chair

Bernabe Dorronsoro	University of Cadiz, Spain

Organizing Committee

Rachid Ellaia	EMI & Mohammed V University, Morocco
Thomas Firmin	University of Lille, France
Gabriel Luque	University of Malaga, Spain

Jamal Toutouh	University of Malaga, Spain
Zakaria Dahi	University of Malaga, Spain

Publicity Chairs

Juan J. Durillo	Leibniz Supercomputing Center, Germany
Grégoire Danoy	University of Luxembourg, Luxembourg
Javier Ferrer	University of Malaga, Spain

Program Committee

Lionel Amodeo	Université de Technologie de Troyes, France
Taha Arbaoui	Université de Technologie de Troyes, France
Mehmet-Emin Aydin	University of Bedfordshire, UK
Mathieu Balesdent	ONERA, France
Abou El Hassan Benyamina	University of Oran, Algeria
Pascal Bouvry	University of Luxembourg, Luxembourg
Claudia Cavallaro	University of Catania, Italy
Krisana Chinnasarn	Burapha University, Thailand
Christian Cintrano	University of Malaga, Spain
Zakaria Dahi	University of Malaga, Spain
Grégoire Danoy	University of Luxembourg, Luxembourg
Patrick de Causmaecker	KU Leuven, Belgium
Javier del Ser	TECNALIA, Spain
Bernabe Dorronsoro	Universidad de Cadiz, Spain
Rachid Ellaia	EMI, Morocco
Bogdan Filipic	Jožef Stefan Institute, Slovenia
José Manuel García Nieto	University of Málaga, Spain
Domingo Jimenez	University of Murcia, Spain
Hamamache Kheddouci	University of Lyon, France
Peter Korosec	Jožef Stefan Institute, Slovenia
Andrew Lewis	Griffith University, Australia
Francisco Luna	University of Malaga, Spain
Gabriel Luque	University of Malaga, Spain
Teodoro Macias Escobar	Technological Institute of Ciudad Madero, Mexico
Roberto Magán-Carrión	University of Granada, Spain
Renzo Massobrio	Universidad de la República, Uruguay
Gonzalo Mejia Delgadillo	Pontificia Universidad Católica de Valparaíso, Chile

Nouredine Melab	University of Lille, France
Edmondo Minisci	University of Strathclyde, UK
José Ángel Morell	University of Malaga, Spain
Amir Nakib	University of Paris-Est Créteil, France
Antonio J. Nebro	University of Málaga, Spain
Sergio Nesmachnow	Universidad de la República, Uruguay
Eneko Osaba Icedo	TECNALIA, Spain
Gregor Papa	Jožef Stefan Institute, Slovenia
Mario Pavone	University of Catania, Italy
Helena Ramalhinho	Universitat Pompeu Fabra, Spain
Rubén Saborido	University of Malaga, Spain
Roberto Santana	University of the Basque Country, Spain
Alejandro Santiago	Polytechnic University of Altamira, Mexico
Nadiya Schvai	Cyclope.ai, France
Andrei Tchernykh	CICESE, Mexico
Jamal Toutouh	University of Malaga, Spain
Alice Yalaoui	Université de Technologie de Troyes, France
Farouk Yalaoui	Université de Technologie de Troyes, France
Xin-She Yang	Middlesex University London, UK

Contents

Learning Methods to Enhance Optimization Tools

Optimization Applied to Learning Methods

Real-World Applications

Advanced Optimization

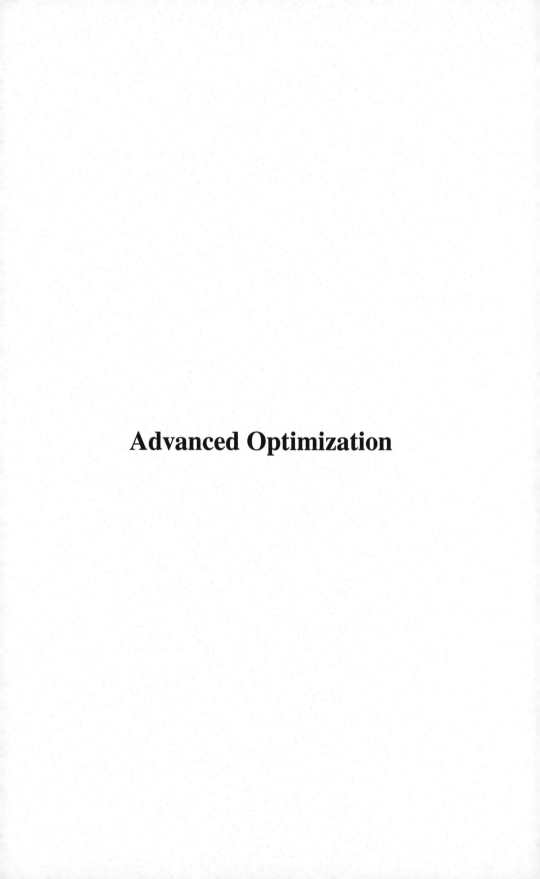

A Comparative Study of Fractal-Based Decomposition Optimization

T. Firmin[1(✉)] and E-G. Talbi[2]

[1] Centre de Recherche en Informatique, Signal et Automatique de Lille, Lille, France
thomas.firmin@univ-lille.fr

[2] University of Lille & INRIA, Lille, France
el-ghazali.talbi@univ-lille.fr

Abstract. In this work, we present a comparative study of 24 different and unique decomposition-based algorithms derived from Fractal Decomposition Algorithm and Simultaneous Optimistic Optimization. These algorithms were built within a generic, flexible and unified algorithmic framework named fractal-based decomposition algorithms. This generic framework is issued from previous works and is succinctly described in this paper. A software, called Zellij, based on this methodology was used to instantiate the 24 algorithms. Under our generic framework, fractal-based decomposition algorithms are made of five independent and well-defined search components: a type of fractal, a tree search algorithm, a scoring method, an exploration, and exploitation strategies. This new family of algorithms, hierarchically decomposes an initial search space using a generic geometrical object, named fractal. The decomposition forms a rooted tree, where fractals are nodes, and the root corresponds to the initial search space. The tree is explored, exploited and expanded using the four other search components. The proposed algorithms were tested and compared to each other on the CEC2020 benchmark. Obtained performances emphasize the impact of each search component, and pointed out the scalability capacity of certain algorithms. Our results strongly suggest that some search components have major impact on FDA and SOO-based algorithms for large-scale problems, whereas others are used to fine tune performances in terms of convergence.

Keywords: Continuous optimization · Hierarchical decomposition · Fractals

1 Introduction

We consider a black-box, non-convex, derivative free and non-linear continuous optimization problem defined by $f : \mathcal{S} \subset \mathbb{R}^n \to \mathbb{R}$:

$$\hat{x} \in \underset{x \in \mathcal{S}}{\operatorname{argmin}} f(x) \tag{1}$$

B. Dorronsoro et al. (Eds.): OLA 2023, CCIS 1824, pp. 3–20, 2023.
https://doi.org/10.1007/978-3-031-34020-8_1

where $\hat{x}$ is the global optima, f the objective function, and $\mathcal{S}$ the search space made of upper and lowers bounds.

Generally, the most popular optimization algorithms used to tackle these problems are of various families. One can distinguish evolutionary algorithms, making a population of solutions converge toward optima, such as differential evolution, genetic algorithm or swarm intelligence [2]. Surrogate-based optimization algorithms, for example, Bayesian optimization [6], can be used to tackle such problems when the objective function gets computationally expensive. Nonetheless, in this paper, we focus on a taxonomic group of optimization techniques, inspired and inherited from *divide-and-conquer* methods. We claim that these flexible and scalable algorithms can overcome some bottlenecks encountered with previous algorithms when tackling high dimensional problems.

In prior works, we generalized algorithms based on the decomposition of the decision space. These algorithms were from the optimization, machine learning and computational intelligence research communities. Such as DIRECT [8], SOO [18], FDA [19] or FRACTOP [4]. We call this family of metaheuristics *fractal-based decomposition* algorithms. Our generalized, flexible and unified framework is made of five different independent and well-defined search components, their combination allows instantiating various *fractal-based decomposition* algorithms. We built a software called *Zellij*[1] which allows to easily instantiate and modify decomposition based metaheuristics. We reproduced previous algorithms and some of their variations, such as Locally Biased DIRECT or DIRECT-Restart [9].

Along these lines, we focused on the comparison of two algorithms, FDA and SOO. We were able to instantiate 24 different versions by modifying the five search components. The results highlight the behaviors and significance of all 24 fractal decomposition-based algorithms in terms of sensitivity to dimension, convergence, and to search components. Furthermore, we noticed that we can adapt search components according to the problem difficulty, so to obtain various behaviors.

The paper is organized as follows. In Sect. 2, a recall of our flexible and generalized framework for fractal-based decomposition algorithm is presented, as well as the five search components. In Sect. 3, we describe FDA and SOO, which are the methods behind the 22 other algorithms instantiated with Zellij. Then, in Sect. 4, all 24 algorithms are presented in terms of search components, some of their properties are also discussed. In Sects. 5 and 6, we explained the selection of the benchmark, the experimental setup, and we present and discuss the performances of the algorithms on the CEC2020 benchmark. Finally, Sect. 7, concludes this work by summarizing our contributions and discussing future works.

[1] https://github.com/ThomasFirmin/zellij.

2 A Recall on Generalized Fractal-Based Decomposition Algorithm

Fractal-based decomposition algorithms can be divided into five generic, unique and independent search components: fractal, tree search, scoring, exploration and exploitation strategies. Their combination allows to quickly instantiates fractal-based decomposition algorithms.

For instance, to instantiate FRACTOP [4], one can use the following combination of search components:

- *Fractal*: Hypercube.
- *Tree search*: Best First Search [3].
- *Scoring*: Belief.
- *Exploration*: Genetic algorithm.
- *Exploitation*: Simulated annealing.

This family of algorithms is based on a hierarchical partition of the search space, using a self-similar object named *fractal*. In our framework a fractal is a generalized abstract object describing a high dimensional *geometrical object*, a *subset* of an initial search space or of another fractal, and a *node* of a tree. Fractals are stored in a k-ary rooted tree, where the initial search space corresponds to the root. Each fractal can be decomposed into k smaller fractals, named children, by using a decomposition function F. Hence, all fractals contain references to their children, and children are a partition of their parents. The tree search component, τ, allows selecting non-expanded fractal within the k-ary rooted tree. A fractal is considered as expanded when the exploration search component, $Explor$, is applied within the selected fractal, and when its children are created. This search component allows to quickly gather information about a fractal, it can be sampling methods, metaheuristics or other optimization algorithms. To determine how promising a fractal is, a scoring component, γ, using gathered information computes the quality value for each fractal. Moreover, the k-ary rooted tree, has a maximum depth D, so, once a leave of this tree reaches level D, instead of the exploration, an exploitation algorithm, $Exploi$, is applied to emphasize the search within a promising area. The exploitation is not restricted by the boundaries of a fractal, so it can freely converge toward local optima.

The pseudocode, with the five search components F, τ, γ, $Explor$ and $Exploi$ is resumed in Algorithm 1. In Fig. 1, the workflow of fractal-based decomposition algorithms is presented. Search components are depicted in blue. The two tests, in orange, correspond to the stopping criterion and the depth test of a fractal (*line 13* and *line 18* in Algorithm 1). There are two additional inputs for fractal-based decomposition algorithms, which are the initial search space $\mathcal{S}$ and the maximum depth D of the k-ary rooted tree.

The following subsections will dive deep into the five search components, their functions, and behaviors.

2.1 Fractal Component

Considering a fractal-based decomposition algorithm, a fractal, in this context, is a self-similar and self-contained object, used to perform a hierarchical partition of the search space. The self-contained property of a fractal ensure that all necessary information describing the object is locally contained within the fractal. Thus, a fractal, is an independent subspace of the initial search space $\mathcal{S}$. In Algorithm 1, the function F takes a fractal, and returns its k children. We can identify several fractal types according to the literature such as hypercubes [4], hyperspheres [19], hyperrectangles from a trisection [8,18] or even Voronoï cells [11,13].

Fractals are categorized by five properties describing their behavior within our framework. We distinguish the *coverage*, which describes the space covered by the hierarchical partition. We then have the *overlapping* between fractals, the *building* and *memory* complexities. These complexities have a major impact on the scalability of fractal-based decomposition algorithms, for example, Voronoï cells are hard to build in high dimensions [10,14]. Finally, we have the *partition size* property, which describes the number of children per fractal. This property has also an impact on the scalability, such as, in FRACTOP [4] where the number of smaller hypercubes of equal size needed to partition their parent has an exponential complexity of 2^n, as the dimension n increases. Or even simplices, where each fractal has $n!$ children [21]. These five properties are summarized in Table 1. The coverage is said to be complete when the partition fully covers the initial search space, and partial when only a part of $\mathcal{S}$ is covered. We can see in this table that there is no dominant, universal, fractal. When selecting a fractal, one has to make concessions on some properties.

2.2 Tree Search Component

Generated fractals are nodes of a $k-$ary rooted tree, where the root corresponds to the entire initial search space. The tree search algorithm allows manipulating the rooted tree and efficiently expend promising fractals. Within this tree, a fractal is characterized by its level (i.e. depth). In Algorithm 1, this component is noted τ, it takes a list of fractals and their quality values, so to returns Q non-expanded fractals.

This search component determines some behaviors of fractal-based decomposition algorithms, such as the tradeoff between exploration and exploitation. In DIRECT, this problem is tackled by using the selection of all *potentially optimal rectangles* [8]. In FDA, a *sorted depth first search* quickly allows exploiting promising deep fractals. Other tree search algorithms can be worthless, such as Breadth First Search, as it will explore all fractals of a level before exploring fractals of the next level. We lose the notion of hierarchy within the partition. The same applied to Depth First Search, as the criterion to select the next fractal is its level and not its quality value. These algorithms can be replaced by Best First Search [3] and some of its variations, such as Beam Search [5] which allows tackling memory issues by pruning the tree. Epsilon Greedy Search [22] or Diverse Best First Search [7] add stochasticity to the tree search component. Different

Table 1. Properties of fractals

Fractal	Hypercube	Trisection	Hypersphere	Voronoï	Simplex
Partition size (k)	2^n	3	$2n$	c*	$n!$
Building complexity	$\mathcal{O}(2^n)$	$\mathcal{O}(n)$	$\mathcal{O}(n)$	$\mathcal{O}(2^n)$**	$\mathcal{O}(n!)$
Coverage of $\mathcal{S}$ by the partition	complete	complete	partial	complete	complete
Overlapping	no	no	yes	no	no
Data structure	2 points of size n	2 points of size n	2 points of size n	See***	n points of size n

* Number of centroids defined by the user
** Valid for usual algorithms, we can reduce this complexity by approximating the Voronoï diagram in high dimensions. Here we consider the complexity depending on the dimension n, but it also depends on the number of centroids.
*** It can be a set of vertices for the QuickHull algorithm or a set of hyperplanes for sampling methods.

algorithms allow selecting the exploration-exploitation tradeoff, such as Cyclic Best First Search [16]. One can also tackle multi-objective problems by using Pareto front selection [15]. These algorithms share a common structure called the `OPEN-CLOSED` list algorithm [3]. The `OPEN` list contains all non-expanded fractals, and the `CLOSED` list contains all expanded (i.e. explored) fractals.

2.3 Scoring Component

To introduce a notion of hierarchy between fractals, we need to assign a quality value to all of them. This component can be seen as an acquisition function used in Bayesian optimization [6], with the difference that it provides information about how promising a fractal is. This value is determined by information obtained by the exploration component, and used by the tree search algorithm to select non-expanded fractals within the `OPEN` list. Measures can be of different natures, some will be statistics about solutions sampled by the exploration component within a fractal, such as the minimum objective value [8,18], the median or the mean. Some use global information, such as in FDA [19] with the distance to the best solution found so far. Others introduce inheritance or uncertainty, such as Belief in FRACTOP [4]. In Algorithm 1, γ takes a fractal, a list of sampled points, their objective values, and, it returns the quality value of the given fractal.

2.4 Exploration and Exploitation Components

The exploration search component defines how to sample within a fractal so that the fractal-based decomposition algorithm can efficiently get information on the landscape and quality of this fractal. One can use sampling methods, metaheuristics or other optimization algorithms. Once, done, the quality of the fractal is computed thanks to the scoring component which uses prior sampled information. For some fractal-based algorithms, this component can be very basic such as in DIRECT or SOO, where centers of all fractals are computed. In FDA, the *Promising Hypersphere Selection* computes three fixed points inside each hypersphere. Other fractal-based decomposition algorithms use active methods. In FRACTOP, a genetic algorithm is used in each fractal.

Some decomposition-based algorithms can suffer from a lack of exploitation (e.g., DIRECT [9]). The exploitation search component is applied to nodes of maximum depth. This search component is not restricted to the fractal boundaries, so it can search within non generated fractals. For instance, in FRACTOP, a simulated annealing is used, whereas in FDA, a coordinate local search is applied. In Algorithm 1, *Explor* and *Exploi* take a fractal and return a list of solutions and their objective values.

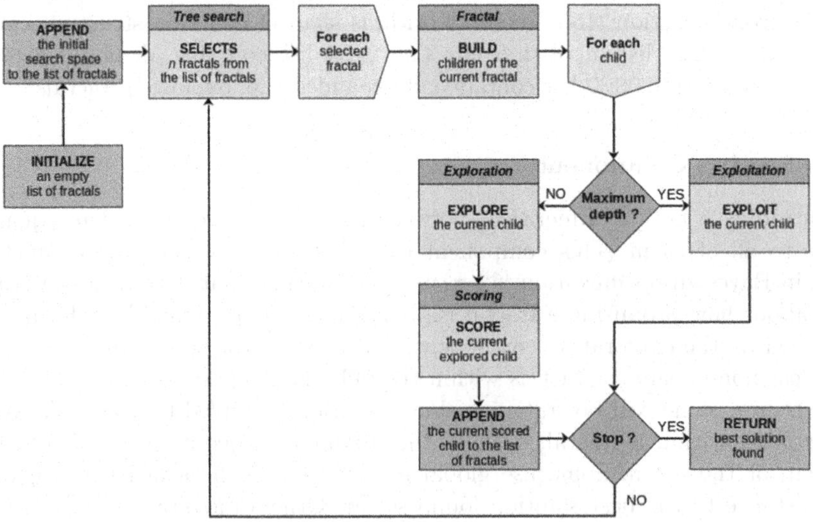

Fig. 1. Workflow of fractal-based decomposition algorithms

Algorithm 1. Fractal-based decomposition algorithm

```
Inputs:
  S                                           Initial search space
  D                                           Maximum depth
  F                                           Function to build fractals
  Explor                                      Exploration strategy
  Exploi                                      Exploitation strategy
  τ                                           Tree search
  γ                                           Scoring
Outputs: x̂                                    Best solution found
  x̂ ← ∞
  OPEN ← [S]                                  List of non-expanded fractals
  CLOSED ← [·]                                List of expanded fractals
  current ← S
  scores ← [+∞]
  while stopping criterion not reached do
      for each leaf ∈ current do
          children ← F(leaf)                  Decompose the leaf
          for each child ∈ children do
              Append child to OPEN
              if level(child) < D then
                  P, values ← Explor(child)
                  score ← γ(child, P, values)
                  Append child to OPEN
                  Append score to scores
                  if min(values) < x̂ then
                      x̂ ← min(values)
              else
                  values ← Exploi(child)
                  if min(values) < x̂ then
                      x̂ ← min(values)
          Append leaf to CLOSED
          index ← Index of leaf in OPEN
          Remove element at index from OPEN
          Remove element at index from scores
      current ← τ(OPEN, scores)
  return x̂
```

3 Related Works

This section describes two popular fractal-based optimization algorithms according to the five search components. In this work, FDA and SOO were used and modified to create new fractal-based decomposition algorithms to better understand the functionalities and behaviors of their search components.

3.1 FDA: Fractal Decomposition Algorithm

The FDA algorithm solves the *curse of dimensionality* problem of FRACTOP [19]. Instead of a hypercubes-based decomposition, it uses hyperspheres. By using such fractals, the decomposition has a lower complexity, but at the cost of overlapping fractals due to an inflation ratio. This ratio partly reduces the lack of space coverage implied by hyperspheres decomposition. The exploration component, called *promising hypersphere selection*, computes three points: the center of the hypersphere and two opposite points equidistant to the center. The heuristic, used to score a fractal, is the *distance-to-the-best* solution found so far. Finally, a leaf at the maximum depth level of the tree, is exploited with an *Intensive Local Search*, which is a coordinate descent algorithm with adaptive step size.

3.2 SOO: Simultaneous Optimistic Optimization

The DOO (Deterministic Optimistic Optimization) and SOO algorithms assume the existence of a semi-metric l, and simplify the Lipschitz-continuous property by only using a *local smoothness* assumption around the global optimum $\hat{x}$ [18]:

$$f(\hat{x}) - f(x) \leq l(\hat{x}, x), \quad \forall x \in \mathcal{S}$$

DOO is used when l is known; otherwise, SOO is more adapted. Both algorithms are deterministic. At each iteration and at each level of the partition tree, the best fractal is selected according to the evaluation of a representative solution inside it (e.g. center). Here, the balance between exploration and exploitation relies on a particular tree search algorithm, and on a heuristic value computed for each fractal, according to one representative solution. In addition, a stochastic version called Sto-SOO has been designed for noisy loss function, where each fractal has to be evaluated multiple times [23].

4 Instantiation of Fractal-Based Decomposition Algorithms

Five properties describe the selected algorithms. They are:

- **Deterministic:** two different runs of the same algorithm on the same noiseless function should give the same results.
- **Axis-aligned:** they sample solutions or set up fractals in an axis-aligned fashion.
- **Symmetrical:** they sample solutions or set up fractal symmetrically.
- **Structure:** they have the same algorithmic structure as described in *Zellij*. They can be divided into the five search components of our generalized and flexible framework.
- **Non-distributed:** We use the non-parallel version of these algorithms, as it can alter their behaviors [20].

Therefore, we choose FDA [19] and SOO [18] as a base for the other 22 new versions obtained by modifying some of their five search components. These two algorithms come from two different communities, and have drastically different behaviors and purposes. Moreover, our choice is supported by the fact that the author in [18] claims that SOO is a generalization of DIRECT. All 24 algorithms are described by search components in Table 1, these search components are described in depth within the following subsections.

4.1 Extensions of Fractal Components

Because the 22 instantiated new algorithms derivate from FDA and SOO, we can distinguish two groups of algorithms, those using hyperspheres to decompose the search space, and those using trisections. Compared to trisection, hyperspheres have the advantage of simultaneously reducing all dimensions, whereas trisections only reduce the longest side of the parent fractal. However, hyperspheres suffer from a low coverage capability. An inflation ratio tries to overcome this behavior by increasing the surface of hyperspheres to the detriment of overlapping fractals. It is important to mention that both hypersphere and trisection suffer from the curse of dimensionality, and so, do not scale well when the dimension increases. Indeed, the Hausdorff measure (n-volume) of a hypersphere tends to 0 as the dimension tends to infinity, meaning that a hypersphere covers less and less space. Concerning the trisection, the reduction of the search space by the children becomes insignificant, as only one dimension is reduced at a time, meaning that SOO needs a deeper and deeper tree to significantly reduce the search space. Figure 2 depicts visual explanations of fractals in FDA and SOO. However, one should be careful and not infer what happens in high dimensions by only looking at these figures, as high dimensional geometry is counterintuitive.

4.2 Extensions of Tree Search Components

As described before, FDA uses the Move-up tree search procedure, and SOO selects the best fractal at each level of the tree. We switched these two algorithms from FDA to SOO. The new versions are named FDA-SOO, and SOO-MoveUp. The goal, here, is to determine if the performances of SOO and FDA can be explained by their respective tree search component. We also tried the Potentially Optimal Rectangle (POR) from DIRECT, which was adapted to Hypersphere by using the radius as a measure of the size of the fractal. Algorithms, using POR, are named FDA-POR and SOO-POR. Moreover, we tried some algorithms coming from the A^* family, such as, Best First Search (BFS) to emphasize the search on the most promising fractal, or Beam Search (BS)[2] introducing a pruning technique. We have also implemented Cyclic Best First Search (CBFS) for SOO and FDA.

[2] Here the beam length was set to 3000.

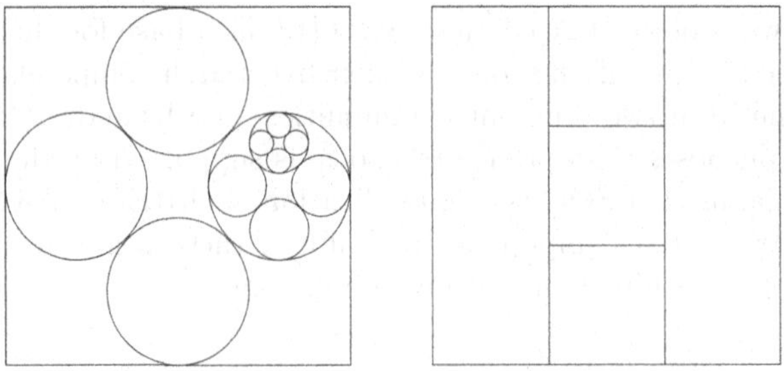

(a) FDA hyperspheres (b) SOO trisections

Fig. 2. Examples of fractals in 2 dimensions

4.3 Extensions of Exploration and Exploitation Components

The FDA algorithm explores a fractal by using the Promising Hypersphere Selection (PHS). The PHS samples three points within a given hypersphere; its center at level l, $\overrightarrow{C}^l$, and two other points $s_1 = \overrightarrow{C}^l + \alpha\frac{r_l}{n}$ and $s_2 = \overrightarrow{C}^l + \alpha\frac{r_l}{n}$, with r_l the radius at level l, α the inflation ratio[3] and n the dimension. SOO only samples the center of each fractal. So, once again, here, the objective is to determine if the exploration component explains the performance of FDA and SOO. Because we cannot directly apply the PHS to SOO, we decided to compute the center and two fixed points on the diagonal of each hyperrectangle. This version is called SOO-Diagonal. We have also instantiated FDA-Center, where only the center of the hypersphere is computed. We also tried to sample $10.n$ solutions using chaos with a Henon map sampling [1]. Algorithms using this method are named FDA-Chaos and SOO-Chaos.

Concerning the exploitation search component, SOO does not use any. So, we decided to instantiate the Intensive Local Search (ILS) from FDA to SOO, by using the longest side of the hyperrectangle as the radius in the ILS. Thus, we have two new algorithms SOO-ILS, and FDA-NoILS[4] for which no ILS is used. Here, we want to test the impact of a local optimizer on fractal-based decomposition algorithms.

4.4 Extensions of Scoring Components

To analyze the behaviors of the scoring component, we tried the Distance-To-The-Best (DTTB) solution found so far from FDA, the mean and the minimum objective value from sampled points within a fractal. For SOO, because it only computes the center, the minimum, and the mean were also used with SOO-Diagonal and SOO-DMean. Moreover, we applied a fuzzy measure named Belief from FRACTOP, the new algorithms are named FDA-Belief and SOO-Belief.

[3] α was set to 1.75.

[4] The maximum depth of the tree was set to 600.

5 Experimental Setup

We have tested the 24 algorithms on the CEC2020 mono-objective continuous benchmark [12], for dimensions 10, 15, 20, 30, 50 and 100. Functions of this benchmark are characterized by five important properties for fractal-based decomposition algorithms. Indeed, functions should not have their global optimum in the center or on an axis of the search space, if so, then the function should be shifted and/or rotated. For example, SOO, at the first iteration, samples one point directly into the center of the initial search space. Then, most of the CEC2020's functions are multimodal, it creates multiple local optima, and so allows analyzing the exploration capability of optimization algorithms. Moreover, functions should not be separable. Indeed, SOO reduces only one dimension at a time, so a focus can be given to only a few dimensions. We want to evaluate if fractal-based optimization algorithms can optimize all dimensions at the same time. Finally, functions have to be asymmetrical because SOO and FDA, sample points and fractals symmetrically. In our experiments, the CEC2020's functions are considered as black box. The budget is set to $5000n$ calls to the objective function.

To statistically compare performances of instantiated algorithms, we applied a two-sided Wilcoxon signed-rank test on the regrets, $r = f(\hat{x}) - f(x^*)$, where $\hat{x}$ is the global known optimum of the evaluated objective function, and x^* is the best solution found by the optimization algorithm. We applied an error rate $\alpha = 0.05$ for the statistical test.

6 Results Analysis

The results, showed in Fig. 3, can be read column by column. For example, the first column represents the performances of FDA, compared to the 23 other algorithms (rows). If the color is gray, then there is no statistical evidence that FDA is better than the algorithm at the current row. If it is green, then $\alpha <$ %5, and the rank of the algorithm at the current column is higher than the rank of the selected row, and conversely if the color is red. Two representative convergence graphs are shown in Fig. 4 and Fig. 5 (on dimensions 50 and 100 for the Composition 3 function), which allow to better understand the classification depicted in Fig. 3 (Tables 2 and 3).

Table 2. Instantiated algorithms using Zellij

#	Algorithms	Geometry	Tree search	Exploration	Exploitation	Scoring
1	FDA	Hyperpshere	MoveUp [19]	Promising Hypersphere Selection [19]	Intensive Local Search [19]	Distance to the best [19]
2	FDA-BFS	Hyperpshere	Best First Search	Promising Hypersphere Selection [19]	Intensive Local Search [19]	Distance to the best [19]
3	FDA-BS	Hyperpshere	Beam Search	Promising Hypersphere Selection [19]	Intensive Local Search [19]	Distance to the best [19]
4	FDA-CBFS	Hyperpshere	Cyclic Best First Search	Promising Hypersphere Selection [19]	Intensive Local Search [19]	Distance to the best [19]
5	FDA-POR	Hyperpshere	Potentially Optimal Rectangle* [8]	Promising Hypersphere Selection [19]	Intensive Local Search [19]	Distance to the best [19]
6	FDA-SOO	Hyperpshere	Best fractal at each level [18]	Promising Hypersphere Selection [19]	Intensive Local Search [19]	Distance to the best [19]
7	FDA-Belief	Hyperpshere	MoveUp [19]	Promising Hypersphere Selection [19]	Intensive Local Search [19]	Belief [4]
8	FDA-mean	Hyperpshere	MoveUp [19]	Promising Hypersphere Selection [19]	Intensive Local Search [19]	Mean
9	FDA-min	Hyperpshere	MoveUp [19]	Promising Hypersphere Selection [19]	Intensive Local Search [19]	Minimum
10	FDA-center	Hyperpshere	MoveUp [19]	Promising Hypersphere Selection [19]	Intensive Local Search [19]	Distance to the best [19]
11	FDA-Chaos	Hyperpshere	MoveUp [19]	Henon map sampling**	Intensive Local Search [19]	Distance to the best [19]
12	FDA-NoILS	Hyperpshere	MoveUp [19]	Promising Hypersphere Selection [19]	∅	Distance to the best [19]
13	SOO	Trisection	Best fractal at each level [18]	Hyperrectangle Center	∅	Minimum
14	SOO-BFS	Trisection	Best fractal at each level [18]	Hyperrectangle Center	∅	Minimum
15	SOO-BS	Trisection	Beam Search	Hyperrectangle Center	∅	Minimum
16	SOO-CBFS	Trisection	Cyclic Best First Search	Hyperrectangle Center	∅	Minimum
17	SOO-MoveUp	Trisection	MoveUp [19]	Hyperrectangle Center	∅	Minimum
18	SOO-POR	Trisection	Potentially Optimal Rectangle [8]	Hyperrectangle Center	∅	Minimum
19	SOO-Belief	Trisection	Best fractal at each level [18]	Hyperrectangle Center	∅	Belief [4]
20	SOO-DTTB	Trisection	Best fractal at each level [18]	Hyperrectangle Center	∅	Distance to the best [19]
21	SOO-Chaos	Trisection	Best fractal at each level [18]	Henon map sampling	∅	Minimum
22	SOO-Diagonal	Trisection	Best fractal at each level [18]	3 points on a Diagonal	∅	Minimum
23	SOO-DMean	Trisection	Best fractal at each level [18]	3 points on a Diagonal	∅	Mean
24	SOO-ILS	Trisection	Best fractal at each level [18]	Hyperrectangle Center	Intensive Local Search [19]	Minimum

* Adapted to hyperspheres, by using the radius as the measure of the size of the fractal.
** Adapted to hyperspheres by using the Box-Muller method [17].

Table 3. Functions of the CEC2020 benchmark [12]

#	Function	Shifted	Rotated	Unimodality	Separability	Symmetrical
1	Bent Cigar	Yes	Yes	Yes	No	Yes
2	Schwefel	Yes	Yes	No	No	No
3	Lunacek Bi-Rastrigin	Yes	Yes	No	No	No
4	Rosenbrock + Griewangk	No	No	Yes	No	Yes
5	Hybrid 1	No	Yes	No	No	No
6	Hybrid 2	No	Yes	No	No	No
7	Hybrid 3	No	Yes	No	No	No
8	Composition 1	Yes	Yes	No	No	No
9	Composition 2	Yes	Yes	No	No	No
10	Composition 3	Yes	Yes	No	No	No

6.1 Sensitivity to the Fractal Search Component

Results suggest that there is an unequivocal difference between FDA and SOO-based algorithms. There is a clear dominance of FDA-based versions on SOO ones for dimensions 50 and 100. For lower dimensions FDA and SOO appear to be equivalent solutions on the CEC2020 benchmark. More fractals types should be tested to have a better idea of their impacts, as FDA and SOO were specifically designed for these geometrical objects.

6.2 Sensitivity to the Tree Search and Scoring Search Components

There is no clear evidence that modifying the tree search from FDA and SOO, by other similar and efficient tree search algorithms, radically improves performances. We can notice, in dimension 100, that using POR decreases performances of the original SOO. And, using the tree search from SOO with FDA also decreases FDA's performances. In low dimensions, instantiating BFS, BS, Move up from FDA, and POR, makes SOO worse.

Concerning the scoring search component FDA and SOO appear to be robust to the modification of this component (except for SOO-DMean), so performances of FDA and SOO might be explained by the fractal, exploration, and exploitation search components. Tree search and scoring components seem to refine FDA and SOO performances, as it can have a little impact on the convergence.

6.3 Sensitivity to the Exploitation and Exploration Search Components

In high dimensions, FDA-Chaos appears to be the worse FDA version. One can explain these performances by looking at convergence plots in Fig. 4 and Fig. 5. Indeed, for dimension 50 we can see that the ILS is applied later compared to other algorithms. One can observe a fast decrease in the error starting at approximately, 200000 evaluations. Whereas for dimension 100, the ILS is never applied.

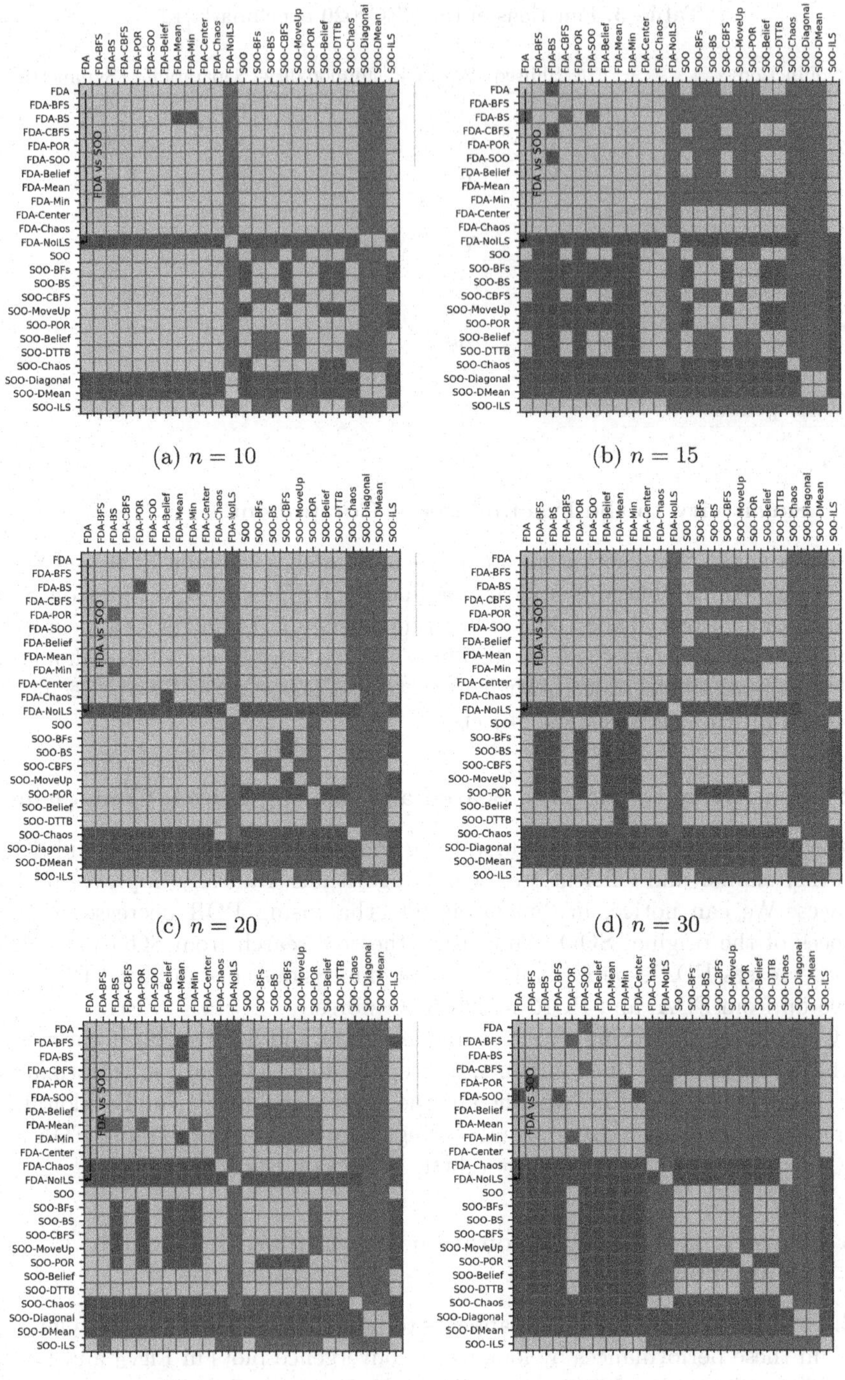

(a) $n = 10$ (b) $n = 15$

(c) $n = 20$ (d) $n = 30$

(e) $n = 50$ (f) $n = 100$

Fig. 3. Pair Wise Wilcoxon test comparison on CEC2020. **Gray:** Statistically unsignificative ($\alpha > 0.05$). **Green:** Better. **Red:** Worse (Color figure online)

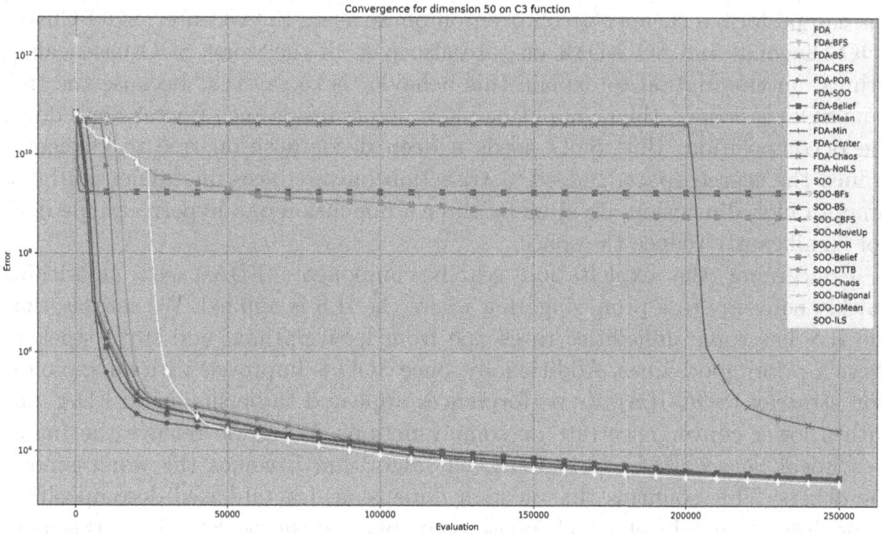

Fig. 4. Convergence plot of the 24 algorithms for dimension 50 with Composition 3 function.

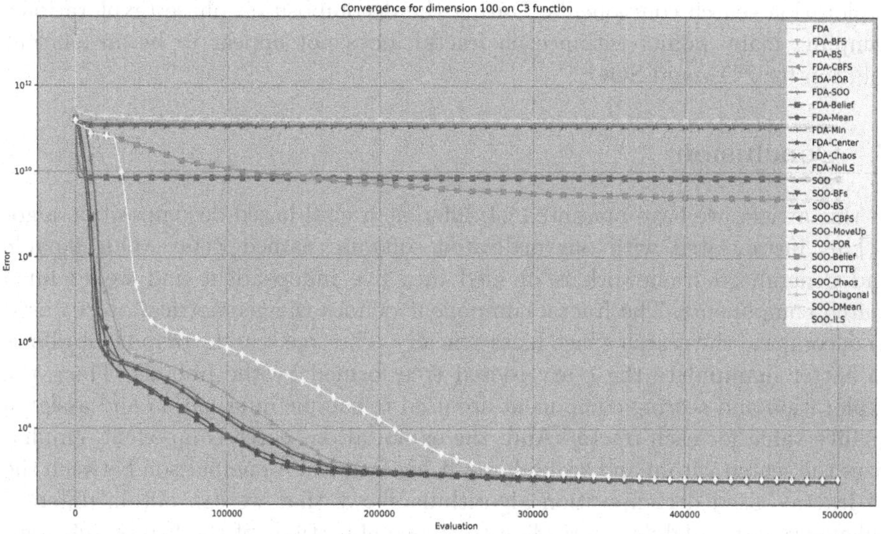

Fig. 5. Convergence plot of the 24 algorithms for dimension 100 with Composition 3 function.

Indeed, the number of sampled points gets higher when the dimension increases, so in dimension 100 for FDA-Chaos, all the budget is consumed on exploring fractals. We can notice that, computing only the center of the hypersphere with FDA-Center, does not decrease in any dimension performances of FDA. Thus, one can perform a cheaper decomposition phase using FDA-Center. SOO-Chaos, SOO-Diagonal and SOO-DMeans, are also part of the worst SOO-based algorithms, an assumption to explain this behavior is to say that because the tree from SOO is deeper, then computing more points inside each fractal slows down the decomposition. But, SOO needs a deep decomposition tree to efficiently reduce the search space. Indeed, a trisection only reduces one dimension at a time. In high dimension, dividing by three a dimension of a hyperrectangle does not significantly reduce the space.

Concerning the exploitation search component, FDA-based algorithms quickly converge to a promising area where the ILS is applied. We can say that the ILS has some difficulties to escape from local optima, and stays stuck in these a priori good areas. Additionally, once SOO is improved with an exploitation strategy (SOO-ILS), its performances appeared to be similar to FDA, but with a lower convergence rate on some functions. When we remove the Intensive Local Search (ILS) from FDA, the algorithm becomes the worst among the others. This confirms that in high dimension, fractal-based decomposition algorithms need a local search phase to emphasize the search within the most promising areas found by the decomposition.

Finally, we can say that FDA and SOO performances are mainly explained by first, the presence or not of an exploitation search component, and then by the exploration search components. During the decomposition phase (exploration), sampling many points within each fractal, does not appear to be an efficient solution for FDA and SOO.

7 Conclusion

In this article, we have presented 24 different fractal-based decomposition algorithms instantiated with our dedicated software named Zellij[5]. Our flexible and generalized framework is divided into five independent and well-defined search components. The fractal component defines the geometrical object used to decompose the search space hierarchically. The tree search component allows to better manipulate the k-ary rooted tree formed by the fractals. Then, the exploration and scoring component are used to sample information and assign a quality value for each fractal. And, the exploitation search component, emphasizes the search within and around promising fractals. A comparison between the 24 fractal-based decomposition algorithms shows that we can obtain different behaviors and scalability according to the combinations of the five search components. Results highlight that some search components (e.g. exploration and exploitation) have a higher impact on the algorithm performances compared to their original versions. The analysis strongly suggests, that in high dimensions, a local optimizer appears to be necessary to improve fractal-based decomposition algorithm performances.

Future work will focus on the development of new scalable search components, particularly the exploration and exploitation component, to reduce the

[5] https://github.com/ThomasFirmin/zellij.

allocated budget and improve the exploration capability. Furthermore, a massively asynchronous framework for fractal-based decomposition algorithms is under development for multi-nodes distributed environments.

Acknowledgment. Experiments presented in this paper were carried out using the Grid'5000 testbed (see https://www.grid5000.fr) supported by a scientific interest group hosted by Inria and including CNRS, RENATER and several Universities as well as other organizations.

This work has been supported by the University of Lille and the ANR-20-THIA-0014 program AI_PhDLille.

References

1. Aslimani, N., El-ghazali, T., Ellaia, R.: A new chaotic-based approach for multi-objective optimization. Algorithms **13**(9) (2020). https://doi.org/10.3390/a13090204. https://www.mdpi.com/1999-4893/13/9/204
2. Bansal, J.C., Singh, P.K., Pal, N.R. (eds.): Evolutionary and Swarm Intelligence Algorithms. SCI, vol. 779. Springer, Cham (2019). https://doi.org/10.1007/978-3-319-91341-4
3. Dechter, R., Pearl, J.: Generalized best-first search strategies and the optimality of A*. J. ACM **32**(3), 505–536 (1985). https://doi.org/10.1145/3828.3830
4. Demirhan, M., Özdamar, L., Helvacıoğlu, L., Birbil, I.: FRACTOP: a geometric partitioning metaheuristic for global optimization. J. Glob. Optim. **14**(4), 415–436 (1999). https://doi.org/10.1023/A:1008384329041
5. Frohner, N., Gmys, J., Melab, N., Raidl, G.R., Talbi, E.G.: Parallel beam search for combinatorial optimization (extended abstract). In: Proceedings of the International Symposium on Combinatorial Search, vol. 15, no. 1, pp. 273–275 (2022). https://doi.org/10.1609/socs.v15i1.21783. https://ojs.aaai.org/index.php/SOCS/article/view/21783
6. Garnett, R.: Bayesian Optimization. Cambridge University Press, Cambridge (2023)
7. Imai, T., Kishimoto, A.: A Novel Technique for Avoiding Plateaus of Greedy Best-First Search in Satisficing Planning, vol. 2 (2011)
8. Jones, D., Perttunen, C., Stuckman, B.: Lipschitzian optimisation without the Lipschitz constant. J. Optim. Theory Appl. **79**, 157–181 (1993). https://doi.org/10.1007/BF00941892
9. Jones, D.R., Martins, J.R.R.A.: The DIRECT algorithm: 25 years later. J. Glob. Optim. **79**(3), 521–566 (2021). https://doi.org/10.1007/s10898-020-00952-6
10. Khachiyan, L., Boros, E., Borys, K., Elbassioni, K., Gurvich, V.: Generating all vertices of a polyhedron is hard. Discret. Comput. Geom. **39**(1–3), 174–190 (2008). https://doi.org/10.1007/s00454-008-9050-5
11. Khodabandelou, G., Nakib, A.: H-polytope decomposition-based algorithm for continuous optimization. Inf. Sci. **558**, 50–75 (2021). https://doi.org/10.1016/j.ins.2020.12.090. https://linkinghub.elsevier.com/retrieve/pii/S0020025521000232
12. Liang, J., Suganthan, P., Qu, B., Gong, D., Yue, C.: Problem definitions and evaluation criteria for the CEC 2020 special session on multimodal multiobjective optimization (2019). https://doi.org/10.13140/RG.2.2.31746.02247
13. Liu, H., Xu, S., Wang, X., Wu, J., Song, Y.: A global optimization algorithm for simulation-based problems via the extended direct scheme. Eng. Optim. **47** (2014). https://doi.org/10.1080/0305215X.2014.971777

14. Mitchell, S.A., et al.: Spoke-darts for high-dimensional blue-noise sampling. ACM Trans. Graph. **37**(2) (2018). https://doi.org/10.1145/3194657
15. Mockus, J.: On the pareto optimality in the context of Lipschitzian optimization. Informatica Lith. Acad. Sci. **22**, 521–536 (2011). https://doi.org/10.15388/Informatica.2011.340
16. Morrison, D., Sauppe, J., Zhang, W., Jacobson, S., Sewell, E.: Cyclic best first search: using contours to guide branch-and-bound algorithms. Nav. Res. Logist. Q. **64**(1), 64–82 (2017). https://doi.org/10.1002/nav.21732
17. Muller, M.E.: A note on a method for generating points uniformly on N-dimensional spheres. Commun. ACM **2**(4), 19–20 (1959). https://doi.org/10.1145/377939.377946
18. Munos, R.: Optimistic optimization of a deterministic function without the knowledge of its smoothness. In: Shawe-Taylor, J., Zemel, R., Bartlett, P., Pereira, F., Weinberger, K. (eds.) Advances in Neural Information Processing Systems, vol. 24. Curran Associates, Inc. (2011)
19. Nakib, A., Ouchraa, S., Shvai, N., Souquet, L., Talbi, E.G.: Deterministic metaheuristic based on fractal decomposition for large-scale optimization. Appl. Soft Comput. **61**, 468–485 (2017). https://doi.org/10.1016/j.asoc.2017.07.042. https://www.sciencedirect.com/science/article/pii/S1568494617304623
20. Nakib, A., Souquet, L., Talbi, E.G.: Parallel fractal decomposition based algorithm for big continuous optimization problems. J. Parallel Distrib. Comput. **133**, 297–306 (2019). https://doi.org/10.1016/j.jpdc.2018.06.002. https://www.sciencedirect.com/science/article/pii/S0743731518304003
21. Paulavičius, R., Žilinskas, J.: Simplicial Lipschitz optimization without the Lipschitz constant. J. Global Optim. **59**(1), 23–40 (2013). https://doi.org/10.1007/s10898-013-0089-3
22. Valenzano, R.A., Xie, F.: On the completeness of best-first search variants that use random exploration. In: AAAI Conference on Artificial Intelligence (2016)
23. Valko, M., Carpentier, A., Munos, R.: Stochastic simultaneous optimistic optimization. In: 30th International Conference on Machine Learning, ICML 2013, vol. 28 (2013)

Diagonal Barzilai-Borwein Rules in Stochastic Gradient-Like Methods

Giorgia Franchini[1(✉)], Federica Porta[1], Valeria Ruggiero[2], Ilaria Trombini[3], and Luca Zanni[1]

[1] Dipartimento di Scienze Fisiche, Informatiche e Matematiche, Università di Modena e Reggio Emilia, Via Campi, 213A, 41125 Modena, Italy
{giorgia.franchini,federica.porta,luca.zanni}@unimore.it

[2] Dipartimento di Matematica e Informatica, Università di Ferrara, via Machiavelli, 30, 44121 Ferrara, Italy
rgv@unife.it

[3] Dipartimento di Scienze Matematiche, Fisiche e Informatiche, Università di Parma, Parco Area delle Scienze, 7/A, 43124 Parma, Italy
ilaria.trombini@unife.it

Abstract. Minimization problems involving a finite sum as objective function often arise in machine learning applications. The number of components of the finite-sum term is typically very large, by making unfeasible the computation of its gradient. For this reason stochastic gradient methods are commonly considered. The performance of these approaches strongly relies on the selection of both the learning rate and the mini-batch size employed to compute the stochastic direction. In this paper we combine a recent idea to select the learning rate as a diagonal matrix based on stochastic Barzilai-Borwein rules together with an adaptive subsampling technique to fix the mini-batch size. Convergence results of the resulting stochastic gradient algorithm are shown for both convex and non-convex objective functions. Several numerical experiments on binary classification problems are carried out to compare the proposed method with other state-of-the-art schemes.

Keywords: Stochastic gradient methods · Diagonal Barzilai-Borwein rules · Variance reduced methods

1 Introduction

In this paper we consider the following optimization problem

$$\min_{x\in\mathbb{R}^d} F(x) \equiv \frac{1}{N}\sum_{i=1}^{N} f_i(x), \tag{1}$$

G. Franchini, F. Porta, V. Ruggiero, I. Trombini and L. Zanni—These authors contributed equally to this work.

B. Dorronsoro et al. (Eds.): OLA 2023, CCIS 1824, pp. 21–35, 2023.
https://doi.org/10.1007/978-3-031-34020-8_2

where each $f_i : \mathbb{R}^d \rightarrow \mathbb{R}$ is a smooth function. This problem arises in many machine learning applications where it is known as empirical risk minimization. The definition of the empirical risk $F(x)$ is based on a random sample $\xi^{(N)} = \{\xi_1^{(N)}, ..., \xi_N^{(N)}\}$ of size N of a random variable ξ whose probability distribution is unknown. Each $f_i(x) \equiv f(x, \xi_i^{(N)})$ denotes the loss function related to the instance $\xi_i^{(N)}$ of this sample, called train set. We are especially interested in the case when the number of components N is very large, and, hence, the adoption of stochastic gradient methods is convenient. Indeed they exploit either a single gradient ∇f_i or a very limited number of them at each iteration, rather than the entire gradient ∇F. Recently a number of stochastic methods has been developed with the aim to adaptively estimate the hyper-parameters introduced in these iterative schemes and to improve the effectiveness of the approach. In particular, we consider the iteration of the standard stochastic gradient method

$$x^{(k+1)} = x^{(k)} - \alpha g_{\mathcal{N}_k}^{(k)}, \tag{2}$$

where, given a randomly selected subset $\mathcal{N}_k \subseteq \mathcal{N} \equiv \{1, \ldots, N\}$ of size $N_k || N$, the stochastic gradient is defined as follows

$$g_{\mathcal{N}_k}^{(k)} = \frac{1}{N_k} \sum_{i \in \mathcal{N}_k} \nabla f_i(x^{(k)})$$

and α is a positive learning rate. In [1,2] the authors devise a technique to estimate the learning rate α by a version of the well-known Barzilai-Borwein (BB) rules, tailored for the stochastic approach and in [3] the authors used BB rules in a stochastic framework to threshold the learning rate. We recall that the standard BB rules are the solutions of the following problems:

$$\alpha_k^{BB1} = \underset{\alpha}{\operatorname{argmin}} \, \|\frac{1}{\alpha} s^{(k-1)} - y^{(k-1)}\|^2 = \frac{{s^{(k-1)}}^T s^{(k-1)}}{{s^{(k-1)}}^T y^{(k-1)}}, \tag{3}$$

$$\alpha_k^{BB2} = \underset{\alpha}{\operatorname{argmin}} \, \|s^{(k-1)} - \alpha y^{(k-1)}\|^2 = \frac{{s^{(k-1)}}^T y^{(k-1)}}{{y^{(k-1)}}^T y^{(k-1)}}, \tag{4}$$

where $s^{(k-1)} = x^{(k)} - x^{(k-1)}$ and $y^{(k-1)} = \nabla F(x^{(k)}) - \nabla F(x^{(k-1)})$. In the stochastic framework, the basic idea is to keep fixed the estimate of the learning rate in a cycle of m inner steps and, at the start of a new cycle, to update the value of α by using the iterates at the end of the last two subsequent cycles and their related gradients (or estimates of these). Variants of this approach are developed for variance reduced methods, such as SVRG, SAG [2] or mS2GD [4]. Since these schemes perform a full gradient evaluation over the whole dataset per epoch, the full gradients in two subsequent outer iterations are involved in the computation of BB approximation. More recently [5], in the context of variable metric proximal gradient iterative methods, the authors propose to replace the learning rate with a diagonal scaling matrix U_k^{-1} derived by properly adjusting the secant conditions (3) and (4). In particular, the minimization is carried out with respect to a diagonal matrix, the entries of the solution are constrained to the BB rules (as safeguarding policy) and a consistency term with the previous metric U_{k-1}, weighted by a positive parameter μ, is included in the objective function. The resulting updating rules for the diagonal matrices $(U_k^{-1})^{BB1}$ and $(U_k^{-1})^{BB2}$ are given as follows:

$$(U_k^{-1})_j^{BB1} = \text{mid}(\alpha_k^{BB2}, \frac{(s_j^{(k-1)})^2 + \mu}{y_j^{(k-1)} s_j^{(k-1)} + \mu (U_{k-1})_j^{BB1}}, \alpha_k^{BB1}), \quad j = 1, \ldots, d, \tag{5}$$

$$(U_k^{-1})_j^{BB2} = \text{mid}(\alpha_k^{BB2}, \frac{y_j^{(k-1)} s_j^{(k-1)} + \frac{\mu}{(U_{k-1})_j^{BB2}}}{(y_j^{(k-1)})^2 + \mu}, \alpha_k^{BB1}), \quad j = 1, \ldots, d, \tag{6}$$

where $\text{mid}(a, b, c) = \min(\max(a, b), c)$. This idea has been borrowed and adapted to the stochastic context in [6, 7], giving rise to the methods named **mS2GD-DBB** and **SRG-DBB-YOU**, respectively. Also in this case the diagonal scaling matrix is kept fixed along the internal iterations of a variance reduced method, as mS2GD or SARAH, so that two full gradients are involved in the updating rule of the scaling matrices. It is crucial to recall that to perform a full gradient evaluation over the dataset per epoch is very expensive and it is not employed in the practical deep learning applications and, more in general, in the big data framework, for its computational cost, the hardware memory constraints and the resulting high inefficiency when training a deep neural network. Moreover, a stochastic gradient algorithm which avoids the computation of the full gradient becomes necessary in the online learning scenarios, where data are not entirely available at the beginning of the training process as well as the full gradient [8, Sec. 2.2.2].

The aim of this paper is to introduce the scaling matrix technique in those stochastic gradient methods which control the variance of the stochastic gradients by means of adaptive subsampling strategies based on a suitable increasing of the mini-batch size employed for their computation, as for example in the method described in [9], known as **ASM** (see also [10]). In more detail, when suitable conditions, assuring that the negative of the current stochastic gradient is a descent direction in expectation, are not satisfied, the size of the mini-batch is increased until the aforementioned conditions are meet. The idea is to keep fixed the scaling matrix for all the iterations based on mini-batches of the same size; the scaling matrix is updated when an increase of the mini-batch size is required or the whole train set has been visited. In Sect. 2, we state the conditions on the stochastic scaled direction $-U^{-1} g_{\mathcal{N}_k}^{(k)}$ which enable to obtain theoretical convergence results. Under suitable assumption on U^{-1}, linear convergence is proved for $\mathbb{E}[F(x^{(k)}) - F^*]$ when F has gradient Lipschitz continuous gradient and satisfies the Polyak-Lojasiewicz condition, while sublinear convergence is obtained for convex F. Finally, we discuss convergence results for general functions. Section 3 is devoted to detail the practical implementation of an algorithm, named **ASM-DIAG**, which combines the adaptive subsampling technique suggested in [9] with the updating rules for the scaling matrices (5)–(6). Finally, in Sect. 4 we describe the results of an extensive numerical experimentation. The conclusions are drawn in Sect. 5.

Notation. We recall that, given two symmetric square matrices, $A \preceq B$ means that $\frac{x^T Ax}{x^T x} \leq \frac{x^T Bx}{x^T x}$ for all $x \neq 0$. Furthermore, we denote by $\lambda_{min}(A)$ and $\lambda_{max}(A)$ the minimum and the maximum eigenvalue of A respectively. Given a vector x, we denote by $\|x\|$ the standard Euclidean norm; if A is a symmetric and positive definite matrix, $\|x\|_A$ denotes the norm with respect to A, defined as $\sqrt{x^T Ax}$.

2 Scaled Stochastic Gradient Methods

We consider a stochastic gradient (SG) iteration where the role of the standard learning rate is taken by a scaling matrix U^{-1} which is assumed symmetric and positive definite:

$$x^{(k+1)} = x^{(k)} - U^{-1} g_{\mathcal{N}_k}^{(k)}, \tag{7}$$

where $g_{\mathcal{N}_k}^{(k)} = \frac{1}{N_k} \sum_{i \in \mathcal{N}_k} \nabla f_i(x^{(k)})$ is a stochastic gradient. As a standard assumption, we assume that the stochastic gradient at the current iterate is an unbiased estimate of the full gradient, i.e., $\mathbb{E}_k[g_{\mathcal{N}_k}] = \nabla F(x^{(k)})$, where $\mathbb{E}_k[\cdot]$ denotes the conditional expected value with respect to the σ-algebra generated by the information collected before iteration k, i.e., assuming $x^{(0)}, \ldots, x^{(k)}$ given.

Following the suggestions in [9–11], the size of the current mini-batch is selected so that suitable conditions are satisfied. In particular, it is required that $-U^{-1} g_{\mathcal{N}_k}^{(k)}$ is a descent direction at least in expectation, that is

$$\mathbb{E}_k \left[\nabla F(x^{(k)})^T U^{-1} g_{\mathcal{N}_k}^{(k)} \right] = \nabla F(x^{(k)})^T U^{-1} \nabla F(x^{(k)}) > 0. \tag{8}$$

Thus, to control the variance of the term on the left hand side, the value of N_k has to be large enough to assure that the following condition is satisfied:

$$\mathbb{E}_k \left[\left(\nabla F(x^{(k)})^T U^{-1} g_{\mathcal{N}_k}^{(k)} - \nabla F(x^{(k)})^T U^{-1} \nabla F(x^{(k)}) \right)^2 \right] \leq \theta^2 \left(\nabla F(x^{(k)})^T U^{-1} \nabla F(x^{(k)}) \right)^2, \tag{9}$$

for a prefixed value $\theta^2 > 0$. Moreover, we observe that, in view of the positive definiteness of U^{-1}, it is required that, at least in expectation, $g_{\mathcal{N}_k}^{(k)}$ and $\nabla F(x^{(k)})$ should not be U^{-1} conjugate; thus, N_k has to be large enough to ensure that the following condition holds:

$$\mathbb{E}_k [{w^{(k)}}^T U^{-1} w^{(k)}] \leq \nu^2 \nabla F(x^{(k)})^T U^{-1} \nabla F(x^{(k)}), \tag{10}$$

where $w^{(k)} = g_{\mathcal{N}_k}^{(k)} - \frac{{g_{\mathcal{N}_k}^{(k)}}^T U^{-1} \nabla F(x^{(k)})}{\nabla F(x^{(k)})^T U^{-1} \nabla F(x^{(k)})} \nabla F(x^{(k)})$ and $\nu^2 > 0$ is a prefixed value.

Now we perform the following additional assumptions:

A ∇F is L-Lipschitz continuous;

B the Polyak-Lojasiewicz (P-L) condition holds

$$\|\nabla F(x)\|^2 \geq 2c(F(x) - F^*), \quad \forall x \in \mathbb{R}^d, \tag{11}$$

where c is a positive constant and $F^* = \inf_{x\in\mathbb{R}^d} F(x)$;
C U^{-1} is a symmetric positive definite matrix such that $U^{-1} \preceq \frac{1}{L(1+\theta^2+\nu^2)I}$ (or equivalently $U \succeq L(1+\theta^2+\nu^2)I$), for given constants θ, ν in (9) and (10).

We remark that assumption B holds when F is c-strongly convex, but it is also satisfied for other functions that are not convex (see [12]). In addition we observe that assumptions A and B do not guarantee the existence of a stationary point for F; nevertheless, under the two assumptions, any stationary point x^* for F is a global minimizer and $F^* = F(x^*)$.

Lemma 1. *Under the Assumptions A and C, we have*

$$\mathbb{E}_k[F(x^{(k+1)})] \leq F(x^{(k)}) - \frac{1}{2}\nabla F(x^{(k)})^T U^{-1} \nabla F(x^{(k)}). \tag{12}$$

Proof. From the general equality $\mathbb{E}[\|z - \mathbb{E}[z]\|^2] = \mathbb{E}[\|z\|^2] - \|\mathbb{E}[z]\|^2$, we obtain from (9) the inequality

$$\mathbb{E}_k[(\nabla F(x^{(k)})^T U^{-1} g_{\mathcal{N}_k}^{(k)})^2] \leq (1+\theta^2)(\nabla F(x^{(k)})^T U^{-1} \nabla F(x^{(k)}))^2. \tag{13}$$

From (10), we can write

$$\begin{aligned} \mathbb{E}_k[{g_{\mathcal{N}_k}^{(k)}}^T U^{-1} g_{\mathcal{N}_k}^{(k)}] &\leq \frac{\mathbb{E}_k[(\nabla F(x^{(k)})^T U^{-1} g_{\mathcal{N}_k}^{(k)})^2]}{\nabla F(x^{(k)})^T U^{-1} \nabla F(x^{(k)})} + \nu^2 \nabla F(x^{(k)})^T U^{-1} \nabla F(x^{(k)}) \\ &\leq (1+\theta^2+\nu^2) \nabla F(x^{(k)})^T U^{-1} \nabla F(x^{(k)}), \end{aligned} \tag{14}$$

where the last inequality follows from (13). In view of (7) and the L-Lipschitz continuity of ∇F, we have that

$$F(x^{(k+1)}) \leq F(x^{(k)}) - \nabla F(x^{(k)})^T U^{-1} g_{\mathcal{N}_k}^{(k)} + \frac{L}{2}\|U^{-1} g_{\mathcal{N}_k}^{(k)}\|^2. \tag{15}$$

By taking the conditional expectation on both sides of the last inequality, recalling (8) and the assumption $U^{-1} \preceq \frac{1}{L(1+\theta^2+\nu^2)} I$, we can write

$$\begin{aligned} \mathbb{E}_k[F(x^{(k+1)})] &\leq F(x^{(k)}) - \nabla F(x^{(k)})^T U^{-1} \nabla F(x^{(k)}) \\ &\quad + \frac{1}{2(1+\theta^2+\nu^2)} \mathbb{E}_k[g_{\mathcal{N}_k}^{(k)} U^{-1} U U^{-1} g_{\mathcal{N}_k}^{(k)}] \\ &\leq F(x^{(k)}) - \left(1 - \frac{1+\theta^2+\nu^2}{2(1+\theta^2+\nu^2)}\right) \nabla F(x^{(k)})^T U^{-1} \nabla F(x^{(k)}) \end{aligned} \tag{16}$$

where the last inequality follows from (14). □

Theorem 2. *Suppose Assumptions A and B hold. Let $\{x^{(k)}\}$ be the sequence generated by (7), where the size N_k of any sub-sample is chosen so that the conditions (9) and (10) are fulfilled and U^{-1} satisfies the assumption C. Then, we have that*

$$\mathbb{E}[F(x^{(k)}) - F^*] \leq (1-\rho)^k (F(x^{(0)}) - F^*), \tag{17}$$

where $\rho = c\,\lambda_{min}(U^{-1}) < 1$.

Proof. From the properties of symmetric and positive definite matrices and Assumption B we can write

$$\begin{aligned} \nabla F(x^{(k)})^T U^{-1} \nabla F(x^{(k)}) &\geq \lambda_{min}(U^{-1}) \|\nabla F(x^{(k)})\|^2 \\ &\geq 2c\lambda_{min}(U^{-1})(F(x^{(k)}) - F^*). \end{aligned} \tag{18}$$

Consequently, by subtracting F^* from both members of inequality (12) in Lemma 1 and using (18), we can write

$$\begin{aligned} \mathbb{E}_k[F(x^{(k+1)}) - F^*] &\leq (F(x^{(k)}) - F^*) - \frac{1}{2}\nabla F(x^{(k)})^T U^{-1} \nabla F(x^{(k)}) \\ &\leq (1 - c\lambda_{min}(U^{-1}))(F(x^{(k)}) - F^*). \end{aligned} \tag{19}$$

We set $\rho = c\lambda_{min}(U^{-1})$; then, by taking the total expectation in the last inequality, we obtain the linear convergence to 0 of $\{\mathbb{E}[F(x^{(k+1)}) - F^*]\}$

$$\mathbb{E}[F(x^{(k+1)}) - F^*] \leq (1-\rho)^{k+1}(F(x^{(0)}) - F^*).$$

□

In the case of a convex function F, we can state the following theorem.

Theorem 3. *Suppose Assumption A holds. Let $\{x^{(k)}\}$ be the sequence generated by (7), where the size N_k of any sub-sample is chosen so that the conditions (9) and (10) are fulfilled and U^{-1} satisfies Assumption C, with $\lambda_{max}(U^{-1}) < \frac{1}{L(1+\theta^2+\nu^2)}$. Assume that $X^* = \underset{x}{\operatorname{argmin}} F(x) \neq \emptyset$ and the function F is convex. Then, we have that*

$$\min_{0 \leq k \leq K} \mathbb{E}[F(x^{(k)}) - F(x^*)] \leq \frac{1}{2(1-\gamma)K}\|x^{(0)} - x^*\|_U^2, \tag{20}$$

where $x^* \in X^*$ and $\gamma = \lambda_{max}(U^{-1})L(1+\theta^2+\nu^2)$.

Proof. Assume $x^* \in X^*$. We have

$$\begin{aligned} &\mathbb{E}_k[\|x^{(k+1)} - x^*\|_U^2] = \\ &= \|x^{(k)} - x^*\|_U^2 + \mathbb{E}_k[\|x^{(k+1)} - x^{(k)}\|_U^2] + 2\mathbb{E}_k[(x^{(k+1)} - x^{(k)})^T U (x^{(k)} - x^*)] \\ &= \|x^{(k)} - x^*\|_U^2 - 2\mathbb{E}_k[g_{N_k}^{(k)}]^T (x^{(k)} - x^*) + \mathbb{E}_k[{g_{N_k}^{(k)}}^T U^{-1} g_{N_k}^{(k)}] \\ &\leq \|x^{(k)} - x^*\|_U^2 - 2\nabla F(x^{(k)})^T (x^{(k)} - x^*) + (1+\theta^2+\nu^2)\nabla F(x^{(k)})^T U^{-1} \nabla F(x^{(k)}) \\ &\leq \|x^{(k)} - x^*\|_U^2 - 2\nabla F(x^{(k)})^T (x^{(k)} - x^*) + (1+\theta^2+\nu^2)\lambda_{max}(U^{-1})\|\nabla F(x^{(k)})\|^2 \\ &\leq \|x^{(k)} - x^*\|_U^2 - 2\nabla F(x^{(k)})^T (x^{(k)} - x^*) + \frac{\gamma}{L}\|\nabla F(x^{(k)})\|^2 \end{aligned} \tag{21}$$

where we set $\gamma = L(1+\theta^2+\nu^2)\lambda_{max}(U^{-1})$; in the first equality we use (7); the second inequality follows from (14). From the assumption on $\lambda_{max}(U^{-1})$, $\gamma < 1$. In view of the convexity of F and the Lipschitz continuity of its gradient, the inequality $\|\nabla F(x^{(k)})\|^2 \leq 2L(F(x^{(k)}) - F(x^*))$ holds [13]. Thus, in view of this last inequality and again the convexity of F, we obtain

$$\mathbb{E}_k[\|x^{(k+1)} - x^*\|_U^2] \leq \|x^{(k)} - x^*\|_U^2 - 2(1-\gamma)(F(x^{(k)}) - F(x^*)). \quad (22)$$

By taking the total expectation, we can write

$$\mathbb{E}[F(x^{(k)}) - F(x^*)] \leq \frac{1}{2(1-\gamma)}(\mathbb{E}[\|x^{(k)} - x^*\|_U^2] - \mathbb{E}[\|x^{(k+1)} - x^*\|_U^2]). \quad (23)$$

Summing up for $k = 0, ..., K-1$ both the members of this last inequality, we obtain

$$\min_{0\leq k\leq K-1} \mathbb{E}[F(x^{(k)}) - F(x^*)] \leq \frac{1}{2(1-\gamma)K}\|x^{(0)} - x^*\|_U^2.$$

□

Finally, we consider the case of non-convex objective function. In this case, $\{\nabla F(x^{(k)})\}$ converges to zero in expectation, with a sub-linear rate of convergence of the smallest gradient arising after K iterations.

Theorem 4. *Suppose Assumption A holds and F is bounded below by F^*. Let $\{x^{(k)}\}$ be the sequence generated by (7), where the size N_k of any sub-sample is chosen so that the conditions (9) and (10) are fulfilled and U^{-1} satisfies Assumption C. Assume that $X^* = \operatorname{argmin}_x F(x) \neq \emptyset$. Then, we have that*

$$\lim_{k\to\infty} \mathbb{E}[\|\nabla F(x^{(k)})\|^2] = 0. \quad (24)$$

Proof. From Lemma 1, by taking the total expectation, we can write

$$\mathbb{E}[\|U^{-1}\nabla F(x^{(k)})\|_U^2] \leq 2\mathbb{E}[F(x^{(k)}) - F(x^{(k+1)})].$$

Summing up for $k = 0, ..., K-1$ both the members of this last inequality, we obtain

$$\begin{aligned}\sum_{k=0}^{K-1} \mathbb{E}[\|U^{-1}\nabla F(x^{(k)})\|_U^2] &\leq 2\mathbb{E}[F(x^{(0)}) - F(x^{(K)})] \\ &\leq 2(F(x^{(0)}) - F^*) < \infty.\end{aligned}$$

Thus, we conclude that

$$\begin{aligned}\min_{0\leq k\leq K-1} \mathbb{E}[\|\nabla F(x^{(k)})\|^2] &\leq \frac{1}{K\lambda_{min}(U^{-1})}\sum_{k=0}^{K-1} \mathbb{E}[\|U^{-1}\nabla F(x^{(k)})\|_U^2] \\ &\leq \frac{2}{K\lambda_{min}(U^{-1})}(F(x^{(0)}) - F^*). \quad (25)\end{aligned}$$

□

3 Practical Implementation

For the convergence results, conditions (9) and (10) have a crucial role. We observe that the left hand side terms of (9) and (10) are bounded from above by the true expectations of individual gradient $\nabla f_i(x^{(k)})$ of the sum in (1), so that (9) and (10) are satisfied when the following conditions involving the mini-batch size N_k hold:

$$\frac{\mathbb{E}_k[\nabla F(x^{(k)})^T U^{-1} \nabla f_i(x^{(k)}) - \|\nabla F(x^{(k)})\|^2_{U^{-1}}]}{N_k} \leq \theta^2 \|\nabla F(x^{(k)})\|^4_{U^{-1}}, \quad (26)$$

$$\frac{\mathbb{E}_k[{w_i^{(k)}}^T U^{-1} w_i^{(k)}]}{N_k} \leq \nu^2 \|\nabla F(x^{(k)})\|^2_{U^{-1}}, \quad (27)$$

where $w_i^{(k)} = \nabla f_i(x^{(k)}) - \frac{\nabla f_i(x^{(k)})^T U^{-1} \nabla F(x^{(k)})}{\nabla F(x^{(k)})^T U^{-1} \nabla F(x^{(k)})} \nabla F(x^{(k)})$. In order to implement the above conditions, the expectation values can be approximated by the sample expectations and the gradient $\nabla F(x^{(k)})$ on the right side by a sample gradient, so that the above conditions can be replaced by the following tests:

$$\frac{\sum_{i \in \mathcal{N}_k} ({g_{\mathcal{N}_k}^{(k)}}^T U^{-1} \nabla f_i(x^{(k)}) - \|g_{\mathcal{N}_k}^{(k)}\|^2_{U^{-1}})^2}{N_k(N_k - 1)} \leq \theta^2 \|g_{\mathcal{N}_k}^{(k)}\|^4_{U^{-1}}, \quad (28)$$

$$\frac{\sum_{i \in \mathcal{N}_k} (\tilde{w}_i^{(k)})^T U^{-1} \tilde{w}_i^{(k)}}{N_k(N_k - 1)} \leq \nu^2 \|g_{\mathcal{N}_k}^{(k)}\|^2_{U^{-1}}, \quad (29)$$

where $\tilde{w}_i^{(k)} = \nabla f_i(x^{(k)}) - \frac{\nabla f_i(x^{(k)})^T U^{-1} g_{\mathcal{N}_k}^{(k)}}{{g_{\mathcal{N}_k}^{(k)}}^T U^{-1} g_{\mathcal{N}_k}^{(k)}} g_{\mathcal{N}_k}^{(k)}$. When these conditions are not satisfied by the current sample size, the sample size is increased until (28) and (29) are satisfied.

Now we specify how to define the scaling matrix which multiplies the stochastic gradient. We call "cycle" the set of m_k steps ($i = 0, ..., m_k - 1$), where the mini-batch is fixed, i.e., the tests (28) and (29) are meet by the selected current mini-batch of size N_k. We impose that any cycle has at most a number of steps corresponding to a visit of the whole dataset (epoch). For any i-th step of a cycle ($i = 0, ..., m_k$), the following basic iteration is repeated

$$x^{(k,i+1)} = x^{(k,i)} - U_k^{-1} g_{\mathcal{N}_{k,i}}^{(k,i)}, \quad i = 0, ..., m_k - 1,$$

where $x^{(k,0)} = x^{(k)}$, $\mathcal{N}_{k,i}$ are subset of size N_k and $x^{(k+1)} = x^{(k,m_k-1)}$. For a whole cycle U_k^{-1} is kept fixed and it is selected by means of either (5) or (6) where $s^{(k-1)}$ and $y^{(k-1)}$ are defined as follows:

$$\begin{aligned} s^{(k-1)} &= x^{(k)} - x^{(k-1)} = x^{(k,0)} - x^{(k-1,0)} \\ y^{(k-1)} &= v^{(k)} - v^{(k-1)}. \end{aligned} \quad (30)$$

The vectors $\{v^{(k)}\}$ are approximations of the full gradient and are computed as in [2]; particularly, starting by $v^{(k)} = 0$,

$$v^{(k)} = \beta g_{\mathcal{N}_{k,i}}^{(k,i)} + (1 - \beta) v^{(k)},$$

where $\beta \in (0, 1)$ is a prefixed parameter.

Furthermore, as explained in [2], the standard BB rules are redefined in a stochastic framework as

$$\alpha_k^{BB1} = \text{mid}\left(\alpha_{min}, \frac{1}{m_{k-1}} \frac{s^{(k-1)^T} s^{(k-1)}}{s^{(k-1)^T} y^{(k-1)}}, \alpha_{max}\right) \tag{31}$$

$$\alpha_k^{BB2} = \text{mid}\left(\alpha_{min}, \frac{1}{m_{k-1}} \frac{s^{(k-1)^T} y^{(k-1)}}{y^{(k-1)^T} y^{(k-1)}}, \alpha_{max}\right) \tag{32}$$

where m_{k-1} is the number of steps of cycle where the size of the mini-batches is N_{k-1}.

For the initial two cycles, U_0^{-1} and U_1^{-1} are set as $\alpha_{ini} I$, with $\alpha_{ini} > 0$ chosen as a small value; furthermore, we set α_{min} and $\tilde{\alpha}_{max}$ as bound values (i.e., 10^{-5}, 10^4); $\alpha_{max} = \tilde{\alpha}_{max}\gamma_\ell$, where ℓ is the counter of the steps and $\{\gamma_\ell\} \subset \mathbb{R}$ is a decreasing sequence as $\mathcal{O}(\frac{1}{\ell})$ [1]. When $\alpha_k^{BB1} = \alpha_k^{BB2}$ or one of the two values is equal to α_{max}, a recovery cycle with $U_k^{-1} = \alpha_{ini} I$ is executed. We denote the described method as **ASM-DIAG-BB1** and **ASM-DIAG-BB2** in according to the rules (5) and (6) respectively used to update the diagonal matrix after a cycle with fixed mini-batch size.

4 Numerical Results

To evaluate the effectiveness of **ASM-DIAG** method, in this section we report the results of a set of numerical experiments aimed to obtain a binary classifier for some datasets with respect to different loss functions. In particular we study the behaviour of **ASM-DIAG** method in both convex and non-convex contexts with respect to the following competitor methods: the standard SG method with a fixed mini-batch size (50 elements) and optimal hand-tuned value for the learning rate, named **SG-mini 50**, the **ASM** method in [9], the **mS2GD-DBB** method in [6], the **SRG-DBB-YOU** method in [7].

We remark that **mS2GD-DBB** and **SRG-DBB-YOU** are hybrid methods since they use cyclically full gradient computations.

We compare the methods by considering the accuracy of the classification measured on the test set and the behaviour of the optimality gap with respect to the epochs. The optimality gap is defined as $|F(x^{(j)}) - F^*|$, where $F(x^{(j)})$ is the objective function computed at the epoch j on the whole train set and F^* is a ground truth value for the exact minimum of F, obtained by a huge number of iterations of a stochastic method.

For the experiments we consider three datasets with four different loss functions: two convex and two non-convex. Table 1 show the details of these datasets and the cardinality of the train and the test sets.

In the following we list the functions used in the objective function in (1):

- logistic regression (LR) loss:

$$f_i(x) = \ln\left[1 + e^{-b_i a_i^T x}\right];$$

Table 1. Features of each dataset.

Dataset	d	#train set (N)	#test set
w8a	300	44774	4975
IJCNN	22	49990	91701
RCV1	47236	20242	10000

- smooth hinge (SH) loss:

$$f_i(x) = \begin{cases} \frac{1}{2} - b_i a_i^T x, & \text{if } b_i a_i^T x \leq 0; \\ \frac{1}{2}(1 - b_i a_i^T x)^2, & \text{if } 0 < b_i a_i^T x < 1\ ; \\ 0, & \text{if } b_i a_i^T x \geq 1; \end{cases}$$

- sigmoid (SIG) loss:

$$f_i(x) = 1 - \tanh(b_i a_i^T x);$$

- logistic difference (LD) loss:

$$f_i(x) = \ln(1 + e^{-b_i a_i^T x}) - \ln(1 + e^{-b_i a_i^T x - 1});$$

where $a_i \in \mathbb{R}^d$ is the sample and $b_i \in \{+1, -1\}$ is the label of the i-th element of the dataset.

Hyper-Parameters Setting

Each of the five methods considered has different hyper-parameters to be set up as best as possible.

1. For **SG-mini 50** the mini-batch size is fixed to 50 and the best learning rate has been obtained by successive trials. We emphasize that this phase is computationally expensive, as it requires a run for each option tried.
2. For **ASM** the initial mini-batch size is set to 3, the initial learning rate α_0 is set to 10 and, using the notation in [9], $\theta = 0.7$, $\nu = 5.84$, $r = 10$, $\gamma = 0.38$, $\eta = 2$ and $\zeta_k = \zeta = 2$. For the dataset *IJCNN* combined with the SH loss, $\theta = 0.9$.
3. For **mS2GD-DBB** we set $\mu = 10^{-3}$ in (5) and the mini-batch size as 50; the initial value of the learning rate and the maximum dimension m of the internal cycle are tuned by a trial procedure.
4. For **SRG-DBB-YOU** we set $\mu = 10^{-3}$ in (6) and the mini-batch size as 50; the initial value of the learning rate and the maximum dimension m of the internal cycle are tuned by a trial procedure.
5. For **ASM-DIAG** we set $\beta = 0.9$, $\alpha_{ini} = 1$, $\alpha_{min} = 10^{-5}$, $\tilde{\alpha}_{max} = 10^4$, $\mu = 10^{-4}$, $\theta = 0.5$, $\nu = 3.82$ and $\gamma_\ell = 1/(0.1l + 1)$.

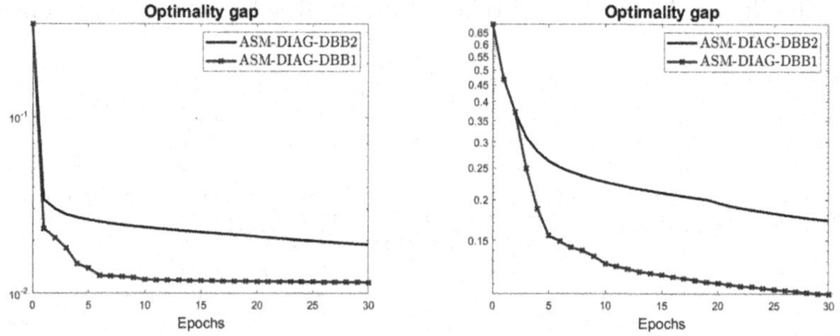

Fig. 1. Comparison between **ASM-DIAG-BB1** (rule (5)) and **ASM-DIAG-BB2** (rule (6)) in the case of the *w8a* dataset with the LD loss (left panel) and the *RCV1* dataset with the LR loss (right panel).

First Experiment

In Fig. 1, the two versions **ASM-DIAG-BB1** and **ASM-DIAG-BB2** of the proposed method are compared in the case of two test problems, i.e., the dataset *w8a* combined with the convex LR loss and the dataset *RCV1* combined with the non-convex LD loss. We highlight the greater performance of **ASM-DIAG-BB1** with respect to the version **ASM-DIAG-BB2**: this behaviour can be observed for all the considered test problems. For this reason, the version **ASM-DIAG-BB1** will always be used when comparing with the other methods.

We underline that to check conditions (28) and (29) can be very expensive from a computational point of view. Therefore, in the implementation of **ASM-DIAG**, this check is not performed at each iteration, but periodically also according to the size of the dataset.

Second Experiment

In this section we present the results of the comparison among all the considered methods. Due to the stochastic nature of the methods, we compute 10 runs with different pseudo-random number generators. Specifically for each combination dataset/loss, in Tables 2, 3, 4 and 5 we present the following metrics:

- average and STandard Deviation (STD) of the optimality gap $|F(\overline{x}) - F^*|$ evaluated on the train set, where $\overline{x}$ is the iterate at the end of the 30th epoch;
- average and STD of the accuracy $A(\overline{x})$ evaluated on the test set, at the end of the 30th epoch;
- the averaged execution time.

Table 2. Results for the LR loss function.

Method		*w8a*	*IJCNN*	*RCV1*
SG-mini 50				
	$\|F(\bar{x}) - F^*\|$	0.0009	0.0002	0.0543
	$\pm STD$	±0.0008	±0.0001	±0.0001
	$A(\bar{x})$	0.9062	0.9199	0.9640
	$\pm STD$	±0.0008	±0.0005	±0.0002
	Time	10.5714	4.9242	13.2564
ASM				
	$\|F(\bar{x}) - F^*\|$	0.0061	0.0002	0.0076
	$\pm STD$	±0.0009	$\pm 5.41e^{-5}$	±0.0006
	$A(\bar{x})$	0.9052	0.9193	0.9632
	$\pm STD$	±0.0014	±0.0004	±0.0008
	Time	28.8524	9.2412	328.0100
mS2GD-BB				
	$\|F(\bar{x}) - F^*\|$	0.0271	0.0038	0.2687
	$\pm STD$	±0.0019	±0.0006	±0.0004
	$A(\bar{x})$	0.8995	0.9168	0.9435
	$\pm STD$	±0.0012	±0.0012	±0.0005
	Time	10.8022	4.0593	10.2900
SRG-DBB-YOU				
	$\|F(\bar{x}) - F^*\|$	0.0654	$8.85e^{-6}$	0.0089
	$\pm STD$	±0.0096	$\pm 1.20e^{-7}$	0.0011
	$A(\bar{x})$	0.8930	0.9202	0.9648
	$\pm STD$	±0.0030	$\pm 2.83e^{-5}$	±0.0005
	Time	10.2375	8.6645	11.2300
ASM-DIAG				
	$\|F(\bar{x}) - F^*\|$	0.0109	0.0013	0.0960
	$\pm STD$	±0.0056	±0.0012	±0.0043
	$A(\bar{x})$	0.9043	0.9185	0.9616
	$\pm STD$	±0.0017	±0.0020	±0.0008
	Time	16.3327	4.6291	27.4500

Table 3. Results for the SH loss function.

Method		*w8a*	*IJCNN*	*RCV1*
SG-mini 50				
	$\|F(\bar{x}) - F^*\|$	0.0010	0.0005	0.0045
	$\pm STD$	±0.0005	±0.0002	±0.0003
	$A(\bar{x})$	0.9072	0.9225	0.9638
	$\pm STD$	±0.0012	±0.0011	±0.0003
	Time	9.7397	7.3017	86.1983
ASM				
	$\|F(\bar{x}) - F^*\|$	0.0134	0.0025	0.0506
	$\pm STD$	± 0.0011	±0.0005	±0.0028
	$A(\bar{x})$	0.9012	0.9169	0.9600
	$\pm STD$	±0.0007	±0.0006	±0.0006
	Time	18.1085	10.5134	89.3124
mS2GD-BB				
	$\|F(\bar{x}) - F^*\|$	0.0196	0.0030	0.0666
	$\pm STD$	±0.0011	±0.0015	±0.0003
	$A(\bar{x})$	0.8996	0.9203	0.9569
	$\pm STD$	±0.0010	±0.0024	±0.0007
	Time	18.9906	6.6724	190.8110
SRG-DBB-YOU				
	$\|F(\bar{x}) - F^*\|$	0.0051	0.0002	0.0020
	$\pm STD$	±0.0116	±0.0001	±0.0004
	$A(\bar{x})$	0.8833	0.9204	0.9628
	$\pm STD$	±0.0077	±0.0004	±0.0007
	Time	18.1700	6.2644	60.6231
ASM-DIAG				
	$\|F(\bar{x}) - F^*\|$	0.0039	0.0006	0.0212
	$\pm STD$	±0.0021	±0.0009	±0.0014
	$A(\bar{x})$	0.9061	0.9204	0.9654
	$\pm STD$	±0.0009	±0.0022	±0.0010
	Time	21.4431	5.5339	156.6770

The execution time reported in the tables does not take into account the preliminary tuning of the hyper-parameters. This phase is very expensive for **SG**, **mS2GD-DBB** and **SRG-DBB-YOU** while **ASM** and **ASM-DIAG** are more robust whit respect to the initial setting. Furthermore, the hybrid methods require the computation of the full gradient several times during the training; this may not be practicable in online learning or in contexts where the dataset is too large compared to the available hardware memory resources. Regarding the comparison between **ASM** and **ASM-DIAG**, we observe that in **ASM** the increase of the mini-batch size is very quickly. This event, especially with medium and large datasets such as *RCV1*, determines a consequent increase in execution time. Furthermore, a very large mini-batch size can grow memory traffic, possibly causing system crashes. In Figs. 2, 3 the comparison between **ASM-DIAG** and the considered methods is shown. The averaged optimality gap and the increase of the mini-batch size are reported. Figure 2 shows the results obtained for *w8a* with SIG loss. Although the behavior of **ASM-DIAG** does not match that of **ASM**, in **ASM** we observe a very large growth of the mini-batch size. As a consequence, the execution time of **ASM** is higher than that of **ASM-DIAG**. Figure 3 shows the experiment results carried out on *IJCNN* with SH loss. The best final performance is reached by the **SRG-DBB-YOU** method, even if its

Table 4. Results for the SIG loss function.

Method		*w8a*	*IJCNN*	*RCV1*
SG-mini 50				
	$\lvert F(\bar{x}) - F^*\rvert$	0.0075	$6.15e^{-7}$	0.0230
	$\pm STD$	±0.0058	$\pm 1.40e^{-7}$	± 0.0002
	$A(\bar{x})$	0.9046	0.9050	0.9650
	$\pm STD$	±0.0026	±0.0000	± 0.0004
	Time	11.0278	3.3114	18.3694
ASM				
	$\lvert F(\bar{x}) - F^*\rvert$	0.0085	0.0108	0.0054
	$\pm STD$	±0.0018	±0.0343	±0.0007
	$A(\bar{x})$	0.9053	0.8996	0.9643
	$\pm STD$	±0.0009	±0.0171	± 0.0007
	Time	25.3240	7.1951	347.4760
mS2GD-BB				
	$\lvert F(\bar{x}) - F^*\rvert$	0.0330	0.0015	0.1818
	$\pm STD$	±0.0000	±0.0008	± 0.0006
	$A(\bar{x})$	0.8949	0.9050	0.9511
	$\pm STD$	±0.0026	±0.0000	± 0.0005
	Time	8.2558	4.0018	8.7160
SRG-DBB-YOU				
	$\lvert F(\bar{x}) - F^*\rvert$	0.0782	$2.20e^{-5}$	0.0064
	$\pm STD$	±0.0202	$\pm 1.02e^{-5}$	±0.0012
	$A(\bar{x})$	0.8828	0.9050	0.9651
	$\pm STD$	±0.0077	±0.0000	±0.0006
	Time	8.3621	3.9150	7.6961
ASM-DIAG				
	$\lvert F(\bar{x}) - F^*\rvert$	0.0111	$2.91e^{-5}$	0.0372
	$\pm STD$	±0.0018	$\pm 5.51e^{-6}$	±0.0056
	$A(\bar{x})$	0.9046	0.9050	0.9644
	$\pm STD$	±0.0012	±0.0000	± 0.0008
	Time	13.1858	7.7255	28.7361

Table 5. Results for the LD loss function; the symbol * denotes a failure.

Method		*w8a*	*IJCNN*	*RCV1*
SG-mini 50				
	$\lvert F(\bar{x}) - F^*\rvert$	$4.32e^{-5}$	0.0329	0.0088
	$\pm STD$	$\pm 7.91e^{-5}$	±0.0113	± 0.0006
	$A(\bar{x})$	0.9068	0.9085	0.9657
	$\pm STD$	±0.0006	±0.0110	±0.0005
	Time	6.6691	5.7566	286.6630
ASM				
	$\lvert F(\bar{x}) - F^*\rvert$	0.0015	*	0.0031
	$\pm STD$	±0.0096	±*	± 0.0003
	$A(\bar{x})$	0.9065	0.6221	0.9643
	$\pm STD$	±0.0007	±0.2358	± 0.0006
	Time	16.2706	16.8134	747.0070
mS2GD-BB				
	$\lvert F(\bar{x}) - F^*\rvert$	0.0173	0.0380	0.1981
	$\pm STD$	±0.0007	±0.0001	± 0.0001
	$A(\bar{x})$	0.8978	0.9050	0.9324
	$\pm STD$	±0.0008	±0.0000	± 0.0004
	Time	6.1348	4.6669	273.6460
SRG-DBB-YOU				
	$\lvert F(\bar{x}) - F^*\rvert$	0.0063	0.0365	0.0045
	$\pm STD$	±0.0009	$\pm 4.30e^{-5}$	±0.0005
	$A(\bar{x})$	0.9025	0.9050	0.9654
	$\pm STD$	±0.0008	±0.0000	±0.0005
	Time	5.7727	3.7509	159.2970
ASM-DIAG				
	$\lvert F(\bar{x}) - F^*\rvert$	0.0115	0.0368	0.0502
	$\pm STD$	±0.0009	$\pm 3.16e^{-5}$	0.0026
	$A(\bar{x})$	0.8998	0.9050	0.9584
	$\pm STD$	±0.0004	±0.0000	±0.0011
	Time	11.3575	5.9963	369.9070

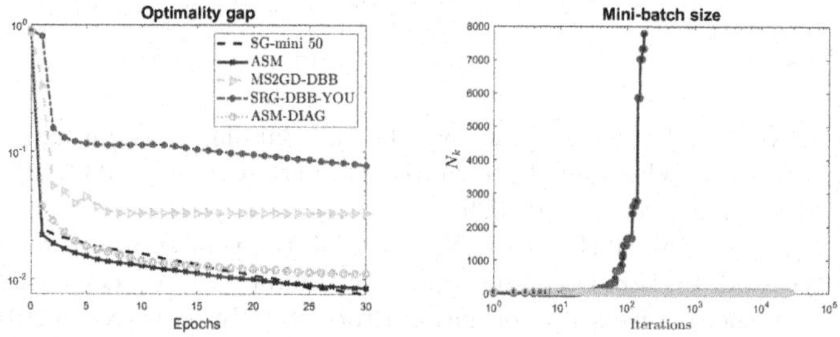

Fig. 2. *w8a* dataset with SIG loss: optimality gap (left panel) and increase of sample size in **ASM** and **ASM-DIAG** (right panel).

behaviour is the worst in the initial 10 epochs. The increase of the mini-batch size for **ASM** is again very large compared to the one for **ASM-DIAG**.

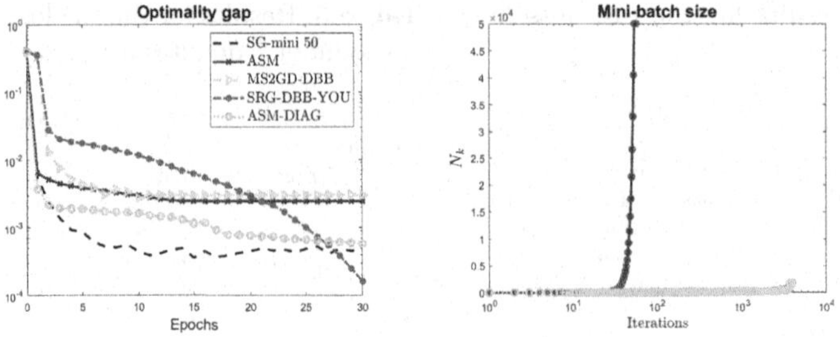

Fig. 3. *IJCNN* dataset with SH loss: optimality gap (left panel) and increase of sample size in **ASM** and **ASM-DIAG** (right panel).

5 Conclusions

In this paper we introduce the idea of selecting the learning rate by means of a diagonal scaling matrix in those stochastic gradient methods which reduce the variance of the stochastic directions through an adaptive increase of the mini-batch size, without any computation of the full gradient of the function to minimize. For the resulting algorithm we prove convergence results in both cases of convex and non-convex objective function. An extensive numerical experimentation has been carried out on binary classification problems in order to evaluate the effectiveness of the proposal. The numerical results show that the suggested approach appears robust with respect to the selection of the hyper-parameters involved in its definition and allows for a less rapid increase of the mini-batch size.

References

1. Liang, J., Xu, Y., Bao, C., Quan, Y., Ji, H.: Barzilai–Borwein-based adaptive learning rate for deep learning. Pattern Recogn. Lett. **128**, 197–203 (2019). https://doi.org/10.1016/j.patrec.2019.08.029
2. Tan, C., Ma, S., Dai, Y.H., Qian, Y.: Barzilai–Borwein step size for stochastic gradient method. In: Lee, D.D., Sugiyama, M., Luxburg, U.V., Guyon, I., Garnett, R. (eds.) Advances in Neural Information Processing Systems (NIPS 2016), vol. 29 (2016)
3. Franchini, G., Ruggiero, V., Trombini, I.: Thresholding procedure via Barzilai–Borwein rules for the steplength selection in stochastic gradient methods. In: Nicosia, G., et al. (eds.) LOD 2021. LNISA, vol. 13164, pp. 277–282. Springer, Cham (2022). https://doi.org/10.1007/978-3-030-95470-3_21
4. Yang, Z., Wang, C., Zang, Y., Li, J.: Mini-batch algorithms with Barzilai–Borwein update step. Neurocomputing **314**, 177–185 (2018)
5. Park, Y., Dhar, S., Boyd, S., Shah, M.: Variable metric proximal gradient method with diagonal Barzilai–Borwein stepsize. In: 2020 IEEE International Conference on Acoustics, Speech and Signal Processing (ICASSP), ICASSP 2020, pp. 3597–3601 (2020). https://doi.org/10.1109/ICASSP40776.2020.9054193

6. Yang, Z., Ma, L.: Adaptive stochastic gradient descent for large-scale learning problems (2022). https://orcid.org/0000-0002-1468-5042
7. Yu, T.T., Liu, X.W., Dai, Y.H., Sun, J.: A mini-batch proximal stochastic recursive gradient algorithm with diagonal Barzilai–Borwein stepsize. J. Oper. Res. Soc. China **11**, 277–307 (2022)
8. Bottou, L.: On-Line Learning and Stochastic Approximations, pp. 9–42. Cambridge University Press, USA (1999)
9. Bollapragada, R., Byrd, R., Nocedal, J.: Adaptive sampling strategies for stochastic optimization. SIAM J. Optim. **28**(4), 3312–3343 (2018)
10. Byrd, R.H., Chin, G.M., Nocedal, J., Wu, Y.: Sample size selection in optimization methods for machine learning. Math. Program. **1**(134), 127–155 (2012)
11. Franchini, G., Ruggiero, V., Zanni, L.: On the steplength selection in stochastic gradient methods. In: Sergeyev, Y.D., Kvasov, D.E. (eds.) NUMTA 2019. LNCS, vol. 11973, pp. 186–197. Springer, Cham (2020). https://doi.org/10.1007/978-3-030-39081-5_17
12. Karimi, H., Nutini, J., Schmidt, M.: Linear convergence of gradient and proximal-gradient methods under the Polyak-Łojasiewicz condition. In: Frasconi, P., Landwehr, N., Manco, G., Vreeken, J. (eds.) ECML PKDD 2016. LNCS (LNAI), vol. 9851, pp. 795–811. Springer, Cham (2016). https://doi.org/10.1007/978-3-319-46128-1_50
13. Nesterov, Y.: Introductory Lectures on Convex Optimization: A Basic Course. Applied Optimization. Kluwer Academic Publication, Boston (2004)

Algorithm Selection for Large-Scale Multi-objective Optimization

Mustafa Mısır[1(✉)] and Xinye Cai[2]

[1] Duke Kunshan University, Kunshan, China
mustafa.misir@dukekunshan.edu.cn
[2] Dalian University of Technology, Dalian, China
xinye@dlut.edu.cn

Abstract. The present study applies Algorithm Selection to automatically specify the suitable algorithms for Large-Scale Multi-objective Optimization. Algorithm Selection has known to benefit from the strengths on multiple algorithm rather than relying one. This trait offers performance gain with limited or no contribution on the algorithm and instance side. As the target application domain, Multi-objective Optimization is a realistic way of approaching any optimization tasks. Most real-world problems are concerned with more than one objective/quality metric. This paper introduces a case study on an Algorithm Selection dataset composed of 4 Multi-objective Optimization algorithms on 63 Large-Scale Multi-objective Optimization problem benchmarks. The benchmarks involve the instances of 2 and 3 objectives with the number of variables changing between 46 and 1006, Hypervolume is the performance indicator used to quantify the solutions derived by each algorithm on every single problem instance. Since Algorithm Selection needs a suite of instance features, 4 simple features are introduced. With this setting, an existing Algorithm Selection system, i.e. ALORS, is accommodated to map these features to the candidate algorithms' performance denoted in ranks. The empirical analysis showed that this basic setting with AS is able to offer better performance than those standalone algorithms. Further analysis realized on the algorithms and instances report similarities/differences between algorithms and instances while reasoning the instances' hardness to be solved.

1 Introduction

Optimization [1] is a process concerned with exploring the best solution regarding some performance criteria. These criteria are referred to objective functions that can measure the solution quality regarding a target problem. The number of objectives determine the nature of the problem. A large group of optimization research focuses on the problems with only one objective, i.e. single-objective optimization. However, the majority of the real-world applications actually come with more than one objective. Those problems are categorized as the multi-objective optimization problems (MOPs) [2]. Further categorization is possible

B. Dorronsoro et al. (Eds.): OLA 2023, CCIS 1824, pp. 36–47, 2023.
https://doi.org/10.1007/978-3-031-34020-8_3

when the number of objectives is exactly two, i.e. bi-objective optimization. If the count exceeds three, then the MOPs are denoted as many-objective optimization. [3,4].

The main challenge of having multiple objectives is that they are likely to be conflicting. Improving one objective can degrade the quality of the remaining objectives. This leads to solution quality evaluation based on various performance indicators utilizing all the objectives. R2 [5,6], Hyper-volume (HV) [7], Generational Distance (GD) [8], Inverted/Inverse GD (IGD) [9], IGD+ [10], Spread [11], and Epsilon [12] are well-known examples of the performance indicators. These indicators are mostly linked to Pareto fronts (PFs) where multiple solutions are maintained. PFs consist of the solutions that do not strictly dominate any other solution, i.e. the solutions that are not worse than the remaining solutions considering all the objectives. In that respect, the algorithms developed for the MOPs mostly operate on the populations of solutions, i.e. the population-based algorithms. Multi-objective Evolutionary Algorithms (MOEAs) [13,14] take the lead in that domain. Non-dominated Sorting Genetic Algorithm II (NSGA-II) [11,15], Pareto Archived Evolution Strategy (PAES) [16], Strength Pareto Evolutionary Algorithm 2 (SPEA2) [17], Pareto Envelope-based Selection Algorithm II (PESA-II) [18] and MOEA based on Decomposition (MOEA/D) [19] are some examples from the literature. There are other population-based algorithms besides MOEAs, using meta-heuristics like Particle Swarm Optimization (MOPSO) [20] and Ant Colony Optimization [21]. It is also possible to see their hybridized variants [22–24].

Despite these immense algorithm development efforts, it is unlikely to see a truly best, i.e. always coming first, algorithm on the existing benchmark scenarios under fair experimental conditions. This practical fact is further supported theoretically by the No Free Lunch (NFL) theorem [25]. This study focuses on automatically determining the algorithm to be applied for each given MOP instance, through Algorithm Selection (AS) [26]. AS is a meta-algorithmic approach offering improved performance through selection. The idea is to automatically choose algorithms from given problem solving scenarios. The selection operations are carried on a given algorithm set [27] consisting of those candidate methods to be picked. The traditional way of approach AS is in the form of performance prediction models. In that respect, a suite of features is needed to characterize the target problem instances. These features are matched with the performance of the candidate algorithms on a group of training instances. While the use of human-engineered features is common for AS, Deep Learning (DL) has also been used for automatically extracting features [28].

AS has been applied to a variety of problem domains such as Boolean Satisfiability (SAT) [29] Constraint Satisfaction (CSP) [30], Blackbox Optimization [31], Nurse Rostering (NRP) [32], Graph Coloring (GCP) [33], Traveling Salesman (TSP) [34] Traveling Thief Problem (TTP) [35], and Game Playing [36]. AS library (ASlib) [37] provides a diverse and comprehensive problem sets for AS. There have been development efforts of new AS systems for addressing these problems. SATzilla [29] is a well known AS method, particularly popularized due to its success in the SAT competitions. Hydra [38] is an example aiming at constructing algorithm sets, a.k.a. Algorithm Portfolios [27], via configuring the given algorithms. The portfolio building task has been studied for different selection tasks [39–42]. 3S [43] delivers algorithm schedules, assigning runtime to the algorithms for each given problem instance. Unlike these AS level contributions, Autofolio [44] takes the search to a higher level by seeking the best AS setting of varying components and parameter configurations. As another high-level approach, AS is used for performing per-instance selection across Selection Hyper-heuristics (SHHs) [45].

The present study performs AS to identify suitable algorithms for the given MOP instance. To be specific, the problem targeted here is the Large-scale MOP (LSMOP) where the number of decision variables can reach up to the vicinity of thousands. The instance set is based on 9 LSMOP benchmarks. Those base benchmarks are varied w.r.t. the number of objectives, i.e. 2 or 3, and the number of decision variables, varies between 46 and 1006, leading to 63 LSMOP instances. The task is to perform per-instance AS using an existing AS system named ALORS [46], among 4 candidate population-based algorithms. Hypervolume (HV) is used as the performance indicator. Experimental analysis carried out illustrated that AS only with 4 basic features outperforms those constituent multi-objective algorithms.

In the remainder of the paper, Sect. 2 discusses the use of AS. An empirical analysis is reported in Sect. 3. Section 4 comes with the concluding remarks besides discussing the future research ideas.

2 Method

ALORS [46] is concerned with the selection task as a recommender system (RS). ALORS specifically uses Collaborative Filtering (CF) [47] in that respect. Unlike the existing AS systems, ALORS is able to perform with the sparse/incomplete performance data, M, while maintaining high, comparable performance to the complete data. The performance refers to running a set of algorithms, A, on a group of instances, I. Thus, the performance data is a matrix of $M_{|I|\times|A|}$. For decreasing the data generation cost of such sparse data has been further targeted in [48,49]. While the entries of the performance data vary from problem to problem, ALORS generalizes them by using the rank data, $\mathcal{M}$. Thus, any given performance data is first converted into rank data. Unlike the traditional AS systems, ALORS builds a prediction model with an intermediate feature-to-feature mapping step, instead of providing a direct rank prediction. The initial, hand-picked/designed features are referenced to a set of latent (hidden) features. These features are extracted directly from the rank performance data by using Singular Value Decomposition (SVD) [50]. SVD is a well-known Matrix Factorization (MF) strategy, used in various CF based RS applications [51]. SVD returns two matrices, U and V besides a diagonal matrix accommodating the singular values as $\mathcal{M} = U\Sigma V^t$. U represents the rows of $\mathcal{M}$, i.e. instances, while V displays its columns, i.e. algorithms, similarly to [52,53]. Beyond representing those data elements, the idea is the reduce the dimensions, $r \leq min(|I|, |A|)$, hopefully eliminating the possible noise in $\mathcal{M}$.

$$\mathcal{M} \approx U_r \Sigma_r V_r^t$$

ALORS maps a given initial set of instance features F to U_r. The predicted performance ranks are calculated by multiplying U_r with the remaining matrices of Σ_r and V_r^t. In that respect, for a new problem instances, ALORS essentially determines an array of values, i.e. a new row for U_r. Its multiplication with Σ_r and V_r^t delivers the expected performance ranks of the candidate algorithms on this new problem instance.

3 Computational Results

Despite the capabilities of ALORS as the sole Algorithm Selection (AS) approach, on working with incomplete performance data, the instance $\times$ algorithm rank data here has the complete performance entries. The AS data is directly derived from [54]. The data on the Large-Scale Multi-objective Optimisation Problem (LSMOP) consists of 4 algorithms. The candidate algorithms are Speed-constrained Multi-objective Particle Swarm Optimization (SMPSO) [55], Multi-objective Evolutionary Algorithm based on Decision Variable Analysis (MOEA/DVA) [56], Large-scale Many-objective Evolutionary Algorithm (LMEA) [57] and Weighted Optimization Framework SMPSO (WOF-SMPSO) [58]. The hypervolume (HV) indicator [59] is used as the performance metric.

Table 1. The base LSMOP instances

Problem	Modality	Separability
LSMOP1	Unimodal	Fully Separable
LSMOP2	Mixed	Partially Separable
LSMOP3	Multi-modal	Mixed
LSMOP4	Mixed	Mixed
LSMOP5	Unimodal	Fully Separable
LSMOP6	Mixed	Partially Separable
LSMOP7	Multi-modal	Mixed
LSMOP8	Mixed	Mixed
LSMOP9	Mixed	Fully Separable

Table 1 shows the specifications of the LSMOP benchmark functions [60]. The functions differ in terms of modality and separability. The 2-objective and 3-objective variants of each function are considered. Besides that further variations on the functions are achieved using different number of decision variables. In total, 63 LSMOP instances are present. The instances are encoded as LSMOPX_m=a_n=b where X is the base LSMOP index, m refers to the number of objectives and n is for the number of decision variables. All these instances are represented using just 4 features. Besides the modality and separability characteristics, the number of objectives and the number of variables are as the instance features.

Table 2. The average ranks of each constituent algorithm besides ALORS where the best per-benchmark performances are in bold (AVG: the average rank considering the average performance on each benchmark function; O-AVG: the overall average rank across all the instances)

Benchmark	SMPSO	MOEA/DVA	LMEA	WOF-SMPSO	ALORS
LSMOP1	4.57	**1.71**	3.71	3.07	1.93
LSMOP2	4.43	**2.43**	3.14	2.5	2.5
LSMOP3	4.86	2.43	3	**2.36**	**2.36**
LSMOP4	4.43	2.79	3.71	2.57	**1.5**
LSMOP5	3.14	**1.86**	5	2.93	2.07
LSMOP6	3.86	3.5	3.64	**1.21**	2.79
LSMOP7	3.71	3.79	4.29	**1.29**	1.93
LSMOP8	4	2.93	4.43	2.14	**1.5**
LSMOP9	2.86	**1.79**	4.43	3.86	2.07
AVG	3.98 ± 0.67	2.58 ± 0.74	3.93 ± 0.66	2.44 ± 0.84	**2.07 ± 0.43**
O-AVG	3.98 ± 1.04	2.58 ± 1.30	3.93 ± 1.19	2.44 ± 1.19	**2.07 ± 0.89**

Table 2 reports the performance of all the candidate algorithms besides ALORS as the automated selection method. Average performance on all the instances show that ALORS offers the best performance with the average rank of 2.07. The closest approach that is the single best method, i.e. WOF-SMPSO, comes with the average rank of 2.44 while SMPSO shows the overall worst performance with the average rank of 3.98. Referring to the standard deviations, ALORS also comes with the most robust behaviour.

Figure 1 reports the selection frequencies of each constituent algorithm. Oracle denotes the optimal selection, i.e. choosing the best algorithm for each instance. The graph shows that ALORS shows similar behaviour to Oracle with minor variations. MOEA/DVA and WOF-SMPSO are the most frequently selected algorithms. Besides the pure selection frequencies, ALORS does not utilize SMPSO at all while it is preferred for two instances by Oracle.

Figure 2 illustrates the importance of each single feature in terms of Gini Index, derived by Random Forest (RF). All four features happen to contribute to the selection model. Being said that *separability* comes as the most critical feature while *modality* is the least important one.

Figure 3 reports the dis/-similarities of the LSMOP benchmark function instances. Linking to the feature importance analysis in Fig. 2, there is no a

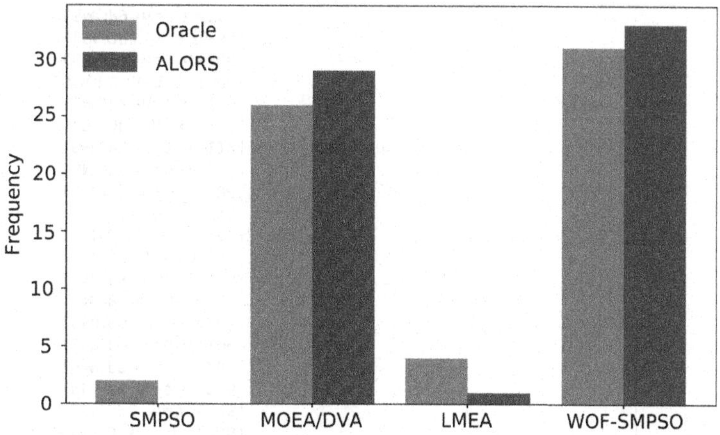

Fig. 1. The selection frequencies of each algorithm by Oracle and ALORS

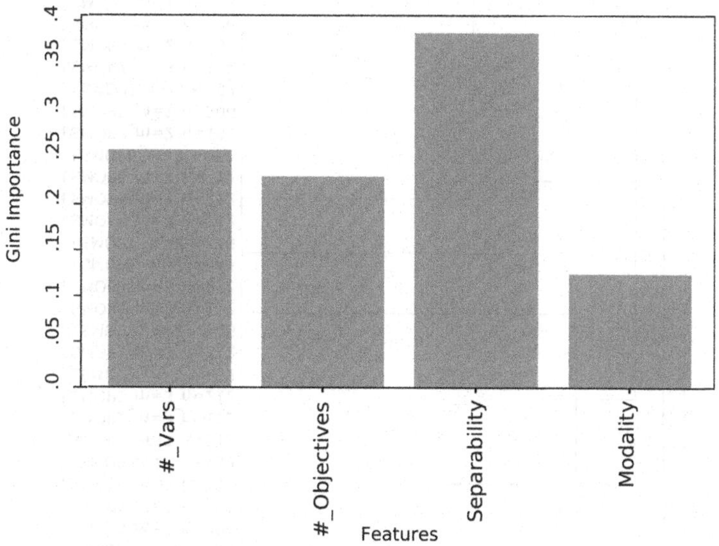

Fig. 2. Importance of the initial, hand-picked LSMOP benchmark function instance features, using Gini Index/Importance

single criterion/feature to emphasize instance dis/-similarity, yet it is still possible to see the effects of separability. As an example, consider the 10 most similar instances provided on the right bottom of the clustering figure. The instances are LSMOP1_m=2_n=46, LSMOP1_m=2_n=106, LSMOP5_m=3_n=212, LSMOP5_m=3_n=112, LSMOP5_m=3_n=52, LSMOP2_m=2_n=106, LSMOP5_m=2_n=1006, LSMOP1_m=2_n=206, LSMOP8_m=3_n=52 and LSMOP9_m=3_n=112. 8 of them are fully separable. The remaining 2 instances are partially separable and mixed, respectively. Referring to the second best fea-

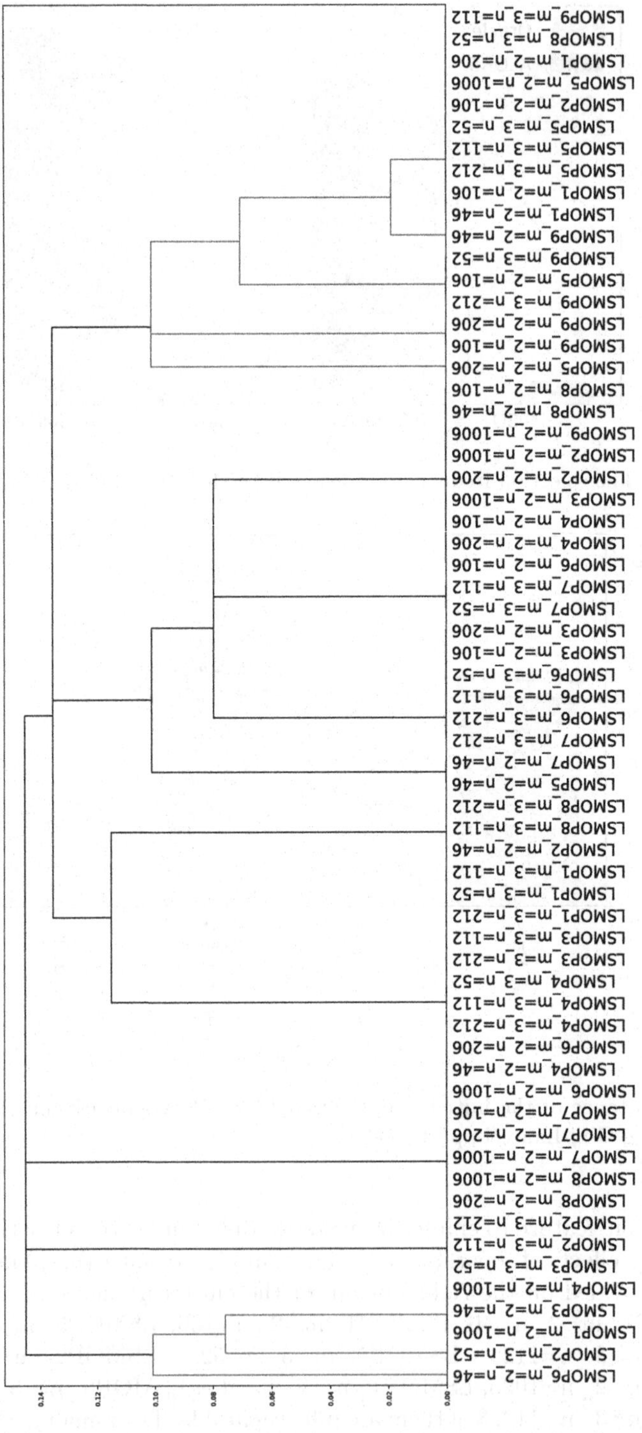

Fig. 3. Hierarchical clusters of instances using the latent features extracted from the performance data by SVD ($k = 3$)

tures, i.e. number of variables, the values change from 52 to 1006. Being said that 1006 occurs only once, thus The half of the instances have 2 objectives while the other half is with 3 objectives. As 2 out of 3 fully separable benchmark functions are unimodal, 7 instances happen to be unimodal. The other 3 instances are mixed in terms of modality.

Figure 4 illustrates the candidate algorithms which are hierarchically clustered. Referring to the best performing standalone algorithm, i.e. WFO-SMPSO, there is resemblance to SMPSO which is the base approach of WFO-SMPSO. Although their performance levels differ, their performance variations across the tested instances are similar.

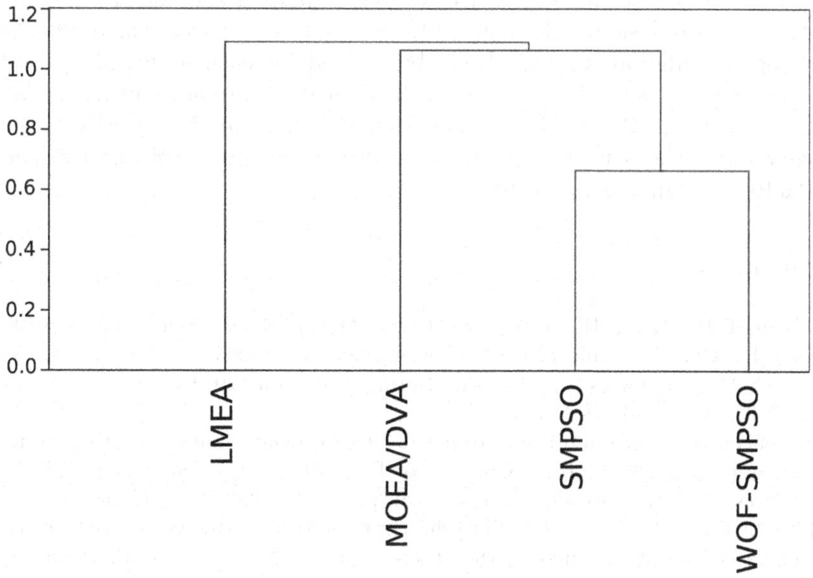

Fig. 4. Hierarchical clusters of algorithms using the latent features extracted from the performance data by SVD ($k = 3$)

4 Conclusion

This study utilizes Algorithm Selection (AS) for Large-Scale Multi-objective Optimization, using Hyper-volume (HV) as the performance criterion. Multi-objective optimization is concerned with the majority of the real-world optimization tasks. In that respect, there have been immense efforts both problem modelling and algorithm development for multi-objective optimization. However, there is no ultimate multi-objective optimization algorithm that can outperform the competing algorithms under fair experimental settings. This practical fact reveals a clear performance gap that can be filled by AS. AS suggests a way to automatically determine the best algorithms for given problem instances.

The present work performs on 4 multi-objective optimization algorithm for 63 benchmarks originated from 9 base problems. For the instance characterization required to use AS, 4 simple instance features are determined. The corresponding computational analysis showed that AS is able to suppress those candidate algorithms. Further analysis carried on the algorithm and instance spaces delivered insights on the instance hardness, instance similarity and algorithm resemblance.

As the first study of using AS for multi-objective optimization, there are a variety of research tasks to be tackled as future research. The initial follow-up work is concerned with extending both the algorithm and instance space. Additionally, the well-known multi-objective performance indicators will be incorporated. The analysis on the algorithm and instance spaces will be extended accordingly. While an AS model will be derived for each indicator, the selection will be also achieved by taking all the indicators into account like a Pareto frontier. The idea will then be reversed to devise AS a multi-objective selection problem where the performance measures are the common AS metrics such as the Par10 score and success rate.

References

1. Chong, E.K., Zak, S.H.: An Introduction to Optimization. Wiley, Hoboken (2004)
2. Deb, K., Deb, K.: Multi-objective optimization. In: Burke, E., Kendall, G. (eds.) Search Methodologies, pp. 403–449. Springer, Boston (2014). https://doi.org/10.1007/978-1-4614-6940-7_15
3. Ishibuchi, H., Tsukamoto, N., Nojima, Y.: Evolutionary many-objective optimization: a short review. In: IEEE Congress on Evolutionary Computation (IEEE World Congress on Computational Intelligence), pp. 2419–2426. IEEE (2008)
4. Li, K., Wang, R., Zhang, T., Ishibuchi, H.: Evolutionary many-objective optimization: a comparative study of the state-of-the-art. IEEE Access **6**, 26194–26214 (2018)
5. Hansen, M.P., Jaszkiewicz, A.: Evaluating the quality of approximations to the non-dominated set. Citeseer (1994)
6. Brockhoff, D., Wagner, T., Trautmann, H.: On the properties of the R2 indicator. In: Proceedings of the 14th Annual Conference on Genetic and Evolutionary Computation, pp. 465–472 (2012)
7. Zitzler, E., Thiele, L.: Multiobjective optimization using evolutionary algorithms—a comparative case study. In: Eiben, A.E., Bäck, T., Schoenauer, M., Schwefel, H.P. (eds.) PPSN 1998. LNCS, pp. 292–301. Springer, Cham (1998). https://doi.org/10.1007/bfb0056872
8. Van Veldhuizen, D.A., Lamont, G.B.: Multiobjective evolutionary algorithm research: a history and analysis. Technical report, Department of Electrical and Computer Engineering Air Force Institute of Technology, OH, Technical Report TR-98-03 (1998)
9. Coello Coello, C.A., Reyes Sierra, M.: A study of the parallelization of a coevolutionary multi-objective evolutionary algorithm. In: Monroy, R., Arroyo-Figueroa, G., Sucar, L.E., Sossa, H. (eds.) MICAI 2004. LNCS (LNAI), vol. 2972, pp. 688–697. Springer, Heidelberg (2004). https://doi.org/10.1007/978-3-540-24694-7_71

10. Ishibuchi, H., Masuda, H., Tanigaki, Y., Nojima, Y.: Modified distance calculation in generational distance and inverted generational distance. In: Gaspar-Cunha, A., Henggeler Antunes, C., Coello, C.C. (eds.) EMO 2015. LNCS, vol. 9019, pp. 110–125. Springer, Cham (2015). https://doi.org/10.1007/978-3-319-15892-1_8
11. Deb, K., Pratap, A., Agarwal, S., Meyarivan, T.: A fast and elitist multiobjective genetic algorithm: NSGA-II. IEEE Trans. Evol. Comput. **6**, 182–197 (2002)
12. Fonseca, C.M., Knowles, J.D., Thiele, L., Zitzler, E.: A tutorial on the performance assessment of stochastic multiobjective optimizers. In: Proceedings of the 3rd International Conference on Evolutionary Multi-Criterion Optimization (EMO), vol. 216, p. 240 (2005)
13. Deb, K.: Multi-objective optimisation using evolutionary algorithms: an introduction. In: Wang, L., Ng, A., Deb, K. (eds.) Multi-objective Evolutionary Optimisation for Product Design and Manufacturing, pp. 3–34. Springer, London (2011). https://doi.org/10.1007/978-0-85729-652-8_1
14. Deb, K.: Multi-objective evolutionary algorithms. In: Kacprzyk, J., Pedrycz, W. (eds.) Springer Handbook of Computational Intelligence, pp. 995–1015. Springer, Heidelberg (2015). https://doi.org/10.1007/978-3-662-43505-2_49
15. Deb, K., Agrawal, S., Pratap, A., Meyarivan, T.: A fast elitist non-dominated sorting genetic algorithm for multi-objective optimization: NSGA-II. In: Schoenauer, M., et al. (eds.) PPSN 2000. LNCS, vol. 1917, pp. 849–858. Springer, Heidelberg (2000). https://doi.org/10.1007/3-540-45356-3_83
16. Knowles, J.D., Corne, D.W.: Approximating the nondominated front using the pareto archived evolution strategy. Evol. Comput. **8**, 149–172 (2000)
17. Zitzler, E., Laumanns, M., Thiele, L.: SPEA2: improving the strength pareto evolutionary algorithm. Technical Report 103, Computer Engineering and Networks Laboratory (TIK), Swiss Federal Institute of Technology (ETH), Zurich, Switzerland (2001)
18. Corne, D.W., Jerram, N.R., Knowles, J.D., Oates, M.J.: PESA-II: region-based selection in evolutionary multiobjective optimization. In: Proceedings of the 3rd Annual Conference on Genetic and Evolutionary Computation (GECCO), pp. 283–290. Morgan Kaufmann Publishers Inc. (2001)
19. Zhang, Q., Li, H.: MOEA/D: a multiobjective evolutionary algorithm based on decomposition. IEEE Trans. Evol. Comput. **11**, 712–731 (2007)
20. Coello, C.C., Lechuga, M.S.: MOPSO: a proposal for multiple objective particle swarm optimization. In: Proceedings of the IEEE Congress on Evolutionary Computation (CEC), vol. 2, pp. 1051–1056. IEEE (2002)
21. Ding, L.P., Feng, Y.X., Tan, J.R., Gao, Y.C.: A new multi-objective ant colony algorithm for solving the disassembly line balancing problem. Int. J. Adv. Manuf. Technol. **48**, 761–771 (2010)
22. Mashwani, W.K.: MOEA/D with DE and PSO: MOEA/D-DE+PSO. In: Bramer, M., Petridis, M., Nolle, L. (eds.) SGAI 2011, pp. 217–221. Springer, London (2011). https://doi.org/10.1007/978-1-4471-2318-7_16
23. Ke, L., Zhang, Q., Battiti, R.: MOEA/D-ACO: a multiobjective evolutionary algorithm using decomposition and antcolony. IEEE Trans. Cybern. **43**, 1845–1859 (2013)
24. Alhindi, A., Zhang, Q.: MOEA/D with tabu search for multiobjective permutation flow shop scheduling problems. In: Proceedings of the IEEE Congress on Evolutionary Computation (CEC), pp. 1155–1164. IEEE (2014)
25. Wolpert, D., Macready, W.: No free lunch theorems for optimization. IEEE Trans. Evol. Comput. **1**, 67–82 (1997)

26. Kerschke, P., Hoos, H.H., Neumann, F., Trautmann, H.: Automated algorithm selection: survey and perspectives. Evol. Comput. **27**, 3–45 (2019)
27. Gomes, C., Selman, B.: Algorithm portfolios. Artif. Intell. **126**, 43–62 (2001)
28. Loreggia, A., Malitsky, Y., Samulowitz, H., Saraswat, V.A.: Deep learning for algorithm portfolios. In: Proceedings of the 13th Conference on Artificial Intelligence (AAAI), pp. 1280–1286 (2016)
29. Xu, L., Hutter, F., Hoos, H., Leyton-Brown, K.: SATzilla: portfolio-based algorithm selection for SAT. J. Artif. Intell. Res. **32**, 565–606 (2008)
30. Yun, X., Epstein, S.L.: Learning algorithm portfolios for parallel execution. In: Hamadi, Y., Schoenauer, M. (eds.) LION 2012. LNCS, pp. 323–338. Springer, Heidelberg (2012). https://doi.org/10.1007/978-3-642-34413-8_23
31. Kerschke, P., Trautmann, H.: Automated algorithm selection on continuous black-box problems by combining exploratory landscape analysis and machine learning. Evol. Comput. **27**, 99–127 (2019)
32. Messelis, T., De Causmaecker, P., Vanden Berghe, G.: Algorithm performance prediction for nurse rostering. In: Proceedings of the 6th Multidisciplinary International Scheduling Conference: Theory and Applications (MISTA 2013), pp. 21–38 (2013)
33. Musliu, N., Schwengerer, M.: Algorithm selection for the graph coloring problem. In: Nicosia, G., Pardalos, P. (eds.) LION 2013. LNCS, vol. 7997, pp. 389–403. Springer, Heidelberg (2013). https://doi.org/10.1007/978-3-642-44973-4_42
34. Kotthoff, L., Kerschke, P., Hoos, H., Trautmann, H.: Improving the state of the art in inexact TSP solving using per-instance algorithm selection. In: Dhaenens, C., Jourdan, L., Marmion, M.-E. (eds.) LION 2015. LNCS, vol. 8994, pp. 202–217. Springer, Cham (2015). https://doi.org/10.1007/978-3-319-19084-6_18
35. Wagner, M., Lindauer, M., Mısır, M., Nallaperuma, S., Hutter, F.: A case study of algorithm selection for the traveling thief problem. J. Heuristics **24**, 295–320 (2018)
36. Stephenson, M., Renz, J.: Creating a hyper-agent for solving angry birds levels. In: AAAI Conference on Artificial Intelligence and Interactive Digital Entertainment (2017)
37. Bischl, B., et al.: ASlib: a benchmark library for algorithm selection. Artif. Intell. **237**, 41–58 (2017)
38. Xu, L., Hoos, H., Leyton-Brown, K.: Hydra: automatically configuring algorithms for portfolio-based selection. In: Proceedings of the 24th AAAI Conference on Artificial Intelligence (AAAI), pp. 210–216 (2010)
39. Mısır, M., Handoko, S.D., Lau, H.C.: OSCAR: online selection of algorithm portfolios with case study on memetic algorithms. In: Dhaenens, C., Jourdan, L., Marmion, M.-E. (eds.) LION 2015. LNCS, vol. 8994, pp. 59–73. Springer, Cham (2015). https://doi.org/10.1007/978-3-319-19084-6_6
40. Mısır, M., Handoko, S.D., Lau, H.C.: ADVISER: a web-based algorithm portfolio deviser. In: Dhaenens, C., Jourdan, L., Marmion, M.-E. (eds.) LION 2015. LNCS, vol. 8994, pp. 23–28. Springer, Cham (2015). https://doi.org/10.1007/978-3-319-19084-6_3
41. Lau, H., Mısır, M., Xiang, L., Lingxiao, J.: ADVISER$^+$: toward a usable web-based algorithm portfolio deviser. In: Proceedings of the 12th Metaheuristics International Conference (MIC), Barcelona, Spain, pp. 592–599 (2017)
42. Gunawan, A., Lau, H.C., Mısır, M.: Designing and comparing multiple portfolios of parameter configurations for online algorithm selection. In: Festa, P., Sellmann, M., Vanschoren, J. (eds.) LION 2016. LNCS, vol. 10079, pp. 91–106. Springer, Cham (2016). https://doi.org/10.1007/978-3-319-50349-3_7

43. Kadioglu, S., Malitsky, Y., Sabharwal, A., Samulowitz, H., Sellmann, M.: Algorithm selection and scheduling. In: Lee, J. (ed.) CP 2011. LNCS, vol. 6876, pp. 454–469. Springer, Heidelberg (2011). https://doi.org/10.1007/978-3-642-23786-7_35
44. Lindauer, M., Hoos, H.H., Hutter, F., Schaub, T.: AutoFolio: an automatically configured algorithm selector. J. Artif. Intell. Res. **53**, 745–778 (2015)
45. Misir, M.: Cross-domain algorithm selection: algorithm selection across selection hyper-heuristics. In: 2022 IEEE Symposium Series on Computational Intelligence (SSCI), pp. 22–29. IEEE (2022)
46. Mısır, M., Sebag, M.: ALORS: an algorithm recommender system. Artif. Intell. **244**, 291–314 (2017)
47. Su, X., Khoshgoftaar, T.M.: A survey of collaborative filtering techniques. Adv. Artif. Intell. **2009**, 4 (2009)
48. Mısır, M.: Data sampling through collaborative filtering for algorithm selection. In: The 16th IEEE Congress on Evolutionary Computation (CEC), pp. 2494–2501. IEEE (2017)
49. Mısır, M.: Active matrix completion for algorithm selection. In: Nicosia, G., Pardalos, P., Umeton, R., Giuffrida, G., Sciacca, V. (eds.) LOD 2019. LNCS, vol. 11943, pp. 321–334. Springer, Cham (2019). https://doi.org/10.1007/978-3-030-37599-7_27
50. Golub, G.H., Reinsch, C.: Singular value decomposition and least squares solutions. Numer. Math. **14**, 403–420 (1970)
51. Koren, Y., Bell, R., Volinsky, C.: Matrix factorization techniques for recommender systems. Computer **42**, 30–37 (2009)
52. Mısır, M.: Matrix factorization based benchmark set analysis: a case study on HyFlex. In: Shi, Y., et al. (eds.) SEAL 2017. LNCS, vol. 10593, pp. 184–195. Springer, Cham (2017). https://doi.org/10.1007/978-3-319-68759-9_16
53. Mısır, M.: Benchmark set reduction for cheap empirical algorithmic studies. In: Proceedings of the IEEE Congress on Evolutionary Computation (CEC) (2021)
54. Zille, H., Mostaghim, S.: Comparison study of large-scale optimisation techniques on the LSMOP benchmark functions. In: IEEE Symposium Series on Computational Intelligence (SSCI), pp. 1–8. IEEE (2017)
55. Nebro, A., Durillo, J., García-Nieto, J., Coello Coello, C., Luna, F., Alba, E.: SMPSO: a new PSO-based metaheuristic for multi-objective optimization. In: IEEE Symposium on Computational Intelligence in Multicriteria Decision-Making (MCDM 2009), pp. 66–73. IEEE Press (2009)
56. Ma, X., et al.: A multiobjective evolutionary algorithm based on decision variable analyses for multiobjective optimization problems with large-scale variables. IEEE Trans. Evol. Comput. **20**, 275–298 (2015)
57. Zhang, X., Tian, Y., Cheng, R., Jin, Y.: A decision variable clustering-based evolutionary algorithm for large-scale many-objective optimization. IEEE Trans. Evol. Comput. **22**, 97–112 (2016)
58. Zille, H., Ishibuchi, H., Mostaghim, S., Nojima, Y.: A framework for large-scale multiobjective optimization based on problem transformation. IEEE Trans. Evol. Comput. **22**, 260–275 (2017)
59. Auger, A., Bader, J., Brockhoff, D., Zitzler, E.: Hypervolume-based multiobjective optimization: theoretical foundations and practical implications. Theor. Comput. Sci. **425**, 75–103 (2012)
60. Cheng, R., Jin, Y., Olhofer, M., et al.: Test problems for large-scale multiobjective and many-objective optimization. IEEE Trans. Cybern. **47**, 4108–4121 (2016)

Solving a Multi-objective Job Shop Scheduling Problem with an Automatically Configured Evolutionary Algorithm

Jesús Para[1], Javier Del Ser[3,4], and Antonio J. Nebro[1,2](✉)

[1] Departamento de Lenguajes y Ciencias de la Computación, University of Málaga, 29071 Málaga, Spain
ajnebro@uma.es

[2] ITIS Software, University of Málaga, 29071 Málaga, Spain

[3] TECNALIA, 48160 Derio, Spain
javier.delser@tecnalia.com

[4] University of the Basque Country (UPV/EHU), 48013 Bilbao, Spain

Abstract. In this work we focus on optimizing a multi-objective formulation of the Job Shop Scheduling Problem (JSP) which considers the minimization of energy consumption as one of the objectives. In practice, users experts in the problem domain but with a low knowledge in metaheuristics usually take an existing algorithm with default settings to optimize problem instances but, in this context, the use of automatic parameter configuration techniques can help to find ad-hoc configurations of algorithms that effectively solve optimization problems. Our aim is to study what improvement in results can be obtained by applying an autoconfiguration approach versus using a set of well-known multi-objective evolutionary algorithms (NSGA-II, SPEA2, SMS-EMOA and MOEA/D) for different instances of the JSP, with varying dimensionality. Our experiments showcase the potential of automated algorithmic configuration for energy-efficient production scheduling, producing better balanced solutions than the multi-objective solvers considered in the study.

Keywords: Multi-Objective Optimization · Job Shop Scheduling · Automatic Algorithm Configuration

1 Introduction

The Job Shop Scheduling problem (JSP) is a combinatorial optimization problem where a set of jobs requiring different processing times have to be scheduled on a set of machines having different processing power [9,15]. The main goal of JSP is typically to minimize the makespan, namely, the minimum time to complete all pending jobs in the production commit, but in the context of Industry 4.0 objectives such as the minimization of the consumed energy during production

B. Dorronsoro et al. (Eds.): OLA 2023, CCIS 1824, pp. 48–61, 2023.
https://doi.org/10.1007/978-3-031-34020-8_4

are of paramount importance. When accounting for this objective, it is intuitive to infer that productivity can be a conflicting goal with the consumption of energy, especially when dealing with energy-demanding assets whose production rate correlate tightly with the amount of required energy (e.g., electric arc furnaces). In this paper, we are interested in a bi-objective formulation of the JSP by considering makespan and production costs as objectives to minimize. The production costs are assumed to be driven by different energy and non-energy related concepts, including the cost of human workforce, which may vary depending on the shift.

A common situation in practice is that an expert in the domain of a optimization problem is interested in solving instances of it, but frequently that user is not an expert in metaheuristics, so the adopted approach usually is to take a well-known algorithm (typically, NSGA-II [10] in the case of multi-objective optimization) configured with defaults settings. Although the results obtained may be good enough, it seems obvious that they could be improved if the algorithm were properly configured. Traditionally, the parameter adjustment of metaheuristics has been addressed by manual adjustment of parameters and conducting pilot tests, which is a tedious and not rigorous process. In this work, we are interested in exploring the potential of automated algorithm design tools for the multi-objective JSP. Given a set of algorithmic components that can be combined together and a set of problems used as a training set, automated algorithm configuration tools can autonomously discover an evolutionary algorithm that is tailored to deal with unseen instances of the same problem [6,7].

This work relies on the jMetal framework [12,16], which has recently released functionalities for the automated design of multi-objective evolutionary algorithms [17], combined with the irace package for automated algorithm configuration and design [14]. We define several training and test instances of a multi-objective formulation of the JSP which, in addition to the factors considered in [19], personnel costs by shifts and different electricity costs by hour are included as new cost concepts. Then, a multi-objective evolutionary algorithm (hereafter denoted as AutoMOEA) is designed from the training set of problem instances by using the aforementioned optimization engine. Finally, we measure the quality of the solutions obtained by AutoMOEA over the test instances to inform a discussion about the performance gaps that automatic algorithm configuration can bring to this family of problems. We must note that our goal is not to find an algorithm capable of outperforming state-of-the-art of techniques for the JSP, but to assess whether using auto-configuration, which is a computing-intensive process, can be worth when compared to known multi-objective evolutionary algorithms with default parameter settings.

The rest of the manuscript is divided as follows: Sect. 2 first contextualizes and exposes the novelty of this research work compared to existing studies. Section 3 poses the multi-objective JSP under consideration, and describes how solutions to this problem can be numerically encoded. Next, Sect. 4 details how to automatic design a multi-objective evolutionary solver. Sections 5 and 6 respectively presents the experimental setup and discusses on the results obtained therefrom. Finally, Sect. 7 draws conclusions and outlines research directions.

2 Previous Work and Motivation

The starting point of this research work is the recent survey contributed in [19], in which a detailed study of the literature related to multi-objective JSP formulations with energy as one of the objectives was performed. The survey analyzes the most relevant publications that tries to solve the JSP problem with energy as one of their objectives, setting the good practices detected and pointing out areas of improvement. This review also comprised a use case made of public synthetic instances aimed to showcase how a principled experimentation with multi-objective metah euristic algorithms should be made. Such synthetic instances were produced by considering machine `on` and `off` times, the time to `idle` and start from `idle`, the power consumption at `startup`, `idle`, and `producing` times, and the manufacturing speed (which relates to the energy consumption while production). This work introduces two new considerations in the problem formulation: *variable electricity cost per hour of the day* and *personnel cost per hour*. These two novel ingredients shed further realism to the use case formulated in [19], as electricity provision contracts in energy-intensive industries are often subject to this modeled feature.

Besides the novel ingredients of the problem statement, this work steps beyond one of the main conclusions arising from the survey in [19]: most existing proposals in the literature related to energy-aware multi-objective JSP develop are small modifications of a classical multi-objective algorithm, which is experimentally justified by comparing its results for a specific problem instance against the classical multi-objective algorithm. Furthermore, the reduced set of problem instances chosen for evaluation makes it difficult to ascertain if the performance gains claimed for the newly proposed algorithm generalize to new problem instances or variants of the already evaluated without requiring a major parameter tuning effort. In practice a plant manager would be willing to have an scheduler that does not require long running latencies to elicit different energy-aware schedules. Therefore, the generalization capability of the algorithmic configuration is essential for the usability of any production scheduling software embracing a multi-objective solver at its core. This calls for further insights on the possibilities of automatic algorithmic configuration tools for multi-objective JSP, examining its relative performance w.r.t. classical solvers and to reflect on whether such performance differences hold when solving new problem instances.

3 Problem Formulation, Solution Encoding and Instances

In short, the JSP aims to assign production jobs (or tasks) to industrial production assets over time. This production schedule must be optimized by taking into account variables and restrictions that impact on the production of the plant under consideration. Among the many objectives that have been regarded in the literature related to the JSP, producing a given set of jobs (production commit) within the minimum total production time possible (i.e., *makespan*) is arguably the most widely adopted optimization goal. Makespan can be defined as the minimum time to complete all the scheduled jobs. There are several investigations in the literature trying to improve the quality of solutions for makespan problems [8,20,26]. It can be described as a sorting task, searching for an optimal arrangement such that production is concentrated in the shortest possible period of time, avoiding gaps between tasks.

We define mathematically the JSP by adopting the notation in [19]. As such, N independent production jobs $\{j_n\}_{n=1}^{N}$ are to be processed through M production machines $\mathcal{M} = \{m_i\}_{i=1}^{M}$. Each job j_n consists of O_n ordered tasks $[T_1^n, \ldots, T_{O_n}^n]$. Task T_k^n is *processed* on a predefined machine $m_j \in \mathcal{M}$ (as in the naive JSP), or on any machine that is qualified for the task, given by $\mathcal{M}_j \subseteq \mathcal{M}$. The assignment of task T_k^n to production machine m_j yields a cost $C_{j,n,k}$. The most simplistic formulation of the JSP considers that overall processing time of the entire production commit is the only cost objective to be minimized, whereas other formulations can pursue the minimization of other conflicting cost goals (e.g. energy) or leverage the flexibility of deciding which qualified machine from $\mathcal{M}_j$ can serve every task. When considering different objectives, the goal of the multi-objective JSP is to find a number of production schedules that differently balance between the considered objectives.

Table 1. Synthetic energy-aware JSP instances under consideration

Instances	LA04 [2]	LA10 [2]	FT06 [3]	ORB01 [2]	ABZ5 [1]	SWV20 [22]	TA12 [4]	DMU11 [11]	YN03 [18]	DMU40 [11]	TA77 [23]
M (machines)	5	5	6	10	10	10	15	15	20	20	20
J (jobs)	10	15	6	10	10	50	20	30	20	50	100

Based on the above description, an instance of the energy-aware JSP is defined by 1) the number of machines; 2) the number of jobs; 3) the tasks that comprise every job; 4) the sequence of machines through which the job has to be processed; and 5) the time taken by every machine to process the job. To realistically produce instances for this problem, we depart from the benchmark JSP instances proposed in [23], which defines a JSP problem as several jobs that have to be processed through a number of machines in an orderly fashion. The time taken by each of these machines to process a given job is predefined. It is important to note that while this information is enough to define a single-objective JSP seeking to minimize the makespan, further information must be enclosed to these instances to model the energy consumption while producing the jobs: the processing speed of the machine, the time and energy cost of turning a given machine `on` and `off` or leaving it `idle`, the electrical hourly tariff and personnel costs depending on the time of the day in which the commit is produced. Problem instances augmented with all these relevant modifications have been made available in a repository [13], from which we select a representative subset of them for the study. Such selected instances are listed in Table 1.

A crucial decision when undertaking the problem defined above with metaheuristic algorithms is the solution encoding strategy, namely, the methodology to numerically represent any solution to the problem. To this end we partially embrace the encoding approach presented in [24], albeit with some modifications. To begin with, an integer list encoding will be used, representing every solution as a list of integers with length $J \times M \times Z$, where J is the number of jobs and M is the number of machines. To include the capability to not work in one or more hours during the day, this length of the array is multiplied by a factor Z, divisible by 24, so that every entry in the list is now one hour in a block of 24 consecutive entries. The algorithm searches for the hours of the day to work with this hourly cost, since it generates an array Z times replicated (multiple of 24), whose binary encoding indicates the working/not working hours. As a result, this directly affects the makespan (more time to completion if the job spans hours whose encoded gen is set to 0) and the cost (due to peak and off-peak hours of electricity cost). With this, we can satisfy the condition that the plant might operate in hours of cheaper energy or instead, decide to produce during periods of lower personnel costs. With each value in this string being between 1 and J, and by imposing that each value can only be repeated J times, the list is traversed sequentially, so that each time a certain value appears over the list a step forward in the job is taken, until job is finished.

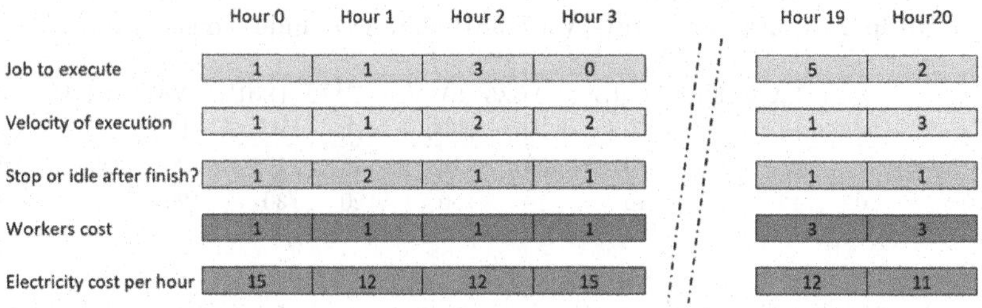

Fig. 1. Example of the adopted solution encoding strategy.

To introduce the energy concept, lists of equal length are generated for machine speed, machine idle, shift and hourly cost, so that the sum of the associated costs of the scalar product of these vectors will give us the production cost. Figure 1 illustrates this encoding strategy, showing the different lists included in the solution's genotype. To begin with, *job to execute* denotes the number of the job to be executed, encoded as an integer between 0 and $J-1$. The *velocity of execution* indicates the speed at which the job is executed, which is a multiplier ($\times 1$, $\times 2$ or $\times 3$) of the baseline processing velocity of the machine. Elements in the list denoted as *stop or idle at finish?* are equal to 1 or 2, depending on whether the machine is set to `idle` state after finishing its job or, instead, is stopped. The *workers cost* list indicates the shift during which the job is processed: morning, afternoon or night shift, each featuring an average labour cost per hour. Finally, the *electricity cost per hour* indicates the average hourly cost of electricity incurred in production as per the hours in which the jobs are processed and the status and processing speed of the machines.

Table 2. Parameter space of AutoMOEA for solving JSP. The population size is 50.

Parameter	Search domain
offspringPopulationSize	[1, 400]
algorithmResult	{externalArchive, population}
populationSizeWithArchive	[10, 200] (subject to algorithmResult == externalArchive)
ranking	{dominanceRanking, strengthRanking}
densityEstimator	{crowdingDistance, kNN}
kValueForKNN	[1, 3] (subject to densityEstimator == kNN)
selection	{tournament, random}
selectionTournamentSize	[2, 10] (subject to selection == tournament)
crossover	JSPCrossover
crossoverProbability	[0.0, 1.0]
mutation	JSPMutation

The above encoding strategy is suitable for the crossover and mutation operators defined in [19], which will be used in our experiments. Specifically, the JSPCrossover operator proceeds as follows: after selecting two parent solutions from the population, a dimension of the encoded solution is chosen uniformly at random between schedule, velocity or idle is chosen uniformly at random. After that, two random positions are selected again within the list corresponding to the selected dimension and they are interchanged between the two parents, leading to two children. In the mutation operator, a random choice is done between (schedule, velocity and idle). Then, a random interchange is done between two positions chosen at random. Mutation is always applied.

4 Automated Design of a Multi-objective Evolutionary Algorithm for the JSP

The process of the automatic design of an multi-objective evolutionary algorithms for our energy-aware JSP variant requires first to define the parameter space, i.e. the parameters and components that can be combined to produce certain algorithm. The JSP is implemented in jMetal, including the aforementioned specific variation operators labeled as JSPMutation and JSPCrossover, so that any of the multi-objective evolutionary algorithms available in jMetal can use them to solve the problem. The combined use of jMetal and irace for auto-parameter tuning is based on the proposal presented in [17], where a study involving the automatic configuration of NSGA-II [10] was presented.

The current design space is detailed in Table 2. Starting by a fixed population size of 50 solutions, the offspring population size can take a value in the range 1 (i.e., steady-state) to 400, and an external archive can be incorporated; in such case, the size of the population can vary within the integer range [10, 200] and the result of the search will be the solutions contained in the archive instead of the population. The archive has a maximum size of 50 individuals, and the crowding distance density estimator [10] is used to prune it when it becomes full.

Table 3. Parameters of NSGA-II and AutoMOEA for JSP (the population size is 100).

Parameter	NSGA-II	AutoMOEA
algorithmResult	population	population
offspringPopulationSize	100	36
ranking	dominanceRanking	strengthRanking
densityEstimator	crowdingDistance	crowdingDistance
crossover	JSSPCrossover	JSSPCrossover
crossoverProbability	0.90	0.955
selection	tournament	tournament
selectionTournamentSize	2	5

The algorithmic template assumes that a ranking method and a density estimator are used for discriminating solutions both in the selection and the replacement steps. These components can be the non-dominance ranking and crowding distance of NSGA-II and the strength ranking and k-nearest neighbour (KNN) density estimator of SPEA2 [27]. If KNN is selected, the value of K ranges between 1 and 3. The selection can be random or by tournament with a tournament size between 2 and 10. Finally, the probability of applying the JSPCrossover operator ranges between 0.0 and 1.0 (the JSPMutation is always applied).

The second step is to choose the training set, which is a key decision because it affects the ability of the designed algorithm to generalize and effectively solve other problems and it also impacts on the computing time required by irace to find the best combination of components. As this is our first study on the capabilities of automatic algorithmic design for tackling the JSP with metaheuristics, we have chosen the three smallest JSP instances, namely, *LA04*, *LA10*, and *FT06*. Thus, we prioritize a faster discovery of the final algorithmic design, hoping that it generalizes nicely even if it is only evaluated in simple problem instances. In this sense, the stopping condition of the AutoMOEA is reduced to compute a maximum of 25,000 function evaluations, while in the experimentation section this number will be increase to 200,000.

The irace tool requires a measure that, given two different algorithmic configurations, determine which one is the best. As the subject of comparison are Pareto front approximations, a quality indicator should be applied. We have selected the hypervolume [28], which requires a reference point. As the optimum of the considered instances are unknown, to define such reference points we have examined for each problem instance all fronts produced after a number of pilot tests performed off-line. After registering the extreme points spanned by all these fronts for every objective, we have added a conservative offset to ensure that the fronts found by AutoMOEA are likely to dominate them.

The last step is to run irace to find a configuration for AutoMOEA. The found one is included in Table 3, where we include in the left column the configuration of NSGA-II as a reference. Interestingly, we observe that AutoMOEA does not use an external archive, as opposed to the AutoNSGA-II algorithm reported in [17], which did make use of it. Furthermore, the offspring population size is reduced from 100 to 36, the ranking scheme is strength ranking, the crossover probabilities are similar than that of NSGA-II, and the tournament size is 5 instead of 2. We must note that we used a 64 cores virtual machine and that the autoconfiguration required about six hours of computing time.

5 Experiments and Results

Once irace has found a potentially good solver over the considered training JSP instances, the next step is to assess whether our auto-designed AutoMOEA is capable of improving the results obtained by other algorithms over the considered JSP instances. Concretely, we have selected representative techniques of the three categories of multi-objective evolutionary algorithms: Pareto-dominance based (NSGA-II [10] and SPEA2 [28]), decomposition-based (MOEA/D [25]), and indicator-based (SMS-EMOA [5]). All the algorithms – including AutoMOEA – are configured with a population size of 50 individuals, whereas the stopping condition is set to a maximum of 200,000 function evaluations. All of them use the same variation operators: JSPCrossover, with a probability of 0.9 (excepting AutoMOEA, which uses 0.955), and JSPMutation. MOEA/D applies the Tschebyscheff aggregation, the neighborhood selection probability is 0.9, the neighbour size is 20, and the maximum number of replaced solutions is 2. GWASFGA requires an epsilon parameter, which is set to 0.01.

To measure the quality of the Pareto front approximations achieved we have used the normalized hypervolume (NHV), which is computed as 1.0 minus the ratio between the hypervolume of the front to be evaluated and the hypervolume of the reference front. The hypervolume metric [28] calculates the volume (in the space of objectives) covered by members of a given set of non-dominated solutions with respect to a reference point. To quantify the results for each algorithm, we perform 25 independent runs per each pair of problem-algorithm, reporting the median and interquartile range (IQR) for each quality indicator.

The results of this comparison are collected in Table 4, where the best and second best results are highlighted in dark and light grey background, respectively. To check for the statistical significance of the performance gaps, we have applied the Wilcoxon rank-sum test [21], at 95% confidence level, with respect to AutoMOEA, whose results are included in the last column. The reference fronts to compute the NHV have been obtained by aggregating, for each problem, the solutions returned by all the runs of all the algorithms, so that the reference front is composed by all non-dominated solutions in the aggregated set.

When inspecting these results, we observe that AutoMOEA gets the best (lowest) indicator and second best values in 7 and 3, respectively, out of the 11 JSP instances under consideration. If we focus on the first three problem instances - which were used for the auto-design process (training) - SPEA2 results to be the solver that yields the overall best indicator values. In fact, SPEA2 is the algorithm that presents the best figures in the 5 smallest instances,

Table 4. Median and interquartile range (IQR) of the results of the NHV quality indicator. Cells with dark and light gray background highlights, respectively, the best and second best indicator values. The algorithm in the last column is the reference algorithm, and the symbols +, − and ≈ indicate that the differences with the reference algorithm are significantly better, worse, or there is no difference according to the Wilcoxon rank sum test (confidence level: 95%).

NSGAII	SPEA2	MOEA/D	SMS-EMOA	AutoMOEA	
LA04	6.13e-02(6.65e-02) ≈	1.18e-02(6.11e-02) ≈	8.01e-01(2.16e-01)−	7.55e-02(6.26e-02)−	4.23e-02(7.28e-02)
LA10	6.67e-02(3.06e-02)+	8.64e-02(3.93e-02) ≈	6.49e-01(1.42e-01)−	8.12e-02(4.08e-02) ≈	9.65e-02(3.18e-02)
FT06	1.05e-01(2.33e-01) ≈	5.53e-02(2.16e-02)+	1.00e+00(0.00e+00)−	1.40e-01(2.81e-01)−	9.76e-02(1.56e-01)
ORB01	3.38e-01(1.51e-01) ≈	2.95e-01(5.58e-02) ≈	1.00e+00(7.82e-02)−	3.16e-01(2.08e-01) ≈	3.10e-01(1.87e-01)
ABZ5	1.65e-01(1.30e-01) ≈	1.30e-01(7.47e-02) ≈	8.12e-01(1.91e-01)−	1.70e-01(6.18e-02)−	1.27e-01(7.70e-02)
SWV20	2.26e-01(6.18e-02)−	2.64e-01(7.06e-02)−	8.79e-01(1.26e-01)−	2.67e-01(6.31e-02)−	1.44e-01(1.11e-01)
TA12	2.51e-01(1.09e-01)−	2.83e-01(7.75e-02)−	8.69e-01(1.99e-01)−	3.49e-01(1.72e-01)−	1.68e-01(9.21e-02)
DMU11	3.20e-01(1.71e-01)−	3.36e-01(1.10e-01)−	1.00e+00(0.00e+00)−	3.23e-01(1.01e-01)−	2.04e-01(1.77e-01)
YN03	2.80e-01(1.49e-01)−	3.57e-01(1.60e-01)−	1.00e+00(4.40e-02)−	3.45e-01(1.75e-01)−	2.27e-01(1.42e-01)
DMU40	6.14e-01(2.60e-01)−	7.33e-01(3.54e-01)−	1.00e+00(0.00e+00)−	6.45e-01(1.91e-01)−	2.86e-01(1.89e-01)
TA77	1.00e+00(0.00e+00)−	1.00e+00(0.00e+00)−	1.00e+00(0.00e+00)−	1.00e+00(0.00e+00)−	5.41e-01(5.70e-01)
+/≈/−	1/4/6	1/4/6	0/0/11	0/2/9	−

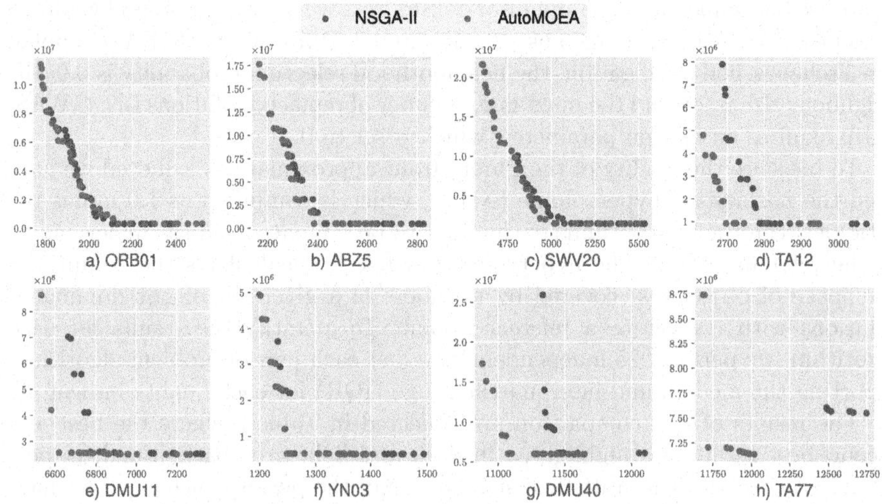

Fig. 2. Median fronts of NSGA-II and AutoMOEA for different test instances.

but the Wilcoxon rank-sum tests indicates that the differences with AutoMOEA are not statistically significant but instance FT06. NSGA-II performs best on instance LA10 and second-best in the 6 largest instances. The other algorithms compared, MOEA/D and SMS-EMOA, do not achieve the best NHV on any problem.

For illustration purposes, we include in Fig. 2 the fronts found by NSGA-II and AutoMOEA, corresponding those with the median value of the NHV indicator, on the 8 largest problem instances. We can observe that in general, the fronts of AutoMOEA have a higher degree of convergence than those obtained by NSGA-II.

6 Discussion

In the light of the results presented in the previous section, we can claim that the use of auto-configuration can lead to an AutoMOEA algorithm that outperforms the rest of classical solvers for most of the synthetic JSP instances under consideration. It is clear that the comparison is not fair, as the other algorithms are configured by default and we have spent an important amount of resources to find the settings of AutoMOEA. However, our aim has not been to make a fair comparison, but to consider a hypothetical real-world scenario in which a JSP-savvy user who is not an expert in metaheuristics is interested in finding an ad-hoc algorithm using a methodology and tools that automatically produce such an algorithm. Our experimentation shows that, from a quantitative point of view (i.e., finding an algorithm with best average performance on the selected JSP instances), the goal has been obtained; what remains is to determine whether the obtained solutions are qualitatively better. The response to this question largely depends on the cost model of the production line under study.

We delve into this matter with a practical example: let us assume that we deal with three scenarios for every JSP instance, depending on the share between fixed and variable production costs. Fixed costs include the cost of the building, maintenance/amortization of the fixed assets, depreciation of the use of machines, indirect personnel costs, etc. Such costs can be can be extrapolated to the hourly production, thus entailing a fixed share of the hourly production cost. On the other hand, variable production costs include the cost of energy, production personnel and electricity tariffs. Since NSGA-II is the best classical algorithm in the benchmark and autoMOEA has been shown to outperform the rest of its counterparts in most instances, the example only considers these algorithms. For each of such scenarios we will inspect the production schedules with least consumed energy among the global set of dominated solutions found by these algorithms for the different JSP instances. Together with their associated makespan values, we proceed by analyze the implications in terms of total economical cost of the production for the three scenarios anticipated above. To this end, the three scenarios will assume a fixed share of the cost equal to 10, 100 and $1,000$ € per hour, respectively. The overall fixed cost C_{fixed} required to complete the production commit is given by the product between the makespan (in hours) and the fixed hourly cost rate assumed for each scenario. Likewise, the variable cost C_{var} associated to the production commit will driven by the cost of the energy and the working shift during which jobs are processed.

Intuitively, a production schedule with a higher makespan would increase the fixed costs, but would surely reduce the variable costs, as the machines work at a lower speed, avoiding night shifts (more expensive) or avoiding hours during the day when the electricity cost is high. On the contrary, a lower makespan

Table 5. Differences between fixed and variable production costs of the minimum-energy solutions found by NSGA-II and AutoMOEA for each JSP instance, assuming a fixed hourly cost rate of 10, 100 and 1,000 € per hour. All quantities are in €.

Instance	LA04	LA10	FT06	ORB01	ABZ5	SWV20	TA12	DMU11	YN03	DMU40	TA77
10 € per hour											
ΔC_{fixed}	300.00	260.00	0.00	300.00	−420.00	855.00	−210.00	1,940.00	3,500.00	1,285.00	−5,070.00
ΔC_{var}	−860.00	−1,240.00	0.00	49.50	−729.00	5,275.00	1,977.50	3,509.00	5,400.00	29,563.00	363,816.00
ΔC_T	−560.00	−980.00	0.00	349.50	−1,149.00	6,130.00	1,767,50	5,449.00	8,900.00	30,848.00	358,746.00
Best?	NSGA-II	NSGA-II	Equal	AutoMOEA	NSGA-II	AutoMOEA	AutoMOEA	AutoMOEA	AutoMOEA	AutoMOEA	AutoMOEA
100 € per hour											
ΔC_{fixed}	1,200.00	5,200.00	0.00	6,000.00	−8,400.00	17,100.00	−4,200.00	38,800.00	14,000.00	25,700.00	−101,400.00
ΔC_{var}	−860.00	−1,240.00	0.00	49.50	−729.00	5,275.00	1,977.50	3,509.00	5,400.00	29,563.00	363,816.00
ΔC_T	340.00	3,960.00	0.00	6,049.50	−9,129.00	22,375.00	−2,222.50	42,309.00	19,400.00	55,263.00	262,416.00
Best?	AutoMOEA	AutoMOEA	Equal	AutoMOEA	NSGA-II	AutoMOEA	NSGA-II	AutoMOEA	AutoMOEA	AutoMOEA	AutoMOEA
1,000 € per hour											
ΔC_{fixed}	12,000.00	52,000.00	0.00	60,000.00	−84,000.00	171,000.00	−42,000.00	388,800.00	140,000.00	257,000.00	−1,014,000.00
ΔC_{var}	−860.00	−1,240.00	0.00	49.50	−729.00	5,275.00	1,977.50	3,509.00	5,400.00	29,563.00	363,816.00
ΔC_T	11,140.00	50,760.00	0.00	60,049.50	−84,729.00	176,275.00	−40,022.50	391,509.00	145,500.00	286,563.00	−650,184.00
Best?	AutoMOEA	AutoMOEA	Equal	AutoMOEA	NSGA-II	AutoMOEA	NSGA-II	AutoMOEA	AutoMOEA	AutoMOEA	NSGA-II

would reduce the fixed part of the cost, but would increase the variable share, with machines working at a higher speed and the production commit concentrated in more expensive working shifts, leading to an overall increase of the production cost. This intuitive reasoning is validated in Table 5. Columns in these tables denote economical differences between the minimum-energy solutions of the Pareto fronts approximated by NSGA-II and AutoMOEA. To begin with, $\Delta C_{fixed} = C_{fixed}(\text{NSGA-II}) - C_{fixed}(\text{AutoMOEA})$ stands for the difference between the fixed production costs of solutions found by both algorithms for each JSP instance, which are intrinsically related to their associated makespan: longer production runs will entail a higher fixed share of the production expenditure. On the other hand, $\Delta C_{var} = C_{var}(\text{NSGA-II}) - C_{var}(\text{AutoMOEA})$ reflect the difference in terms of variable production costs. Finally, $\Delta C_T = \Delta C_{fixed} + \Delta C_{var}$ stands for the total cost difference between the schedules of both algorithms. It is important to note that when any of these gaps is positive, the schedule optimized by AutoMOEA yields an economical profit when compared to that of NSGA-II. Conversely, negative values of these gaps unveil that the schedule of NSGA-II is more economically convenient.

As can be observed in this table, the economical convenience of adopting the minimum-energy production schedule found by AutoMOEA depend on the type of product manufactured by the plant and its associated production costs. Longer makespans can be detrimental for the economical viability of the plant as fixed costs become higher, so the variable share of the production costs play a crucial role in this regard. This can be noted in instance LA10: if fixed hourly costs are low, the lower variable share of the minimum-energy schedule evolved by NSGA-II results to be critical for the overall economical balance of the commit. Conversely, when fixed costs increase, the shorter makespan associated to the minimum-energy solution of AutoMOEA gives rise to a higher surplus of fixed costs, making this solution more profitable than that of NSGA-II. This, together with non-functional restrictions that often hold in practical industrial scenarios (e.g., availability of personnel for night shifts, or regulatory constraints in terms of sustainability), suggests that no universal answer can be given to the question of whether in real scenarios it makes sense to resort to automatic algorithm configuration tools to improve the cost-productivity balance of production lines.

7 Conclusions and Future Work

In this work we have presented a study about the adoption of an automatic algorithm design approach to efficiently solve a multi-objective JSP formulation that includes energy considerations as an objective. Our proposal is based on combining two software tools: the jMetal optimization framework and the irace package for automatic algorithm configuration. The result is a multi-objective solver (AutoMOEA), which results from an automatic design process over a training set composed of three small instances of the JSP, as well as a parameter space comprising 9 dimensions of a multi-objective metaheuristic algorithm.

Our proposal has been experimentally validated in a benchmark consisting of representative multi-objective optimizers (NSGA-II, SPEA2, MOEA/D, and SMS-EMOA) and 11 JSP instances with varying number of machines (from 5 to 20) and jobs (from 10 to 100). The results of the experiments in terms of the normalized hypervolume quality indicator have revealed that AutoMOEA outperforms the rest techniques according, achieving the best overall results with statistical significance in most of the cases. Performance gaps are not significant in the JSP instances where AutoMOEA does not achieve the best indicator values. Our discussion has also delved into the relevance of performance gaps in practical industrial settings, exposing that they must be examined further in terms of the fixed and variable shares of the cost of evolved schedules.

Future work will elaborate in this last concluding remark, devising new formulations of the energy-aware JSP problem that incorporate non-functional aspects that may affect the quality and economical viability of the produced schedules. To this end, we plan to investigate how to efficiently deal with changes over time in the availability of staff and machinery. Other cost models tailored for different industries will be also explored by the proposed optimization framework.

Acknowledgements. This work has been partially funded by the Spanish Ministry of Science and Innovation (grant PID2020-112540RB-C41, AEI/FEDER, UE) and the Basque Government (IT1456-22).

References

1. Adams, J., Balas, E., Zawack, D.: The shifting bottleneck procedure for job shop scheduling. Manag. Sci. **34**(3), 391–401 (1988)
2. Applegate, D., Cook, W.: A computational study of the job-shop scheduling problem. ORSA J. Comput. **3**(2), 149–156 (1991)
3. Balas, E.: Machine sequencing via disjunctive graphs: an implicit enumeration algorithm. Oper. Res. **17**(6), 941–957 (1969)

4. Balas, E., Vazacopoulos, A.: Guided local search with shifting bottleneck for job shop scheduling. Manag. Sci. **44**(2), 262–275 (1998)
5. Beume, N., Naujoks, B., Emmerich, M.: SMS-EMOA: multiobjective selection based on dominated hypervolume. Eur. J. Oper. Res. **181**(3), 1653–1669 (2007)
6. Bezerra, L.C.T., López-Ibáñez, M., Stützle, T.: Automatic design of evolutionary algorithms for multi-objective combinatorial optimization. In: Bartz-Beielstein, T., Branke, J., Filipič, B., Smith, J. (eds.) PPSN 2014. LNCS, vol. 8672, pp. 508–517. Springer, Cham (2014). https://doi.org/10.1007/978-3-319-10762-2_50
7. Bezerra, L.C.T., López-Ibáñez, M., Stützle, T.: Automatic component-wise design of multiobjective evolutionary algorithms. IEEE Trans. Evol. Comput. **20**(3), 403–417 (2016). https://doi.org/10.1109/TEVC.2015.2474158
8. Blythe, J., et al.: Task scheduling strategies for workflow-based applications in grids. In: IEEE International Symposium on Cluster Computing and the Grid, vol. 2, pp. 759–767 (2005)
9. Davis, L., et al.: Job shop scheduling with genetic algorithms. In: International Conference on Genetic Algorithms and their Applications, vol. 140 (1985)
10. Deb, K., Pratap, A., Agarwal, S., Meyarivan, T.: A fast and elitist multiobjective genetic algorithm: NSGA-II. IEEE Trans. Evol. Comput. **6**(2), 182–197 (2002)
11. Demirkol, E., Mehta, S., Uzsoy, R.: Benchmarks for shop scheduling problems. Eur. J. Oper. Res. **109**(1), 137–141 (1998)
12. Durillo, J.J., Nebro, A.J.: jMetal: a Java framework for multi-objective optimization. Adv. Eng. Softw. **42**(10), 760–771 (2011)
13. van Hoorn, J.J.: Job shop instances and solutions (2015). https://jobshop.jjvh.nl/
14. López-Ibáñez, M., Dubois-Lacoste, J., Cáceres, L.P., Birattari, M., Stützle, T.: The irace package: iterated racing for automatic algorithm configuration. Oper. Res. Perspect. **3**, 43–58 (2016)
15. Manne, A.S.: On the job-shop scheduling problem. Oper. Res. **8**(2), 219–223 (1960)
16. Nebro, A.J., Durillo, J.J., Vergne, M.: Redesigning the jMetal multi-objective optimization framework. In: Genetic and Evolutionary Computation Conference, pp. 1093–1100 (2015)
17. Nebro, A.J., López-Ibáñez, M., Barba-González, C., García-Nieto, J.: Automatic configuration of NSGA-II with jMetal and irace. In: Genetic and Evolutionary Computation Conference, pp. 1374–1381 (2019)
18. Nowicki, E., Smutnicki, C.: An advanced tabu search algorithm for the job shop problem. J. Sched. **8**(2), 145–159 (2005)
19. Para, J., Del Ser, J., Nebro, A.J.: Energy-aware multi-objective job shop scheduling optimization with metaheuristics in manufacturing industries: a critical survey, results, and perspectives. Appl. Sci. **12**(3) (2022)
20. Reza Hejazi, S., Saghafian, S.: Flowshop-scheduling problems with makespan criterion: a review. Int. J. Prod. Res. **43**, 2895–2929 (2005)
21. Sheskin, D.J.: Handbook of Parametric and Nonparametric Statistical Procedures. Chapman & Hall/CRC (2007)
22. Storer, R.H., Wu, S.D., Vaccari, R.: New search spaces for sequencing problems with application to job shop scheduling. Manag. Sci. **38**, 1495–1509 (1992)
23. Taillard, E.: Benchmarks for basic scheduling problems. Eur. J. Oper. Res. **64**(2), 278–285 (1993)
24. Weisse, T.: An Introduction to Optimization Algorithms (2020)
25. Zhang, Q., Li, H.: MOEA/D: a multiobjective evolutionary algorithm based on decomposition. IEEE Trans. Evol. Comput. **11**(6), 712–731 (2007)

26. Zhou, X., Zhang, G., Sun, J., Zhou, J., Wei, T., Hu, S.: Minimizing cost and makespan for workflow scheduling in cloud using fuzzy dominance sort based heft. Futur. Gener. Comput. Syst. **93**, 278–289 (2019)
27. Zitzler, E., Laumanns, M., Thiele, L.: SPEA2: improving the strength pareto evolutionary algorithm. Technical report. 103, Swiss Federal Institute of Technology (ETH), Zurich, Switzerland (2001)
28. Zitzler, E., Thiele, L.: Multiobjective evolutionary algorithms: a comparative case study and the strength pareto approach. IEEE Trans. Evol. Comput. **3**(4), 257–271 (1999)

Solving the Nurse Scheduling Problem Using the Whale Optimization Algorithm

Mehdi Sadeghilalimi, Malek Mouhoub(✉), and Aymen Ben Said

Department of Computer Science, University of Regina, Regina, Canada
{msv368,mouhoubm,abb549}@uregina.ca

Abstract. Managing human resources is crucial in organizations, and this can be best done through optimal workforce scheduling. Workforce scheduling is conducted regularly in transportation, manufacturing, retail stores, academic institutions, and health care units. For the latter, healthcare personnel must be assigned required shifts to satisfy hospital requirements, while optimizing costs and quality of service. In this context, we propose a nature-inspired technique based on the Whale Optimization Algorithm (WOA) for solving the Nurse Scheduling Problem (NSP). More precisely, we have redefined the WOA to deal with this combinatorial problem efficiently. To assess the performance of different variants of our discrete WOA, we conducted several experiments on randomly generated NSP instances. In the experiments, our WOA has been compared with variants of the Branch & Bound (B&B) and the Stochastic Local Search (SLS) algorithms. B&B uses constraint propagation at different levels while SLS starts with an initial configuration obtained with a backtrack search technique. Overall, the results of the comparative experiments demonstrate the superiority of the proposed WOA in terms of quality of the solution returned and the related running time.

Keywords: Combinatorial Optimization · Nature-Inspired Techniques · Metaheuristics · Nurse Scheduling Problem (NSP)

1 Introduction

The Nurse Scheduling Problem (NSP) consists of assigning nurses to a set of shifts such that all the hospital requirements are satisfied while defined costs are minimized. Due to the challenge and complexity of this problem, the NSP is among the hard combinatorial applications to solve. This has motivated several researchers to investigate different facets of the NSP and the outcomes have been reported in the literature. In order to overcome the difficulty for solving the NSP in practice, we propose a nature inspired technique, based on the Whale Optimization Algorithm (WOA). We consider a variant of the NSP where a minimum and a maximum number of nurses per shift is defined to meet hospital requirements. In addition, to ease the burden on nurses, some constraints are defined accordingly. We have redefined the WOA operators so that it can efficiently deal with this discrete optimization problem. We also propose different functions for the exploration strategy in order to diverse the search process and escape local minima. To assess the performance of different variants of our discrete WOA, we conducted several experiments on randomly generated NSP instances. Our WOA has been

B. Dorronsoro et al. (Eds.): OLA 2023, CCIS 1824, pp. 62–73, 2023.
https://doi.org/10.1007/978-3-031-34020-8_5

compared with variants of the Branch & Bound(B&B) and the Stochastic Local Search (SLS) algorithms. B&B [2] uses constraint propagation [7] to remove some locally inconsistent values, which will reduce the size of the search space. SLS starts with an initial configuration obtained with a Depth-First Search (DFS) technique. SLS then attempts to improve it, at each iteration, while maintaining satisfiability. Like for B&B, the backtrack search technique is enhanced with constraint propagation. The results of the comparative experiments demonstrate the superiority of the proposed WOA in terms of quality of the solution returned and the related running time. Despite the use of constraint propagation, both B&B and DFS still suffer from their inherited exponential time cost.

2 Related Works

There are many classifications for NSP, and different methods have been used to solve this problem, including constraint programming, metaheuristic methods, and mathematical programming. In the NSP, constraints can be categorized as hard or soft. Hard constraints are the ones that cannot be violated, while soft constraints can be violated and are often associated a penalty function to minimize [20]. In the following, we will report on the main NSP solving methods.

Metaheuristics use random search to discover near-optimal answers in a proper time. Basically, these techniques trade the quality of the solution returned for the processing time. Jan et al. [11] defined the NSP as a Multi-Objective Optimization (MOO) problem and used genetic algorithms to solve it. The method proposed in [8] uses the Ant Colony Optimization (ACO) technique to tackle the NSP in a dynamic environment. In this study, according to the soft and hard constraints defined for the problem, such as the priorities of the hospital and nurses, etc., a schedule is created. Despite the limitations of the problem and the various scenarios considered, the simulation results showed that this method does provide acceptable results. In 2013, Wu et al. [23] used ACO to solve the NSP. Indeed, ACO has been shown to perform reasonably well for over-constrained problems. In 2015, Jafari and Salmasi [10] define the NSP using several constraints and objectives. The focus is on maximizing nurses' priorities and hospital policies. Mathematical programming has been adopted to model the problem, and simulated annealing to solve it. In [19], the authors used the Bee Colony Optimization algorithm to maximize nurses' priorities and minimize penalties related to violating some soft constraints.

Exact methods where used to solve the NSP and guarantee to find an optimal solution if one exists. These methods, such as branch and bound [1,22], are however time consuming and suffer from their exponential time cost, especially when the size of the problem increases. In [1], the NSP is defined as a MOO where nurses' priorities and fairness need to be optimized. Here, the NSP is defined into subproblems where each is tackled with the branch and bound algorithm.

In [15], the authors used Bayesian optimization algorithm to solve the NSP through a learning mechanism. The proposed method relies on human behavior to learn the main NSP scenarios. A goal programming method was adopted in [21] to solve the NSP as a set of hard and soft constraints.

Hybrid methods have also been adopted to solve the NSP. These methods combine two or more algorithms with the goal to take advantage of each. In [24], Zhang et al. used a genetic algorithm and variable neighborhood search to find good schedules for the NSP. In [4], the authors used a Memetic Approach to solve the NSP. The latter uses Tabu search for small-scale problem, combined with GAs and the steepest descent improvement heuristic. It has been demonstrated that a combination of these techniques (which complement each others) performs better than when using these algorithms separately.

3 Problem Formulation

The Formulation of our NSP is listed below. Basically the main goal is to assign nurses to daily shifts such that a set of constraints are met (following hospital personnel policies) while an overall cost is minimized.

Decision Variables

$$x_{ijk} = \begin{cases} 1, & \text{if nurse } i \text{ works in shift } j \text{ on day } k. \\ 0, & \text{otherwise.} \end{cases} \tag{1}$$

We assume that we have d days and 3 shifts per day: morning shift ($j = 1$), evening shift ($j = 2$), and night shift ($j = 3$).

Constraints

1. **Minimum and Maximum number of nurses per shift.** The following constraint expresses the minimum and maximum assigned number of nurses per shift j in day k.
$$Q_{jk} \leq \sum_{i} x_{ijk} \leq S_{jk} \tag{2}$$
Q_{jk} and S_{jk} are respectively the minimum and the maximum number of nurses needed for shift j in day k.
2. **Maximum number of shifts for a given nurse during the schedule.** The following constraint sets the maximum number of shifts w_i, for a given nurse i during the schedule.
$$\sum_{j}\sum_{k} x_{ijk} \leq w_i \tag{3}$$
3. **Maximum number of consecutive shifts.** The following constraint sets the maximum number of consecutive shifts L for a given nurse i during the schedule. A consecutive shift corresponds to $j = 3$ for day k, followed by $j = 1$ for day $k+1(mod\,d)$.
$$\sum_{k}^{d} (x_{i3k} + x_{i1(k+1\,mod\,d)}) \leq L \tag{4}$$

4. **Maximum number of night shifts.** Each nurse i should not work more than n_i night shifts in the schedule.

$$\sum_k x_{ijk} \leq n_i \quad j = 3 \tag{5}$$

Objective: Hospital Costs to Minimize

$$\min(\sum_i \sum_j \sum_k c_{ij} \cdot x_{ijk}) \tag{6}$$

c_{ij} is the cost of nurse i working in shift j for any day.

4 Proposed Solving Approach

We propose a new solving method based on the Whale Optimization Algorithm (WOA) [16]. WOA is considered as a combination of two variants of PSO, namely the moth flame and the grey wolf techniques [5]. WOA is inspired by the behavior of humpback whales and has been effective in solving optimization problems. WOA uses shrinking encircling as well as spiral motions for exploitation. Exploration is achieved by having whales moving randomly. Given the discrete nature of the NSP, we have defined WOA operators as follows.

4.1 Individual Representation and Fitness Function

Each whale corresponds to a potential solution (schedule) and is expressed by a matrix where rows list nurses' shifts while columns represent the different days. Each entry (i, j) will then correspond to a given shirt $(1 \leq j \leq 3)$ assigned to a particular nurse i, as depicted in Fig. 1. Note that this representation implicitly represents the fact that each nurse cannot have more than one shift per day. The fitness function of a given whale corresponds to the objective function we defined in Eq. 6.

Shift Days	Day (1, 2, .., d)						
Nurse (1,2,3, ..., i)	j (0,1,2,3)	j (0,1,2,3)	j (0,1,2,3)	j (0,1,2,3)	j (0,1,2,3)	j (0,1,2,3)	...

Fig. 1. A solution representation in WOA

4.2 Example

Let us consider the NSP instance defined with the following parameters: $n = 3$, $k = 7$, $Q_{jk} = 1$, $S_{jk} = 3$, $w_i = 7$, $L = 0$, and $n_i = 3$. The costs are listed in Table 1. An example of potential schedule (whale) is depicted in Fig. 2. The fitness function (objective) of this whale is computed as follows.

$$X_1 = \sum_{i=1}^{3}\sum_{j=1}^{3}\sum_{k=1}^{7} c_{ij} \cdot x_{ijk} = 1.63$$

Moreover, the whale in Fig. 2 satisfies all the constraints, as shown below.

$$Q_{jk} \leq \sum_{i} x_{ijk} \leq S_{jk} \quad \Rightarrow \quad 1 \leq \sum_{i=1}^{3} x_{ij1} = 2 \leq 3$$

$$\sum_{j}\sum_{k} x_{ijk} \leq w_i \quad \Rightarrow \quad \sum_{j=0}^{3}\sum_{k=1}^{7} x_{1jk} = 6 \leq 7$$

$$\sum_{a=1}^{k} x_{ijk} \leq n_i \quad j = 3 \quad \Rightarrow \quad \sum_{a=1}^{7} x_{13k} = 1 \leq 3$$

Shift Days	Day 1	Day 2	Day 3	Day 4	Day 5	Day 6	Day 7
Whale 1	0	2	1	1	3	2	1
	1	3	2	3	0	1	2
	2	1	0	2	1	3	2

Fig. 2. A solution representation

Table 1. Cost information for the NSP instance.

Nurse no.	$c(i, j)$		
	Shift 1	Shift 2	Shift 3
1	0.1	0.1	0.11
2	0.095	0.095	0.105
3	0.07	0.07	0.08

4.3 Exploitation and Exploration in Discrete WOA

Before we define the discrete versions of the exploration functions (spiral motion and shrinking encircling), let us first presents the definition of distance in discrete WOA. The distance between two whales (representing two potential schedules) is equal to the Hamming distance between the related matrices. This basically corresponds to the number of entries that are different in both matrices.

Shrinking Encircling. The following equations were defined in [16] to guide the shrinking encircling exploration function. In this search process, each whale (represented by $X(t)$) approaches its prey (best whale, X^*) by rotating around it.

$$D = |C \cdot X^*(t) - X(t)| \tag{7}$$

$$X(t+1) = X^*(t) - A \cdot D \tag{8}$$

$$A = 2a \cdot r - a \tag{9}$$

$$C = 2 \cdot r \tag{10}$$

a and r are random parameters in [0,2] and [0,1] respectively.

We have redefined the above equations as follows. We first set parameter C to 1. Then, we defined the distance D as the Hamming distance we defined earlier. Therefore, Eq. 7 will compute the Hamming distance between $X(t)$ and $X(t)^*$. Equation 8 will then allow whale $X(t)$ to move closer to X^* by reducing (according to A) the number of entries that are different in both whales.

For example, we assume that the best whale, X^*, and whale $X(t)$ are as depicted in Fig. 3. The different values in both whales (identified in grey) are used to calculate the Hamming Distance, which is equal to 5. Let us assume that $a = 0.5$ and $r = 0.9$, hence, A equals 0.4. This means that from the entries identified as different, by the Hamming Distance, we randomly select 40% (2 entries) from the best whale and assign them to the current whale $X(t)$. In Fig. 4, we assume that the entries shown in green are those that are randomly selected for whale $X(t)$. These 2 values (3 and 2) will be replaced with the corresponding entries in X^* (1 and 1 respectively). This will allow $X(t+1)$ to move closely to X^* by reducing 40% of the different entries.

Shift Days	Day 1	Day 2	Day 3	Day 4	Day 5	Day 6	Day 7
X*(t)	0	2	1	1	3	2	1
	1	3	2	3	0	1	2
	2	1	0	2	1	3	2

Shift Days	Day 1	Day 2	Day 3	Day 4	Day 5	Day 6	Day 7
X(t)	0	1	2	1	3	2	1
	1	3	2	2	0	2	2
	2	3	0	2	1	3	2

Fig. 3. Hamming distance between $X(t)$ and $X(t)^*$ (5 different entries)

Shift Days	Day 1	Day 2	Day 3	Day 4	Day 5	Day 6	Day 7
X(t)	0	1	2	1	3	2	1
	1	3	2	2	0	2	2
	2	3	0	2	1	3	2

Fig. 4. Selecting a percentage of entries (in green) based on the value of A (Color figure online)

Spiral Attack. The following equations were defined in [16] to express the spiral attack exploration function. Here, each whale (represented by $X(t)$) approaches its prey (best whale, X^*) by following a spiral curve.

$$X(t+1) = D' \cdot e^{bl} \cdot \cos(2\pi l) + X^*(t) \tag{11}$$

In Eq. 11, b is a constant and l is a random variable between $[-1, 1]$. We have used a similar equation, with distance D' defined as the Hamming distance between the two whales. The following equations define the spiral attack for the NSP.

$$X(t+1) = X^*(t) - A \cdot D' \tag{12}$$

$$A = e^{bl} \cdot \cos(2\pi l) \tag{13}$$

To balance shrinking and spiral attacks, a random parameter, p, is generated between [0, 1] to choose between the two attacks as follows.

$$A = \begin{cases} 2a \cdot r - a & p < 0.5 \\ e^{bl} \cdot \cos(2\pi l) & p \geq 0.5 \end{cases} \tag{14}$$

Exploration. For the exploration phase, we take a similar approach to the exploitation phase to redefine the operators. The only difference is that we choose a random whale (X_{rand}) instead of the best whale (X^*). The following equation guides the exploration process.

$$X(t+1) = X_{rand}(t) - A \cdot D \tag{15}$$

It should be noted that in Eq. 15, A is calculated from Eq. 9. In addition to Eq. 15 allowing a given whale to perform a shrinking motion towards a random whale, we also consider the following techniques for exploration. Each of these methods alter some values of a given whale $X(t)$.

- **Random Resetting Mutation (RRM).** In this method, a number of entries of $X(t)$ are randomly selected and their values will be randomly changed.
- **Swap Mutation (SwM).** Pairs of values are randomly selected. Then their values are interchanged.
- **Scramble Mutation (ScM).** A subset of contiguous entries are selected from $X(t)$. Then these values are randomly scrambled.
- **Inverse Mutation (IM).** A subset of contiguous entries are selected and inverted.

5 Experimentation

To evaluate the performance of our proposed WOA, we conducted comparative experiments against variants of B&B [2] and SLS techniques. We updated the B&B in [2] to reflect the minimization variant of the NSP. B&B relies on the Depth First Search (DFS) strategy to generate and explore all candidate solutions. B&B uses the Upper Bound (UB) and Lower Bound (LB) parameters to minimize the overall hospital cost by pruning the non-optimal solutions as well as the sub-branches that are unlikely to lead to an optimal solution (the overall hospital cost is computed according to the partial costs related to shift assignments as shown in Table 2). Given that we are dealing with a minimization problem, we use the UB to track the best most-recent solution found during search, and use the LB as an optimistic estimation for the best solution during the exploration of sub-branches. The LB is computed based on the cost of the current sub-branch plus the minimum shift patterns' costs that can possibly be assigned to the remaining nurses to over-estimate the best possible solution at any search stage. The purpose of estimating the LB is to mainly forward-check if the current decision may lead to a better solution or not, as early as possible during search, to avoid the exploration of nodes that may not lead to optimal solutions. The UB and LB parameters contribute to the pruning process in such a way that if the LB becomes greater or equal to the UB at any point during search, there will be no need to further explore the remaining nodes because the current sub-branch will definitely not lead to a better solution than the optimal one that has already been found.

Although B&B guarantees the optimality of the solution returned, it may come with an exponential running time cost with respect to the domain size and the number of nurses ($O(d^n)$, where d designate the domain size and n the number of nurses). Therefore, we adopt constraint propagation techniques [7] to overcome this challenge and minimize the B&B execution time. Constraint propagation [7] may be considered as a pre-processing step before the B&B execution, and it mainly consists in optimizing the domain size by simply eliminating the locally inconsistent values that does not satisfy some set of constraints including unary, binary, and k-ary constraints. Enforcing constraint propagation may lead to two main situations. The first situation is removing all the domain values and ending up with an empty domain which confirms the inconsistency of the problem (non-existence of feasible solutions). The second situation is ending up with a reduced domain of values which will reduce the size of the search space to be used by B&B for finding the optimal solution. Considering the scope of the constraints presented in our problem formulation, and given the fact that constraints 3, 4, and 5 (in Sect. 3) are unary constraints, we apply Node Consistency (NC) [13] using these constraints before search for the purpose of minimizing the domain size and consequently reduce the B&B running time (node consistency is simply the process of eliminating the values that violates the unary constraints). We call, B&B + NC, the search method using NC as a pre-processing phase. Furthermore, we adopt the Generalized Arc Consistency (GAC) algorithm [6,14] using global constraint 2 (in Sect. 3) to eliminate some inconsistent domain values that are not part of any feasible solution. We call, B&B + NC + GAC, the method using GAC, in addition to NC.

For SLS, we use three variants. These variants simply work by getting an initial feasible solution and then trying to tune the solution while maintaining the feasibility of the solution by sequentially looking for an alternative value for each variable from its respective domain such that replacing this value would minimize our objective when paired with the remaining values in the solution. Note that improving the solution may not be guaranteed. The main difference between our proposed SLS variants is the method of obtaining the initial solution. The first variant, that we call SLS, consists of a random search to generate the initial solution. The second variant may be considered as an alternative to B&B. Given that B&B is very slow since it may require exploration of the entire search space to return the optimal solution, we propose a different method that simply returns the first feasible solution found following a Depth-first search (DFS) instead of searching for the optimal solution. After getting the initial solution, the algorithm iterates by improving the initial solution. This process works by filtering the variables' domains (i.e. removing the values with a greater cost that would not improve the solution even if they do not violate any constraint, if selected) and then systematically looking for a better value that improves the objective for each variable, using a brute-force search. We call, DFS + SLS, this second method. Since we know that DFS comes with an exponential time cost, our third SLS variant involves constraint propagation as a pre-processing step, before search, to optimize the search space and to avoid exploring the values that are not part of any feasible solution. Like for B&B, we use NC and GAC to eliminate some inconsistent values from the variables' domains which will reduce the size of the search space. We call DFS + NC + GAC + SLS this third method.

All the algorithms are implemented in MATLAB software on a computer with a Core i5-6200U processor at 2.3 GHz and 8 GB of RAM.

Table 2. The cost table

Nurse no.	$c(i, j)$		
	Shift 1	Shift 2	Shift 3
1	0.81	0.16	0.64
2	0.90	0.79	0.37
3	0.12	0.31	0.81
4	0.91	0.52	0.53
5	0.63	0.16	0.35
6	0.09	0.60	0.93
7	0.27	0.26	0.87
8	0.54	0.65	0.55
9	0.95	0.68	0.62
10	0.96	0.74	0.58
11	0.15	0.45	0.20
12	0.97	0.08	0.30
13	0.95	0.22	0.47
14	0.48	0.91	0.23
15	0.80	0.22	0.84

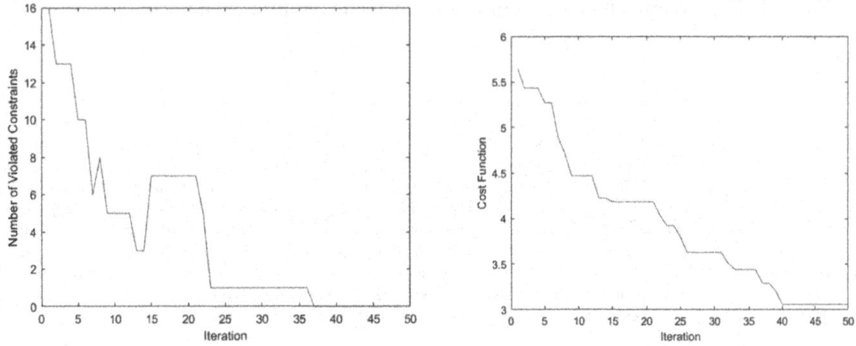

Fig. 5. The number of violated constraints (left) and cost function improvement (right) in each iteration

Table 3. The NSP parameters used in the experiments for different numbers of nurses

n	d	Q_{jk}	S_{jk}	w_i	L	n_i
5	7	1	4	5	2	3
10	7	1	7	5	2	3
15	7	1	12	5	2	3
20	7	1	15	5	2	3
30	7	1	25	5	2	3
50	7	1	35	5	2	3
60	7	1	45	5	2	3
80	7	1	65	5	2	3

The first experiments are conducted in order to depict the convergence trend of WOA. We randomly generate NSP instances with the following parameters: $n = 15$, $k = 7$, $Q_{jk} = 1$, $S_{jk} = 12$, $w_i = 5$, $L = 2$, and $n_i = 3$. The cost of nurses for different shifts is considered as a uniform distribution according to Table 2. According to the parameters listed above and the cost information in Table 2, WOA seeks to find a suitable schedule for nurses with respect to all constraints, and optimizing the overall cost function. In this regard, the left chart of Fig. 5 shows the convergence trend in terms of solved constraints. As noticed, all constraints are satisfied with less than 40 iterations. While WOA is satisfying the constraints, the cost function is decreasing at each iteration, and finally, after 40 iterations, the algorithm has been able to optimize the cost. This process is depicted in the right chart of Fig. 5. The NSP instances parameters used to conduct the comparative experiments are depicted in Table 3.

Table 4 reports on the experiment results comparing variants of the WOA algorithm, as described in the previous section, with B&B and SLS, as described previously. The quality of the best solution returned (BS) and the corresponding running time (RT) where used as comparison criteria. All the results are averaged over 10 run. The experiments were conducted on several NSP instances with the number of nurses varying from 5 to 80. While it is hard to distinguish between the different variants for small size

Table 4. The Experimental results in various methods for different number of nurses

Method	Number of Nurses															
	5		10		15		20		30		50		60		80	
	BS	RT (s)	BS	RT (s)	BC	RT (s)	BS	RT (s)	BS	RT (s)	BS	RT (s)	BS	RT (s)	BS	RT (s)
WOA	10.22	1.01	21.29	1.62	33.39	3.03	40.83	2.90	69.93	37.86	106.25	244.34	124.75	2221.78	177.66	3393.23
WOA + RRM	10.82	1.50	21.72	1.46	30.56	1.50	43.09	9.98	65.04	44.33	105.04	412.05	129.19	653.03	170.18	4196.58
WOA + SwM	10.29	0.95	22.81	1.53	32.48	**0.97**	41.85	14.96	65.28	24.40	103.26	340.61	127.09	1073.41	175.55	959.06
WOA + ScM	9.78	1.73	19.56	**0.11**	29.51	3.49	40.68	6.01	63.38	42.05	102.01	332.04	**123.40**	240.24	**169.480**	1301.61
WOA + IM	10.45	1.85	22.19	0.21	33.01	1.60	41.21	15.41	63.49	4.14	103.78	277.68	127.66	599.18	173.491	2005.55
SLS	11.57	**0.69**	24.10	0.50	35.82	1.10	47.81	**1.18**	74.72	**1.46**	114.40	**1.21**	143.03	**1.85**	190.09	**3.40**
DFS + SLS	14.86	18.90	28.68	164.11	38.91	345.57	43.62	404.75	76.08	1060.31	119.21	3873.79	148.34	4351.57	189.96	6901.90
DFS + NC + GAC + SLS	12.34	5.49	25.81	89.24	32.28	100.16	49.33	246.71	67.98	1394.28	109.96	3830.05	140.86	4781.31	185.54	5813.59
B&B + NC	**9.68**	1894	**18.86**	14415	**25.55**	22689	**38.58**	28137	**52.34**	34259	**89.35**	41459	–	–	–	–
B&B + NC + GAC	**9.68**	534	**18.86**	11400	**25.55**	16211	**38.58**	23418	**52.34**	29768	**89.35**	37108	–	–	–	–

NSPs, WOA + ScM is superior to the other methods in terms of solution quality, for large number of nurses. ScM consists of randomly scrambling a subset of contiguous entries which is effective given the nature of our problem. B&B is an exact method and always return the optimal solution (except for large instances). The algorithm does however suffer from its inherited exponential time const. The same can be said when adding DFS as an initial step for the SLS algorithm. In both B&B and SLS, constraint propagation does help lowering the running time (as a consequence of reducing the running time). However, this effort is still not enough to compete with WOA variants.

6 Conclusion and Future Work

The NSP is crucial in clinics and hospitals. Providing a schedule traditionally requires a lot of time and effort. We propose a discrete variant of the WOA algorithm to efficiently tackle the NSP. In order to evaluate the efficiency of our method, we conducted several experiments on different NSP instances. The results obtained are promising.

In the near future, we plan to explore other nature-inspired techniques such as the PSO [12] and GAs [9], and will conduct an experimental analysis on real-world scenarios. We anticipate that the latter will require dealing with the challenging task of solving the NSP in a dynamic environment. An example is the Physician Scheduling in Emergency Rooms (PSER) which consists in finding a good (ideally the best) assignment covering all the required shifts and duties, meeting work regulations and hospital policies, and maximizing individual preferences as much as possible. This task becomes even more difficult when schedules need to be re-planned in real time, due to an unexpected change in demand or physicians (or nurses) call in sick. In this context, we will adopt a nature-inspired solution that we have reported in [3, 17, 18]. Indeed, dynamic changes require an algorithm that works in an iterative manner which is consistent with the nature of metaheuristics.

References

1. Baskaran, G., Bargiela, A., Qu, R.: Integer programming: using branch and bound to solve the nurse scheduling problem. In: 2014 International Conference on Artificial Intelligence and Manufacturing Engineering (IIE ICAIME2014) (2014)
2. Ben Said, A., Mouhoub, M.: A constraint satisfaction problem (CSP) approach for the nurse scheduling problem. In: Symposium Series on Computational Intelligence (2022)

3. Bidar, M., Mouhoub, M.: Nature-inspired techniques for dynamic constraint satisfaction problems. Oper. Res. Forum **3**(2), 1–33 (2022)
4. Burke, E., Cowling, P., De Causmaecker, P., Berghe, G.V.: A memetic approach to the nurse rostering problem. Appl. Intell. **15**(3), 199–214 (2001)
5. Camacho-Villalón, C.L., Dorigo, M., Stützle, T.: Exposing the grey wolf, moth-flame, whale, firefly, bat, and antlion algorithms: six misleading optimization techniques inspired by bestial metaphors. Int. Trans. Oper. Res. (2022)
6. Cheng, K.C., Yap, R.H.: An MDD-based generalized arc consistency algorithm for positive and negative table constraints and some global constraints. Constraints **15**(2), 265–304 (2010)
7. Dechter, R., Cohen, D.: Constraint Processing. Morgan Kaufmann, Burlington (2003)
8. Gutjahr, W.J., Rauner, M.S.: An ACO algorithm for a dynamic regional nurse-scheduling problem in Austria. Comput. Oper. Res. **34**(3), 642–666 (2007)
9. Holland, J.H.: Genetic algorithms. Sci. Am. **267**(1), 66–73 (1992)
10. Jafari, H., Salmasi, N.: Maximizing the nurses' preferences in nurse scheduling problem: mathematical modeling and a meta-heuristic algorithm. J. Ind. Eng. Int. **11**(3), 439–458 (2015)
11. Jan, A., Yamamoto, M., Ohuchi, A.: Evolutionary algorithms for nurse scheduling problem. In: Proceedings of the 2000 Congress on Evolutionary Computation, CEC00 (Cat. No. 00TH8512), vol. 1, pp. 196–203. IEEE (2000)
12. Kennedy, J., Eberhart, R.: Particle swarm optimization. In: Proceedings of ICNN 1995 International Conference on Neural Networks, vol. 4, pp. 1942–1948. IEEE (1995)
13. Larrosa, J.: Node and arc consistency in weighted CSP. In: AAAI/IAAI, pp. 48–53 (2002)
14. Lecoutre, C., Szymanek, R.: Generalized arc consistency for positive table constraints. In: Benhamou, F. (ed.) CP 2006. LNCS, vol. 4204, pp. 284–298. Springer, Heidelberg (2006). https://doi.org/10.1007/11889205_22
15. Li, J., Aickelin, U.: A Bayesian optimization algorithm for the nurse scheduling problem. In: The 2003 Congress on Evolutionary Computation, CEC 2003, vol. 3, pp. 2149–2156. IEEE (2003)
16. Mirjalili, S., Lewis, A.: The whale optimization algorithm. Adv. Eng. Softw. **95**, 51–67 (2016)
17. Mouhoub, M.: Dynamic path consistency for interval-based temporal reasoning. In: Hamza, M.H. (ed.) The 21st IASTED International Multi-conference on Applied Informatics (AI 2003), Innsbruck, Austria, 10–13 February 2003, pp. 393–398. IASTED/ACTA Press (2003)
18. Mouhoub, M., Sukpan, A.: Conditional and composite temporal CSPs. Appl. Intell. **36**(1), 90–107 (2012). https://doi.org/10.1007/s10489-010-0246-z
19. Rajeswari, M., Amudhavel, J., Pothula, S., Dhavachelvan, P.: Directed bee colony optimization algorithm to solve the nurse rostering problem. Comput. Intell. Neurosci. **2017** (2017)
20. Russell, S., Norvig, P.: Artificial Intelligence: A Modern Approach. Prentice Hall, Hoboken (2002)
21. Topaloglu, S.: A multi-objective programming model for scheduling emergency medicine residents. Comput. Ind. Eng. **51**(3), 375–388 (2006)
22. Woeginger, G.J.: Exact algorithms for NP-hard problems: a survey. In: Jünger, M., Reinelt, G., Rinaldi, G. (eds.) Combinatorial Optimization—Eureka, You Shrink! LNCS, vol. 2570, pp. 185–207. Springer, Heidelberg (2003). https://doi.org/10.1007/3-540-36478-1_17
23. Wu, J., Lin, Y., Zhan, Z., Chen, W., Lin, Y., Chen, J.: An ant colony optimization approach for nurse rostering problem. In: 2013 IEEE International Conference on Systems, Man, and Cybernetics, pp. 1672–1676. IEEE (2013)
24. Zhang, Z., Hao, Z., Huang, H.: Hybrid swarm-based optimization algorithm of GA & VNS for nurse scheduling problem. In: Liu, B., Chai, C. (eds.) ICICA 2011. LNCS, vol. 7030, pp. 375–382. Springer, Heidelberg (2011). https://doi.org/10.1007/978-3-642-25255-6_48

A Hierarchical Cooperative Coevolutionary Approach to Solve Very Large-Scale Traveling Salesman Problem

Rui Zhong[1(✉)], Enzhi Zhang[1], and Masaharu Munetomo[2]

[1] Graduate School of Information Science and Technology, Hokkaido University, Sapporo, Japan
{rui.zhong.u5,enzhi.zhang.n6}@elms.hokudai.ac.jp
[2] Information Initiative Center, Hokkaido University, Sapporo, Japan
munetomo@iic.hokudai.ac.jp

Abstract. In this paper, we propose a hierarchical Cooperative Coevolution framework (hCC) to deal with the Very Large-scale Traveling Salesman Problem (VLSTSP). Due to the existence of the curse of dimensionality, it is difficult to find an acceptable solution for VLSTSP with conventional Evolutionary Algorithms (EAs). Cooperative Coevolution (CC) framework, which divides the problems into multiple subcomponents and optimizes them independently, offers a potential opportunity to find suitable solutions. However, conventional CC with large-scale sub-size decomposition will still be affected by the curse of dimensionality, and small-scale sub-size decomposition ignores many interactions between subcomponents. Although the initial sub-size of the proposed hCC is small-scale, the ignored interactions will be reconsidered in the higher layer of optimization. Another issue is how to decompose the VLSTSP. In the numerical experiments, we design two strategies: (1). Random decomposition. (2). Decomposition based on the greedy solution. 10 symmetric instances of VLSTSP ranging from 38,478 to 238,025 are employed to evaluate our proposal, and the basic optimizer is greedy Local Search (gLS). Experimental results show that our proposal has great potential and scalability to deal with VLSTSP, which can be easily extended to deal with various types of TSP.

Keywords: hierarchical Cooperative Coevolution (hCC) · greedy Local Search (gLS) · Very Large-scale Traveling Salesman Problem (VLSTSP)

1 Introduction

The Traveling Salesman Problem (TSP) [8] is one of the most famous NP-hard combinatorial optimization problems and has been widely studied in the fields of operations research [14,17] and theoretical computer science [12] in the past decades. Many real-world problems based on TSP also exist, such as planning

B. Dorronsoro et al. (Eds.): OLA 2023, CCIS 1824, pp. 74–84, 2023.
https://doi.org/10.1007/978-3-031-34020-8_6

and logistics [4], arranging school bus paths [9], transportation of farming equipment [10], and so on.

The time complexity of solving TSP by brute-force search is $O(n!)$. Owing to the existence of the curse of dimensionality [6], it is difficult to deal with Large-scale TSP (LSTSP) and almost impossible to find the global optimal tour of Very Large-scale TSP (VLSTSP) under the limited computational time. Meanwhile, researchers notice that evolutionary algorithms (EAs) can achieve an acceptable solution by iteration, and many algorithms have been proposed to deal with LSTSP [2,7,11], which achieve great success. However, VLSTSP makes the performance of EAs degrade rapidly, and the curse of dimensionality is a huge obstacle in solving VLSTSP [5].

Cooperative Coevolution (CC) [13] is a flexible and efficient framework to deal with Large-scale optimization problems (LSOPs). Based on divide and conquer, the original LSOPs are decomposed into several non-separable subcomponents and optimized alternately. This strategy can alleviate the influence of the curse of dimensionality directly and accelerates the convergence of optimization.

However, the conventional CC framework is also defective to solve VLSTSP: The decomposition with large-scale subcomponents will still degenerate the performance of EAs, and the decomposition with small-scale subcomponents will ignore many interactions between subcomponents. Therefore, this paper proposes a hierarchical Cooperative Coevolution (hCC) framework and combined it with greedy Local Search (gLS) to solve VLSTSP, our proposal is named hCC-gLS. Although hCC-gLS has a small-scale decomposition in the initial stage, the repeat of subcomponents optimization and combination will reconsider the neglected interactions in the higher layer of optimization.

The rest of the paper is organized as follows: Sect. 2 covers the related works. Section 3 introduces our proposal hCC-gLS in detail. In Sect. 4, we show the experimental results of our proposal. Section 5 discusses the analysis and future direction of research. Finally, we conclude our paper in Sect. 6.

2 Related Works

2.1 Traveling Salesman Problem (TSP)

The simplest instance of TSP can be described as: A salesman tries to find the shortest closed route to visit a set of cities under the conditions that each city is visited exactly once. The salesman assumes the distances between any pair of cities are assumed to be known. In this work, we concentrate on the 2-D symmetric TSP. Given a list of N city coordinates $\{x_1, x_2, ..., x_N\} \in \mathbb{R}^2$, we wish to find an optimal permutation σ over the cities that minimizes the tour length:

$$L(\sigma, \mathbf{X}) = \sum_{i=1}^{N} \|\mathbf{x}_{\sigma(i)} - \mathbf{x}_{\sigma(i+1)}\|^2 \tag{1}$$

where $\sigma(1) = \sigma(N+1)$, $\sigma(i) \in \{1, ..., N\}$, $\sigma(i) \neq \sigma(j)$ for any $i \neq j$, and $\mathbf{X} = \{\mathbf{x}_1, \mathbf{x}_2, ..., \mathbf{x}_i\}$ is a matrix consisting of all city coordinates $\mathbf{x}_i$.

2.2 Cooperative Coevolution (CC)

Inspired by divide and conquer, CC decomposes the VLSTSPs into multiple subcomponents, which can be described as:

$$\min f(\mathbf{x}_1, \mathbf{x}_2, ..., \mathbf{x}_n) = (\min_{c_1} f_1(..., ...), ..., \min_{c_m} f_m(..., ...)) \quad (2)$$

where $\mathbf{x}_i$ denotes the coordinate of city i, and $f(\cdot)$ represents the objective function. The c_j means a subcomponent in the CC framework, and $f_j(\cdot)$ stands for the objective function in subcomponent j. Different from the CC in continuous LSOPs, it is unnecessary to maintain a context vector [1] to form a complete solution, each subcomponent can be directly evaluated in TSP. In a study for solving VLSTSPs, the CC deals with the problem by dividing it into a set of smaller and simpler subcomponents and optimizing them separately. In summary, there are three steps consisting of the CC framework.

Problem decomposition: Decomposing the VLSTSPs into multiple subcomponents.
Subcomponent optimization: Optimization methods are applied to each subcomponent.
Cooperative combination: Combining the solutions of all subcomponents to construct the complete solution.

3 Our Proposal: hCC-gLS

In this section, we will introduce our proposal: hCC-gLS in detail. Figure 1 demonstrates the main steps.

In the decomposition period, we design two strategies to form the subcomponents: random decomposition and knowledge-based decomposition. In random decomposition, we randomly shuffle the city list $[0, 1, 2, ..., n]$ and divide the cities with pre-defined subcomponents' size. In knowledge-based decomposition, we first execute the greedy search (GS) to find an initial solution, and decomposition is also implemented based on the order of this solution.

In the optimization stage, we optimize the subcomponents with gLS iteratively, and each optimized subcomponent is merged with the nearest subcomponent to form a larger subcomponent. The procedure of optimization and combination is repeated until all subcomponents are merged.

3.1 Hierarchical Cooperative Coevolution (hCC)

hCC framework is proposed to deal with VLSTSP. Figure 2 is an example to demonstrate how hCC works.

In Fig. 2(a), we divide 4 cities as a subcomponent, and the interactions between subcomponents are neglected. However, these ignored interactions are reconsidered in Fig. 2(d) and (f). This strategy endows the ability of the optimizer to get rid of the local optima.

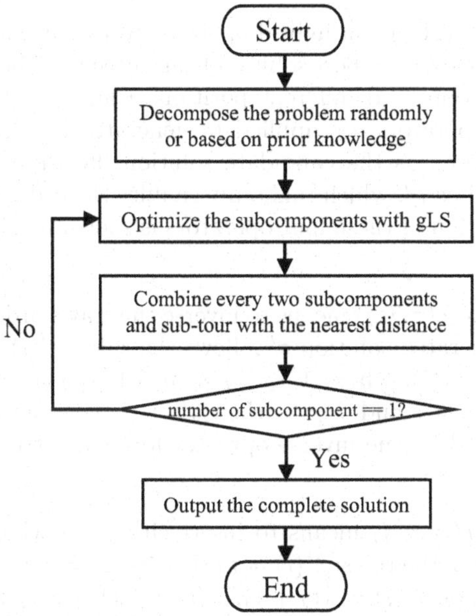

Fig. 1. The flowchart of hCC-gLS.

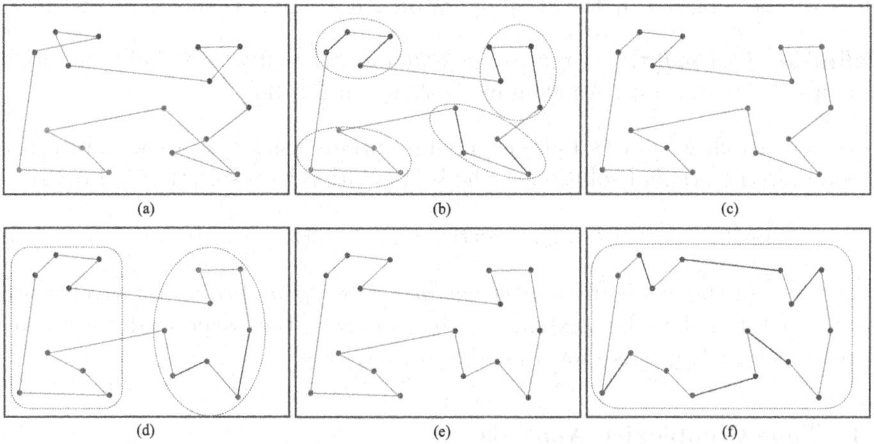

Fig. 2. The demonstration of hCC-gLS. (a) The tour and decomposition are found by the GS. (b) gLS is employed to optimize each subcomponent. (c) Subcomponents and sub-tours are merged. (d) Repeat the subcomponents optimization. (e) Repeat the combination of subcomponents and sub-tours. (f) Repeat the subcomponents optimization until sub-tours form a complete solution.

3.2 Greedy Local Search (gLS)

Greedy Local Search (gLS) is a hybrid operator to produce candidate solutions [15]. This operator selects the best solution from three neighbors greedily. Specifically, after two random and different positions i and j are selected, *inverse*, *insert*, and *swap* operators are applied to generate three neighbor solutions. And the best is chosen as the candidate solution. Furthermore, we develop a new operator named *rand*, which randomly shuffles the order between city i and city j. A demonstration of these four operators is shown in Fig. 3 and defined as follows:

Definition 1. $inverse(\pi, i, j)$ means to inverse the travel order of cities between i and j, and the generated solution π' follows the rules: $\pi'(i) = \pi(j), \pi'(i+1) = \pi(j-1), ..., \pi'(j) = \pi(i)$, where $1 \leq i, j \leq n \wedge 1 \leq j - i \leq n - 1$. n is the tour length. When $i = 1$ and $j = n$, then $\pi'(i) = \pi(j)$ and $\pi'(j) = \pi(i)$. Two edges will be replaced by the inverse operator for symmetric TSP. An example is shown in Fig. 3(b).

Definition 2. $insert(\pi, i, j)$ means to insert city j into the position i. And a new solution π' follows the rules: $\pi'(i) = \pi(j), \pi'(i+1) = \pi(i), ..., \pi'(j) = \pi(j-1)$ in the case of $i < j$, or $\pi'(j) = \pi(j+1), ..., \pi'(i-1) = \pi(i), \pi'(i) = \pi(j)$ in the case of $i > j$. In general, three edges will be replaced by the insert operator. An example is shown in Fig. 3(c).

Definition 3. $swap(\pi, i, j)$ means to swap the city i and city j in travel tour, which follows simple principle: $\pi'(i) = \pi(j)$ and $\pi'(j) = \pi(i)$. In general, four edges will be replaced by the swap operator. An example is shown in Fig. 3(d).

Definition 4. $rand(\pi, i, j)$ means to randomize the city order between city i and city j in travel tour. An example is shown in Fig. 3(e).

Once the search generates four neighbor solutions using the above strategies, considering the original solution π, the best solution π' is selected by Eq. (3).

$$\pi' = \min(\pi, inverse(\pi, i, j), insert(\pi, i, j), swap(\pi, i, j), rand(\pi, i, j)) \tag{3}$$

gLS repeat the procedure of *inverse*, *insert*, *swap*, and *rand* iteratively until the computational budget exhausted. In summary, the pseudocode of greedy search-based hCC-gLS is shown in Algorithm 1.

3.3 Time Complexity Analysis

This section analyzes the time complexity of greedy search-based hCC-gLS. There are two stages of our proposal which need to be analyzed: (1). Initialization (greedy search) stage (2). hCC-gLS stage.

Suppose the dimension of VLSTSP is n, in the initial solution generation (greedy search), we first build the adjacent matrix to save the adjacent information of the cities, and the time complexity is $O(n(n-1)/2) = O(n^2)$. To find the

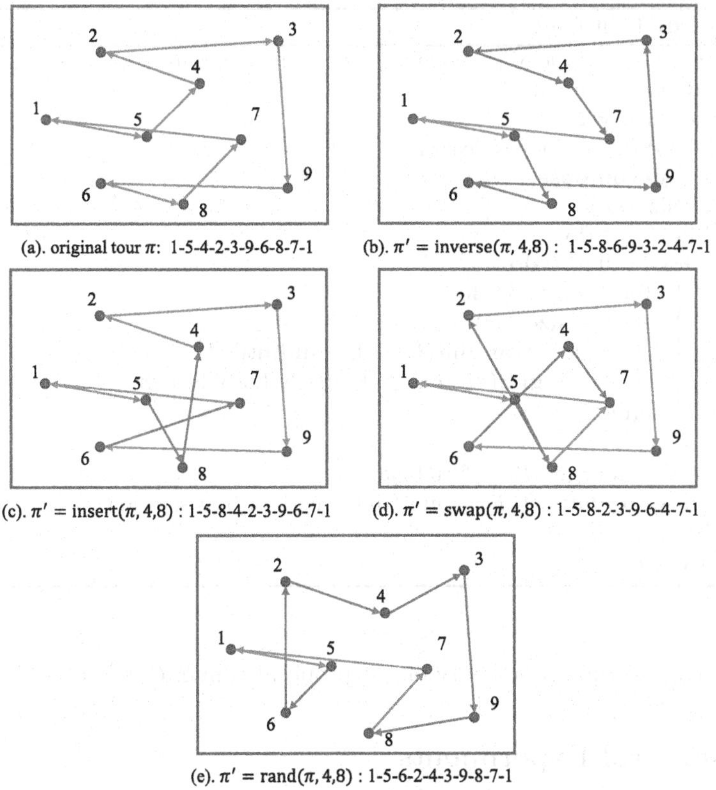

(a). original tour π: 1-5-4-2-3-9-6-8-7-1

(b). π' = inverse(π, 4,8) : 1-5-8-6-9-3-2-4-7-1

(c). π' = insert(π, 4,8) : 1-5-8-4-2-3-9-6-7-1

(d). π' = swap(π, 4,8) : 1-5-8-2-3-9-6-4-7-1

(e). π' = rand(π, 4,8) : 1-5-6-2-4-3-9-8-7-1

Fig. 3. The *inverse*, *insert*, *swap*, and *rand* operators generate neighbor solutions. (a) Original tour π. (b) Inverse π with $(i, j) = (4, 8)$. (c) Insert π with $(i, j) = (4, 8)$. (d) Swap π with $(i, j) = (4, 8)$. (e) Randomize π with $(i, j) = (4, 8)$.

greedy tour, each city has to find the closest city which has not been allocated, thus the time complexity is also $O(n(n-1)/2) = O(n^2)$. In summary, the time complexity of initialization is $O(n^2)$.

In the hCC-gLS stage, suppose the optimization iteration for each subcomponent is k, and the minimal scale of the subcomponent is m. Figure 4 further demonstrates this tree-structure optimization.

In layer 1, the number of subcomponents is $\mathbf{int}((n+m-1)/m)$, where $\mathbf{int}()$ only keeps an integral part of a value, and optimization complexity is $4k$ because four neighbor solutions are generated for once search. Therefore, the time complexity of bottom layer is $O(4k \cdot \mathbf{int}((n+m-1)/m))$, which approximately equals to $O(4kn/m))$. And the time complexity of the second layer is approximately equal to $O(4kn/2m))$, the third layer is $O(4kn/4m))$, and so on. Thus the time complexity of hCC-gLS is:

$$4k \cdot O(n/m + n/2m + n/4m...) := 4k \cdot O(n/m) = O(4kn/m) \tag{4}$$

Algorithm 1: hCC-gLS

Input: cities : C, scale of subcomponents : s, Maximum iteration : M
Output: Best solution : E

Function hCC-gLS(C, n, M):
  $E \leftarrow$ **GS**(C) # Greedy search
  $sT \leftarrow$ **decompose**(E, s)
  $L \leftarrow$ **size**(sT)
  while $L \neq 1$ **do**
    for $i = 0$ *to* L **do**
      for $j = 0$ *to* M **do**
        $l \leftarrow$ **size**(sT_i)
        $r_1, r_2 \leftarrow$ **randint**$(1, l-1)$, **randint**$(1, l-1)$ # $r_1 \neq r_2$
        $sT_i \leftarrow$ **gLS**(sT_i, r_1, r_2) # greedy Local Search
      end
    end
    $sT \leftarrow$ **merge**(sT) # Combination
    $E \leftarrow$ **update**(sT, E) # update the current best solution
  end
  return E

In summary, the time complexity of our proposal is $\max(O(n^2), O(4kn/m))$.

4 Numerical Experiments

In this section, a set of numerical experiments are executed for the algorithm investigation. In Sect. 4.1, we introduce the experiment settings, including the experiment environment, benchmark instances, and comparing methods. Section 4.2 provides the experimental result of benchmarks.

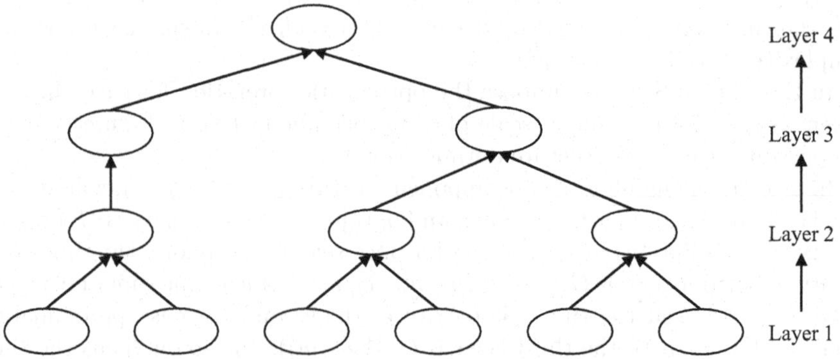

Fig. 4. A demonstration of the structure of hCC framework.

4.1 Experiment Settings

Experiment Environment. All algorithms are programmed with Python 3.7 and implemented in Hokkaido University's high-performance intercloud supercomputer equipped with CentOS operating system, Intel Xeon Gold 6148 CPU, and 384 GB RAM.

Benchmark Instances. In this paper, we apply 10 symmetric VLSTSP instances to evaluate our proposal, they are bby34656, pba38478, ics39603, rbz43748, fht47608, fna52057, bna56769, dan59296, sra104815, and ara238025. The number in the name of instances is the number of cities. All instances are provided by VLSI Data Sets [3].

Compared Methods. Table 1 shows the shortened name and description of compared methods. The size of subcomponents and maximum iteration of a subcomponent are 100. To compare these methods fairly, the evaluation times of the complete solution among all compared methods are identical. Notice that this description is to complete solution, which means all mutually exclusive subcomponents optimized 1 time are equal to optimizing a complete problem 1 time. Thus, from Fig. 4, we can calculate the maximum iteration times without hCC framework is equal to $100 \times p$, and p is the number of layers.

Table 1. The shortened name and description of compared methods

Name	Description
gLS	The shortened name of greedy Local Search, the initial tour is randomly generated
hCC-gLS	Our proposed hCC framework with gLS optimizer, the initial tour is randomly generated
gLS-G	The shortened name of greedy Local Search, the initial tour is found by greedy search
hCC-gLS-G	Our proposed hCC framework with gLS optimizer, the initial tour is found by greedy search

4.2 Experimental Results

In this section, the performance of our proposal is studied. Experiments are conducted on the benchmark functions presented in Sect. 4.1. Table 2 shows the experimental results within 10 independent trial runs with compared methods.

5 Discussion

This section consists of two parts: Sect. 5.1 analyzes the experimental results, and Sect. 5.2 lists some future topics.

Table 2. The experimental results of 10 trial runs among 4 compared methods.

Inst	gLS		hCC-gLS		gLS-G		hCC-gLS-G	
	mean dis.	time (sec.)	mean dis.	time (sec.)	mean dis.	time (sec.)	mean dis.	time (sec.)
bby34656	15,822,156	228	**13,820,836**	495	125,990	1674	**125,602**	1985
pba38478	19,182,518	236	**16,749,710**	508	134,730	1824	**134,414**	2162
ics39603	20,648,522	301	**17,561,793**	652	133,045	2014	**132,624**	2373
rbz43748	23,281,293	330	**20,294,489**	627	157,154	2126	**156,725**	2314
fht47608	26,678,349	364	**22,924,353**	683	155,180	2546	**154,895**	2747
fna52057	31,080,146	404	**27,072,440**	833	186,428	3044	**185,971**	3315
bna56769	32,034,954	443	**28,271,489**	908	199,225	3607	**198,776**	3902
dan59296	36,224,847	464	**31,725,212**	955	208,295	3949	**207,667**	4260
sra104815	80,233,169	796	**70,242,189**	1846	329,426	11813	**328,548**	12468
ara238025	284,079,330	1905	**246,528,008**	4655	760,400	58652	**758,527**	60351

5.1 Experimental Results Analysis

Comparison Between gLS and hCC-gLS. Experimental results between the gLS and hCC-gLS in Table 2 show that the introduction of hCC can support the gLS to find a better solution, and the extra time consumption is affordable. Although the random decomposition cannot capture the interactions with high accuracy, paper [18] provides a mathematical proof that random grouping has a high probability to capture some correct interactions, and these limited correct interactions and divide-and-conquer strategy can accelerate the optimization, while pure gLS cannot detect the superior connections in VLSTSP due to the curse of dimensionality.

Comparison Between gLS-G and hCC-gLS-G. Greedy Search (GS) is a naive but efficient method, especially in VLSTSPs, improvements in Table 2 further verify the efficiency of GS in high-dimensional problems. This greedy scheme connects the nearest city from candidates to construct the solution, which is at the cost of computational time and easily trapped into the local optimum. However, this greedy strategy is consistent with the decomposition principle that the closer cities have stronger relationships, thus, the solution found by GS not only an initial solution but also contains the correct linkage information, which can help the hCC framework to form the subcomponent suitably and accelerate the optimization.

5.2 Potential Topics

The above experimental results and analysis show that our proposal hCC-gLS has great potential to solve VLSTSP, and there are still some open topics for our future research.

More Powerful Optimizer. Greedy Local Search is a simple but efficient optimizer for exploitation. However, as the dimension of the problem increases, the concept of local is affected by the curse of dimensionality, and the scale of local also increases exponentially. No Free Lunch Theorem [16] proves that all optimization algorithms have the same averaging performance on all possible problems, and algorithm A performs better on exploration may perform worse on exploitation. Thus, it is necessary to apply different types of optimizers for specific problems with various scales. For example, local search is employed in the early stage when the scale of subcomponents is quite small, and Genetic Algorithm (GA) or Ant Colony Optimization (ACO) is applied in the late stage when the scale of subcomponents is quite large. And how to make this decision is an interesting topic. Reinforcement Learning (RL) is a good choice. The environment of RL in this situation is the scale of subcomponents, the agent is a decision maker, the action is the execution of exploitation or exploration, and the reward is the performance of a certain decision.

Parallelization. VLSTSP is a very time-consuming task. From the execution time in Table 2, a trial run of the instance with 100 million cities is unaffordable without parallelization, even in the high-performance supercomputer. Luckily, parallelization is a friendly and flexible approach to dealing with VLSTSP. It is unnecessary to form a complete solution in optimization to be evaluated, each subcomponent can be optimized separately and form an independent sub-tour to calculate the distance. Thus, parallelization and GPU-based programming are suitable approaches to accelerate the implementation of VLSTSP.

6 Conclusion

This paper proposes a hierarchical Cooperative Coevolution framework (hCC) to deal with VLSTSP based on divide and conquer, and greedy Local Search (gLS) as a basic optimizer is applied to optimize the subcomponents. We also emphasize the importance of the initial solution. A well-performed solution contains effective linkage information which can help the hCC framework to form subcomponents.

At the end of this paper, we list some interesting and potential topics which can improve our algorithm. Finally, our proposal is a promising study for addressing VLSTSP.

Acknowledgement. This work was supported by JSPS KAKENHI Grant Number JP20K11967.

References

1. Van den Bergh, F., Engelbrecht, A.P.: A cooperative approach to particle swarm optimization. IEEE Trans. Evol. Comput. **8**(3), 225–239 (2004)
2. Chen, J., Wang, Y., Xue, X., Cheng, S., El-Abd, M.: Cooperative co-evolutionary metaheuristics for solving large-scale TSP art project. In: 2019 IEEE Symposium Series on Computational Intelligence (SSCI), pp. 2706–2713. IEEE (2019)

3. Cook, W.: Traveling salesman problem. http://www.math.uwaterloo.ca/tsp/index.html/
4. Huang, B., Yao, L., Raguraman, K.: Bi-level GA and GIS for multi-objective TSP route planning. Transp. Plan. Technol. **29**(2), 105–124 (2006)
5. Ismkhan, H.: Effective heuristics for ant colony optimization to handle large-scale problems. Swarm Evol. Comput. **32**, 140–149 (2017)
6. Köppen, M.: The curse of dimensionality. In: 5th Online World Conference on Soft Computing in Industrial Applications (WSC5), vol. 1, pp. 4–8 (2000)
7. Krasnogor, N., Smith, J.: A memetic algorithm with self-adaptive local search: TSP as a case study. In: Proceedings of the 2nd Annual Conference on Genetic and Evolutionary Computation, pp. 987–994 (2000)
8. Larranaga, P., Kuijpers, C.M.H., Murga, R.H., Inza, I., Dizdarevic, S.: Genetic algorithms for the travelling salesman problem: a review of representations and operators. Artif. Intell. Rev. **13**(2), 129–170 (1999)
9. Lawler, E.L., Lenstra, J.K., Kan, A.H.R., Shmoys, D.B.: The traveling salesman problem: a guided tour of combinatorial optimization. J. Oper. Res. Soc. **37**(5), 535–536 (1986). https://doi.org/10.1057/jors.1986.93
10. Lenstra, J., Shmoys, D.: In pursuit of the traveling salesman: mathematics at the limits of computation. Not. Am. Math. Soc. **63**, 635–638 (2016). https://doi.org/10.1090/noti1397
11. Meuth, R.J., Wunsch, D.C.: Divide and conquer evolutionary TSP solution for vehicle path planning. In: 2008 IEEE Congress on Evolutionary Computation (IEEE World Congress on Computational Intelligence), pp. 676–681 (2008). https://doi.org/10.1109/CEC.2008.4630868
12. Pihera, J., Musliu, N.: Application of machine learning to algorithm selection for TSP. In: 2014 IEEE 26th International Conference on Tools with Artificial Intelligence, pp. 47–54 (2014). https://doi.org/10.1109/ICTAI.2014.18
13. Potter, M.A., De Jong, K.A.: A cooperative coevolutionary approach to function optimization. In: Davidor, Y., Schwefel, H.-P., Männer, R. (eds.) PPSN 1994. LNCS, vol. 866, pp. 249–257. Springer, Heidelberg (1994). https://doi.org/10.1007/3-540-58484-6_269
14. Uluçınar, Ş.: Coevolutionary memetic algorithms for solving traveling salesman problem (TSP). Ph.D. thesis, Eastern Mediterranean University (EMU) (2013)
15. Wang, C., Lin, M., Zhong, Y., Zhang, H.: Solving travelling salesman problem using multiagent simulated annealing algorithm with instance-based sampling. Int. J. Comput. Sci. Math. **6**(4), 336–353 (2015). https://doi.org/10.1504/IJCSM.2015.071818
16. Wolpert, D., Macready, W.: No free lunch theorems for optimization. IEEE Trans. Evol. Comput. **1**(1), 67–82 (1997). https://doi.org/10.1109/4235.585893
17. Yan, L., Kongyu, Y.: Immunity genetic algorithm based on elitist strategy and its application to the TSP problem. In: 2008 International Symposium on Intelligent Information Technology Application Workshops, pp. 3–6. IEEE (2008)
18. Yang, Z., Tang, K., Yao, X.: Large scale evolutionary optimization using cooperative coevolution. Inf. Sci. **178**(15), 2985–2999 (2008). https://doi.org/10.1016/j.ins.2008.02.017. Nature Inspired Problem-Solving

Tornado: An Autonomous Chaotic Algorithm for High Dimensional Global Optimization Problems

Nassime Aslimani[1(✉)], El-Ghazali Talbi[1], and Rachid Ellaia[2]

[1] University of Lille and INRIA, Lille, France
n.aslimani@yahoo.fr, el-ghazali.talbi@univ-lille.fr

[2] Engineering for Smart and Sustainable Systems Research Center (E3S), Mohammadia School of Engineers, Mohammed V University in Rabat, Rabat, Morocco
ellaia@emi.ac.ma

Abstract. In this paper we propose an autonomous chaotic optimization algorithm, called Tornado, for high dimensional global optimization problems. The algorithm introduces advanced symmetrization, levelling and fine search strategies for an efficient and effective exploration of the search space and exploitation of the best found solutions. To our knowledge, this is the first accurate and fast autonomous chaotic algorithm solving large scale optimization problems.

A panel of various benchmark problems with different properties was used to assess the performance of the proposed chaotic algorithm. The obtained results have shown the scalability of the algorithm in contrast to chaotic optimization algorithms encountered in the literature. Moreover, in comparison with some state-of-the-art metaheuristics (e.g. evolutionary algorithms, swarm intelligence), the computational results revealed that the proposed Tornado algorithm is an effective and efficient optimization algorithm.

A panel of various benchmark problems with different properties was used to assess the performance of the proposed chaotic algorithm. The obtained results have shown the scalability of the algorithm in contrast to chaotic optimization algorithms encountered in the literature. Moreover, in comparison with some state-of-the-art metaheuristics (e.g. evolutionary algorithms, swarm intelligence), the computational results revealed that the proposed Tornado algorithm is an effective and efficient optimization algorithm.

Keywords: Global optimization · Chaos optimization algorithm · Levelling · Symmetrization · Fine search · Large scale optimization

1 Introduction

Chaos theory is a branch of mathematics dealing on the study of dynamical systems whose apparently-random states of disorder and irregularities are often

B. Dorronsoro et al. (Eds.): OLA 2023, CCIS 1824, pp. 85–113, 2023.
https://doi.org/10.1007/978-3-031-34020-8_7

governed by deterministic laws [1]. Chaotic behavior exists in many natural systems, including fluid flow, weather and climate. It also occurs spontaneously in some systems with artificial components, such as stock market and road traffic. Chaotic systems are characterized by high sensitive dependence to initial conditions, an effect which is popularly known as the butterfly effect [2]. As a result of this sensitivity, the behaviour of such systems appears to be stochastic, even though the model of the system is deterministic, meaning that their future behaviour is fully determined by their initial conditions, with no random elements involved. Another consequence of the butterfly effect is unpredictability. Small differences in initial inputs yield widely divergent solutions results after several cycles through the system. In recent years, chaos has gained increasing attention and have been widely investigated in various disciplines such as control [3,4] and optimization [5].

Nowadays, there is a need for more effective and efficient optimization techniques, able to solve high dimensional problems. State-of-the-art chaos based optimization algorithms (COAs) are not efficient for high dimensional optimization problems [6]. They are not even operational for a dimension greater than 5 [6]. Existing COAs are deficient in terms of:

- *Exploration of the search space*: indeed, the irregularity of the chaos dynamics grows quickly with the problem dimension. This is due to the intrinsic imprevisibility of chaotic dynamics [6].
- *Exploitation of the best found solutions*: the main search mechanism used in COAs is not adapted for a good exploitation. It selects in a random way the direction around the current solution [7].

This paper is the culmination of an approach that leads to an autonomous COA algorithm. First, the following strategies have been introduced in a gradient-based chaotic algorithm to improve the regularity and the flexibility of the algorithm [8,9]:

- *Symmetrization*: on the one hand, Symmetrization induces a regular structure into the chaotic dynamics for a better exploration. On the other hand, based on a stochastic decomposition strategy, it enables an efficient and scalable alternative search mechanisms for a better exploitation in the search space.
- *Levelling*: in fact, the chaotic dynamics has been restructured by a leveling approach. This allows to generate different flexible chaotic levels to improve the exploration and the exploitation of the search space.
- *Hybridization with local search*: a combination with gradient based algorithm has been carried out for continuous differentiable functions.

In this paper, an autonomous Chaos is introduced which speed-ups the convergence and improves the accuracy of the search for high dimensional problems. An autonomous and pure chaotic algorithm has been developed, in which the combination with a local search algorithm (e.g. gradient descent) has been replaced by a chaotic fine search. The computational results for many test functions with different properties and levels of complexity has shown the effectiveness, efficiency, and scalability of the autonomous chaotic algorithm in tackling high dimensional optimization problems.

This paper is organized as follows. In Sect. 2, the related work on chaos optimization algorithms and state-of-the art global optimization algorithms (e.g. evolutionary algorithms, swarm intelligence) are presented. Section 3 details the novel autonomous chaos optimization, the Tornado algorithm. Section 4 shows the computational experiments of the proposed algorithm. A comparison has been carried out as well with popular chaos optimization algorithms and state-of-the-art stochastic metaheuristics (e.g. evolutionary algorithms, swarm intelligence). The conclusion and the perspectives of this work are made in Sect. 5.

2 Related Work

Consider an optimization problem with bounding constraints[1]:

$$\text{Minimize } f(X) \text{ subject to } X \in [L, U], \tag{1}$$

where

- $f : \mathbb{R}^n \longrightarrow \mathbb{R}$, denotes the objective function,
- $X = (x_1, .., x_n) \in \mathbb{R}^n$, the decision vector whose components x_i are bounded by lower bounds l_i and upper bounds u_i. and $[L, U] = \prod_{i=1}^{n} [l_i, u_i]$.

Chaos is a universal nonlinear phenomenon with stochastic, ergodic, and regular properties. Ergodicity can be used as a search mechanism for optimization. The sequence of solutions is generated by means of a chaotic map. Different chaotic maps exist in the literature [10]. The most popular ones are:

- The logistic map: $x_k = \mu . x_k(1 - x_k), \quad 0 < x_0 < 1, \quad 0 \leqslant \mu \leqslant 4$
- The Kent map: $x_{k+1} = \begin{cases} x_k/\beta & \text{if } 0 < x_k < \beta \\ (1 - x_k)/(1 - \beta) & \text{if } \beta < x_k < 1 \end{cases},$
- The Henon map: $\begin{cases} x_{k+1} = 1 - a x_k^2 + y_k \\ y_{k+1} = b x_k \end{cases} \quad (x_0, y_0) \in \mathbb{R}^2, \ a, b > 0$

Chaos has been embedded in the development of novel search strategies for global optimization known as chaos optimization algorithms (COAs). COAs have the properties of easy implementation, reduced execution time and robust mechanisms of escaping from local optimum. COA has been used in many applications such as optimization of power flow problems [11], control systems [12], neural networks [13], cryptography [14] and image processing [15]. In [10], the best chaotic sequences generated by sixteen different chaotic maps have been analysed.

Chaos based optimization has been originally proposed in 1997 [5]. It includes generally two main stages:

- *Global search*: an exploration of the global search space is carried out. A sequence of chaotic solutions is generated using a chaotic map. Then, the objective functions are evaluated and the solution with the best objective function is chosen as the current solution.

[1] Without loss of generality, we consider only minimization problems.

- *Local search*: the current solution is assumed to be close to the global optimum after a given number of iterations, and it is viewed as the centre on which a little chaotic perturbation, and the global optimum is obtained through local search. The above two steps are iterated until some specified stopping criterion is satisfied.

Observations from existing COA algorithms reveal that COA still presents some drawbacks especially with problems involving high dimensional spaces. Furthermore, the exploration ability of the COA decreases with the increase of the dimension space particularly because of the irregularity and the rigidity of the chaos dynamic which does not always authorize the exploration of some isolated regions containing the global optimum. Moreover, chaotic search has poor fine search ability, and then existing COAs suffer from the exploitation aspect. Most of the efficient chaos based optimization algorithms (COAs), proposed in the literature, are hybrid algorithms. Used generally as a global search strategy, COA is combined with local search efficient procedures such as gradient descent [9], grey-wolf [16], golden section search [17], and stochastic metaheuristics (e.g. butterfly [18], particle swarm [19,20], cucko search [21], firefly [22], genetic algorithms [23]).

Hence, few articles proposed an autonomous COA algorithm for global optimization [7,24,25]. Rather, COA has been widely involved in hybridization strategies, and by contrast, these few autonomous COA approaches involve only low-dimensional problems, and that reveals their limited efficiency and especially their incapacity in handling higher-dimensional problems [6]. According to the aforementioned difficulties, this paper presents a new COA approach based on new strategies including symmetrization, levelling, and fined local search.

In the last two decades, many efficient metaheuristics have been developed for tackling continuous optimization problems. Most of state-of-the-art algorithms are stochastic metaheuristics:

- *Differential evolution (DE)*: DE has two main control parameters that are required to be fixed by a user before the evolutionary process starts: the scaling factor F, and the crossover control parameter CR. Many adaptive and self-adaptive DE variants have been developed (e.g. L-SHADE [26]). jSO [27] and SHADE-cnEpSin [28] are DE-based winners of the CEC'2017 competition. SALSHADE-cnEpSin [29] and LSHADE-RSP [30] are the DE-based winners of the CEC'2018 competition.
- *Evolution strategies (ES)*: CMA-ES (Covariance Matrix Adaptation-Evolution Strategy) represents an efficient algorithm for global optimization [31] CMA-ES is a population based multivariate sampling algorithm, in which new candidate solutions are sampled using the multivariate normal distribution, based on the adaptation of covariance matrix.
- *Particle swarm optimization (PSO)*: designing learning methods that can use previous search information more efficiently was one of the most salient PSO research topics. The Orthogonal Learning PSO (OLPSO) [32] and the heterogeneous CLPSO [33] represent one of the most efficient PSO-based algorithms to solve global optimization problems. In OLPSO, orthogonal learning (OL)

strategy is used to discover useful information and guide particles to fly in better directions by constructing a much promising and efficient exemplar [32]. In CLPSO, the swarm population is divided into two subpopulations. Each subpopulation is assigned to focus solely on either exploration or exploitation. Comprehensive learning (CL) strategy is used to generate the exemplars for both subpopulations [33].

- *Estimation of distribution algorithms (EDA)*: the principle of EDA is to explore the space of potential solutions by generating and sampling promising solutions [34]. The main stage is the construction of an explicit probabilistic model that tries to capture the probability distribution of the promising solutions by using tree-structured or Bayesian networks [35]. As the univariate EDAs assume that all the variables are independent, it is widely used to solve separable problems [36]. It has been shown that univariate EDAs such as univariate marginal distribution algorithm continuous (UMDAc) is efficient for solving some multimodal nonseparable problems [37,38].
- *Hybrid metaheuristics*: the hybrid metaheuristic LSHADE_SPACMA (Semi-Parameter Adaptation Hybrid with CMA-ES) shows its efficiency for solving the CEC'2017 benchmark problems [39]. The HS-ES (Hybrid Sampling Evolution Strategy) is the general winner of the CEC'2018 competition on real parameter bound-constrained optimization [40]. It combines CMA-ES and univariate sampling UMDAc algorithms. Univariate sampling is very effective for solving multimodal nonseparable problems. As the CMA-ES has obvious advantages for solving unimodal nonseparable problems, the proposed HS-ES tries to take advantages of these two complementary algorithms to improve the performance of the search.

3 The Tornado Algorithm

The proposed Tornado algorithm is composed of three main procedures:

- The chaotic global search (CGS): CGS is a full exploration-based Chaotic search procedure. Its goal is to produce initial solutions that will be improved and refined by other exploitation-based chaotic search procedures.
- The chaotic local search (CLS): CLS is an exploitation-based Chaotic search procedure. Starting from an initial solution given by CGS, it exploits the neighbourhood of the solution. By focusing on successive promising solutions, CLS allows also the exploration of promising neighbouring regions.
- The chaotic fine search (CFS): CFS is a full exploitation-based Chaotic procedure. It uses a coordinate adaptive zoom strategy to intensify the search around the current optimum.

The structure of the proposed Tornado approach is given in Algorithm 1. In this work, we use the Henon map as a generator of a chaotic sequence. We consider a sequence $(Z_k)_{1\leq k\leq N_h}$ of normalized Henon vectors $Z_k =$

$(z_{k,1}, z_{k,2}, .., z_{k,n}) \in \mathbb{R}^n$ through the following linear transformation of the standard Henon map (2) (Fig. 1):

$$z_{k,i} = \frac{y_{k,i} - \alpha_i}{\beta_i - \alpha_i}, \quad \forall\,(k,i) \in [\![1, N_h]\!] \times [\![1, n]\!], \tag{2}$$

where $\alpha_i = \min_k(y_{k,i})$ and $\beta_i = \max_k(y_{k,i})$.

Thus, we get $\forall\,(k,i) \in [\![1, N_h]\!] \times [\![1, n]\!], \quad 0 \leqslant z_{k,i} \leqslant 1$.

In this work, the sequence of normalized Henon map vectors (Z_k) is defined as: $a = 1.5, \quad b = 0.2, \quad \forall k \in [\![1, n]\!], \quad (x_{k,0}, y_{k,0}) = (r_k, 0),\ r_k \sim U(0,1)$.

Algorithm 1 : The Tornado algorithm structure

```
Initialisation of the Henon chaotic sequence ;
Set k = 1 ;
Repeat
   Chaotic Global Search (CGS);
   Set s = 1 ;
   Repeat;
      Chaotic Local Search (CLS);
      Chaotic Finest Search (CFS);
      s = s + 1;
   Until s = Ml ; /* Ml is the number of CLS/CFS by cycle */
   k = k + 1 ;
Until k = M ; /* M is maximum number of cycles of Tornado */
```

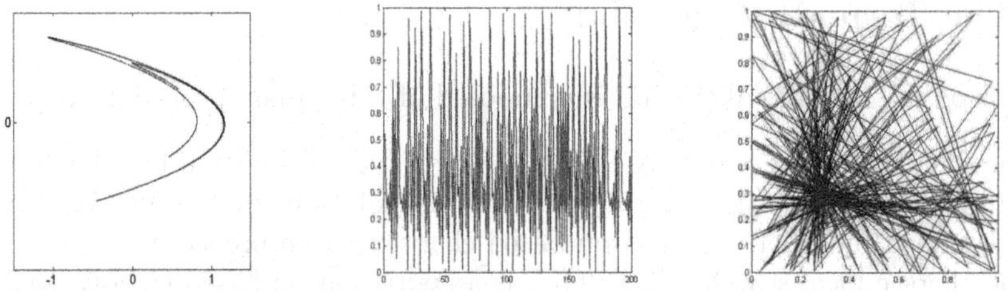

(a) *Henon map attractor* (b) *Henon map (200 iterations)* (c) *Henon map dynamic in 2D (200 iterations)*

Fig. 1. Illustration of Henon Map.

In general, chaos dynamics suffer from irregularity and rigidity, which induces a deficient exploration [?]. Indeed, the chaos dynamics does not always enable to cover some isolated regions of the search space. For a better exploration of the search space, the proposed *Tornado* algorithm uses symmetrization and levelling strategies to better control the flexibility and the orientation of the chaotic search process. In fact, those proposed strategies provide a distribution of chaos variables that contains several layers of symmetric solutions.

Chaotic global search and Chaotic local search will be briefly explained in next sections, while further details can be found in

3.1 Chaotic Global Search (CGS)

In order to improve the exploration ability of the chaos dynamic, the CGS proceeds by restructuring the chaos dynamics using to approaches: Levelling and Symmetrization.

- *Levelling approach:* In order to provide more diversification in the chaotic distribution, CGS proceeds by levelling with N_c chaotic levels. More precisely, the CGS procedure generates three chaotic variables for each iteration k, and in each level, $l \in [\![1, N_c]\!]$ according to:

$$X_1 = L + Z_l Z_k \times (U - L) \tag{3}$$
$$X_2 = \theta + Z_l Z_k \times (U - \theta) \tag{4}$$
$$X_3 = U - Z_l Z_k \times (U - \theta). \tag{5}$$

  Note that we drop k from the subscript in the notation $X_{i,k}$ for sake of simplicity.
- *Symmetrization approach:* As the exploration of all the dimensions in a high-dimensional space is not practical because of combinatorial explosion, we have introduced a new strategy consisting of a stochastic decomposition of the search space $\mathbb{R}^n$ into two vectorial subspaces: a vectorial line $\mathcal{D}$ and its corresponding hyperplane $\mathcal{H}$:

$$\mathrm{IR}^n = \mathcal{D} \oplus \mathcal{H}, \quad \mathcal{D} = \mathrm{IR} \times e_p, \quad \mathcal{H} = \mathrm{vect}(e_i)_{i \neq p}. \tag{6}$$

  By consequence,

$$\forall X = (x_1, x_2, \ldots, x_n) \in \mathrm{IR}^n: \quad X = X_d + X_h, \tag{7}$$

  where

$$X_d = (0, \cdot\cdot, 0, x_p, 0, \cdot\cdot, 0) \in \mathcal{D}, X_h = (x_1, \cdot\cdot, x_{p-1}, 0, x_{p+1}, \cdot\cdot, x_n) \in \mathcal{H}. \tag{8}$$

The symmetrization approach based on this stochastic decomposition of the design space provides two main advantages:

- It Reduces significantly the complexity of a high dimensional problem in a way as if we were dealing with a 2D space with four directions.
- The symmetric chaos is consequently more regular and more ergodic than the initial one (Fig. 2).

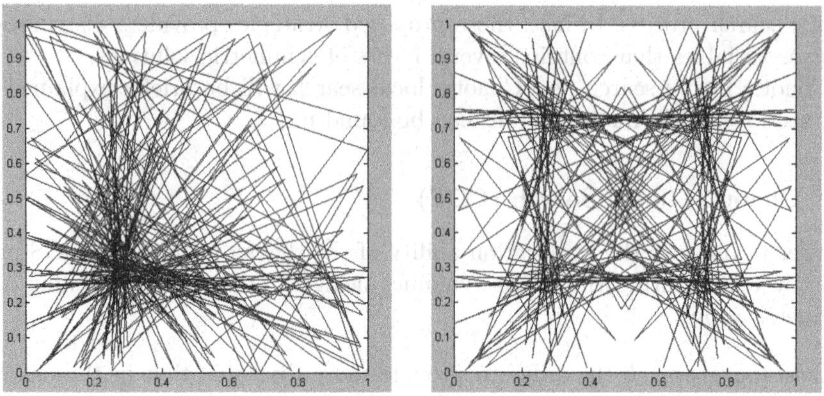

(a) *Henon map dynamic in 2D (200 iterations)* (b) *Symmetrized Henon map (200 iterations)*

Fig. 2. Illustration of symmetrisation approach in 2D.

Therefore, by using the stochastic decomposition (6), at each chaotic level $l \in [\![1, N_c]\!]$, CGS generates four symmetric chaotic points using axial symmetries $\mathcal{S}_{\theta+\mathcal{D}}$, $\mathcal{S}_{\theta+\mathcal{H}}$:

$$\begin{array}{ll} X_{i,1} = X_i, & X_{i,2} = \mathcal{S}_{\theta+\mathcal{D}}(X_{i,1}), \\ X_{i,3} = \mathcal{S}_{\theta+\mathcal{H}}(X_{i,2}), & X_{i,4} = \mathcal{S}_{\theta+\mathcal{D}}(X_{i,3}) = \mathcal{S}_{\theta+\mathcal{H}}(X_{i,1}). \end{array} \tag{9}$$

where the axial symmetries $\mathcal{S}_{\theta+\mathcal{D}}$, $\mathcal{S}_{\theta+\mathcal{H}}$ are defined as follows:

$$\mathcal{S}_{\theta+\mathcal{D}}(X) = X_d + (2\theta_h - X_h) \tag{10}$$

$$\mathcal{S}_{\theta+\mathcal{H}}(X) = (2\theta_d - X_d) + X_h \tag{11}$$

At last, the best solution among these all generated chaotic points as illustrated by Algorithm 3 (Fig. 3).

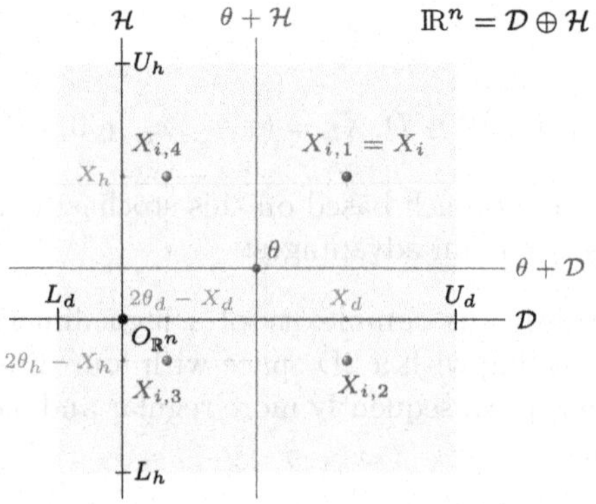

Fig. 3. Generation of chaotic variables by the symmetrization approach in CGS

Algorithm 2 : Chaotic global search (CGS).

Input: f, U, Z, N_c, k
Output: X_c
$Y = +\infty$; $\theta = \frac{1}{2}(U + L)$
for $l = 1$ **to** N_c
    Generate three chaotic variables X_1, X_2, and X_3 according to the following:
    $X_1 = \theta + Z_l\, Z_k \times (U - \theta)$, $X_2 = U - Z_l\, Z_k \times (U - \theta)$, $X_3 = L + Z_l\, Z_k \times (U - L)$
    for $i = 1$ **to** 3
        Select **randomly** an index $p \in \{1, \cdots, n\}$ and **decompose** X_i according to (79)
        Generate the four corresponding symmetric points $(X_{i,j})_{1 \leqslant j \leqslant 4}$ according to (9) and (11)
        for $j = 1$ **to** 4
            if $Y > f(X_{i,j})$
                $X_c = X_{i,j}$; $Y = f(X_{i,j})$
            end if
        end for
    end for
end for

3.2 Chaotic Local Search (CLS)

The Chaotic local search proceeds by exploiting the neighbourhood of the solution ω found by the chaotic global search CGS. However, CLS contributes also to the exploration of the decision space by looking for potential solutions relatively far from the current solution. In fact, the CLS conducts the search process near the current solution ω within a local search area $\mathcal{S}_l$ of radius $\mathcal{R}_l = r \times \mathcal{R}$ focused on ω, where $r \sim U(0,1)$ is a random parameter corresponding to the reduction rate, and $\mathcal{R}$ denotes the radius of the search area $\mathcal{S} = \prod_{i=1}^{n} [l_i, u_i]$ defined as follows:

$$\mathcal{R} = \frac{1}{2}(U - L) = \left(\frac{1}{2}(u_1 - l_1), \ldots, \frac{1}{2}(u_n - l_n)\right) \tag{12}$$

Like the CGS, the CLS also uses a levelling approach by creating N_l chaotic levels focused on ω. In each chaotic level $\eta \in [\![0, N_l - 1]\!]$, the local search process is limited to a local area $\mathcal{S}_{l,\eta}$ focused on ω characterized by its radius $\mathcal{R}_\eta$ defined by the following:

$$\mathcal{R}_\eta = \gamma_\eta \times \mathcal{R}_l = r \times \gamma_\eta \times \mathcal{R}, \tag{13}$$

where γ_η is a decreasing parameter trough levels which we have formulated in this work as follows:

$$\gamma_\eta = \frac{10^{-2r\eta}}{1 + \eta} \tag{14}$$

where $r \sim U(0,1)$ is a random number distributed uniformly within the range $[0, 1]$.

In fact, the levelling approach used by the CLS corresponds to a progressive zoom focus on the current solution ω carried out through N_l chaotic levels, and γ_η is the factor (decreasing throughout the chaotic levels η) that controls the speed of this zoom process ($\gamma_\eta \searrow 0$).

Moreover, once the CGS provides an initial solution ω, the CLS intensifies the search around this solution, through several chaotic layers. In each cycle of the Tornado algorithm, a given number (i.e. M_l) of CLS procedures is applied. Hence, the CLS participates also to the exploration of neighbouring regions by following the zoom dynamic as shown in Fig. 4.

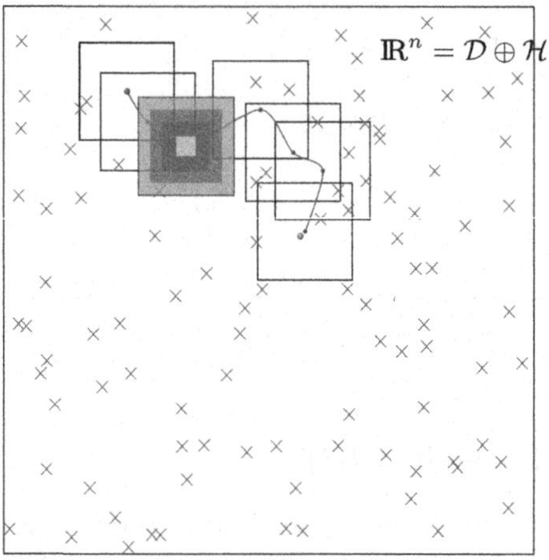

Fig. 4. Illustration of the selection of symmetric chaotic variables in CLS.

Moreover, in each chaotic level η, CLS generates two symmetric chaotic variables X_1, X_2 according to Fig. 5:

$$X_1 = Z \times \mathcal{R}_\eta, \quad X_2 = (1 - Z) \times \mathcal{R}_\eta = \mathcal{R}_\eta - X_1. \tag{15}$$

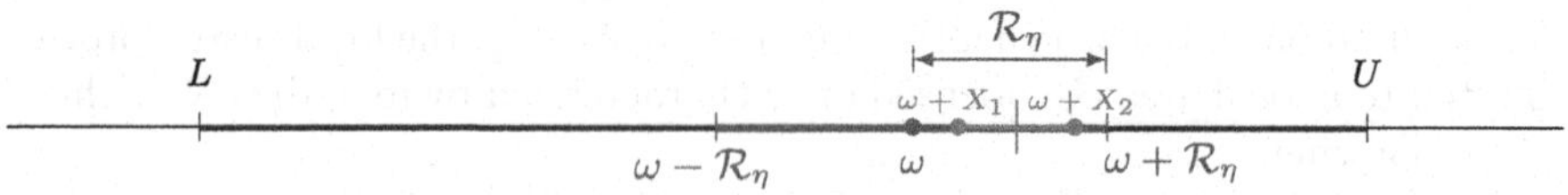

Fig. 5. Selection of symmetric chaotic variables in CLS.

We select randomly an index $p \in \{1, .., n\}$ and generate the corresponding stochastic decomposition of $\mathbb{R}^n$:

$$\mathbb{R}^n = \mathcal{D} \oplus \mathcal{H}, \quad \mathcal{D} = \mathbb{R} \times e_p, \quad \mathcal{H} = \text{vect}(e_i)_{i \neq p}. \tag{16}$$

Then, we get the corresponding decomposition of each chaotic variable $X_{i,(i=1,2)}$:

$$X_i = X_{i,d} + X_{i,h}. \tag{17}$$

Finally we generate from each chaotic variable $X_{i,\,(i=1,2)}$, N_p symmetric chaotic points $(X_{i,j})_{1 \leqslant j \leqslant N_p}$ using the polygonal model (Fig. 6):

$$X_{i,j} = \omega + X_i = \omega + \cos(2\pi.j/N_p)X_{i,d} + \sin(2\pi.j/N_p)X_{i,h}, \tag{18}$$

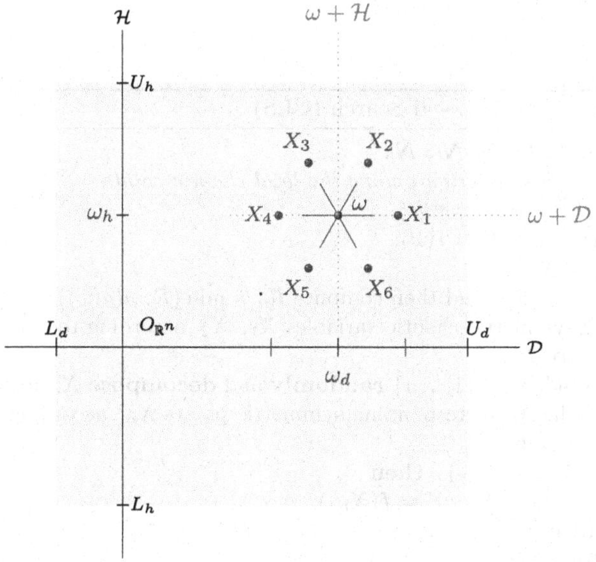

Fig. 6. Illustration of the generation of $N_p = 6$ symmetric chaotic points in CLS.

When ω is close enough to the borders of the search area $\mathcal{S}$, the search process can leave it and then it may give an infeasible solution localized outside $\mathcal{S}$.

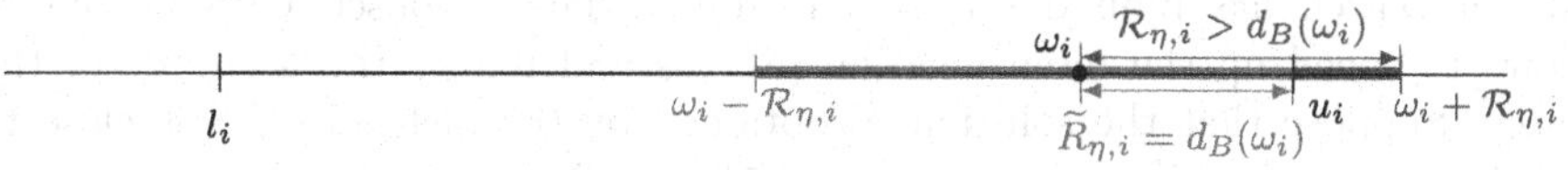

Fig. 7. Illustration of overflow:$\mathcal{R}_{\eta,i} > d_B(\omega_i)$

In fact, that occurs when $\mathcal{R}_{\eta,i} > d_B(\omega_i)$ for at least one component ω_i (Fig. 7), where $d_{\mathrm{B}}(\omega_i)$ denotes the distance of the component ω_i to borders l_i, u_i defined as follows:

$$d_{\mathrm{B}}(\omega_i) = \min(u_i - \omega_i, \omega_i - l_i). \tag{19}$$

To prevent this overflow, we consider the improved radius $\widetilde{R}_\eta$ instead of $\mathcal{R}_\eta$, given by the following:

$$\widetilde{R}_\eta = \min\left(\mathcal{R}_\eta, d_{\mathrm{B}}(\omega)\right), \tag{20}$$

where $d_{\mathrm{B}}(\omega) = (d_{\mathrm{B}}(\omega_1), \ldots, d_{\mathrm{B}}(\omega_n))$.

This guarantees $\widetilde{R}_{\eta,i} \leqslant d_{\mathrm{B}}(\omega_i)$, $\forall i \in [\![1, n]\!]$.

Hence, Eqs. (15) become

$$X_1 = Z \times \widetilde{R}_\eta, \quad X_2 = (1 - Z) \times \widetilde{R}_\eta. \tag{21}$$

Finally, the algorithm of the chaotic local search (CLS) is described in Algorithm 3.

Algorithm 3 : Chaotic Local Search (CLS)

Input:$\boldsymbol{f}$, $\boldsymbol{\omega}$, $\boldsymbol{L}$, $\boldsymbol{U}$, $\boldsymbol{Z}$, $\boldsymbol{N_l}$, $\boldsymbol{N_p}$
Output: X_l: *best solution among the local chaotic points*
$\mathcal{R} = \frac{1}{2}(U - L)$; $\mathcal{R}_l = r \times \mathcal{R}$;
$X = \omega$; $X_l = \omega$; $Y = f(\omega)$;
for $\eta = 0$ **to** $N_l - 1$
  Set $\mathcal{R}_\eta = \gamma_\eta \times \mathcal{R}_l$, and then compute $\widetilde{R}_\eta = \min\left(\mathcal{R}_\eta, d_B(\omega)\right)$
  Generate 2 symmetric chaotic variables X_1, X_2 according to (21)
  for $i = 1$ **to** 2
    Select an index $p \in \{1, .., n\}$ **randomly** and **decompose** X_i according to (17)
    Generate the N_p corresponding symmetric points $X_{i,j}$ according to (18)
      for $j = 1$ **to** N_p
        if $Y > f(X_{i,j})$ **then**
          $X_l = X_{i,j}; Y = f(X_{i,j})$;
        end if
      end for
  end for
end for

3.3 Chaotic Fine Search (CFS)

Chaotic search has limited fine search ability. The proposed CFS procedure allows to speed up the intensification process and refines the accuracy of the search. Suppose that the solution X obtained by the method *CLS* is close to the global optimum X_o with precision 10^{-p}, $p \in \mathbb{N}$. Then, we have:

$$X = X_o + \varepsilon, \quad \|\varepsilon\| < 10^{-p} \tag{22}$$

Thus, the distance ε can be interpreted as a parasitic signal of the solution, which is sufficient to filter in a suitable way to reach the global optimum, or the distance to which is the global optimum of its approximate solution. We carry out a chaotic search in a local area in which the radius adapts to the distance $\varepsilon = X - X_o$, component by component. However, in practice, the global optimum is not known a priori. To work around this difficulty, knowing that as the search process proceeds the resulting solution X is supposed to be close enough to the overall optimum, the trick found is to consider instead of the relation (22) the difference between the current solution X and its decimals fractional parts of order η,($\eta \in \mathbb{N}$):

$$\varepsilon_\eta = |X - X_\eta|$$

where the fractional of order η, X_η is the closest point of X to the precision $10^{-\eta}$ defined by: $X_\eta = 10^{-\eta} round(10^\eta X)$

For instance, Table 1 illustrates the fifth fractional parts as well as the corresponding errors for $X = (2.854732, 1.384527)$ (Fig. 8).

Table 1. Illustration of 5 first fractional and their corresponding errors.

Order k	$k-$ Fractional part	Error of order k
0	$X_0 = (3, 1)$	$\varepsilon_0 = (0.145268, 0.384127)$
1	$X_1 = (2.9, 1.4)$	$\varepsilon_1 = (0.045268, 0.084127)$
2	$X_2 = (2.85, 1.38)$	$\varepsilon_2 = (0.004732, 0.004127)$
3	$X_3 = (2.855, 1.384)$	$\varepsilon_3 = (0.000268, 0.000127)$
4	$X_4 = (2.8547, 1.3841)$	$\varepsilon_4 = (0.000032, 0.000027)$

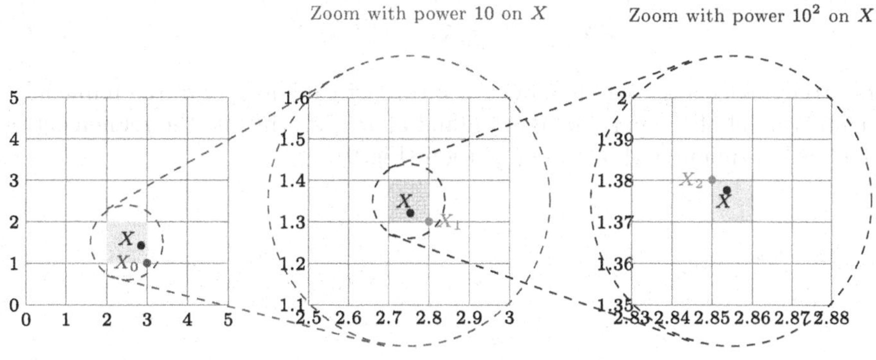

Fig. 8. Illustration of the 10 power zoom via the successive fractional parts.

Moreover, in order to perturb a potential local optima we propose to add a stochastic component in the round process, in fact we consider the stochastic round $[.]_{st}$ formalised as:

$$[X]_{st} = \begin{cases} round(X) + P, & if \quad mod(k,2) = 0 \\ round(X), & otherwise \end{cases} \tag{23}$$

where $P \sim U(-1,1)^d$ is a stochastic perturbation operated on X alternatively during the process. Thus, we get a the new formulation of the η−error of X:

$$\widetilde{\varepsilon}_\eta(X) = |X - 10^{-\eta}[10^\eta X]_{st})| \tag{24}$$

The chaotic fine search CFS has a structure similar to the CLS local chaotic search. Indeed it operates by levelling on N_f levels, except the fact that the local area of level η is defined by its radius $\mathcal{R}_\eta$ proportional to the η−error ε_η and given by:

$$\mathcal{R}_\eta = \frac{1}{1+\eta^2}\widetilde{R}, \quad \eta \in [\![0, N_f - 1]\!] \tag{25}$$

This way the local area search is carried out in a narrow domain that allow a focus adapted coordinate by coordinate unlike the uniform local search in CLS.

This time the modified radius $\widetilde{R}$ is defined by the following:

$$\widetilde{R} = \begin{cases} s \times R \cdot \widetilde{\varepsilon}_\eta, & if \quad r > 0.5 \\ T \cdot R \cdot \widetilde{\varepsilon}_\eta, & otherwise \end{cases} \tag{26}$$

where $r, s \sim U(0,1)$ and $T \sim U(0,1)^d$.

The $\mathcal{R}_\eta$ radius design allows you to zoom at an exponential rate of decimale over the levels. Indeed, we have:

$$\|\mathcal{R}_\eta\| \leqslant \|\varepsilon_\eta\|.\mathcal{R} < 10^{-\eta} \times \mathcal{R}. \tag{27}$$

Thus, the fine chaotic search allows an ultra fast exploitation of the immediate neighbourhood of the current solution and allows in principle the refinement of the global optimum with a good precision (Fig. 9).

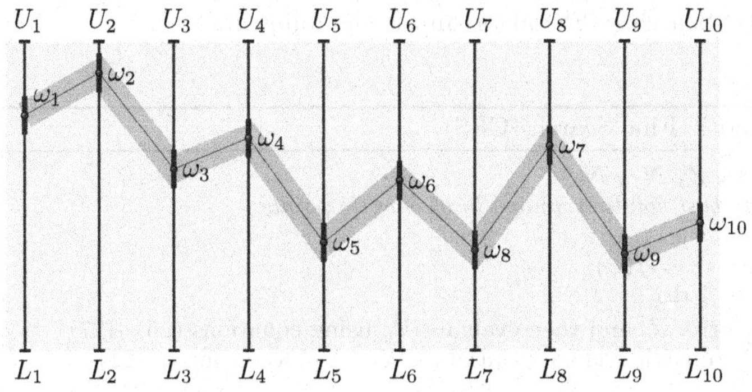

(a) *Illustration of the uniform local search area in CLS using uniform reduction factor*

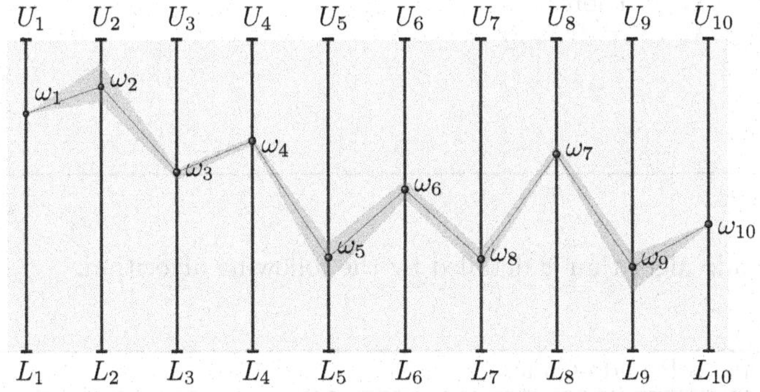

(b) Illustration of the coordinate adaptative local search area in CFS based on the fractionnal error information ε_η

Fig. 9. Illustration of the coordinate adaptative local search in CFS.

The Fine Chaotic Search (CFS) algorithm is the following:

Algorithm 4 : Chaotic Fine Search (CFS)

Input:$\boldsymbol{f}$, $\boldsymbol{\omega}$, $\boldsymbol{L}$, $\boldsymbol{U}$, $\boldsymbol{Z}$, $\boldsymbol{N_f}$, $\boldsymbol{N_p}$
Output: X_l: *the best solution among local chaotic points*
$\mathcal{R} = \frac{1}{2}(U - L)$;
$X = \omega$; $X_l = \omega$; $Y = f(\omega)$;
for $\eta = 0$ **to** $N_l - 1$ **do**
  Compute the $\eta-$error $\widetilde{\varepsilon_\eta}$ and then evaluate $\widetilde{R}_\eta$ using equations (25)$-$(27)
  Generate two symmetrical chaotic variables X_1, X_2 according to (21)
  for $i = 1$ **to** 2
    Choose **randomly** p in $\{1, \cdots, n\}$ and **decompose** X_i using (17)
    Generate N_p symmetrical points $X_{i,j}$ according to (18)
    for $j = 1$ **à** N_p
      if $Y > f(X_{i,j})$ **then**
        $X_l = X_{i,j}; Y = f(X_{i,j})$;
      end if
    end for
  end for
end for

Finally the Tornado algorithm is detailed by the following algorithm:

Algorithm 5 : Tornado Pseudo-Code.

Given :f, L, U, Z, M, M_l, N_c, N_l, N_f, N_p
Output : X, Y
$k = 1$; $Y = +\infty$;
while $k \leqslant M$ **do**
  $X_c =$ **CGS** (f, L, U, Z_k, N_c)
  if $Y > f(X_c)$ **then**
    $X = X_c$; $Y = f(X_c)$;
  end for
  $s = 1$;
  while $s \leqslant M_l$ **do**
    $X_l =$ **CLS** $(f, X, L, U, Z_{s+k}, N_l, N_p)$
    if $Y > f(X_l)$ **do**
      $X = X_l$; $Y = f(X_l)$;
    end if
    $X_f =$ **CFS** $(f, X, L, U, Z_{s+k}, N_f, N_p)$
    if $Y > f(X_l)$ **do**
      $X = X_f$; $Y = f(X_f)$;
    end if
    $s = s + 1$;
  end while
  $k = k + 1$;
fin tant que

---

4 Computational Experiments

In this section, computational experiments are carried out in order to assess the performance of the proposed Tornado algorithm for high dimensional problems (i.e. 50, 100, and 200 variables). All the experiments were run using Intel(R) Core(TM) $i3$ 4005U CPU 1.70 GHz with 4 GB RAM. The implementation of all used algorithms was done in MatLab. Upon recommendation from CEC conference competitions[2], a set of 24 well known benchmark problems were selected with diverse properties and different levels of complexity (i.e. unimodal, multimodal, separable, non separable, shifted, rotated, noisy) as illustrated by Tables 2 and 3. Unimodality shows the exploitation capability of the developed algo-

Table 2. High dimensional Benchmark functions used in our experiments.

B.Function	Expression	C	Search region	Optimum
Shifted Bent Cigar	$f_1 = y_1^2 + 10^6 \sum_{i=2}^{D} y_i^2 + bias,\ y = x - o,\ bias = 100$	US	$[-100, 100]^D$	$bias$
Shifted Rastrigin	$f_2 = \sum_{i=1}^{D} [y_i^2 - 10\cos(2\pi y_i) + 10] + bias,\quad y = x - o,\quad bias = 200$	MS	$[-5, 12, 5, 12]^D$	$bias$
Shifted Non Continuous Rastrigin	$f_3 = \sum_{i=1}^{D} [z_i^2 - 10\cos(2\pi z_i) + 10] + bias,\quad z = y - o,\quad y_i = \begin{cases} x_i & \text{if } \lvert x_i\rvert \leq 0.5 \\ round(2x_i)/2, & \text{if } \lvert x_i\rvert > 0.5 \end{cases}\ bias = 300$	MS	$[-5, 12, 5, 12]^D$	$bias$
Shifted Discuss	$f_4 = 10^6 y_1^2 + \sum_{i=2}^{D} y_i^2 + bias,\ y = x - o,\ bias = 600$	US	$[-100, 100]^D$	$bias$
Shifted Levy	$f_5 = \sin^2(\pi y_1) + \sum_{i=1}^{D-1} (y_i - 1)^2)[1 + 10\sin^2(\pi y_i + 1)] + (y_D - 1)^2[1 + \sin^2(2\pi y_D)] + bias,\quad y = x - o,\ bias = 500,$	MN	$[-50, 50]^D$	$biais$
Shifted Rotated H.C Elliptic	$f_6(x) = \sum_{i=1}^{D} (10^6)^{\frac{i-1}{D-1}} y_i^2 + bias,\ y = M(x - o),\ bias = 400$	UN	$[-100, 100]^D$	$bias$
Shifted Rotated Rosenbrock	$f_7(x) = \sum_{i=1}^{D-1} (100(y_{i+1} - y_i^2)^2 + (y_i - 1)^2) + bias,\ y = M(x - o),\ bias = 700$	MN	$[-30, 30]^D$	$bias$
SR Expended Schaffer F6	$f_8 = g(y_1, y_2) + g(y_1, y_2) + \cdots + g(y_D, y_1) + bias,\ y = M(x - o),\ g(u, v) = 0.5 + \frac{\sin^2(u^2 + v^2) - 0.5}{(1 + 0.001(u^2 + v^2))^2},\ bias = 800,$	MN	$[-100, 100]^2$	$biais$
S.R. HappyCat	$f_9 = \lvert \sum_{i=1}^{D} y_i^2 - D\rvert^{\frac{1}{4}} + \frac{0.5}{D}(\sum_{i=1}^{D} y_i)^2 - \sum_{i=1}^{D} y_i) + 0.5,\ y = M(x - o)$	MN	$[-5, 10]^D$	$biais$
S.R. Zakharov	$f_{10} = \sum_{i=1}^{D} x_i^2 + (\sum_{i=1}^{D} 0.5ix_i)^2 + (\sum_{i=1}^{D} 0.5ix_i)^4 + biais\ y = M(x - o)\ biais = 1000$	UN	$[-5, 10]^D$	0
S.R. Ackley	$f_{11} = -20\exp(-0.2\sqrt{\frac{1}{D}\sum_{i=0}^{D} y_i^2}) - \exp(\frac{1}{D}\sum_{i=0}^{D}\cos(2\pi y_i)) + 20 + e + biais,\ y = M(x - o),\ bias = 1100$	MN	$[-32.768, 32.768]^D$	$biais$
S.R HGBat	$f_{12} = \lvert(\sum_{i=1}^{D} x_i^2)^2 - (\sum_{i=1}^{D} x_i)^2\rvert^{0.5} + 0.5(\sum_{i=1}^{D} x_i)^2 - \sum_{i=1}^{D} x_i)/D + 0.5 + biais\ y = M(x - o)\ bias = 1200$	UN	$[-5, 10]^D$	$biais$

C: Characteristic, U: Unimodal, M: Multimodal, S: Separable, N: Non-Separable.

[2] Competition on single objective real-parameter numerical optimization.

Table 3. High dimensional Benchmark functions used in our experiments (continued).

B.Function	Expression	C	Search region	Optimum
Quartic	$f_{13} = \sum_{i=1}^{D} i x_i^4 + rand(0,1)$	MS	$[-1.28, 1.28]^D$	0
Inverted cosine wave	$f_{14} = -\sum_{i=1}^{D-1} \exp\left(-y_i/8\right)\cos(4\sqrt{y_i}),\ y_i = x_i^2 + x_{i+1}^2 + 0.5x_i x_{i+1}$	MN	$[-5, 5]^D$	$-n+1$
Penalized 1	$f_{15} = \frac{\pi}{D}\{10\sin^2(3\pi x_1) + \sum_{i=1}^{D-1}(y_i - 1)^2)[1 + 10\sin^2(\pi y_{i+1})] + (y_D - 1)^2\} + \sum_{i=1}^{D} u(x_i, 10, 100, 4)\ (*),\quad y_i = 1 + \frac{1}{4}(x_i + 1)$	MN	$[-50, 50]^D$	0
Himmelblau	$f_{16} = \frac{1}{D}\sum_{i=1}^{D}(x_i^4 - 16x_i^2 + 5x_i)$	MS	$[-5, 5]^D$	-78.3323
Alpine	$f_{17} = \sum_{i=1}^{D} \lvert x_i \sin(x_i) + 0.1x_i \rvert$	MS	$[-10, 10]^D$	0
PowerSum	$f_{18} = \sum_{i=1}^{D}\left(\sum_{k=1}^{4} x_i^k - b_k\right)^2 \ /\ b = (8, 18, 44, 114)$	MN	$[0, n]^D$	0
Cosine Mixture	$f_{19} = \sum_{i=1}^{D} x_i^2 - 0.1\sum_{i=1}^{D} cos(5\pi x_i)$	MS	$[-1; 1]^D$	$-0.1n$
Schwefel 2.22	$f_{20} = \sum_{i=1}^{D} \lvert x_i \rvert + \prod_{i=1}^{D} \lvert x_i \rvert$	UN	$[-10, 10]^D$	0
Powell sum	$f_{21} = \sum_{i=1}^{D} \lvert x_i \rvert^{i+1}$	MS	$[-100, 100]^D$	0
Easom	$f_{22} = -(-1)^D\left(\prod_{i=1}^{D}\cos(x_i)\right)\exp\left(-\sum_{i=1}^{D}(x_i - \pi)^2\right)$	UN	$[-10, 10]^D$	-1
Mishra 2	$f_{23} = (1 + \chi_D)^{\chi_D},\quad \chi_D = D - \frac{1}{2}\sum_{i=1}^{D-1} x_i + x_{i+1}$	MN	$[0, 1]^D$	2
Brown	$f_{24} = \sum_{i=1}^{D-1}(x_i^2)^{(x_{i+1}^2+1)} + (x_{i+1}^2)^{(x_i^2+1)}$	UN	$[-1, 4]^D$	0

C: Characteristic, U: Unimodal, M: Multimodal, S: Separable, N: Non-Separable.

rithms, while multi-modality confirms the exploration capabilities. The shifted global optimum for all the functions is provided as $o = (o_1, o_2, .., o_D)$ and the functions are defined as $z = x - o$ for shifted functions and $z = (x - o).M$ for shifted rotated functions where M is the transformation matrix for the rotating matrix. For instance, $F_1 - F_6$ are shifted functions and $F_7 - F_{12}$ are shifted and rotated functions.

For the proposed Tornado algorithm, we have used the same set of values of the parameters (e.g. number of chaotic levels) for all experiments. The algorithm is not very sensitive to those parameters. In the current study, the parameters were set as follows:

- The number of CGS chaotic levels (N_c): $N_c = 5$.
- The number of CLS chaotic levels (N_l): $N_l = 5$.
- The number of CFS chaotic levels (N_f): $N_f = 10$.
- The number of CLS-CFS per cycle (M_l): $M_l = 100$.

4.1 Comparison with Other Chaotic Optimization Algorithms

In order to show the effectiveness of our new chaotic optimization strategy, this section presents a comparison of the Tornado algorithm with state-of-the-art

COA variants such as ICOLM [7] and ICOMM [25]. We have adopted the suggested parameters by the authors of those algorithms, as illustrated in Table 4. Three set of parameters have been suggested by the authors.

Table 4. The set of parameters used by ICOLM and ICOMM approaches.

Configuration	M_g	M_l	M_{gl1}	M_{gl2}	λ	λ_{gl1}	λ_{gl2}
C1	800	400	6	6	0.1	0.04	0.01
C2	800	400	6	6	0.01	0.04	0.01
C3	800	400	6	6	0.001	0.04	0.01

Where M_g is the maximum number of iterations of chaotic global search, M_{gl1} is maximum number of iterations of first chaotic Local search in global search, M_{gl2} is the maximum number of iterations of second chaotic local search in global search, M_l is the maximum number of iterations of chaotic local search, λ_{gl1} is the step size in first global-local search, λ_{gl2} is step size in second global-local search, and λ is the step size in chaotic local search. The other specific parameters of algorithms are given below:

- ICOLM uses Lozi map with: $a = 1.7,\ \ b = 0.5$.
- ICOMM uses Henon map with: $a = 4,\ \ b = 0.9$.

We choose the number of function evaluations (FEs) as a stopping criteria. The maximum number of function evaluations was 10^4 for all functions. Since the algorithms are stochastic in nature, 30 independent runs of each algorithm are carried out. The performance indicators used are the mean and the standard deviation. The comparison results for functions $(f_1 - f_{15})$ on moderate dimension $D = 10$ are shown in Table 5.

It is observed from the results presented in Table 5 that the performance of our Tornado algorithm strongly dominates the existing COA approaches for all functions. Indeed, the computational results show clearly the deficiency of the classical COA approaches (here ICOLM and ICOMM) to even deal with moderate 10-dimensional problems whereas Tornado succeeds systematically. Therefore, it is needless to show the carried comparisons for high dimensional problems.

4.2 Comparison with State-of-the-Art Algorithms

We have also compared the obtained results with three state-of-the-art algorithms from different families of stochastic optimization algorithms:

- CMA-ES[3]: a Covariance Adaptation Evolution Strategy (ES) based algorithm [31]. It is a ES algorithm in which the Covariance matrix is deterministically adapted from the last move of the algorithm.

[3] Available in the MATLAB library Yarpiz.

- L-SHADE[4]: SHADE is an adaptive Differential Evolution (DE) which incorporates succes-history based parameter adaptation and one of the state-of-the-art DE algorithm. L-SHADE is an extension of SHADE using Linear Population Size Reduction (LPSR) [26].
- CLPPSO[5]: a comprehensive learning particle swarm optimizer (CLPPSO) embedded with local search (LS) is proposed to pursue higher optimization performance by taking the advantages of CLPPSO's strong global search capability and LS's fast convergence ability [33].

Table 5. Comparison results for $f_1 - f_{12}$ on dimension $D = 10$ over 30 runs

No	Stats	Tornado	ICOLM(1)	ICOLM(2)	ICOLM(3)	ICOMM(1)	ICOMM(2)	ICOMM(3)
F_1	Mean	5,30E-08	2.58E+08	2.28E+06	3.21E+04	2.44E+08	3.19E+06	8.98E+06
	Std	6.10E-08	6.23E+07	9.28E+05	1.13E+04	9.28E+07	3.66E+06	2.40E+07
F_2	Mean	1.24E+00	4.76E+02	2.99E+01	1.87E+03	1.36E+03	4.14E+02	1.04E+03
	Std	4.97E-01	9.73E+01	7.04E+00	7.64E+02	4.88E+02	9.11E+01	2.94E+02
F_3	Mean	2.75E+00	1.93E+01	7.05E+01	8.39E+01	2.82E+01	1.74E+01	2.21E+01
	Std	9.58E-01	8.05E+00	2.15E+01	2.18E+01	9.00E+00	9.14E+00	1.11E+01
F_4	Mean	5.62E-09	6.13E+02	9.92E+00	1.89E+04	1.99E+04	1.15E+04	8.33E+03
	Std	8.96E-09	4.47E+02	3.44E+00	9.36E+03	9.79E+03	3.26E+03	5.34E+03
F_5	Mean	0.00E+00	3.85E+00	3.31E+00	4.30E+00	4.99E-01	1.29E-01	1.26E-01
	Std	1.14E-13	9.93E-01	1.76E+00	1.40E+00	1.71E-01	9.85E-02	1.05E-01
F_6	Mean	3.14E-04	2.35E+05	3.91E+03	3.77E+01	7.84E+04	7.97E+03	2.35E+02
	Std	7.54E-04	1.51E+05	2.55E+03	1.89E+01	2.96E+04	9.05E+03	2.29E+02
F_7	Mean	8.51E+00	7.39E+01	1.82E+01	1.70E+01	1.93E+02	4.13E+01	1.21E+01
	Std	2.09E+01	4.28E+01	2.90E+01	2.89E+01	1.21E+02	3.22E+01	2.26E+01
F_8	Mean	5.10E-05	1.45E-01	4.46E-02	4.65E-03	8.78E-02	3.38E-02	6.78E-04
	Std	1.55E-04	1.01E-01	8.97E-02	8.41E-03	5.14E-02	3.39E-02	9.88E-04
F_9	Mean	1.99E+00	8.39E-01	8.80E-01	7.51E-01	1.04E+00	1.01E+00	9.08E-01
	Std	3.25E-02	1.99E-01	1.22E-01	2.24E-01	1.74E-01	1.93E-01	1.61E-01
F_{10}	Mean	8.87E+07	1.45E+04	8.59E+00	1.87E+05	9.09E+05	3.47E+05	9.42E+05
	Std	2.80E+08	2.88E+04	2.90E+00	1.51E+05	7.93E+05	3.83E+05	4.43E+05
F_{11}	Mean	2.00E+01	2.05E+01	2.06E+01	2.03E+01	2.04E+01	2.05E+01	2.04E+01
	Std	5.84E-03	7.42E-02	4.20E-02	1.03E-01	1.05E-01	9.15E-02	5.20E-01
F_{12}	Mean	4.87E-01	4.53E-01	4.75E-01	4.63E-01	4.92E-01	4.94E-01	4.99E-01
	Std	1.89E-02	3.36E-02	1.62E-02	1.58E-02	8.58E-03	6.69E-03	4.75E-03

[4] Available at sites https://google.com/site/tanaberyoji/software.
[5] Available in https://github.com/hmofrad/Adaptative-CLPPSO.

Table 6. Comparison results for $f_1 - f_{24}$ problems on dimension $D = 50$ over 30 runs.

No	CLPSO		CMAES		L-SHADE		Tornado	
	Mean	Std	Mean	Std	Mean	Std	Mean	Std
f_1	7.17E+01	2.28E+01	2.44E+03	1.69E+03	3.48E−03	3.36E−03	**3.00E−12**	**2.46E−12**
f_2	2.51E+03	2.50E+03	3.23E+02	1.20E+01	3.60E+01	8.43E+00	**5.97E−01**	**1.33E+00**
f_3	1.54E+01	1.22E+00	2.82E+02	1.26E+01	1.11E+02	1.29E+01	**2.20E+00**	**8.36E−01**
f_4	1.97E+01	1.97E+01	2.75E+03	1.01E+03	1.00E−11	5.55E−12	**2.73E−12**	**4.40E−13**
f_5	1.87E-02	5.33E-03	9.80E−11	3.16E−11	**4.55E−13**	**1.14E−13**	1.24E−12	1.80E−13
f_6	7.42E-08	1.60E-08	3.33E−06	1.02E−06	9.00E−11	6.72E−11	**9.09E−13**	**1.80E−13**
f_7	4.89E+02	1.55E+01	3.86E+01	5.28E−01	4.38E+01	2.32E−02	**3.00E−02**	**3.16E−02**
f_8	4.08E-01	3.66E-02	1.00E−02	2.10E−03	5.00E−02	1.44E−02	**8.67E−02**	**9.17E−03**
f_9	3.44E+00	0.00E+00	**2.30E−01**	**2.97E−02**	2.95E+00	6.44E−05	2.95E+00	2.62E−04
f_{10}	1.93E+00	1.57E-01	**0.00E+00**	**1.51E−04**	0.00E+00	4.81E−04	2.70E−01	5.61E−02
f_{11}	2.13E+01	1.75E-02	2.12E+01	8.01E−02	2.09E+01	3.04E−01	**2.00E+01**	**1.16E−02**
f_{12}	4.92E-01	2.46E-03	4.90E−01	4.29E−02	**4.90E−01**	**1.12E−03**	5.00E−01	2.32E−03
f_{13}	3.85E-02	6.78E-03	6.51E−03	1.87E−03	3.15E−03	9.47E−04	**6.70E−04**	**3.96E−04**
f_{14}	1.24E+01	1.65E+00	1.37E+01	5.56E−01	3.55E−14	1.23E−14	**0.00E+00**	**1.07E−14**
f_{15}	3.49E-07	2.62E-07	1.81E−10	4.35E−11	2.68E−15	1.44E−15	**9.42E−33**	**0.00E+00**
f_{16}	1.51E-07	6.05E-08	1.47E+00	6.45E−01	9.00E−02	8.79E−02	**1.56E−09**	**3.26E−14**
f_{17}	1.42E-14	1.55E-14	1.22E−04	1.26E−04	**0.00E+00**	**0.00E+00**	**0.00E+00**	**0.00E+00**
f_{18}	3.36E-04	9.86E-05	3.03E−07	1.80E−07	2.96E−02	6.61E−02	**0.00E+00**	**0.00E+00**
f_{19}	2.66E+00	2.74E-02	2.80E+00	1.57E+00	2.56E+00	1.43E+00	**2.04E−15**	**5.96E−16**
f_{20}	8.68E+00	2.45E+00	2.21E−04	6.51E−05	3.26E−06	1.92E−06	**6.04E−170**	**0.00E+00**
f_{21}	2.94E-04	1.51E-05	4.44E+10	7.46E+10	3.32E−16	5.81E−16	**2.52E−196**	**0.00E+00**
f_{22}	1.00E+00	0.00E+00	1.00E+00	0.00E+00	**0.00E+00**	**0.00E+00**	**0.00E+00**	**9.61E−17**
f_{23}	0.00E+00	0.00E+00	1.33E+00	3.18E−08	4.00E−08	1.87E−08	**0.00E+00**	**0.00E+00**
f_{24}	2.54E-07	1.30E-07	2.32E−11	1.09E−11	6.59E−15	5.40E−15	**4.99E−77**	**2.16E−77**

The choice of the algorithms in the computational study is mainly driven by the high-quality of their results and the availability of code. We avoid the risk of non-optimal implementations and hence unfair comparisons. The maximum number of function evaluations is set to $2 \times 10^3 \times D$ for all algorithms and tested functions. We have adopted the suggested parameters by the authors of those algorithms. The computational results (i.e. error mean, standard deviation) are

Table 7. Comparison results for $f_1 - f_{24}$ problems on dimension $D = 100$ over 30 runs.

No	CLPSO		CMAES		L-SHADE		Tornado	
	Mean	Std	Mean	Std	Mean	Std	Mean	Std
F_1	1.86E+06	4.33E+05	1.10E+01	8.96E+00	1.10E+01	8.96E+00	**6.37E−12**	**2.96E−12**
F_2	1.38E+02	1.13E+01	1.97E+02	2.19E+01	1.97E+02	2.19E+01	**1.10E+00**	**4.22E−01**
F_3	2.69E+01	1.27E+01	1.27E+02	7.72E+00	1.27E+02	7.72E+00	**5.61E+00**	**2.10E+00**
F_4	1.63E+00	3.66E-01	3.95E−08	2.34E−08	3.95E−08	2.34E−08	**4.32E−12**	**3.94E−13**
F_5	2.15E+02	4.34E+01	7.92E−07	5.44E−07	7.92E−07	5.44E−07	**1.59E−12**	**2.60E−13**
F_6	1.84E-03	2.24E-04	3.25E−11	1.56E−11	3.25E−11	1.56E−11	**1.59E−12**	**1.97E−13**
F_7	4.89E+02	1.55E+01	8.70E+01	9.60E−01	1.22E+02	1.75E+00	**2.58E−02**	**8.84E−03**
F_8	4.08E-01	3.66E-02	3.45E−03	7.91E−04	2.08E−02	3.51E−03	**1.56E−04**	**3.77E−04**
F_9	3.44E+00	0.00E+00	**2.00E−01**	**2.42E−02**	3.44E+00	2.51E−04	3.44E+00	9.72E−04
F_{10}	1.93E+00	1.57E-01	**1.76E−05**	**7.07E−06**	4.29E−03	1.09E−03	9.48E−01	1.12E−01
F_{11}	2.13E+01	1.75E-02	2.13E+01	1.93E−02	2.13E+01	3.64E−02	**1.98E+01**	**4.31E−02**
F_{12}	4.92E-01	2.46E-03	5.04E−01	3.16E−02	4.92E−01	1.40E−03	**5.00E−01**	**3.73E−02**
F_{13}	5.86E-02	1.02E-02	1.37E−02	2.18E−03	7.72E−03	2.03E−03	**5.40E−04**	**5.52E−05**
F_{14}	3.04E+01	4.94E-01	2.82E+01	1.99E+00	1.33E−11	4.26E−12	**0.00E+00**	**2.25E−14**
F_{15}	5.52E-08	2.69E-08	4.39E−12	1.86E−12	3.54E−12	2.16E−12	**4.71E−33**	**0.00E+00**
F_{16}	4.79E-08	1.47E-08	2.83E+00	1.26E+00	5.35E+00	3.54E−01	**1.41E−06**	**1.00E−14**
F_{17}	1.88E-17	1.59E-17	8.46E−05	8.67E−05	2.02E−28	1.24E−28	**0.00E+00**	**0.00E+00**
F_{18}	1.03E-04	2.37E-05	5.10E−09	1.89E−09	7.98E−01	2.88E−01	**0.00E+00**	**0.00E+00**
F_{19}	3.47E+00	3.12E-02	3.50E+00	8.60E−16	3.50E+00	1.25E−11	**5.33E−15**	**1.54E−15**
F_{20}	9.26E+00	5.79E+00	3.35E−03	8.92E−04	3.02E−01	7.29E−02	**4.20E+00**	**1.59E+00**
F_{21}	2.23E-04	2.59E-05	4.12E−05	7.78E−06	4.56E−04	3.56E−04	**0.00E+00**	**0.00E+00**
F_{22}	1.00E+00	0.00E+00	1.00E+00	0.00E+00	1.00E+00	0.00E+00	**9.99E−16**	**1.57E−16**
F_{23}	0.00E+00	0.00E+00	9.77E−01	2.33E+01	8.53E−05	3.07E−05	**0.00E+00**	**0.00E+00**
F_{24}	2.49E-07	7.74E-08	1.63E−12	9.08E−13	1.47E−10	1.02E−10	**4.99E−148**	**4.20E−148**

presented in Tables 6, 7 and 8 for 30 independent runs. Moreover Tables 9, 10 and 11 show the ranking of the algorithms according to their computational results.

The obtained results show that the Tornado algorithm dominated largely the other algorithms for most functions. The same conclusion has been observed for all dimensions of the functions (i.e. 50, 100, 200).

Table 8. Comparison results for $f_1 - f_{24}$ problems on dimension $D = 200$ over 30 runs.

No	CLPSO		CMAES		L-SHADE		Tornado	
	Mean	Std	Mean	Std	Mean	Std	Mean	Std
F_1	2.44E+03	1.69E+03	1.19E+00	3.52E-01	3.48E-03	3.36E-03	**3.00E-12**	**2.46E-12**
F_2	3.23E+02	1.20E+01	7.86E+01	5.67E+00	3.60E+01	8.43E+00	**5.97E-01**	**1.33E+00**
F_3	2.82E+02	1.26E+01	2.38E+01	4.97E+00	1.11E+02	1.29E+01	**2.20E+00**	**8.36E-01**
F_4	2.75E+03	1.01E+03	1.04E+04	4.40E+03	1.00E-11	5.55E-12	**2.73E-12**	**4.40E-13**
F_5	9.80E-11	3.16E-11	**0.00E+00**	**0.00E+00**	4.55E-13	1.14E-13	1.24E-12	1.80E-13
F_6	3.33E-06	1.02E-06	4.52E-08	9.63E-09	9.00E-11	6.72E-11	**9.09E-13**	**1.80E-13**
F_7	3.86E+01	5.28E-01	1.85E+02	3.90E-01	4.38E+01	2.32E-02	**3.00E-02**	**3.16E-02**
F_8	1.00E-02	2.10E-03	1.00E-05	4.99E-06	5.00E-02	1.44E-02	**8.67E-02**	**9.17E-03**
F_9	**2.30E-01**	**2.97E-02**	2.77E-01	2.64E-02	2.95E+00	6.44E-05	2.95E+00	2.62E-04
F_{10}	**0.00E+00**	**1.51E-04**	4.09E+03	6.24E+03	0.00E+00	4.81E-04	2.70E-01	5.61E-02
F_{11}	2.12E+01	8.01E-02	2.15E+01	1.03E-02	2.09E+01	3.04E-01	**2.00E+01**	**1.16E-02**
F_{12}	4.90E-01	4.29E-02	5.08E-01	5.40E-02	**4.90E-01**	**1.12E-03**	5.00E-01	2.32E-03
F_{13}	6.51E-03	1.87E-03	2.92E-02	2.21E-03	3.15E-03	9.47E-04	**6.70E-04**	**3.96E-04**
F_{14}	1.37E+01	5.56E-01	6.12E+01	1.03E+01	3.55E-14	1.23E-14	**0.00E+00**	**1.07E-14**
F_{15}	1.81E-10	4.35E-11	2.87E-15	1.54E-16	2.68E-15	1.44E-15	**9.42E-33**	**0.00E+00**
F_{16}	1.47E+00	6.45E-01	3.46E+00	1.30E+00	9.00E-02	8.79E-02	**1.56E-09**	**3.26E-14**
F_{17}	1.22E-04	1.26E-04	**0.00E+00**	**0.00E+00**	**0.00E+00**	**0.00E+00**	**0.00E+00**	**0.00E+00**
F_{18}	3.03E-07	1.80E-07	1.26E-12	2.71E-14	2.96E-02	6.61E-02	**0.00E+00**	**0.00E+00**
F_{19}	2.80E+00	1.57E+00	3.50E+00	0.00E+00	2.56E+00	1.43E+00	**2.04E-15**	**5.96E-16**
F_{20}	2.21E-04	6.51E-05	2.14E-03	4.75E-04	3.26E-06	1.92E-06	**6.04E-170**	**0.00E+00**
F_{21}	4.44E+10	7.46E+10	1.23E-06	9.76E-08	3.32E-16	5.81E-16	**2.52E-196**	**0.00E+00**
F_{22}	1.00E+00	0.00E+00	1.00E+00	0.00E+00	0.00E+00	0.00E+00	**0.00E+00**	**9.61E-17**
F_{23}	1.33E+00	3.18E-08	1.33E+02	1.54E+02	4.00E-08	1.87E-08	**0.00E+00**	**0.00E+00**
F_{24}	2.32E-11	1.09E-11	3.63E-16	2.00E-16	6.59E-15	5.40E-15	**4.99E-77**	**2.16E-77**

The final ranking of the evaluated algorithms is performed by using all the obtained results. The algorithms are sorted for each test function. The ranking are summed up and are presented in Tables 9, 10 and 11. Clearly the Tornado algorithm is the winner, while LSHADE is the runner-up. We notice also that the performance of the CMA-ES algorithm decreases function of the dimension of the problem.

Table 9. The rank of the four algorithms for the functions test on $D = 50$.

	f_1	f_2	f_3	f_4	f_5	f_6	f_7	f_8	f_9	f_{10}	f_{11}	f_{12}	f_{13}	f_{14}	f_{15}	f_{16}	f_{17}	f_{18}	f_{19}	f_{20}	f_{21}	f_{22}	f_{23}	f_{24}	Mean Rank
CLPSO	4	4	1	3	4	3	4	4	4	4	4	4	4	3	4	2	4	3	4	3	4	3	3	4	3,52
CMAES	3	2	4	4	3	4	2	1	1	2	3	2	3	4	3	4	1	2	3	4	3	4	4	2	2.76
LSHADE	2	3	3	2	1	2	3	2	2	3	2	1	2	2	2	3	1	4	2	1	2	1	2	3	2.12
Tornado	1	1	2	1	2	1	1	3	3	1	1	3	1	1	1	1	1	1	1	2	1	2	1	1	**1.48**

Table 10. The rank of the four algorithms for the functions test on $D = 100$

	f_1	f_2	f_3	f_4	f_5	f_6	f_7	f_8	f_9	f_{10}	f_{11}	f_{12}	f_{13}	f_{14}	f_{15}	f_{16}	f_{17}	f_{18}	f_{19}	f_{20}	f_{21}	f_{22}	f_{23}	f_{24}	Mean Rank
CLPSO	4	2	4	4	4	4	3	4	4	4	2	2	3	3	4	3	4	3	3	3	4	3	3	4	3.40
CMAES	3	3	3	3	3	2	1	1	1	2	3	3	4	4	3	4	1	2	4	4	3	4	4	2	2.72
LSHADE	2	4	2	2	2	3	2	2	2	3	4	1	2	2	2	2	1	4	2	2	1	1	2	3	2.20
Tornado	1	1	1	1	1	1	4	3	3	1	1	4	1	1	1	1	1	1	1	1	2	2	1	1	**1.56**

Table 11. The rank of the four algorithms for the functions test on $D = 200$

	f_1	f_2	f_3	f_4	f_5	f_6	f_7	f_8	f_9	f_{10}	f_{11}	f_{12}	f_{13}	f_{14}	f_{15}	f_{16}	f_{17}	f_{18}	f_{19}	f_{20}	f_{21}	f_{22}	f_{23}	f_{24}	Mean Rank
CLPSO	4	1	4	3	4	4	2	2	1	1	3	2	3	3	4	3	4	3	4	3	4	3	3	4	3.04
CMAES	3	4	3	4	1	3	4	1	2	4	4	4	4	4	3	4	1	2	3	4	3	4	4	2	3.12
LSHADE	2	3	2	2	2	2	3	3	3	2	2	1	2	2	2	2	1	4	2	1	2	1	2	3	2.12
Tornado	1	2	1	1	3	1	1	4	4	3	1	3	1	1	1	1	1	1	1	2	1	2	1	1	**1.62**

Table 12. Total of Mean time (per run) consumed by the four algorithms on dimensions $D = 50$, $D = 100$ and $D = 200$.

	CLPSO	L-SHADE	CMAES	Tornado
Total of mean Time for $D = 50$	247.2 s	95.1 s	902.2 s	109.2 s
Total of mean Time for $D = 100$	536.4 s	328.3 s	2640.3 s	215.1 s
Total of mean Time for $D = 200$	1415.4 s	1385.4 s	10425.6 s	614.8 s

On the other hand, Table 12 indicates that Tornado releases the shortest execution time whereas we observe that CMAES is so far the algorithm that consumes the most execution time. This is due to the covariance matrix process used by CMAES which is not integrated in the function evaluation (comparison criteria).

Figures 10 and 11 show the convergence of the four algorithms for functions $f_1 - f_{10}$ on dimensions $D = 50$ and $D = 100$. The obtained results show a quick convergence for the Tornado algorithm compared to other algorithms. Other carried experiments show the same trend for problems with $D = 200$.

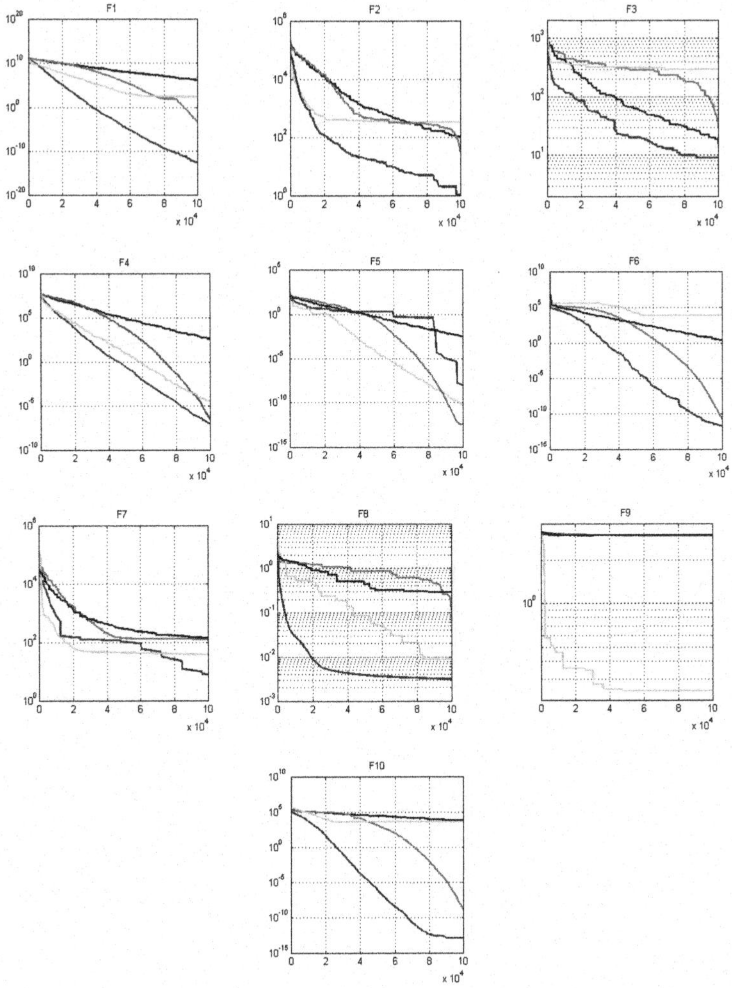

Tornado L-SHADE CMAES CLPSO

Fig. 10. Convergence performance of the four different methods for functions $f_1 - f_{10}$ on dimension $D = 50$.

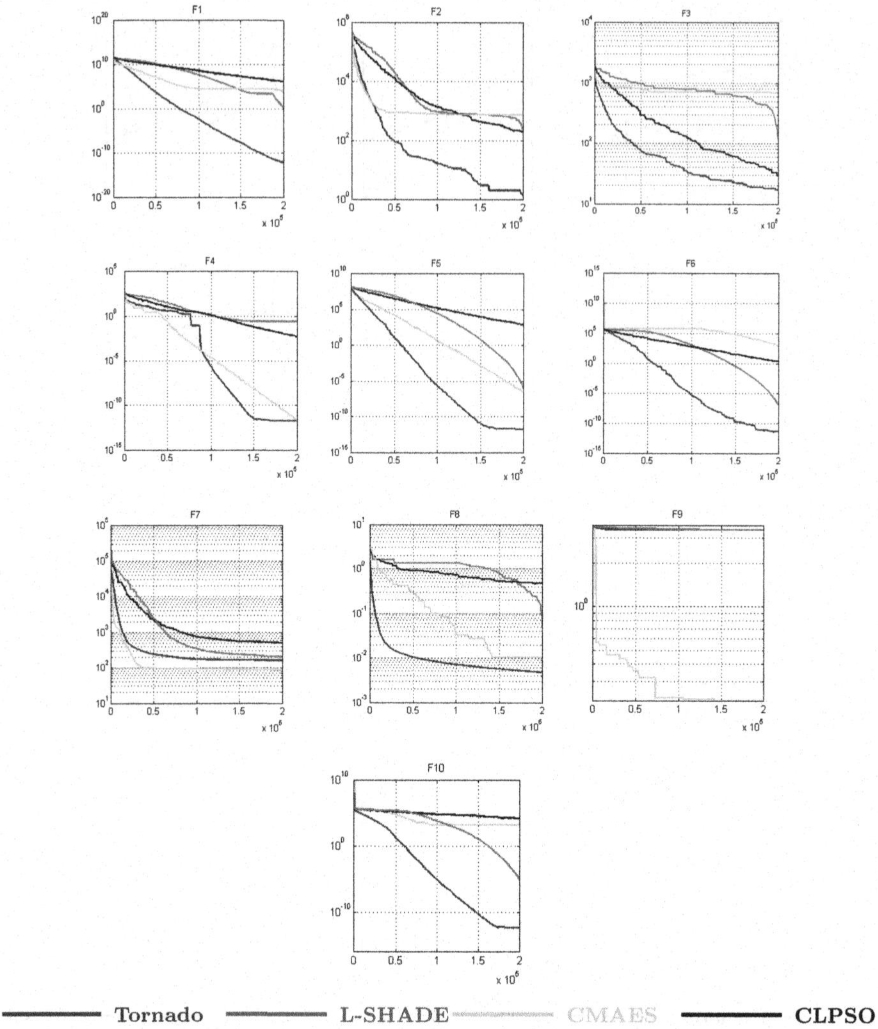

Fig. 11. Convergence performance of the four different methods for functions $f_1 - f_{10}$ on dimension $D = 100$.

5 Conclusions and Perspectives

In the big era, there is a need for developing optimization algorithms able to effectively solve problems with hundreds, thousands, and even millions of variables. In this paper we have proposed an autonomous chaotic optimization algorithm, called Tornado, for high dimensional global optimization problems. The algorithm introduces advanced symmetrization, levelling and fine search strategies for an efficient and effective exploration of the search space and exploitation of the best found solutions. To our knowledge, this is the first accurate and fast autonomous chaotic algorithm solving high dimensional optimization problems.

The obtained results has shown the scalability of the algorithm in contrast to chaotic optimization algorithms encountered in the literature. Moreover, in comparison with some state-of-the-art metaheuristics (e.g. evolutionary algorithms, swarm intelligence), the computational results revealed that the proposed Tornado algorithm is an effective and efficient optimization algorithm.

We will investigate the application of the Tornado algorithm to high dimensional scale real-life optimization problems such as learning of deep neural networks, the optimization of the hyper-parameters of deep convolution neural networks, and demand energy management in smart grids. An extension of the Tornado algorithm to solve multi-objective optimization problems using scalarization and Pareto approaches is also under study.

A parallel implementation of the algorithm on heterogeneous parallel architectures, composed of multi-cores and clusters of GPUs, will be also investigated. We are also interested in the design of Fractals based decomposition strategies. The Chaotic approach will be combined to a Fractal based decomposition model, in which chaotic search is applied in each Fractal. This combination will generate highly parallel approaches to be implemented on exascale parallel architectures composed of millions of GPU cores. The parallel model will also improve the exploration capabilities of the Tornado algorithm.

References

1. Wu, X.X., Chen, Z.: Introduction of Chaos Theory, Shanghai Science and Technology (1996)
2. Lorenz, E.N.: Deterministic nonperiodic flow. J. Atmos. Sci. **20**(2), 130–141 (1963)
3. Auerbach, D., Grebogi, C., Ott, E., Yorke, J.A.: Controlling chaos in high dimensional systems. Phys. Rev. Lett. **69**(24), 3479 (1992)
4. Pecora, L.M., Carroll, T.L.: Synchronization in chaotic systems. Phys. Rev. Lett. **64**(8), 821 (1990)
5. Li, B., Jiang, W.: Chaos optimization method and its application. Control Theory Appl. **14**(4), 613–615 (1997)
6. Yang, D., Li, G., Cheng, G.: On the efficiency of chaos optimization algorithms for global optimization. Chaos Solitons Fractals **34**(4), 1366–1375 (2007)
7. Hamaizia, T., Lozi, R.: Improving chaotic optimization algorithm using a new global locally averaged strategy. In: Emergent Properties in Natural and Artificial Complex Systems, p. 17 (2011)
8. Aslimani, N., Ellaia, R.: A new hybrid algorithm combining a new chaos optimization approach with gradient descent for high dimensional optimization problems. Comput. Appl. Math. **37**(3), 2460–2488 (2018)
9. Aslimani, N., Ellaia, R.: A new chaos optimization algorithm based on symmetrization and levelling approaches for global optimization. Numer. Algorithms **79**(4), 1021–1047 (2018). https://doi.org/10.1007/s11075-018-0471-9
10. Feng, J., Zhang, J., Zhu, X., Lian, W.: A novel chaos optimization algorithm. Multimed. Tools Appl. **76**(16), 17405–17436 (2017)

11. Shengsong, L., Min, W., Zhijian, H.: Hybrid algorithm of chaos optimisation and SLP for optimal power flow problems with multimodal characteristic. IEE Proc.-Gener. Transm. Distrib. **150**(5), 543–547 (2003)
12. Wang, J., Wang, X.: A global control of polynomial chaotic systems. Int. J. Control **72**(10), 911–918 (1999)
13. Ishii, S., Sato, M.: Constrained neural approaches to quadratic assignment problems. Neural Netw. **11**(6), 1073–1082 (1998)
14. Wong, K., Man, K.-P., Li, S., Liao, X.: A more secure chaotic cryptographic scheme based on the dynamic look-up table. Circuits Syst. Signal Process. **24**(5), 571–584 (2005)
15. Gao, H., Zhang, Y., Liang, S., Li, D.: A new chaotic algorithm for image encryption. Chaos Solitons Fractals **29**(2), 393–399 (2006)
16. Ibrahim, R.A., Abd Elaziz, M., Lu, S.: Chaotic opposition-based grey-wolf optimization algorithm based on differential evolution and disruption operator for global optimization. Expert Syst. Appl. **108**, 1–27 (2018)
17. Alikhani Koupaei, J., Hosseini, S.M.M., Maalek Ghaini, F.M.: A new optimization algorithm based on chaotic maps and golden section search method. Eng. Appl. Artif. Intell. **50**, 201–214 (2016)
18. Arora, S., Singh, S.: An improved butterfly optimization algorithm with chaos. J. Intell. Fuzzy Syst. **32**(1), 1079–1088 (2017)
19. Petrović, M., Vuković, N., Mitić, M., Miljković, Z.: Integration of process planning and scheduling using chaotic particle swarm optimization algorithm. Expert Syst. Appl. **64**, 569–588 (2016)
20. Wang, L., Liu, X., Sun, M., Qu, J., Wei, Y.: A new chaotic starling particle swarm optimization algorithm for clustering problems. Math. Probl. Eng. **2018** (2018)
21. Wang, G.-G., Deb, S., Gandomi, A.H., Zhang, Z., Alavi, A.H.: Chaotic cuckoo search. Soft Comput. **20**(9), 3349–3362 (2016)
22. Fister, I., Jr., Perc, M., Kamal, S.M., Fister, I.: A review of chaos-based firefly algorithms: perspectives and research challenges. Appl. Math. Comput. **252**, 155–165 (2015)
23. Yan, H., Zhou, L., Liu, L.: Chaos genetic algorithm optimization design based on permanent magnet brushless DC motor. In: Jia, L., Liu, Z., Qin, Y., Ding, R., Diao, L. (eds.) Proceedings of the 2015 International Conference on Electrical and Information Technologies for Rail Transportation. LNEE, vol. 377, pp. 329–337. Springer, Heidelberg (2016). https://doi.org/10.1007/978-3-662-49367-0_34
24. dos Santos Coelho, L.: Tuning of PID controller for an automatic regulator voltage system using chaotic optimization approach. Chaos Solitons Fractals **39**(4), 1504–1514 (2009)
25. Hamaizia, T., Lozi, R., Hamri, N.: Fast chaotic optimization algorithm based on locally averaged strategy and multifold chaotic attractor. Appl. Math. Comput. **219**(1), 188–196 (2012)
26. Tanabe, R., Fukunaga, A.S.: Improving the search performance of shade using linear population size reduction. In: 2014 IEEE Congress on Evolutionary Computation (CEC), pp. 1658–1665. IEEE (2014)
27. Brest, J., Maučec, M.S., Bošković, B.: Single objective real-parameter optimization: algorithm jSO. In: 2017 IEEE Congress on Evolutionary Computation (CEC), pp. 1311–1318. IEEE (2017)
28. Salgotra, R., Singh, U., Singh, G.: Improving the adaptive properties of LSHADE algorithm for global optimization. In: 2019 International Conference on Automation, Computational and Technology Management (ICACTM), pp. 400–407. IEEE (2019)

29. Salgotra, R., Singh, U., Saha, S., Nagar, A.: New improved SALSHADE-cnEpSin algorithm with adaptive parameters. In: 2019 IEEE Congress on Evolutionary Computation (CEC), pp. 3150–3156. IEEE (2019)
30. Akhmedova, S., Stanovov, V., Semenkin, E.: LSHADE algorithm with a rank-based selective pressure strategy for the circular antenna array design problem. In: ICINCO (1), pp. 159–165 (2018)
31. Loshchilov, I.: CMA-ES with restarts for solving CEC 2013 benchmark problems. In: 2013 IEEE Congress on Evolutionary Computation, pp. 369–376. IEEE (2013)
32. Al-Bahrani, L.T., Patra, J.C.: A novel orthogonal PSO algorithm based on orthogonal diagonalization. Swarm Evol. Comput. **40**, 1–23 (2018)
33. Lynn, N., Suganthan, P.N.: Heterogeneous comprehensive learning particle swarm optimization with enhanced exploration and exploitation. Swarm Evol. Comput. **24**, 11–24 (2015)
34. Larrañaga, P., Lozano, J.A.: Estimation of Distribution Algorithms: A New Tool for Evolutionary Computation, vol. 2. Springer, Heidelberg (2001)
35. Hauschild, M., Pelikan, M.: An introduction and survey of estimation of distribution algorithms. Swarm Evol. Comput. **1**(3), 111–128 (2011)
36. Kabán, A., Bootkrajang, J., Durrant, R.J.: Toward large-scale continuous EDA: a random matrix theory perspective. Evol. Comput. **24**(2), 255–291 (2016)
37. Dong, W., Chen, T., Tiňo, P., Yao, X.: Scaling up estimation of distribution algorithms for continuous optimization. IEEE Trans. Evol. Comput. **17**(6), 797–822 (2013)
38. Larrañaga, P., Etxeberria, R., Lozano, J.A., Peña, J.M.: Optimization in continuous domains by learning and simulation of Gaussian networks (2000)
39. Mohamed, A.W., Hadi, A.A., Fattouh, A.M., Jambi, K.M.: LSHADE with semi-parameter adaptation hybrid with CMA-ES for solving CEC 2017 benchmark problems. In: 2017 IEEE Congress on evolutionary computation (CEC), pp. 145–152. IEEE (2017)
40. Zhang, G., Shi, Y.: Hybrid sampling evolution strategy for solving single objective bound constrained problems. In: 2018 IEEE Congress on Evolutionary Computation (CEC), pp. 1–7. IEEE (2018)

Learning

Neural Network Information Leakage Through Hidden Learning

Arthur Carvalho Walraven da Cunha[1(✉)], Emanuele Natale[1], and Laurent Viennot[2]

[1] Inria d'Université Côte d'Azur, Sophia Antipolis, France
{arthur.carvalho-walraven-da-cunha,emanuele.natale}@inria.fr
[2] Inria de Paris, Paris, France
laurent.viennot@inria.fr

Abstract. We investigate the problem of making an artificial neural network perform hidden computations whose result can be easily retrieved from the network's output. In particular, we consider the following scenario. A user is provided a neural network for a classification task by a third party. The user's input to the network contains sensitive information and the third party can only observe the output of the network. I this work, we provide a simple and efficient training procedure, which we call *hidden learning*, that produces two networks: *(i)* one that solves the original classification task with performance near to state of the art; *(ii)* a second one that takes as input the output of the first, retrieving sensitive information to solve a second classification task with good accuracy. Our result might expose important issues from an information security point of view, as for the use of artificial neural networks in sensible applications.

Keywords: Artificial neural network · Hidden computation · Information security

1 Introduction

In this paper, we investigate the possibility of an attacker training an Artificial Neural Network (ANN) such that, while its behaviour looks legitimate on a given task, it secretly performs an additional task, possibly revealing information it should not. In particular, we investigate the question: when using a model from the shelf, is it possible that it computes and outputs more than supposed?

Such question naturally emerges with the current surge of *machine learning as a service* scenarios (MLaaS) [16], which has motivated plenty of research on the associated privacy and security problems [13]. Within the taxonomy of attacks investigated by previous works, particular attention has been devoted to model inversion (MI) attacks [1], in which an attacker tries to retrieve sensible features about the input data by only accessing the model's output. One can apply this strategy with or without knowledge of the model itself (white-box vs. black-box attacks).

B. Dorronsoro et al. (Eds.): OLA 2023, CCIS 1824, pp. 117–128, 2023.
https://doi.org/10.1007/978-3-031-34020-8_8

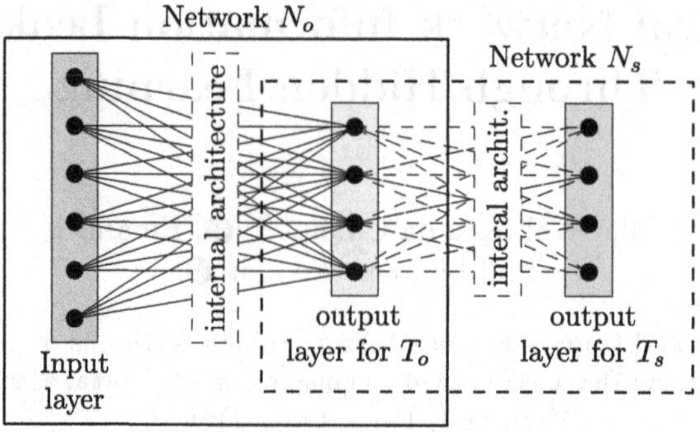

Fig. 1. Diagram illustrating the basic components of the Hidden Learning framework. See Sect. 2 for a description of the components.

In this work, we consider a setting in which the attacker forges the weights of the model based on the training data, thus being in a much more powerful position compared to the MI settings. To ensure that the model looks unsuspicious, we further require the attacker to use a conventional design for the network and that it achieves state-of-the-art accuracy. (We further discuss MI and its relation with the present work in Sect. 3.)

A natural way to perform hidden learning would be to combine two networks with steganographic techniques; however, it is unclear how to do this under the mentioned restrictions without making the model look suspicious.

In this paper, we investigate what may be regarded as the most natural strategy to achieve the mentioned goal. We consider a simple scheme that trains a network for two tasks at the same time, namely, the *official* task, which a user expects it to perform, and a *secret* task, which is achieved by feeding the output of the network to a *secret network* (see Fig. 1). We call this scheme *hidden learning*, and we formally define it in Sect. 3.

To provide some intuition for the proposed framework, consider sets of points on the Euclidean plane sampled from two standard gaussians centred at $(0, 1)$ and $(0, -1)$. The official task is to classify those points according to the gaussian they come from, so it only depends on one of the coordinates of the points. In such set up, the faithful model should use the best separating line, $y = 0$. However, the line $y = x$ would still achieve substantial accuracy on the official task while revealing some information about the input x coordinate.

An example where hidden learning could be problematic would be the scenario where, for better handling the Covid-19 crisis, the government of a country hires a company to develop a smartphone application for estimating how many people are at risk in each region of the country. Each user is asked to feed sensitive health information to a neural network that outputs a probability that the user can develop severe Covid reaction if infected and a probability that the user was already infected. Only these two probabilities and the user's region are communicated to the company's server so it can provide statistics to the government. If the application is open source, independent coders can check that the application does indeed behave as expected. However, by applying hidden learning to set up the weights of the neural network embedded in the application, the company could use a secret (private) network to retrieve additional information from the user's output. Data such as high risk of cardio-vascular accident could be valuable for some insurance companies, which might be tempted to discreetly change their coverage conditions for cardio-vascular risks in certain regions accordingly.

Our main goal is to draw attention to the possibility of an attack on the weights of a model by showing that it can be made effective with a simple approach at a very low computational cost.

After formally defining our framework (Sect. 2) and discussing related works (Sect. 3), we describe and discuss our experiments on several synthetic tasks defined on the CIFAR-10 and Fashion MNIST datasets (Sects. 4 and 5). Finally, we provide our conclusions about the results in Sect. 6.

2 Hidden Learning Framework

In this section, we formally describe the *Hidden Learning* framework, whose main components are represented in Fig. 1.

We start by providing the key definitions. Let S be a generic set and k_o and k_s be two positive integers. Hidden Learning is performed by considering two classification tasks:

- the *official task* T_o, which asks to classify points into S in k_o categories;
- the *secret task* T_s, which asks to classify points into S in k_s categories.

In order to perform those two tasks, the Hidden Learning framework produces two artificial neural networks:

- an *official network* N_o, which assigns each $x \in S$ to a vector $N_o(x) \in [0,1]^{k_o}$ of scores associated to the k_o categories of the task T_o;
- a *secret network* N_s, which classifies vectors in $[0,1]^{k_o}$ into k_s categories.

Remark 1. The only specific constraint in the above framework lies in the co-domain of the official network N_o, namely the space of vectors in $\mathbb{R}^{k_o}$, which are then passed to a softmax function. The latter is a natural choice in many scenarios and is consistent with typical MI attack settings, in which the attacker is assumed to have query access to some model's scores about the possible output categories [15].

The training of the official and secret networks is simultaneous: at each epoch, the updates of the weights of the two networks are computed by back-propagation according to a combination of the loss functions for the respective tasks. As a first simple choice for combining the loss functions, we consider their sum.

More formally, let $L_o(\hat{y}, y)$ and $L_s(\hat{y}, y)$ be the loss functions for the official task T_o and the secret task T_s, respectively. The network is trained by optimizing the combined loss function $L_o(\hat{y}, y)+L_s(\hat{y}, y)$. More details about how we perform the training in our experiments can be found in Sect. 4.

3 Related Work

Our work is closely related to the class of privacy attacks to neural network models known as (*white box*) model inversion (MI) attacks [1]. In the latter setting, given an output $f(x)$ and the model f that produced it, an attacker tries to reconstruct the corresponding input x. We emphasize that, in contrast to the MI setting in which the attacker does not intervene in the creation of the model f, our hidden learning framework assumes that the attacker can forge the model f (our N_o) itself in a disguised fashion that allows, by design, to easily invert it (using N_s). Note also that contrarily to many MI settings, the training data is not considered sensitive here, while the attack concerns input data fed to the model in production use. We also mention here *black box* MI attacks which, as the name suggest, are a more restrictive kind of MI attacks where the attacker only needs to be able to arbitrarily query the model and observe the corresponding output, without any knowledge about the model internals [5]. Contrarily to this setting, we do not assume that the attacker can propose forged inputs and get the corresponding outputs.

Part of our experiments verifies the robustness of the secret network to perturbations of the official one. This can be compared to recent works which investigate the sensitivity of the explainability of a model when the latter is perturbed as a consequence of other procedures, such as the disruption of input attribution that arises when standard neural network compression methods are employed, as recently shown in [11].

The present work investigates a simple approach to produce a neural network (the official network N_o) which performs some *hidden computation* that can be exploited by a third party to extract sensitive information from private inputs. In this respect, it falls in the general area of [12]. While the application of artificial neural networks for standard steganographic tasks (*statically* hiding information in a given object) is being actively investigated [17,20], we are not aware of works which, like the present one, explore how to produce an artificial neural network which tries to hide information in its output through calculations that are entirely transparent to the party who is making use of it. In particular, its architecture should be legitimate for the official task. A concept related to the latter is that of backdoor attacks on deep neural networks, where the goal is to produce a neural network that appears to solve a task, but behaves quite differently when fed specific triggering inputs [3,10]. It has also been shown that the latter triggering inputs can be designed via steganography so that they would not be identifiable by direct inspection [9].

4 Experiments

This section describes our experiments on the Hidden Learning framework, described in Sect. 2.

We perform experiments on the classical CIFAR-10 dataset [7] and Fashion MNIST dataset (FMNIST) [19]. Both of them consist of small-size images (32×32 and 28×28 pixels, respectively) classified in 10 classes:

- airplane, automobile, bird, cat, deer, dog, frog, horse, ship, and truck for CIFAR-10,
- T-shirt/top, trouser, pullover, dress, coat, sandal, shirt, sneaker, bag, and ankle boot for FMNIST.

4.1 Description of Experimental Results

This section describes the experiments summarized in Table 1. All values are rounded to the fourth decimal point.

We have adopted the same architecture for all experiments up to the number of neurons in the output layers of the tasks T_o and T_s. For simplicity, we have opted for a simple convolutional architecture for the official network, based on LeNet5 [8]:

- A convolutional layer with 16 kernels 3×3, stride 1×1, padding of one, and ReLu [4] activation function; which is followed by 2×2 max pooling;
- Two convolutional layers with 32 kernels, and otherwise identical to the previous (including the max pooling);
- A fully connected linear layer.

As for the secret network, we consider a multilayer perceptron with two hidden layer with ReLu activation, the first with 16 and the second with 32 hidden nodes. We remark that the above choices cover all hyperparameters.

We ran two types of experiments.

Hidden Learning Experiments. These experiments, summarized in Table 1, show the accuracy achieved by the official network N_o over several tasks described below. With the expression T_o-and-T_s we refer to the accuracies achieved in the experiments in which the networks N_o and N_s were trained in the hidden learning framework. The row T_o of column T_o-then-T_s and the column T_s-only show the accuracies achieved in the experiments in which the networks N_o and N_s were trained by taking into account, respectively, only the loss function for T_o and for T_s (separately). Finally, the rows T_s of the column T_o-then-T_s show the accuracies achieved by N_s in the experiments in which, first, the network N_o was trained by taking into account only the loss function for T_o and, then, N_s was trained by taking into account only the loss function for T_s, while the weights of N_o are not modified. We observe that the latter experiments resemble black-box MI where the attacker has access to the full training dataset with corresponding model outputs.

Robustness Experiments. These experiments, summarized in Table 2, estimate the secret network's robustness to perturbations of the official network. We do so by adding gaussian noise with zero mean and standard deviation σ to each weight of the official network. Each column of Table 2 shows the accuracies of the official and secret networks, on the train and test sets, for different values of σ, averaged over 10 independent noise injections.

Recall that both CIFAR-10 and FMNIST associate inputs to labels from 10 classes. We have simulated information removal by creating subtasks of classification into[1]

- Two classes ($C2$): one class for the inputs belonging to any of the first 5 original classes, and other for the inputs belonging to any of the last 5. For instance, for CIFAR-10, the first class in this subtask is "*airplane* or *automobile* or *bird* or *cat* or *deer*" while the other is "*dog* or *frog* or *horse* or *ship* or *truck*".
- Five classes ($C5$): we pair original classes to create new ones. Furthermore, we do this while avoiding pairs contained in the classes for the last subtask. This ensures that the solutions to one of those subtasks do not provide any information about the other. Using CIFAR-10 labels as an example, the classes for this subtask are "*airplane* or *dog*", "*automobile* or *frog*", "*bird* or *horse*", "*cat* or *ship*", and "*deer* or *truck*".
- The first n classes (Fn): classification into $n+1$ classes, namely, the first n original classes, and an extra one combining all the other. Exemplifying as before, for $n = 3$ this subtask comprises the classes "*airplane*", "*automobile*", "*bird*", and "*neither an airplane nor an automobile nor a bird*".
- The last n classes (Ln): same as the previous subtask, but for the n last original classes.

We organized the experiments by choosing one of those subtasks as T_o and the other as T_s. We also consider cases where T_s is the original classification into 10 classes.

In the tables, we refer to the original task as C10, to subtasks with 2 and 5 classes as C2 and C5, respectively, to the classification into the first m original classes as Fm, and to the classification into the last n original classes as Ln.

We initialized the weights of all the neural networks using Glorot uniform initialization [2]. and then trained all of them for 40 epochs using the ADAM optimizer [6] with a learning rate of 0.001. over 45,000 training entries for CIFAR-10 and 54,000 for FMNIST, organized into batches of size 64. Even though the actual number of training points in those datasets is, respectively, 50,000 and 60,000, we reserved 10% of those to use as the validation dataset. When training N_o and N_s simultaneously, we chose sets of weights that maximize the sum of the accuracies of both networks. The accuracy values discussed in this work refer to the performance of the networks with these sets of weights on the test set. The test dataset consists of 10,000 data points for both CIFAR-10 and FMNIST. Those do not take any part in the training.

[1] The symbols between parenthesis refer to the one used in the experiment tables.

Furthermore, in subtasks of the type Fn and Ln, some of the classes are the same as in the original task, so each corresponds to 10% of the dataset. On the other hand, the extra class merges all the remaining original classes, corresponding to $10 - n$ tenths of the points. We try to compensate for this unbalance by proportionally under-weighting the loss for these extra classes. More precisely, when computing the loss for subtasks of type Fn or Ln, we divide the loss by $10 - n$ whenever the input belongs to, respectively, the first n or last n original classes.

The results of our experiments are discussed in Sect. 5.

5 Discussion

We start by discussing the experiments summarized in Table 1. Comparing the accuracy achieved by the official network in the *Hidden Learning* experiments (T_o-and-T_s) with its accuracy when trained for T_o only (provided in the T_o row of the T_o-then-T_s column), we can see that the framework does not sensibly decrease accuracy: for CIFAR-10 the two numbers are respectively[2] 68.5 ± 6.4 and 71.2 ± 10.6, while for FMNIST we have 91.5 ± 2.2 and 92.1 ± 2.6.

The corresponding accuracies achieved by the secret network N_s, namely when trained with the framework and when trained after N_o has been trained alone and is not modified, are respectively 58.7 ± 9.5 and 46.5 ± 16.3 on CIFAR-10, and 82.6 ± 10.9 and 65.7 ± 15.1 on FMNIST. Hence, we can see that Hidden Learning drastically improves the accuracy compared to what may be regarded as a black-box MI approach (as mentioned in Sect. 4).

We can furthermore see that, when the entire architecture is trained by uniquely taking into account the loss function of the secret task T_s, N_s achieves accuracies which are only slightly better than those achieved with the Hidden Learning framework, scoring 59.1 ± 8.3 on CIFAR-10 and 83.4 ± 9.9 on FMNIST. The fact that the framework matches the latter results for T_s shows that it is effective in exploiting the whole network despite the interference of the official task.

We observe that the gain in accuracy for T_s is especially significant in the experiments where this task involves fewer classes. This finding is consistent with the fact that, in such cases, the secret network has fewer neurons as input and, thus, when N_s is trained independently (T_s), it should get access to less information in the first place.

Finally, we remark that, since our tasks consisted of different ways to group and split the original dataset classes into different ones, we also verified that our results are not sensitive to the ordering of the original labels.

We now discuss the robustness experiments summarized in Table 2. The goal of these experiments is to provide a first assessment of the sensitivity of the secret network N_s to perturbations of the official network N_o. We remark that

[2] The value reported after the average is the sample standard deviation. All reported statistical values are rounded to the first decimal place.

Table 1. Summary table of experimental results described in Sect. 4.1.

Exp.	Task	CIFAR-10			FMNIST		
		T_o and T_s	T_o then T_s	T_s only	T_o and T_s	T_o then T_s	T_s only
C2-C10	T_o	73.5%	74.9%		94.2%	94.3%	
	T_s	43.5%	21.3%	47.5%	85.3%	50.2%	86.5%
C5-C10	T_o	65.3%	64.8%		90.3%	89.5%	
	T_s	61.4%	49.6%	61.3%	89.8%	87.2%	89.5%
C5-C2	T_o	64.3%	65.0%		89.9%	89.9%	
	T_s	75.7%	66.3%	74.3%	94.3%	90.0%	94.5%
C2-C5	T_o	66.3%	74.7%		94.2%	94.4%	
	T_s	53.0%	27.6%	58.4%	85.6%	51.1%	88.0%
F2-L8	T_o	78.2%	89.8%		94.0%	96.3%	
	T_s	51.7%	22.2%	50.3%	87.0%	38.2%	86.9%
F3-L7	T_o	71.0%	81.4%		90.9%	93.3%	
	T_s	56.3%	41.9%	58.3%	88.4%	68.4%	88.9%
F4-L6	T_o	63.5%	67.0%		90.1%	92.1%	
	T_s	58.0%	45.6%	62.2%	89.9%	74.8%	90.1%
F5-L5	T_o	61.4%	60.4%		90.4%	90.2%	
	T_s	64.4%	49.1%	66.1%	92.2%	78.6%	92.9%
F6-L4	T_o	63.4%	60.4%		89.9%	89.6%	
	T_s	60.6%	51.4%	64.2%	78.9%	74.9%	78.7%
F7-L3	T_o	64.2%	64.0%		90.5%	90.3%	
	T_s	67.1%	63.1%	64.7%	66.1%	65.3%	65.9%
F8-L2	T_o	65.8%	65.8%		87.7%	89.3%	
	T_s	41.3%	45.3%	49.5%	62.0%	64.1%	71.8%
F2-L5	T_o	79.4%	89.3%		95.3%	96.3%	
	T_s	57.3%	33.2%	59.5%	92.3%	48.2%	92.4%
F5-L2	T_o	64.1%	58.9%		90.6%	90.2%	
	T_s	60.8%	73.9%	44.9%	80.3%	69.3%	75.0%
F3-L3	T_o	78.6%	80.8%		92.3%	93.6%	
	T_s	70.9%	60.9%	65.9%	64.7%	59.0%	66.0%

our experiments were not optimized to improve network robustness to weight noise, e.g. by some regularization approach [21].

The table displays the corresponding accuracies obtained for the smallest values we considered for the standard deviation of the gaussian noise applied (σ for short) to the weights of N_o, namely from 0 to 0.1 with a step of 0.025.

When noise is very low ($\sigma = 0.025$), the average test accuracy for N_o drops by 4.7% for CIFAR-10 while we see a 1.1% average decrease for FMNIST. The corresponding percentages for N_s are 5.8% (CIFAR-10) and 1.7% (FMNIST).

Table 2. Accuracies obtained in robustness experiments described in Sect. 4.1.

Exp.	Task	CIFAR-10					FMNIST				
		$\sigma = 0$	0.025	0.05	0.075	0.1	0	0.025	0.05	0.075	0.1
C2-C10	T_o	73.5%	71.0%	65.5%	60.6%	54.9%	94.2%	93.4%	91.6%	83.8%	74.0%
	T_s	43.5%	38.5%	30.6%	21.5%	14.0%	85.3%	80.3%	67.2%	48.8%	37.2%
C5-C10	T_o	65.3%	56.3%	43.5%	33.5%	28.1%	90.3%	88.3%	80.9%	73.1%	50.3%
	T_s	61.4%	51.0%	34.8%	23.4%	17.1%	89.8%	87.7%	77.7%	69.4%	39.9%
C5-C2	T_o	64.3%	56.9%	42.8%	32.5%	29.0%	89.9%	88.4%	84.3%	75.7%	54.4%
	T_s	75.7%	71.0%	61.1%	55.0%	53.9%	94.3%	93.2%	90.6%	83.3%	69.1%
C2-C5	T_o	66.3%	64.5%	59.1%	54.1%	52.4%	94.2%	93.2%	90.9%	86.1%	79.2%
	T_s	53.0%	45.7%	32.0%	25.1%	24.2%	85.6%	82.8%	73.3%	53.6%	48.8%
F2-L8	T_o	78.2%	78.0%	73.9%	64.0%	66.3%	94.0%	93.6%	92.9%	90.1%	86.6%
	T_s	51.7%	44.6%	32.9%	23.7%	18.6%	87.0%	84.4%	72.5%	58.8%	51.4%
F3-L7	T_o	71.0%	70.0%	64.2%	43.5%	41.8%	90.9%	91.1%	86.1%	80.4%	74.9%
	T_s	56.3%	45.4%	35.2%	27.6%	22.4%	88.4%	85.0%	74.8%	64.2%	46.5%
F4-L6	T_o	63.5%	60.0%	53.1%	47.8%	35.0%	90.1%	89.8%	86.4%	79.6%	70.0%
	T_s	58.0%	50.1%	38.6%	27.8%	28.5%	89.9%	88.2%	81.3%	74.5%	56.1%
F5-L5	T_o	61.4%	55.2%	46.9%	39.1%	34.7%	90.4%	89.4%	85.8%	78.6%	74.7%
	T_s	64.4%	60.6%	51.2%	37.0%	30.9%	92.2%	91.3%	86.0%	76.6%	70.5%
F6-L4	T_o	63.4%	56.9%	44.4%	31.8%	28.8%	89.9%	88.8%	85.0%	77.4%	67.2%
	T_s	60.6%	55.4%	43.8%	38.9%	32.3%	78.9%	78.4%	74.4%	70.2%	63.0%
F7-L3	T_o	64.2%	56.1%	42.6%	26.2%	25.9%	90.5%	88.7%	84.6%	73.5%	53.8%
	T_s	67.1%	62.5%	44.9%	40.2%	31.8%	66.1%	65.9%	65.1%	62.3%	52.9%
F8-L2	T_o	65.8%	55.9%	40.4%	29.0%	17.6%	87.7%	84.7%	75.5%	62.0%	48.9%
	T_s	41.3%	39.5%	32.4%	31.2%	27.4%	62.0%	60.1%	67.7%	65.4%	66.6%
F2-L5	T_o	79.4%	78.7%	74.7%	63.1%	58.2%	95.3%	94.9%	93.4%	90.0%	82.8%
	T_s	57.3%	53.4%	43.1%	30.9%	26.8%	92.3%	91.3%	88.4%	83.6%	74.0%
F5-L2	T_o	64.1%	59.2%	47.8%	37.3%	35.1%	90.6%	89.2%	85.9%	79.9%	75.8%
	T_s	60.8%	57.6%	53.9%	39.3%	37.1%	80.3%	79.4%	79.8%	74.6%	74.6%
F3-L3	T_o	78.6%	74.3%	68.0%	62.3%	38.7%	92.3%	91.9%	88.4%	84.6%	75.3%
	T_s	70.9%	64.9%	53.9%	45.9%	33.7%	64.7%	64.6%	63.2%	54.5%	49.9%

In comparison, when $\sigma = 0.5$, N_o achieves average accuracy 31.5% $\pm$ 13.3 on FMNIST and 26.0% $\pm$ 13.6 on CIFAR10. For N_s those values are 28.4% $\pm$ 13.2 and 25.4% $\pm$ 10.9. This indicates that the perturbation in the official output tends not to disturb the computation of the secret network unless it is strong enough to change the official answer.

We can appreciate from the table that a noise level of 0.1 already deteriorates the accuracy of the official network by 29.5% and 22.3% on average for CIFAR-10 and FMNIST, respectively. In particular, the fact that N_o achieves, on across different experiments, higher accuracies (22.9% difference) on FMNIST (average 91.4 $\pm$ 2.2) than on CIFAR10 (average 68.5 $\pm$ 6.4) in the absence of noise corresponds to lower deterioration when $\sigma = 0.1$, namely 69.2 ± 12.3 versus 39.0 ± 14.0.

We remark that the average standard deviation of test accuracies for N_o appears quite low on both datasets despite the heterogeneity of the experiments (especially the number of output classes). The trend is consistent for N_s, where the noiseless averages are 82.6 ± 10.9 for FMNIST and 58.7 ± 9.5 for CIFAR-10, while the corresponding numbers when $\sigma = 0.1$ are respectively 57.2 ± 12.5 and 28.5 ± 9.9.

6 Conclusions

In this work, we have introduced *Hidden Learning*, a simple and efficient training procedure that produces two networks, an official and a secret one, such that the official network solves an official task with performance comparable to state-of-the-art; and the secret network uses the output of the official one to solve a secret task with considerable accuracy. After contextualizing the above framework in the current research on Model Inversion and related attacks on neural networks, we have tested it on several synthetic tasks. In our experiments, the framework shows to be effective in tuning the official network to enable the attacker to better recover information via a secret network which is computationally very light. Thus, the possibility for such attacks should be taken into account when using a model provided by a third party.

Our preliminary investigation demands more sophisticated ones, particularly on possible defence mechanisms against the Hidden Learning framework. Even if the official network is suspected to be produced by such a framework, naive strategies to use the it while preventing information leakage, such as perturbing the network weights, appear ineffective in our robustness experiments[3]. More generally, the fact that the official network is, by design, produced to assist the secret network in extracting information might allow the framework to find ways around defence mechanisms that have been proven successful against similar attacks, such as model inversion ones. On the other hand, differential privacy [18], together with strategies to decouple data from model training [14], should prove successful in protecting against it.

Acknowledgements. The authors are grateful to the OPAL infrastructure from Université Côte d'Azur for providing resources and support.

References

1. Fredrikson, M., Jha, S., Ristenpart, T.: Model inversion attacks that exploit confidence information and basic countermeasures. In: Proceedings of the 22nd ACM SIGSAC Conference on Computer and Communications Security, pp. 1322–1333. ACM, Denver Colorado (2015). https://doi.org/10.1145/2810103.2813677, https://dl.acm.org/doi/10.1145/2810103.2813677

[3] Similarly, we expect the framework to be robust to output truncation.

2. Glorot, X., Bengio, Y.: Understanding the difficulty of training deep feedforward neural networks. In: Proceedings of the Thirteenth International Conference on Artificial Intelligence and Statistics, pp. 249–256. JMLR Workshop and Conference Proceedings (2010). http://proceedings.mlr.press/v9/glorot10a.html. ISSN: 1938-7228
3. Gu, T., Liu, K., Dolan-Gavitt, B., Garg, S.: BadNets: evaluating backdooring attacks on deep neural networks. IEEE Access **7**, 47230–47244 (2019). https://doi.org/10.1109/ACCESS.2019.2909068
4. Hahnloser, R.H.R., Sarpeshkar, R., Mahowald, M.A., Douglas, R.J., Seung, H.S.: Digital selection and analogue amplification coexist in a cortex-inspired silicon circuit. Nature **405**(6789), 947–951 (2000). https://doi.org/10.1038/35016072, https://www.nature.com/articles/35016072
5. He, Z., Zhang, T., Lee, R.B.: Model inversion attacks against collaborative inference. In: Proceedings of the 35th Annual Computer Security Applications Conference, pp. 148–162. ACM, San Juan Puerto Rico (2019). https://doi.org/10.1145/3359789.3359824, https://dl.acm.org/doi/10.1145/3359789.3359824
6. Kingma, D.P., Ba, J.: Adam: a method for stochastic optimization. In: International Conference on Learning Representations (2015). http://arxiv.org/abs/1412.6980, arXiv: 1412.6980
7. Krizhevsky, A.: Learning Multiple Layers of Features from Tiny Images. Master's thesis, Department of Computer Science, University of Toronto, p. 60 (2009)
8. Lecun, Y.: Gradient-based learning applied to document recognition. Proc. IEEE **86**(11), 47 (1998)
9. Li, S., Xue, M., Zhao, B., Zhu, H., Zhang, X.: Invisible backdoor attacks on deep neural networks via steganography and regularization. IEEE Trans. Dependable Secure Comput. (2020). https://doi.org/10.1109/TDSC.2020.3021407
10. Nguyen, T.A., Tran, A.: Input-aware dynamic backdoor attack. In: Larochelle, H., Ranzato, M., Hadsell, R., Balcan, M.F., Lin, H.T. (eds.) Advances in Neural Information Processing Systems, vol. 33. Annual Conference on Neural Information Processing Systems 2020, NeurIPS 2020 (December), pp. 6–12 (2020). https://proceedings.neurips.cc/paper/2020/hash/234e691320c0ad5b45ee3c96d0d7b8f8-Abstract.html
11. Park, G., Yang, J.Y., Hwang, S.J., Yang, E.: Attribution preservation in network compression for reliable network interpretation. arXiv:2010.15054 [cs] (2020). http://arxiv.org/abs/2010.15054, arXiv: 2010.15054
12. Petitcolas, F., Anderson, R., Kuhn, M.: Information hiding-a survey. Proc. IEEE **87**(7), 1062–1078 (1999). https://doi.org/10.1109/5.771065, http://ieeexplore.ieee.org/document/771065/
13. Qayyum, A., et al.: Securing machine learning in the cloud: a systematic review of cloud machine learning security. Front. Big Data **3**, 587139 (2020). https://doi.org/10.3389/fdata.2020.587139, https://www.frontiersin.org/articles/10.3389/fdata.2020.587139/full
14. Ryffel, T., et al.: A generic framework for privacy preserving deep learning. arXiv:1811.04017 [cs, stat] (2018). http://arxiv.org/abs/1811.04017, arXiv: 1811.04017
15. Shokri, R., Stronati, M., Song, C., Shmatikov, V.: Membership inference attacks against machine learning models. In: 2017 IEEE Symposium on Security and Privacy (SP), pp. 3–18 (2017). https://doi.org/10.1109/SP.2017.41. iSSN: 2375-1207
16. Tafti, A.P., LaRose, E., Badger, J.C., Kleiman, R., Peissig, P.: Machine learning-as-a-service and its application to medical informatics. In: Perner, P. (ed.) MLDM

2017. LNCS (LNAI), vol. 10358, pp. 206–219. Springer, Cham (2017). https://doi.org/10.1007/978-3-319-62416-7_15
17. Tao, J., Li, S., Zhang, X., Wang, Z.: Towards robust image steganography. IEEE Trans. Circuits Syst. Video Technol. **29**(2), 594–600 (2019). https://doi.org/10.1109/TCSVT.2018.2881118
18. Wang, Y., Si, C., Wu, X.: Regression model fitting under differential privacy and model inversion attack. In: IJCAI (2015)
19. Xiao, H., Rasul, K., Vollgraf, R.: Fashion-MNIST: a novel image dataset for benchmarking machine learning algorithms. arXiv:1708.07747 [cs, stat] (2017). http://arxiv.org/abs/1708.07747, arXiv: 1708.07747
20. Yang, Z., Guo, X., Chen, Z., Huang, Y., Zhang, Y.: RNN-Stega: linguistic steganography based on recurrent neural networks. IEEE Trans. Inf. Forensics Secur. **14**(5), 1280–1295 (2019). https://doi.org/10.1109/TIFS.2018.2871746
21. Zheng, S., Song, Y., Leung, T., Goodfellow, I.: Improving the robustness of deep neural networks via stability training. In: 2016 IEEE Conference on Computer Vision and Pattern Recognition (CVPR), pp. 4480–4488 (2016). https://doi.org/10.1109/CVPR.2016.485

Mixing Data Augmentation Methods for Semantic Segmentation

Rubén Escobedo(✉) and Jónathan Heras

University of La Rioja, Logroño, Spain
{ruescog,jonathan.heras}@unirioja.es

Abstract. Deep learning models are the state-of-the-art approach to deal with semantic segmentation tasks. However, training deep models require a considerable amount of images that might be difficult to obtain. This issue can be faced by means of data augmentation techniques that generate new images by applying geometric or colour transformations, or more recently by mixing several images using techniques such as CutMix or CarveMix. Unfortunately, mixing strategies are usually implemented as ad-hoc methods and are difficult to incorporate into the pipeline to train segmentation models. In this work, we present a library that implements several mixing strategies for data augmentation in semantic segmentation tasks. In particular, we provide a set of callbacks that can be integrated into the training pipeline of FastAI segmentation models. We have tested the library with a vineyard dataset and show the benefits of combining mixing strategies with traditional data augmentation techniques; namely an improvement of almost 5% was achieved using these methods regarding models trained only with traditional data augmentation methods.

Keywords: Data Augmentation · Semantic Segmentation · Deep Learning

1 Introduction

Semantic segmentation is a computer vision task that aims to classify every pixel of an image in a fixed set of classes. This task has received a lot of attention in recent years due to its multiple applications in contexts such as agriculture, manufacturing, robotics or medicine [4]. The interest in semantic segmentation is partially due to the development of deep learning architectures that provide accurate segmentation models [3]. One of the main drawbacks that hinder the adoption of deep learning models for semantic segmentation tasks is the annotation of a large number of images—a tedious and time-consuming task that might require expert knowledge [7]. A common technique to deal with this issue is data augmentation [14].

This work was partially supported by Ministerio de Ciencia e Innovación [PID2020-115225RB-I00/AEI/10.13039/501100011033].

B. Dorronsoro et al. (Eds.): OLA 2023, CCIS 1824, pp. 129–136, 2023.
https://doi.org/10.1007/978-3-031-34020-8_9

Data augmentation is a set of techniques that generate additional training data from existing data. The most widely used data augmentation methods for computer vision tasks are based on the application of image transformations such as rotations, flips or translations [6]; however, the diversity of images that can be generated with this approach is limited. A more advanced data augmentation strategy consists in generating new images using generative models [18]; but, the application of these generative methods is challenging [13]. A different approach that has emerged to achieve a trade-off between diversity of images and implementation easiness is the mixing of existing data [19]; for example, CutMix overlaps a region from an image with another image (potentially different), making a new image.

Mixing data augmentation methods have been primarily designed for image classification tasks, but they are not generalised to semantic segmentation problems. This is probably due to the fact that, in semantic segmentation, transformations must be applied not only to the image but also to its associated mask. Moreover, existing mixing data augmentation methods for semantic segmentation are implemented as ad-hoc libraries and, therefore, it is challenging to combine and compare them. In this paper, we aim to deal with these two drawbacks by the development of a Python library that facilitates the application of mixing data augmentation methods for semantic segmentation.

The rest of the paper is organised as follows. In the next section, we provide an overview of existing mixing data augmentation methods for semantic segmentation. Subsequently, in Sect. 3, we explain how those methods have been implemented in a Python library, and the results of evaluating such a library in a vineyard dataset are presented in Sect. 4. The paper ends with some conclusions and further work. The developed library is available at https://github.com/ruescog/semantic_segmentation_augmentations.

2 Mixing Data Augmentation

Mixing data augmentation strategies for semantic segmentation can be classified into two classes: CutOut methods and CutMix methods.

Given an image, CutOut methods, first presented in [2], are a family of techniques that drop regions from such an image, and fill them with the result of a mathematical function such as the mean or the mode of the image, or with a constant value, usually zero. CutOut methods for semantic segmentation have been mainly applied in the literature in two different ways. The classical CutOut method [2], from now on CutOut, picks randomly a region from the image and replaces it with 0s; whereas, the HideAndSeek method [15] divides the image into a grid and, randomly, replaces some portions of the grid with 0s. In addition, in this paper, we propose a new method called CutOutSemantic, a method that randomly picks a region and replaces all the pixels associated with a given class inside that region with 0s. It is worth noting that all the aforementioned transformations are not only applied to the images but also to their corresponding masks. An example of each one of these methods is provided in Fig. 1.

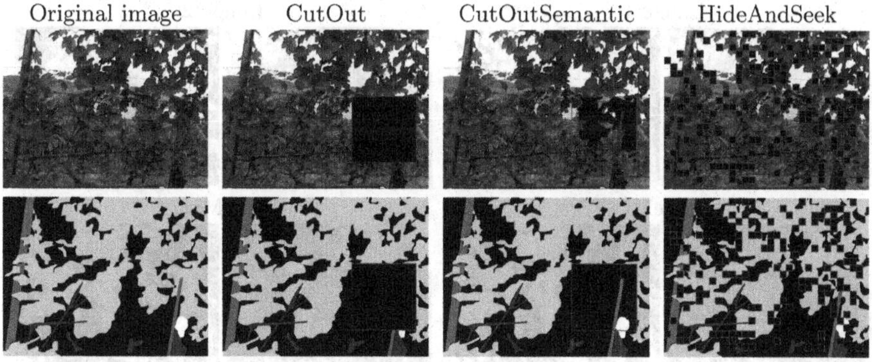

Fig. 1. Samples of different variants of the CutOut method. The region that is modified has been highlighted with a red frame. The same images with a higher resolution have been provided on the project webpage. (Color figure online)

As CutOut methods, CutMix techniques, that first appeared in [17], also drop a region from an image, but instead of replacing it with a fixed value, such a region is filled with another region from either the same image or from a different one—the same transformation is also applied to the mask associated with the image. In the literature, we can find 5 variants of CutMix method called CarveMix [19], CutMix [17], RICAP [16], ResizeMix [11] and SelfMix [20]. The classical CutMix method replaces the picked region with a random region from a different image. Similarly, CarveMix replaces the picked region with a region of interest (that is, a region that contains objects different to the background) from a different image. The RICAP method shuffles 4 image regions from (potentially) different image; and the ResizeMix technique replaces the picked region with a resized version of the given image. Moreover, SelfMix replaces the picked region with a region of interest from a different image, but keeping as much information as possible; that is, picking only the class pixels from the overlapped region.

In addition to the existing CutMix methods available in the literature, in the present paper, we propose three new methods. CutMix+Mod is analogous to CutMix but the replacement region is transformed by applying an operation such as flipping or enhancing its contrast. CutMixSemantic is analogous to CutMixRandom but in the replacement region the pixels of a given class are set to 0. Finally, TransparenceMix is a new method where the background pixels of the replacement region are seen as a transparency; so, the original values of the given image are used for those pixels. Examples of the existing CutMix methods can be found in Fig. 2.

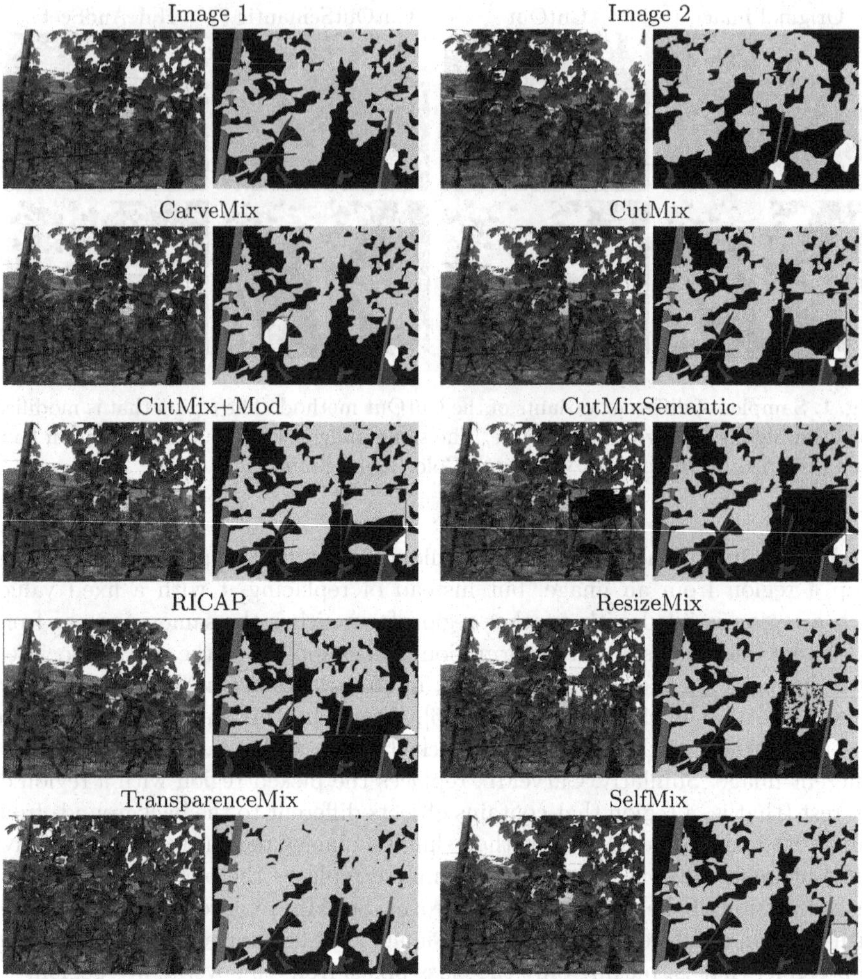

Fig. 2. Samples of different variants of the CutMix method. The region that is modified has been highlighted with a red frame. The same images with a higher resolution have been provided on the project webpage. (Color figure online)

3 Library Description

We have designed, and implemented, an open-source library in Python that implements all the aforementioned methods for the FastAI library [5]. To this aim, the augmentation methods have been implemented using *Callbacks*, which are objects that can perform actions at various stages of the training process. In particular, the Callbacks provided in our library are applied before passing a batch of images to the model during the training process.

All the mixing augmentation methods presented in the previous section consists of two steps: first, a region from an image is chosen; and, then, such a region is filled in some way. Additionally, an optional in-between step can be taken, applying traditional transformations to the selected regions. This abstract separation allowed us to divide the API into three different components called `HoleMaker`, `RegionModifier` and `HolesFilling`, and the different implementations of these components provide the mixing strategies.

In the case of the `HoleMaker` component, several strategies have been designed as classes that implement the `HoleMaker` interface:

- Random: selects a random region with a fixed size.
- Bounded: selects a random region with a fixed size, but this region must be inside the image.
- Attention: selects a random region, but this region must have enough pixels of information (that is, containing non-background elements).
- Point: selects a fixed region given a point and with a fixed aperture.
- ROI: selects a region of interest; which is a region that contains a group of pixels that belong to a fixed class that are separated from the other pixels from that class.

For all these strategies, several parameters can be fixed, such as the size of the region or a threshold in the attention strategy to determine whether a region has enough information. One of these strategies is always provided as input to the classes implementing the `HolesFilling` component.

The `RegionModifier` component can be used after the region has been chosen, allowing the user to apply some traditional augmentations (those transformations are provided by means of the Albumentations library [1]) to the selected region.

The `HolesFilling` component is an interface that has been implemented with the different mixing methods presented in the previous section. All the mixing strategies can be configured by fixing the number of holes that will be taken from the original image, the `HoleMaker` strategy, and the probability for applying the mixing augmentation. In addition, some strategies, such as CutMix, include a parameter with the transformations that will be applied to the region used for filling. Finally, multiple callbacks can be applied during the training process as shown in Fig. 3.

```
# List of callbacks
totalcbs = [
    CutOutRandom(holes_num = 1, p = 0.5,
                 hole_maker = HoleMakerRandom((50, 50))),
    CutOutSemantic(holes_num = 1, occlusion_class = 2, p = 1,
                   hole_maker = HoleMakerRandom((50, 50))),
    HideAndSeek(deactivation_p = 0.25, p = 0.2,
                      hole_maker = HoleMakerPoint((150, 150))
                                                          ),
    CutMixRandom(holes_num = 1, p = 0.5,
                 hole_maker = HoleMakerRandom((50, 50))),
    CutMixSemantic(holes_num = 1, p = 1,
                   hole_maker = HoleMakerRandom((50, 50))),
    RICAP(t = 0.25, p = 0.2),
    ResizeMix(p = 0.5),
    CarveMix(holes_num = 2, random_position = True, p = 1)
]

# Definition of the Learner object with the u-net
# architecture and with the associated callbacks.
# dls corresponds with the dataloader to load the images
# resnet50 is the backbone of the U-net architecture.
learner = unet_learner(dls, resnet50, cbs=totalcbs)
```

Fig. 3. Snippet showing how to define multiple callbacks and add them to a Learner object.

4 Results

In order to test the methods developed in our library, we used a vineyard dataset presented in [8]—a dataset that consists of 85 images (60 for training and 25 for testing) of a vineyard taken from an agricultural robot, see Figs. 1 and 2. Using the vineyard dataset, we trained different models using a U-Net segmentation architecture with a Resnet50 backbone [12], and combining traditional data augmentation methods with each mixing strategy implemented in our library. All the models were trained with the libraries PyTorch [10] and FastAI [5] and using a GPU Nvidia RTX 3080 Ti. The procedure presented in [5] was employed to set the learning rate for the different architectures. Models were trained for 30 epochs and early stopping was applied after 5 epochs without validation improvement. Each data augmentation method was applied to construct deep segmentation models by using a 5-fold strategy where the 60 images of the training set were split using 80% for training and 20% for validation and to apply early stopping. The resulting models were evaluated on the test set using the Averaged Dice metric, also known as DiceMulti [9].

Table 1 shows the results achieved by traditional and mixing data augmentation methods. The mean DiceMulti achieved by the models trained using traditional data augmentation methods was 0.8129, and this result was surpassed by all the models trained with the mixing strategies.

Table 1. Mean (std) DiceMulti of the mixing data augmentation methods. In bold the best results.

Method	Results
Traditional	0.8129 (0.0583)
CarveMix	0.8512 (0.0054)
CutOut	0.8504 (0.0193)
CutOutSemantic	0.8313 (0.0257)
CutMixRandom	0.8431 (0.0217)
CutMix + Mod.	0.8550 (0.0206)
CutMixSemantic	0.8435 (0.0205)
HideAndSeek	0.8546 (0.0121)
RICAP	0.8553 (0.0211)
ResizeMix	0.8474 (0.0287)
TransparenceMix	0.8612 (0.0221)
SelfMix	**0.8619 (0.0193)**

The best results were obtained by using the SelfMix method with a mean DiceMulti of 0.8619, an improvement of about 5% regarding the base models trained with traditional methods. Another method that also achieved a considerable improvement was TransparenceMix. Both these methods combine images but use transparency as background; hence, the integration of the original image and the replacement region is more natural.

5 Conclusions and Further Work

In this paper, we have presented a library for easily applying mixing data augmentation strategies in the construction of deep segmentation models. As shown in the present paper, these techniques might have a positive impact when training those segmentation models. As further work, we plan to integrate this library in a set of tools that facilitate the construction of deep segmentation models.

References

1. Buslaev, A., Iglovikov, V.I., Khvedchenya, E., Parinov, A., Druzhinin, M., Kalinin, A.A.: Albumentations: fast and flexible image augmentations. Information **11**(2), 125 (2020)
2. DeVries, T., Taylor, G.W.: Improved regularization of convolutional neural networks with cutout. arXiv preprint arXiv:1708.04552 (2017)
3. Garcia-Garcia, A., Orts-Escolano, S., Oprea, S., Villena-Martinez, V., Martinez-Gonzalez, P., Garcia-Rodriguez, J.: A survey on deep learning techniques for image and video semantic segmentation. Appl. Soft Comput. **70**, 41–65 (2018)

4. Hao, S., Zhou, Y., Guo, Y.: A brief survey on semantic segmentation with deep learning. Neurocomputing **406**, 302–321 (2020)
5. Howard, J., Gugger, S.: Deep Learning for Coders with fastai and PyTorch. O'Reilly Media, Sebastopol (2020)
6. Isensee, F., Jaeger, P.F., Kohl, S.A., Petersen, J., Maier-Hein, K.H.: nnU-Net: a self-configuring method for deep learning-based biomedical image segmentation. Nat. Methods **18**(2), 203–211 (2021)
7. Lin, T.-Y., et al.: Microsoft COCO: common objects in context. In: Fleet, D., Pajdla, T., Schiele, B., Tuytelaars, T. (eds.) ECCV 2014. LNCS, vol. 8693, pp. 740–755. Springer, Cham (2014). https://doi.org/10.1007/978-3-319-10602-1_48
8. Marani, R., Milella, A., Petitti, A., Reina, G.: Deep neural networks for grape bunch segmentation in natural images from a consumer-grade camera. Precision Agric. **22**(2), 387–413 (2021)
9. Opitz, J., Burst, S.: Macro f1 and macro f1. arXiv preprint arXiv:1911.03347 (2019)
10. Paszke, A., et al.: Pytorch: an imperative style, high-performance deep learning library. In: Advances in Neural Information Processing Systems, vol. 32 (2019)
11. Qin, J., Fang, J., Zhang, Q., Liu, W., Wang, X., Wang, X.: Resizemix: mixing data with preserved object information and true labels. arXiv preprint arXiv:2012.11101 (2020)
12. Ronneberger, O., Fischer, P., Brox, T.: U-Net: convolutional networks for biomedical image segmentation. In: Navab, N., Hornegger, J., Wells, W.M., Frangi, A.F. (eds.) MICCAI 2015. LNCS, vol. 9351, pp. 234–241. Springer, Cham (2015). https://doi.org/10.1007/978-3-319-24574-4_28
13. Salimans, T., et al.: Improved techniques for training GANs. In: 30th International Conference on Neural Information Processing Systems, pp. 2234–2242. Curran Associates Inc. (2016)
14. Simard, P., Victorri, B., LeCun, Y., Denker, J.S.: Tangent prop-a formalism for specifying selected invariances in an adaptive network. In: In Proceedings of Neural Information Processing Systems (NeuriPS 1991), vol. 91, pp. 895–903 (1991)
15. Singh, K.K., Yu, H., Sarmasi, A., Pradeep, G., Lee, Y.J.: Hide-and-seek: a data augmentation technique for weakly-supervised localization and beyond. arXiv preprint arXiv:1811.02545 (2018)
16. Takahashi, R., Matsubara, T., Uehara, K.: RICAP: random image cropping and patching data augmentation for deep CNNs. In: Asian Conference on Machine Learning, pp. 786–798. PMLR (2018)
17. Yun, S., Han, D., Oh, S.J., Chun, S., Choe, J., Yoo, Y.: CutMix: regularization strategy to train strong classifiers with localizable features. In: Proceedings of the IEEE/CVF International Conference on Computer Vision, pp. 6023–6032 (2019)
18. Zeng, Q., Ma, X., Cheng, B., Zhou, E., Pang, W.: GANs-based data augmentation for citrus disease severity detection using deep learning. IEEE Access **8**, 172882–172891 (2020)
19. Zhang, X., et al.: CarveMix: a simple data augmentation method for brain lesion segmentation. In: de Bruijne, M., et al. (eds.) MICCAI 2021. LNCS, vol. 12901, pp. 196–205. Springer, Cham (2021). https://doi.org/10.1007/978-3-030-87193-2_19
20. Zhu, Q., Wang, Y., Yin, L., Yang, J., Liao, F., Li, S.: Selfmix: a self-adaptive data augmentation method for lesion segmentation. In: Wang, L., Dou, Q., Fletcher, P.T., Speidel, S., Li, S. (eds.) MICCAI 2022. LNCS, vol. 13434, pp. 683–692. Springer, Cham (2022)

Real-Time Elastic Partial Shape Matching Using a Neural Network-Based Adjoint Method

Alban Odot(✉), Guillaume Mestdagh, Yannick Privat, and Stéphane Cotin

Mimesis Team, Inria, Strasbourg, France
{alban.odot,stephane.cotin}@inria.fr

Abstract. Surface matching usually provides significant deformations that can lead to structural failure due to the lack of physical policy. In this context, partial surface matching of non-linear deformable bodies is crucial in engineering to govern structure deformations. In this article, we propose to formulate the registration problem as an optimal control problem using an artificial neural network where the unknown is the surface force distribution that applies to the object and the resulting deformation computed using a hyper-elastic model. The optimization problem is solved using an adjoint method where the hyper-elastic problem is solved using the feed-forward neural network and the adjoint problem is obtained through the backpropagation of the network. Our process improves the computation speed by multiple orders of magnitude while providing acceptable registration errors.

Keywords: Optimal control · Artificial neural network · Hyper-elasticity

1 Introduction

We consider an elastic shape-matching problem between a deformable solid and a point cloud. Namely, an elastic solid in its reference configuration is represented by a tridimensional mesh, while the point cloud represents a part of the solid boundary in a deformed configuration. The objective of the procedure is not only to deform the mesh so that its boundary matches the point cloud, but also to estimate the displacement field inside the object.

This situation also arises in computer-assisted liver surgery, where augmented reality is used to help the medical staff navigate the operation scene [3]. Most methods for intra-operative organ shape-matching revolve around a biomechanical model to describe how the liver is deformed when forces are applied to its boundary. Sometimes, a deformation is created by applying forces [13] or constraints [7,11] to enforce surface correspondence. Other approaches prefer to solve an inverse problem, where the final displacement minimizes a cost functional among a range of admissible displacements [5]. However, while living tissues are known to exhibit a highly nonlinear behavior [8], using hyperelastic

B. Dorronsoro et al. (Eds.): OLA 2023, CCIS 1824, pp. 137–147, 2023.
https://doi.org/10.1007/978-3-031-34020-8_10

models in the context of real-time shape matching is prohibited due to high computational costs. For this reason, the aforementioned methods either fall back to linear elasticity [5] or to the linear co-rotational model [13]. In this paper, we perform real-time hyperelastic shape matching by predicting nonlinear displacement fields using a neural network. The network is included in an adjoint-like method, where the backward chain is executed automatically using automatic differentiation.

Neural networks are used to predict solutions to partial differential equations, in compressible aerodynamics [14], structural optimization [15] or astrophysics [6]. Here we work at a small scale, but try to obtain real-time simulations using complex models. Also, the medical image processing literature is full of networks that perform shape-matching in one step [12]. However, the range of available displacement fields is limited by the training dataset of the network, and thus less robust to unexpected deformations. On the other hand, assigning a very generic task to the network results in a very flexible method, where details of the physical model, including the range of forces that can be applied to the liver and the zones where they apply may be chosen after the training. Therefore, our shape-matching approach provides a good compromise between the speed of learning-based methods with the flexibility of standard simulations. We want to mention that for the rest of this article due to how the method is formulated we interchangeably use the terms "shape-matching" and "registration".

We start by presenting the method split into three parts. First, the optimization problem; second, the used neural network and finally, the adjoint method computed using an automatic differentiation framework.

We then present the results considering a toy problem involving a square section beam and a more realistic one involving a liver.

2 Methods

2.1 Optimization Problem

To model the registration problem, we use the optimal control formulation introduced in Mestdagh and Cotin [9]. The deformable object is represented by a tetrahedral mesh, endowed with a hyperelastic model. In its reference configuration, the elastic object occupies the domain Ω_0, whose boundary is $\partial\Omega_0$. When a displacement field $\mathbf{u}$ is applied to Ω_0, the deformed domain is denoted by $\Omega_{\mathbf{u}}$, and its boundary is denoted by $\partial\Omega_{\mathbf{u}}$ as shown in Fig. 1. Applying a surface force distribution $\mathbf{g}$ onto the object boundary results in the elastic displacement $\mathbf{u_g}$, solution to the static equilibrium equation

$$\mathbf{F}(\mathbf{u_g}) = \mathbf{g}, \tag{1}$$

where $\mathbf{F}$ is the residual from the hyperelastic model. Displacements are discretized using continuous piecewise linear finite element functions so that the system state is fully known through the displacement of mesh nodes, stored in $\mathbf{u}$. Note that $\mathbf{g}$ contains the nodal forces that apply on the mesh vertices. As we

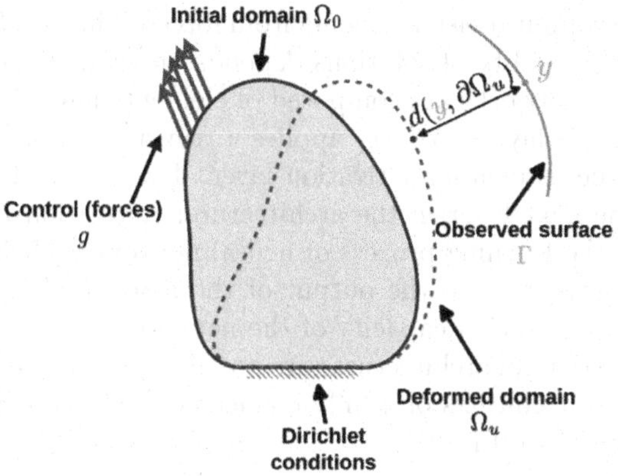

Fig. 1. Schematic of the problem which we are trying to optimize for.

only consider surface loadings, nodal forces are zero for nodes inside the domain. Finally, the observed data are represented by a point cloud $\Gamma = \{y_1, \ldots, y_m\}$.

We compute a nodal force distribution that achieves the matching between $\partial\Omega_{\mathbf{u_g}}$ and Γ by solving the optimization problem

$$\min_{\mathbf{g}\in G} \quad \Phi(\mathbf{g}) + \tfrac{\alpha}{2}\|\mathbf{g}\|^2 \tag{2}$$

$$\text{where} \qquad \Phi(\mathbf{g}) = J(\mathbf{u_g}), \tag{3}$$

where, $\alpha > 0$ is a regularization parameter, G denotes the set of admissible nodal forces distributions, and J is the least-square term

$$J(\mathbf{u}) = \tfrac{1}{2m}\sum_{j=1}^{m} d^2(y_j, \partial\Omega_{\mathbf{u}}). \tag{4}$$

Here, $d(y, \partial\Omega_{\mathbf{u}}) = \min_{x\in\partial\Omega_{\mathbf{u}}} \|y - x\|$ denotes the distance between $y \in \Gamma$ and $\partial\Omega_{\mathbf{u}}$. The functional J measures the discrepancy between $\partial\Omega_{\mathbf{u}}$ and Γ, and it evaluates to zero whenever every point $y \in \Gamma$ is matched by $\partial\Omega_{\mathbf{u}}$.

A wide range of displacement fields $\mathbf{u}$ are minimizers of problem (2), but most of them have no physical meaning. Defining a set of admissible controls G is critical to generate only displacements that are consistent with a certain physical scenario. The set B decides, among others, on which vertices nodal forces may apply, but also which magnitude they are allowed to take. Selecting zones where surface forces apply is useful to obtain physically plausible solutions.

2.2 A Neural Network to Manage the Elastic Problem

Nonlinear elasticity problems are generally solved using a Newton method, which yields very accurate displacement fields at a high computational cost. In this paper, we give a boost to the direct solution procedure by using a pre-trained neural network to compute displacements from forces. This results in much faster estimates, while the quality of solutions depends on the network training.

Artificial neural networks are composed of elements named artificial neurons grouped into multiple layers. A layer applies a transformation on its input data and passes it to the associated activation layer. The result of this operation is then passed to the next layer in the architecture. Activation functions play an important role in the learning process of neural networks. Their role is to apply a nonlinear transformation to the output of the associated layers thus greatly improving the representation capacity of the network.

While a wide variety of architectures are possible we will use the one proposed by Odot et al. [10]. It consists of a fully-connected feed-forward neural network with 2 hidden layers (see Fig. 2).

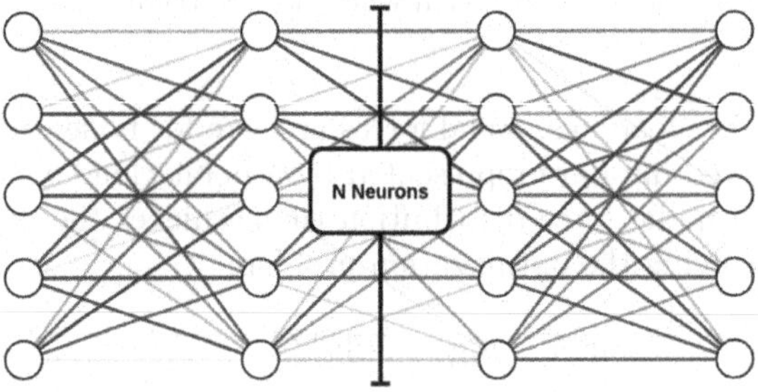

Fig. 2. The proposed architecture is composed of 4 fully connected layers of size the number of degrees of freedom with a PReLU activation function. The input is the nodal forces and the output is the respective nodal displacements.

The connection between two adjacent layers can be expressed as follows

$$\mathbf{z}_i = \sigma_i(\mathbf{W}_i \mathbf{z}_{i-1} + \mathbf{b}_i) \text{ for } 1 \leqslant i \leqslant n+1, \tag{5}$$

where n is the total number of layers, $\sigma(.)$ denotes the element wise activation function, $\mathbf{z}_0$ and $\mathbf{z}_{n+1}$ denotes the input and output tensors respectively, $\mathbf{W}_i$ and $\mathbf{b}_i$ are the trainable weight matrices and biases in the ith layer.

In our case the activation functions $\sigma(.)$ are PReLU [4], which provides a learnable parameter a, allowing us to adaptively consider both positive and negative inputs. From now on, we denote the forward pass operation in the network by

$$\mathbf{u}_\mathbf{g} = \mathbf{N}(\mathbf{g}). \tag{6}$$

2.3 An Adjoint Method Involving the Neural Network

We now give a closer look at the procedure to evaluate Φ and its derivatives. We use an adjoint method, where the only variable controlled by the optimization solver is $\mathbf{g}$. As J only operates on displacement fields, the physical model plays the role of an intermediary between these two protagonists. The adjoint method is well suited to the network-based configuration, as the network can be used as a black box.

In a standard adjoint procedure, a displacement is computed from a force distribution by solving (1) using a Newton method, and it is then used to evaluate $\Phi(\mathbf{g})$. The Newton method is the algorithm of choice when dealing with non-linear materials, it iteratively solves the hyper-elastic problem producing accurate solutions. This method is also known for easily diverging when the load is reaching a certain limit that depends on the problem. To compute the deformation, one requires the application of multiple substeps of load which highly increases the computation times. The backward chain requires solving an adjoint problem to evaluate the objective gradient, namely

$$\nabla\Phi(\mathbf{g}) = \mathbf{p_g} \quad \text{where} \quad \nabla\mathbf{F}(\mathbf{u_g})^{\mathrm{T}}\mathbf{p_g} = \nabla J(\mathbf{u_g}). \tag{7}$$

In (7), the adjoint state $\mathbf{p_g}$ is solution to a linear system involving the hyper-elasticity Jacobian matrix $\nabla\mathbf{F}(\mathbf{u_g})$. When the network is used, however, the whole pipeline is much more straightforward, as the network forward pass is only composed of direct operations. The network-based forward and backward chains read

$$\Phi(\mathbf{g}) = J \circ \mathbf{N}(\mathbf{g}) \quad \text{and} \quad \nabla\Phi(\mathbf{g}) = \mathbf{p_g} = [\nabla\mathbf{N}(\mathbf{g})]^{\mathrm{T}}\, \nabla J(\mathbf{u_g}), \tag{8}$$

respectively. On a precautionnary basis, let us take a brief look at the (linear) adjoint operator $\nabla\mathbf{N}(\mathbf{g})^{\mathrm{T}}$. When $\nabla\mathbf{N}(\mathbf{g})^{\mathrm{T}}$ is applied, the information propagates backward in the network, following the same wires as the forward pass. The displacement gradient $\nabla J(\mathbf{u_g})$ is fed to the output tensor $\mathbf{s}_{n+1}$ and the adjoint state is read at the network entry $\mathbf{s}_0$. In between, the relation between two layers is the adjoint operation to (5). It reads

$$\mathbf{s}_{i-1} = \mathbf{W}_i^{\mathrm{T}}\, \nabla\sigma_i(\mathbf{W}_i\mathbf{z}_{i-1} + \mathbf{b}_i)\, \mathbf{s}_i \quad \text{for} \quad 1 \leqslant i \leqslant n+1, \tag{9}$$

where $\nabla\sigma_i(\mathbf{W}_i\mathbf{z}_{i-1} + \mathbf{b}_i)$ is a diagonal matrix saved during the forward pass.

The network-based adjoint procedure is summarized in Algorithm 1, keeping in mind the backward chain is handled automatically. Given a nodal force vector $\mathbf{g}$, evaluating $\Phi(\mathbf{g})$ and $\nabla\Phi(\mathbf{g})$ requires one forward pass and one backward pass in the network. Then, (2) may be solved iteratively using a standard gradient-based optimization algorithm. Because both network passes consist only of direct operations, the optimization solver is less likely to fail for accuracy reasons, compared to a Φ evaluation based on an iterative method.

Algorithm 1: Network-based adjoint method to evaluate Φ.

Data: Current iterate $\mathbf{g}$
Perform the forward pass $\mathbf{u_g} = \mathbf{N}(\mathbf{g})$
Evaluate $J(\mathbf{u}_g v)$ and $\nabla J(\mathbf{u_g})$
Perform the backward pass $\mathbf{p_g} = [\nabla\mathbf{N}(\mathbf{g})]^{\mathrm{T}}\, \nabla J(\mathbf{u_g})$
Result: $\nabla\Phi(\mathbf{g}) = \mathbf{p_g}$

3 Results

Our method is implemented in Python. To be more specific, we use PyTorch to handle the network and evaluate J on the GPU, while the optimization solver is a limited memory BFGS algorithm [1] available in the Scipy package. Our numerical tests run on a Titan RTX GPU and AMD Ryzen 9 3950x CPU, with 32 GiB of RAM.

3.1 Surface-Matching Tests on a Beam Mesh

To assess the validity of our method, we first consider a toy problem involving a square section beam with 304 hexahedal elements. The network is trained using 20,000 pairs $(\mathbf{g}, \mathbf{u_g})$, computed using a Neo-Hookean material law with a Young modulus $E = 4,500$ Pa and a Poisson ratio $\nu = 0.49$.

We create 10,000 additional synthetic deformations of the beam, distinct from the training dataset, using the SOFA finite element framework [2]. Figure 3 shows three examples of synthetic deformations, along with the sampled point clouds. Generated deformations include bending (Fig. 3a), torsion (Fig. 3c) or a combination of them (Fig. 3b). For each deformation, we sample the deformed surface to create a point cloud. We then apply our algorithm with a relative tolerance of 10^{-4} on the objective gradient norm. We computed some statistics regarding the performance of our method over a series of 10,000 different scenarios and obtained the following results: mean registration error: $6 \times 10^{-5} \pm 6.15 \times 10^{-5}$, mean computation time: 48 ms ± 19 ms and mean number of iterations: 27 ± 11.

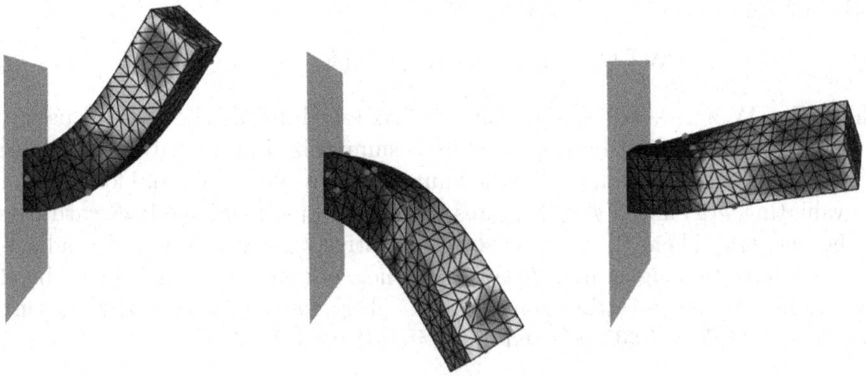

(a) Reg. error: 5.9×10^{-5}, time: 0.07 s, iterations: 13 (b) Reg. error: 6.6×10^{-5} m, time: 0.09 s, iterations: 15 (c) Reg. error: 3.4×10^{-5}, time: 0.115 s, iterations: 19

Fig. 3. Deformations from the test dataset. The red dots represent the target point clouds, and the color map represents the Von Mises stress error of the neural network prediction. (Color figure online)

Using a FEM solver, each sample of the test dataset took between 1 and 2 s to compute. This is mostly due to the complexity of the deformations as shown in Fig. 3. Such displacement fields require numerous costly Newton-Raphson iterations to reach equilibrium. The neural network provides physical deformations in less than a millisecond regardless of the complexity of the force or resulting deformation, which highly improves the computation time of the method. From our analysis, the time repartition of the different tasks in the algorithm is pretty consistent, even with denser meshes. Network predictions and loss function evaluations represent 10% to 15% each, gradient computations represent up to the last 80% of the whole optimization process. This allows us to reach an average registration error of 5.37×10^{-5} in less time than it takes to compute a single simulation of the problem using a classic FEM solver.

Due to the beam shape symmetry, some point clouds may be compatible with several deformed configurations, resulting in wrong displacement fields returned by the procedure. However, our procedure achieved a satisfying surface matching in each case. These results on a toy scenario prove that our algorithm provides fast and accurate registrations.

In the next section, we apply our method in the field of augmented surgery with the partial surface registration of a liver and show that with no additional computation our approach produces with satisfying accuracy the forces that generate such displacements.

3.2 An Application in Augmented Surgery and Robotics

We now turn to another test case involving a more complex domain. The setting is similar to [9, Sect. 3.2]. In this context, a patient-specific liver mesh is generated from tomographic images and the objective is to provide augmented reality by registering, in real-time, the mesh to the deformed organ. During the surgery, only a partial point cloud of the visible liver surface can be obtained. The contact zones with the surgical instruments can also be estimated. In our

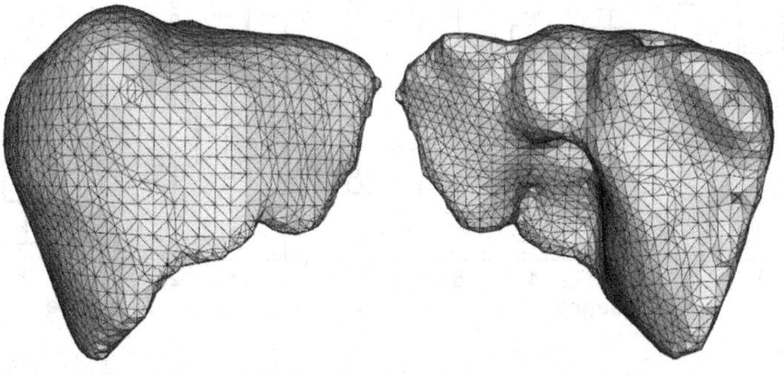

Fig. 4. Mesh of the liver used in this section. Composed of 3,046 vertices and 10,703 tetrahedral elements which represents a challenge compared to the one used in Sect. 3.1

case, the liver mesh contains 3,046 vertices and 10,703 tetrahedral elements. Homogeneous Dirichlet conditions are applied at zones where ligaments hold the liver, and at the hepatic vein entry. Like previously, we use a Neo-Hookean constitutive law with $E = 4,500$ Pa and $\nu = 0.49$, and the network is trained on 20,000 force/displacement pairs. We create 5 series of synthetic deformations by applying a variable local force, distributed on a few nodes, on the liver mesh boundary. For each series, 50 incremental displacements are generated, along with the corresponding point clouds. The network-based registration algorithm is used to update the displacement field and forces between two frames. We also run a standard adjoint method involving the Newton algorithm, to compare with our approach. As the same mesh is used for data generation and reconstruction, the Newton-based reconstruction is expected to perform well (Fig. 4).

3.3 Liver Partial Surface Matching for Augmented Surgery

In this subsection, we present two relevant metrics: target registration error and computation times. In augmented surgery, applications such as robot-aided surgery or holographic lenses require accurate calibrations that rely on registration. One of the most common metrics in registration tasks is the target registration error (TRE), which is the distance between corresponding markers not used in the registration process. In our case we work on the synthetic deformation of a liver, thus, the markers will be the nodes of the deformed mesh. The 5 scenarios present similar results with TRE between 3.5 mm and 0.5 mm. Such errors are entirely acceptable and preserve the physical properties of the registered mesh. We point out that the average TRE for the classic method is around 0.1 mm which shows the impact of the network approximations.

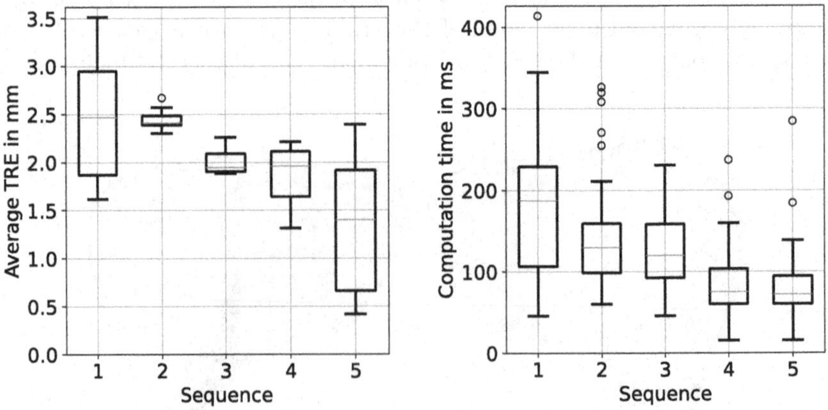

Fig. 5. Average target registration error and computation times of each sequence.

Due to the non-linearity introduced by the Neo-Hookean material used to simulate the liver we need multiple iterations to converge toward the target point cloud. Considering the complexity of the mesh, computing a single iteration of the algorithm using a classical solver takes multiple seconds which leads to an average of 14 min per frame. Our proposed algorithm uses a neural network to improve the computation speed of both the hyper-elastic and adjoint problems. The hyper-elastic problem takes around 4 to 5 ms to compute while the adjoint problem takes around 11 ms. This leads to great improvement in convergence speed as seen in Fig. 5 where on average we reduce the computation time by a factor of 6000.

3.4 Force Estimation for Robotic Surgery

In the context of liver computer-assisted surgery, the objective is to estimate a force distribution supported by a small zone on the liver boundary. Such a local force is for instance applied when a robotic instrument manipulates the organ. In this case, it is critical to estimate the net force magnitude applied by the instrument, to avoid damaging the liver. To represent the uncertainty about the position of the instruments the reconstructed forces are allowed to be nonzero on a larger support than the original distribution. Figure 6 shows the reference and reconstructed deformations and nodal forces for three frames of the same series. While the Newton-based reconstruction looks similar to the reference one,

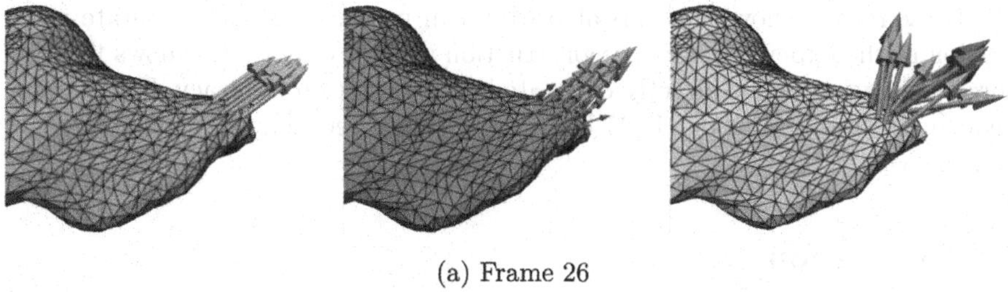

(a) Frame 26

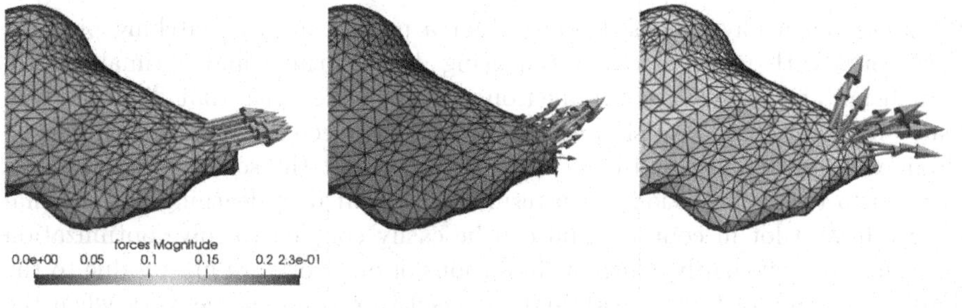

(b) Frame 50

Fig. 6. Synthetic liver deformations and force distributions (left), reconstructed deformations and forces using the Newton method (middle) and the network (right) for test case 3.

network-based nodal forces are much noisier. This is mostly due to the network providing only an approximation of the hyperelastic model.

The great improvement in speed comes at the cost of precision. As shown in Fig. 6 the neural network provides noisy force reconstructions. This is mostly due to prediction errors since the ANN only approximates solutions. These errors also propagate through the backward pass (adjoint problem), thus, accumulate in the final solution. Although the force estimation is noisy for most cases it remains acceptable as displayed in Fig. 7. The red dotted line corresponds to the average error obtained with the classical adjoint method (10.04 %). While we are not reaching such value, some sequences such as 1 and 3 provide good reconstructions. The difference in errors between scenarios is mostly due to training force distribution. This problem can be corrected by simply adding more data to the dataset thus providing better coverage of the force and deformation space.

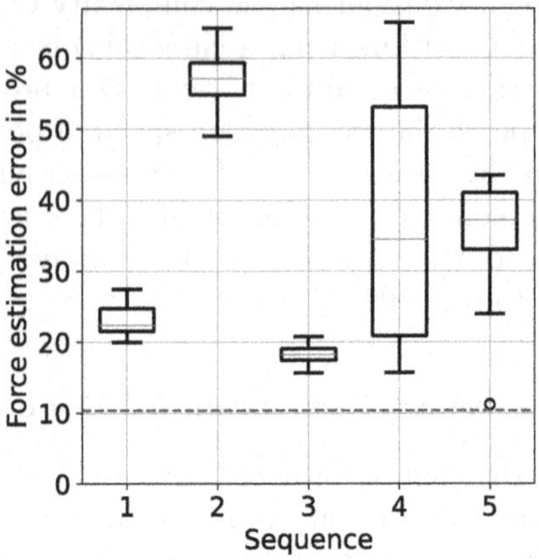

Fig. 7. Force estimation error of the 5 sequences using our method, in red the average force reconstruction error with the classical method.

These results show that this algorithm can produce fast and accurate registration at the expense of force reconstruction accuracy. This also shows that the force estimation is not directly correlated to registration accuracy. For example sequence 1 has the worst TRE but a good force reconstruction compared to sequence 4.

4 Conclusion

We presented a physics-based solution for a partial surface-matching problem that works with non-linear material using deep learning and optimal control formalism. The results are obtained on two main scenarios that differ both in scale and complexity. We showed that a fast and accurate registration can be obtained in both cases and can, in addition, predict the set of external forces that led to the deformation. Such results show that deep learning and optimal control have a lot in common and can be easily coupled to solve optimization problems very efficiently. Current limitations of our work are mostly due to the limited accuracy of the network and the need to retrain the network when the shape or material parameters of the model change.

References

1. Byrd, R.H., Lu, P., Nocedal, J., Zhu, C.: A limited memory algorithm for bound constrained optimization. SIAM J. Sci. Comput. **16**(5), 1190–1208 (1995). https://doi.org/10.1137/0916069
2. Faure, F., et al.: SOFA: a multi-model framework for interactive physical simulation. In: Payan, Y. (ed.) Soft Tissue Biomechanical Modeling for Computer Assisted Surgery. SMTEB, vol. 11, pp. 283–321. Springer, Heidelberg (2012). https://doi.org/10.1007/8415_2012_125
3. Haouchine, N., et al.: Impact of soft tissue heterogeneity on augmented reality for liver surgery. IEEE Trans. Visual. Comput. Graph. **21**(5), 584–597 (2015). https://doi.org/10.1109/TVCG.2014.2377772
4. He, K., Zhang, X., Ren, S., Sun, J.: Delving deep into rectifiers: surpassing human-level performance on ImageNet classification. In: Proceedings of the IEEE International Conference on Computer Vision (ICCV) (2015)
5. Heiselman, J.S., Jarnagin, W.R., Miga, M.I.: Intraoperative correction of liver deformation using sparse surface and vascular features via linearized iterative boundary reconstruction. IEEE Trans. Med. Imaging **39**(6), 2223–2234 (2020). https://doi.org/10.1109/TMI.2020.2967322
6. Khan, S., Green, R.: Gravitational-wave surrogate models powered by artificial neural networks. Phys. Rev. D **103**, 064015 (2021). https://doi.org/10.1103/PhysRevD.103.064015
7. Malti, A., Bartoli, A., Hartley, R.: A linear least-squares solution to elastic shape-from-template. In: Proceedings of the IEEE Conference on Computer Vision and Pattern Recognition, pp. 1629–1637 (2015)
8. Marchesseau, S., Chatelin, S., Delingette, H.: Nonlinear biomechanical model of the liver. In: Payan, Y., Ohayon, J. (eds.) Biomechanics of Living Organs, Translational Epigenetics, vol. 1, pp. 243–265. Academic Press, Oxford (2017). https://doi.org/10.1016/B978-0-12-804009-6.00011-0
9. Mestdagh, G., Cotin, S.: An optimal control problem for elastic registration and force estimation in augmented surgery. In: Wang, L., Dou, Q., Fletcher, P.T., Speidel, S., Li, S. (eds.) MICCAI 2022. LNCS, pp. 74–83. Springer, Cham (2022). https://doi.org/10.1007/978-3-031-16449-1_8
10. Odot, A., Haferssas, R., Cotin, S.: DeepPhysics: a physics aware deep learning framework for real-time simulation. Int. J. Numer. Meth. Eng. **123**(10), 2381–2398 (2022). https://doi.org/10.1002/nme.6943
11. Peterlík, I., et al.: Fast elastic registration of soft tissues under large deformations. Med. Image Anal. **45**, 24–40 (2018). ISSN 1361-8415. https://doi.org/10.1016/j.media.2017.12.006
12. Pfeiffer, M., et al.: Non-rigid volume to surface registration using a data-driven biomechanical model. In: Martel, A.L., et al. (eds.) MICCAI 2020. LNCS, vol. 12264, pp. 724–734. Springer, Cham (2020). https://doi.org/10.1007/978-3-030-59719-1_70
13. Plantefève, R., Peterlik, I., Haouchine, N., Cotin, S.: Patient-specific biomechanical modeling for guidance during minimally-invasive hepatic surgery. Ann. Biomed. Eng. **44**(1), 139–153 (2015). https://doi.org/10.1007/s10439-015-1419-z
14. Renganathan, S.A., Maulik, R., Ahuja, J.: Enhanced data efficiency using deep neural networks and gaussian processes for aerodynamic design optimization. Aerosp. Sci. Technol. **111**, 106522 (2021). https://doi.org/10.1016/j.ast.2021.106522
15. White, D.A., Arrighi, W.J., Kudo, J., Watts, S.E.: Multiscale topology optimization using neural network surrogate models. Comput. Meth. Appl. Mech. Eng. **346**, 1118–1135 (2019). https://doi.org/10.1016/j.cma.2018.09.007

We Won't Get Fooled Again: When Performance Metric Malfunction Affects the Landscape of Hyperparameter Optimization Problems

Kalifou René Traoré[1,2](✉), Andrés Camero[2,3], and Xiao Xiang Zhu[1]

[1] Data Science in Earth Observation, Technical University of Munich, Arcisstrasse 21, Munich, Germany
xiaoxiang.zhu@tum.de

[2] Remote Sensing Institute, German Aerospace Center (DLR), Münchener Strasse 20, Weßling, Germany
{kalifou.traore,andres.camerounzueta}@dlr.de

[3] Helmholtz AI, Neuherberg, Germany

Abstract. Hyperparameter optimization (HPO) is a well-studied research field. However, the effects and interactions of the components in an HPO pipeline are not yet well investigated. Then, we ask ourselves: *Can the landscape of HPO be biased by the pipeline used to evaluate individual configurations?* To address this question, we proposed to analyze the effect of the HPO pipeline on HPO problems using fitness landscape analysis. Particularly, we studied over 119 generic classification instances from either the DS-2019 (CNN) and YAHPO (XGBoost) HPO benchmark data sets, looking for patterns that could indicate evaluation pipeline malfunction, and relate them to HPO performance. Our main findings are: (i) In most instances, large groups of diverse hyperparameters (i.e., multiple configurations) yield the same *ill* performance, most likely associated with majority class prediction models (predictive accuracy) or models unable to attribute an appropriate class to observations (log loss); (ii) in these cases, a worsened correlation between the observed fitness and average fitness in the neighborhood is observed, potentially making harder the deployment of local-search-based HPO strategies. (iii) these effects are observed across different HPO scenarios (tuning CNN or XGBoost algorithms). Finally, we concluded that the HPO pipeline definition might negatively affect the HPO landscape.

Keywords: Hyperparameter Optimization · Fitness Landscape Analysis · Benchmarking

1 Introduction and Related Work

Modern data-driven approaches dealing with large-scale data require domain, data science, and technical expertise. The variety of application tasks (e.g., classification and object detection) often require designing models that are not necessarily reusable in other tasks, and this process is both resource-demanding

B. Dorronsoro et al. (Eds.): OLA 2023, CCIS 1824, pp. 148–160, 2023.
https://doi.org/10.1007/978-3-031-34020-8_11

and error-prone [3,8,12]. Thus, automating the design of ML pipelines, a.k.a. *AutoML* [6], is much more desirable.

AutoML is usually split into four main activities: data preparation, feature engineering, model generation, and model estimation [5]. *Hyperparameter optimization* (HPO [1]) is an important task in model generation. HPO aims at automatically tuning the hyperparameters of learning algorithms, and as with all optimization problems, it is facing the process of minimizing/maximizing a target function (e.g., the performance metric of the model) subject to a set of constraints. HPO is a well-studied field [1], but the effects and interaction between the components of its pipeline are not yet well investigated. Recently, authors [10] have proposed to characterize the search space of AutoML pipelines using *fitness landscape analysis* (FLA [11]). In the same line, [15] proposed a FLA-base framework to characterize NAS problems, and applied it to a multi-sensor data fusion problem [14]. Despite the great results and insights provided by these studies, the relation between HPO and the rest of the HPO pipeline remains barely explored.

Therefore, in this study, we pose the following research question: ***Can the landscape of HPO be biased by the pipeline used to evaluate individual configurations?*** To address this question, we propose to study HPO in the context of AutoML using FLA. Particularly, using *fitness distance correlation* (FDC [7]), *locality* and *neutrality* [2], we aim at patterns that arise from evaluation pipeline issues and assess how they could alter the landscapes of HPO problems. The results on over 119 instances from either the DS-2019 [13] or the YAHPO [9] HPO benchmarks show the existence of large groups of diverse HP configurations that yield the same *ill* fitness value. This *illness* could be explained by the fitness metric selection (e.g., predictive accuracy and log loss), that *induce* various *suboptimal* model behavior, in scenarios of *different natures* (tuning CNN and XGBoost algorithms). More precisely, for the predictive accuracy, we suspect the generation of majority class predictors. In the case of the log loss criterion, the generation of models unable to classify. A complementary analysis of locality shows that the resulting landscapes are more rugged, with lesser correlation between the observed fitness and the fitness in the neighborhood. In other words, these problems are hard to tackle using a local-search strategy.

The rest of the paper is as follows: The next section introduces the methodology used in the study, Sect. 3 presents results of landscape analysis on HPO problems, and Sect. 4 provides conclusions.

2 Methodology

Given a HPO problem, let S be the HP configuration space, f the fitness function that assigns a value $f(x) \in \mathbb{R}$ to all configurations $x \in S$, and $N(x)$ a neighborhood operator that provides a structure to S. Then, the fitness landscape is defined as $\mathcal{L} = (S, f, N)$.

We are interested in exploiting the landscape definition to study the relation between the HPO landscape and the HPO pipeline, and check whether the pipeline may bias the HPO landscape. Particularly, we propose to use the FDC and *locality* to characterize this relation. The motivation is that issues related to the evaluation pipeline should affect the fitness of configurations irrespectively of their configuration, and thus their distance to the optimum. In other words, repetitive or grouping patterns (such as lines) might appear when visualizing distributions of distances to the optimum. Moreover, the locality of the configuration space should be arbitrarily affected, i.e., some configurations should present an unexpected or *random* behavior (in relation to the neighborhood).

Without loss of generality, we consider the problem of tuning the HPs of a fixed model, e.g. neural network architecture or XGBoost ensemble, to perform a task (e.g., classification). Typically, the HP configuration space consists of mixed type features (continuous, discrete or categorical). Thus, we propose to evaluate the distance between individuals using a dedicated similarity function, $\delta(x, y)$, introduced by [4]. Then, we define a neighborhood function $N(x) = \{y \in S \mid \delta(x, y) < \Delta\}$.

The FDC is often interpreted as a measure of the existence of search trajectories from randomly picked configurations to the known global optimum. In practice, the FDC is not collected as a correlation score but visualized as the distribution of fitness versus distance to the global optimum. It writes as: $\text{FDC}(f, x^*, S) = \{(\delta(x^*, y), f(y)) \mid \forall y \in S\}$, where $x^* \in S$ is the global optimum. On the other hand, *locality* corresponds to the relationship between the observed fitness and the distribution of average fitness in the neighborhood [2].

Moreover, the *neutrality degree* [2] provides an additional picture of the interaction between solutions in the landscape. It is defined as $N_d(x) = |\{x' \in N(x) \mid |f(x') - f(x)| < \epsilon\}|$, and is interpreted as the number of neighbors of x that have a *similar* fitness. In this case, we set $\epsilon = \texttt{max}_{\text{fitness}}/C$. Besides, Table 1 and 2 give a description of abbreviations and symbols used in the paper.

Table 1. Table of abbreviations used in the paper.

Abbreviation	Description
HP	Hyperparameter
HPO	Hyperparameter Optimization
CV	Computer Vision
FLA	Fitness Landscape Analysis
FDC	Fitness Distance Correlation
CNN	Convolutional Neural Network
MMCE	Mean Misclassification Error

3 Results

To evaluate the proposed methodology, we propose to analyze the **DS-2019** and **YAHPO** HPO benchmark data sets. DS-2019 consists of a tabular benchmark for the scenario of tuning the HPs of a (fixed) convolutional neural network

(CNN), a ResNet-18, on ten instances of CV classification. For each instance, 15 hyperparameters should be optimized, including the batch size, number of epochs, and momentum, among others.

Table 2. Table of symbols used in the paper.

Symbol	Description
S	Hyperparameter Configuration Space
N	Neighborhood operator
f	Fitness evaluation function
$\mathcal{L}$	Fitness landscape derived from the combination of S, N, f
$\mathtt{max}_{\text{fitness}}$	Maximum fitness value observed in S
$\mathtt{max}_{\text{dist}}$	Maximum pairwise distance (to the optimum) measured in S
$N_d(x)$	Neutrality degree of a HP solution x
$\delta(x, y)$	Gower distance between HP solutions x and y
Δ	Threshold (in Gower distance) used by N to assign neighbors
ϵ	Threshold (in fitness) used by N_d to assign neutral neighbors
C	Constant used to define the fitness neutrality threshold ϵ

YAHPO consists of a tabular benchmark for the scenarios of tuning various learning algorithms (e.g. XGBoost, Neural Networks) for 119 instances of classification, in various domains of applications. Many of the instances were obtained from the collaborative open-source OpenML platform, gathering an ever-growing number of machine learning instances. In this study, we focus on the specific scenario tuning a set of 15 HPs for the XGBoost learning algorithm.

The code used for the experiments is available following the anonymized link: https://github.com/anonymous-for-open-review/late-breaking-automlConf-2022.

3.1 Classification Accuracy

The following paragraphs introduce results when evaluating solutions using the metric of *predictive accuracy* (DS-2019). In the case of YAHPO, the available metric is the *mean misclassification error* (*MMCE = 1 − predictive accuracy*), and we focus on the instances *WDBC, YEAST, MINIBOONE, and ISOLET*. Similar observations are made for the rest of the 119 instances.

Fitness Distance Correlation (FDC)

First, for each instance, we randomly sampled 1000 HP configurations and computed the FDC. Results are shown in Figs. 1 and 2, respectively, for DS-2019 and YAHPO.

Overall, the distances to the global optimum cover a wide range of values: the distribution of distances is wide and uniform in most instances. This is the case for both benchmarks. This suggests a large diversity in the HP configurations (with respect to the optimum), for the sample and potentially the whole configuration space. This is true for both benchmarks, with slightly more narrow distributions of distances for YAHPO. This suggests slightly less diversity within the XGBoost HP configuration space (YAHPO), than one of the CNN classifiers (DS-2019).

In most instances of DS-2019, the fitness also covers a wide range of values, as opposed to relatively more narrow distributions of fitness on YAHPO. This suggests a larger influence on HP configuration on the fitness of CNN classifiers (DS-2019) than on more classical and notoriously robust ensembles of models, i.e XGBoost (YAHPO). Overall, the distributions of fitness all seem to be multimodal, with a principal mode for large fitness values (i.e., good configurations), and another mode for *odd* values, for both benchmarks.

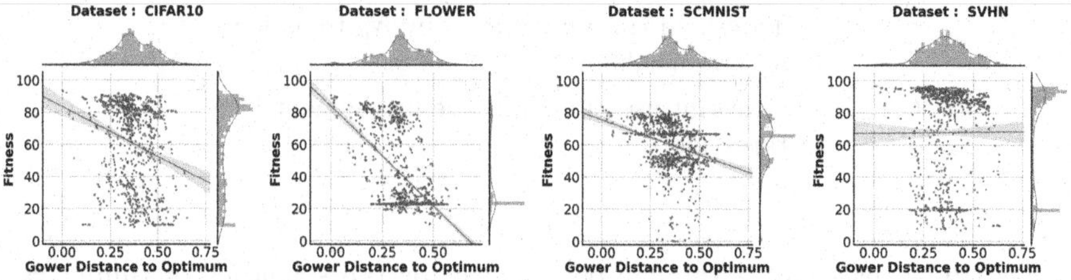

Fig. 1. FDC plot for instances from the DS2019 benchmark, and the corresponding regression line in blue. The fitness function is the *predictive accuracy.* (Color figure online)

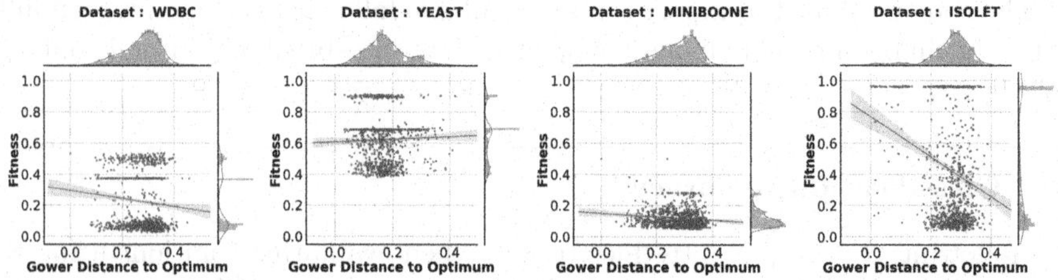

Fig. 2. FDC plot for a few instances from the YAHPO HPO benchmark, and the corresponding regression line in blue. The fitness function is the MMCE. (Color figure online)

Besides, we checked the data distribution for each instance, and we notice that the *odd* modes could be correlated to the majority class. Note that the fitness metric used is predictive accuracy for DS-2019, and its opposite the MMCE for YAHPO. For example in DS-2019, on FLOWER it is around 25%, SCMNIST around 65% and SVHN around 20%. In YAHPO, on WDBC it is around 38%, on YEAST around 70%, and ISOLET around 97%, among others. In particular, configurations are affected regardless of the distance to the optimum. In other words, very diverse configurations yield the same fitness value. This phenomenon could be attached to issues with the learning process, failing to properly fit the data and being stuck in poor local optima (i.e., majority class prediction), preventing them to reach the fitness that their HP configuration would normally yield. Besides, there is no clear global correlation between the observed fitness and distance to the global optimum. This could be caused by the multi-modal nature of the distributions of fitness.

Neighborhood

Next, we seek to identify how the observed artifacts, i.e., the *majority* class predictors, affect the locality of landscapes. Figures 3 and 4 show the distribution of average neighbor fitness as a function of the observed fitness, respectively, for instances from DS-2019 and YAHPO. The black dash-dotted line represents the bisector, i.e., the line connecting all points of equal value on both axis. To generate the plots, we used the previously sampled configurations, and identified the maximal pairwise distance (of any individual) to the optimum $\mathtt{max}_{\text{dist}}$, and maximum observed fitness $\mathtt{max}_{\text{fitness}}$. Given a constant $C = 40$, we discretize the range of fitness values into intervals, where a step is equal to the maximum observed fitness $\mathtt{max}_{\text{fitness}}$ divided by C. In order to decide if a configuration is a neighbor, we set $\Delta = \mathtt{max}_{\text{dist}}/C$.

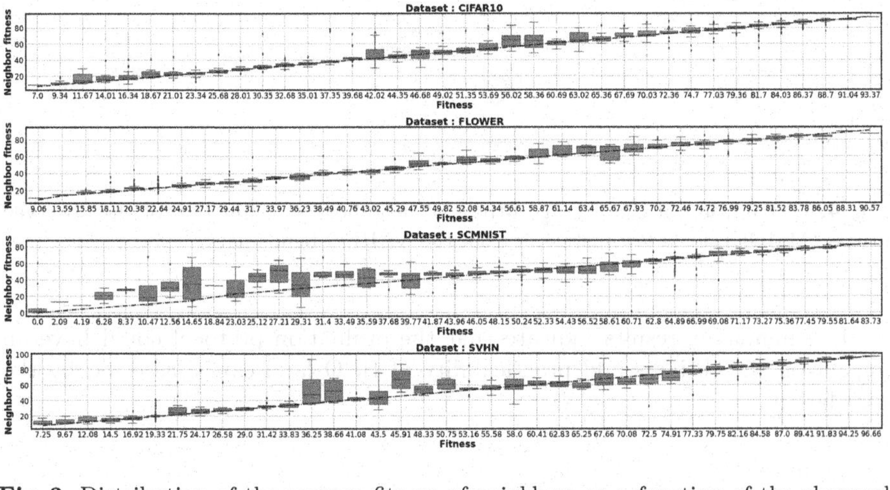

Fig. 3. Distribution of the average fitness of neighbors as a function of the observed fitness (*predictive accuracy*), for a few instances of the DS2019 benchmark.

Overall, we observe in many instances of DS-2019 a strong correlation between the observed fitness and the average fitness in the neighborhood. Indeed, the box-plots are aligned with the bisector. From the perspective of local search, it is easy to navigate the configuration space by consistently improving the fitness, from randomly distant and bad configurations to configurations of high fitness, for instances from DS-2019. This is less the case in YAHPO, as show in Fig. 4. Indeed, we observe a weaker correlation between the two variables. In the presence of unexpected mode in the distributions of fitness (see Fig. 2), e.g. YEAST and ISOLET, we find that a majority of neighbors tend to have the fitness of the observed mode, respectively around 70% and 97%. This suggests that the respective landscapes might have many local optima surrounded by plane areas at the odd fitness value. Thus, the chances for local search of being stuck are higher in such instances. Besides, the instances with more uniform and wider distribution of fitness (Fig. 1) tend to have a near perfect correlation. On the other hand, the more the distributions are multi-modal and with peaky modes, the worse the correlation between the variables of interest. This is the case for both benchmarks.

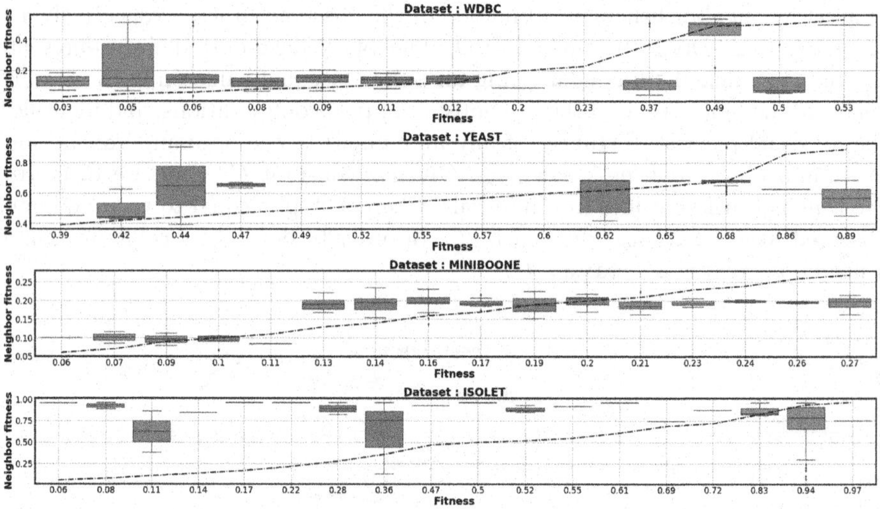

Fig. 4. Distribution of the average fitness of neighbors as a function of the observed fitness (MMCE), for a few instances from the YAHPO benchmark.

To summarize, results indicate that the evaluation protocol could have an impact on the easiness and practicability of HPO landscapes, assessed by the correlation.

Neutrality

Next, we look into the neutrality of the landscape. Figures 5 and 6 show the distribution of neutral neighbor counts as a function of the current fitness, respectively, for instances from DS-2019 and YAHPO.

For instances from DS-2019, the neutrality degree is equal to or greater than one, for most ranges of fitness values. In order words, most configurations have at least one neutral neighbor. Also, note that the FDC and *locality* results for CIFAR-10 are *good*, while for SCMNIST adn SVHN, with a multi-modal distribution of fitness (Fig. 1), coupled with lower *local* correlation (i.e., between the fitness and the fitness in the neighborhood, Fig. 3), the results are *bad*. Regarding CIFAR-10, the neutrality degree is on average consistently greater than two. In other words, most configurations have two or (many) more *neutral* neighbors. On the other hand, for SCMNIST and SVHN, the neutrality degree is inconsistent and with lower values on average. In particular, N_d is lower for fitness values ranging from 6.28 to 43.98% for SCMNIST, i.e., generally bad configurations have fewer neutral neighbors than *mid* and *good* configurations. Also, as expected, there is a huge number of neutral neighbors around the majority class prediction fitness: around 65% for SCMNIST and 20% for SVHN.

In YAHPO, we find that the range of fitness for which one can find solutions with neutral neighbors is limited. In practice, it represents a fraction of the range covered by all evaluated solutions. Besides, the neutrality degree is generally above 2, with larger counts associated to the modes in the distributions of fitness (See Fig. 2). These two facts suggest that the landscapes might be highly rugged with many local optima (no neutral neighbors), with areas centered towards values of the observed modes.

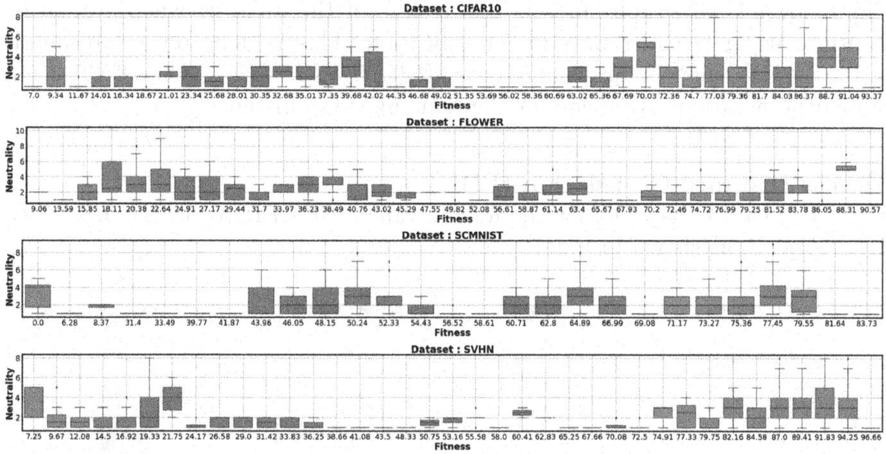

Fig. 5. Neutrality degree as a function of the observed fitness (*predictive accuracy*), for a few instances from the DS2019 benchmark.

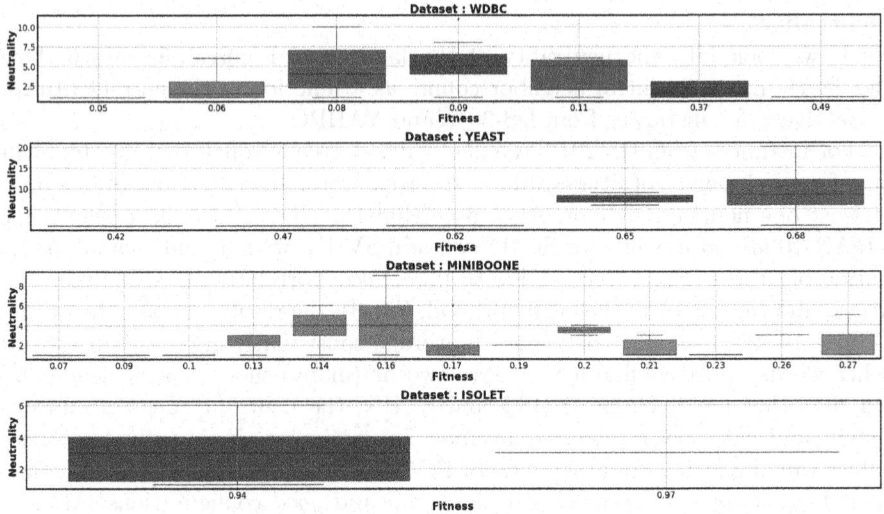

Fig. 6. Neutrality degree as a function of the observed fitness (MMCE), for a few instances from the YAHPO benchmark.

As a summary, the evaluation pipeline *malfunction* is responsible for an *imbalanced* landscape, i.e., the AutoML pipeline generates arbitrary *peaks* of fitness (low N_d) in areas of expected continuous fitness.

3.2 Log Loss

The following paragraphs present the results of the analysis when evaluating solutions using the Log Loss for classification. This is done on a sub-sample of the YAHPO benchmark, namely *SEMEION, VEHICLE, SEGMENT, KC1*. Similar observations are made for the remaining 115 instances.

Fitness Distance Correlation (FDC)

First, we look at the FDC for the four instances *SEMEION, VEHICLE, SEGMENT, KC1*, as shown in Fig. 7.

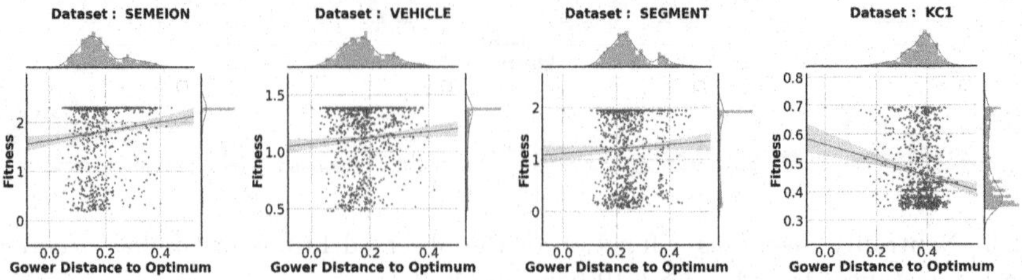

Fig. 7. FDC plot for a few instances from the YAHPO HPO benchmark, and the corresponding regression line in blue. The fitness function is the Log Loss. (Color figure online)

Similarly to results gathered in Sect. 3.1 (*predictive classification accuracy*), we find that the distributions of fitness are also covering a wide range of values, and are multi-modal. Several distributions also have the artifact identified previously: an unexpected mode around large (*poor*) fitness values, associated with HP configurations of highly variable *Gower* dissimilarity to the optimum. For instance, it is around the Log Loss value of 2.25 for SEMEION, 1.55 for VEHICLE, and 1.98 for SEGMENT. It affects HP configurations at Gower distances 0.15 to 0.45, i.e. covering the whole range of dissimilarity to the optimal HP. This phenomenon is also observed for a majority of the 119 analyzed instances of YAHPO. Besides, another mode can exist (around 0.35 for KC1) at low fitness values, i.e *good* HP configurations.

When looking at the nature of the instances (number of classes), we find that the unexpected mode correlates with the fitness value yielded by the Log Loss metric when attributing an *equal* probability of occurring to all classes, for all observations. In other words, solutions associated with the mode are likely solutions with *no classification ability.* This phenomenon could occur since the Log Loss metric does not penalize such behavior, a local optimum to which many solutions could naturally converge to.

Neighborhood

Next, we look at the neighborhood in the landscapes generated by the Log Loss metric. This is shown in Fig. 8. Overall, we find that the correlation between the average fitness of neighbors and the observed fitness, is weak in the case of distributions of fitness with the identified artifact (*peaks*). Most neighbors have a fitness value of the unexpected mode, e.g. a Log Loss value of 2.25 for SEMEION

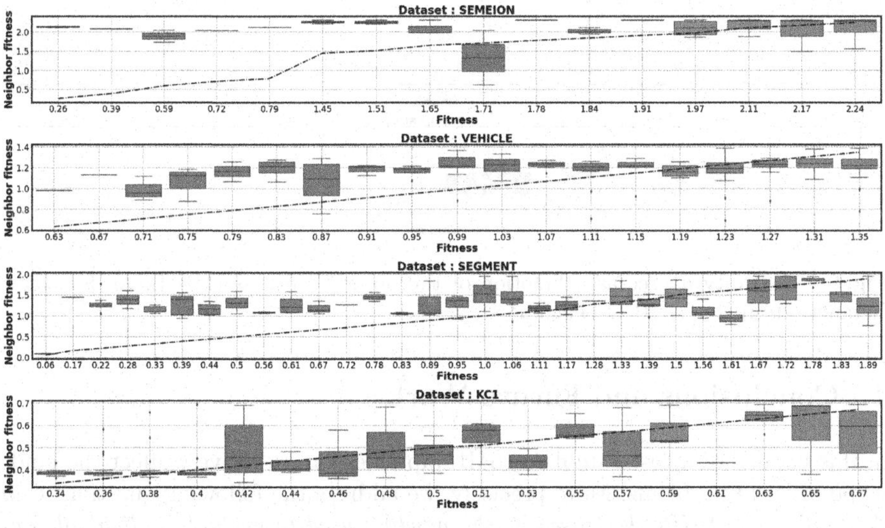

Fig. 8. Distribution of the average fitness of neighbors as a function of the observed fitness (Log Loss), for a few instances from the YAHPO benchmark.

and 1.35 for VEHICLE. This observation is in line with those made when using the metric of MMCE (see Figs. 1 and 2).

To summarize, using the Log Loss as an evaluation metric might also negatively impact the easiness of the associated landscapes, by increasing their ruggedness and decreasing their fitness potential.

Neutrality

Next, we look into the neutrality of the landscapes, as shown in Fig. 9. Similar to the analysis provided when evaluating with the MMCE, we find that few solutions have a neutral neighbor, and these are found within a restricted range of fitness. This suggests highly rugged landscapes. Besides, the highest counts of neutral neighbors are for solutions associated with the modes in the distributions of fitness (see Fig. 7). For instance, at a Log Loss value of 2.2 for SEMEION, and between 1.23 and 1.35 for VEHICLE. In other words, the landscapes are highly rugged (numerous local minima), and surrounded by a plateau of HP configurations associated with the high Log Loss values of the observed mode.

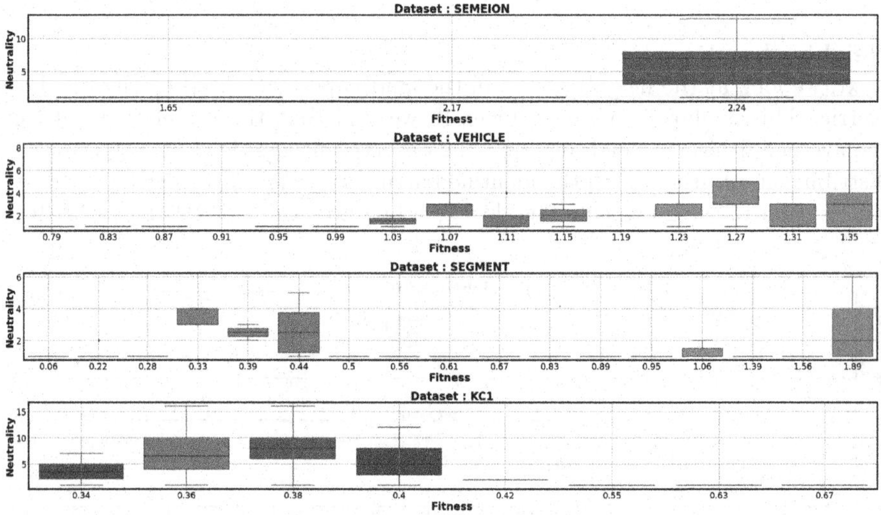

Fig. 9. Neutrality degree as a function of the observed fitness (Log Loss), for a few instances from the YAHPO benchmark.

4 Conclusions and Future Work

In this paper, we investigate if AutoML pipelines can negatively affect the landscape of HPO problems. More precisely, we address the following question: *Can the landscape of HPO be biased by the pipeline used to evaluate individual configurations?* To tackle this question, we have studied the fitness landscape of over 119 HPO instances obtained from either the DS-2019 (CNN) or YAHPO

(XGBoost) HPO benchmark data sets, using the concepts of *fitness distance correlation*, *locality*, and *neutrality*.

The FDC analysis shows unhealthy patterns in many HPO instances, with large groups of very diverse HP configurations with the same *ill* fitness value. These resulting peaks in fitness appear to be outliers in the respective distributions. Looking at the locality (fitness versus fitness in the neighborhood), we observe two things: First, there is a correlation between both variables of interest in healthy landscapes, suggesting that an *easy* path from randomly picked HP configurations could lead to the best performers, i.e., local-search may *do the job*. Second, for HPO problems negatively affected by the mentioned *illnesses* (i.e., the majority class predictors, or the inability to classify), the correlation between the current fitness and fitness in the neighborhood is worsened, indicating more rugged local landscapes.

Even though the *majority class prediction* problem for models trained and evaluated using some metrics (e.g., accuracy) is well known, the results show that the problem may not be taken seriously into account. This is also the case with the *inability to classify* arising when using the Log Loss as a training and evaluation criterion.

Thus, a great amount of resources is wasted when addressing HPO (i.e., many *simple* majority class or *inable* models are evaluated). Furthermore, the evidence shows that *the landscape of HPO problems could be negatively affected by the evaluation pipeline being used.*

Future work will further investigate how such artifacts affect HPO algorithms in practice.

Acknowledgements. Authors acknowledge support by the European Research Council (ERC) under the European Union's Horizon 2020 research and innovation program (grant agreement No. [ERC-2016-StG-714087], Acronym: *So2Sat*), by the Helmholtz Association through the Framework of Helmholtz AI [grant number: ZT-I-PF-5-01] - Local Unit "Munich Unit @Aeronautics, Space and Transport (MASTr)" and Helmholtz Excellent Professorship "Data Science in Earth Observation - Big Data Fusion for Urban Research" (W2-W3-100), by the German Federal Ministry of Education and Research (BMBF) in the framework of the international future AI lab "AI4EO – Artificial Intelligence for Earth Observation: Reasoning, Uncertainties, Ethics and Beyond" (Grant number: 01DD20001) and the grant DeToL. The authors also acknowledge support by DAAD for a Doctoral Research Fellowship.

References

1. Bischl, B., et al.: Hyperparameter optimization: foundations, algorithms, best practices and open challenges (2021). https://doi.org/10.48550/ARXIV.2107.05847, https://arxiv.org/abs/2107.05847
2. Clergue, M., Verel, S., Formenti, E.: An iterated local search to find many solutions of the 6-states firing squad synchronization problem. Appl. Soft Comput. **66**, 449–461 (2018). https://doi.org/10.1016/j.asoc.2018.01.026, https://www.sciencedirect.com/science/article/pii/S1568494618300322

3. Elsken, T., Metzen, J.H., Hutter, F.: Neural architecture search: a survey. J. Mach. Learn. Res. **20**(1), 1997–2017 (2019)
4. Gower, J.C.: A general coefficient of similarity and some of its properties. Biometrics **27**(4), 857–871 (1971). http://www.jstor.org/stable/2528823
5. He, X., Zhao, K., Chu, X.: AutoML: a survey of the state-of-the-art. Knowl.-Based Syst. **212**, 106622 (2021)
6. Hutter, F., Kotthoff, L., Vanschoren, J.: Automated Machine Learning - Methods, Systems, Challenges. Springer, Berlin (2019). https://doi.org/10.1007/978-3-030-05318-5
7. Jones, T., Forrest, S.: Fitness distance correlation as a measure of problem difficulty for genetic algorithms. In: Proceedings of the 6th International Conference on Genetic Algorithms, pp. 184–192. Morgan Kaufmann Publishers Inc., San Francisco (1995)
8. Ojha, V.K., Abraham, A., Snášel, V.: Metaheuristic design of feedforward neural networks: a review of two decades of research. Eng. Appl. Arti. Intell. **60**, 97–116 (2017). https://doi.org/10.1016/j.engappai.2017.01.013, https://www.sciencedirect.com/science/article/pii/S0952197617300234
9. Pfisterer, F., Schneider, L., Moosbauer, J., Binder, M., Bischl, B.: YAHPO gym - an efficient multi-objective multi-fidelity benchmark for hyperparameter optimization (2021)
10. Pimenta, C.G., de Sá, A.G.C., Ochoa, G., Pappa, G.L.: Fitness landscape analysis of automated machine learning search spaces. In: Paquete, L., Zarges, C. (eds.) EvoCOP 2020. LNCS, vol. 12102, pp. 114–130. Springer, Cham (2020). https://doi.org/10.1007/978-3-030-43680-3_8
11. Pitzer, E., Affenzeller, M.: A comprehensive survey on fitness landscape analysis. In: Fodor, J., Klempous, R., Suárez Araujo, C.P. (eds.) Recent Advances in Intelligent Engineering Systems. Studies in Computational Intelligence, vol. 378, pp. 161–191. Springer, Heidelberg (2012). https://doi.org/10.1007/978-3-642-23229-9_8
12. Ren, P., Xiao, Y., Chang, X., Huang, P.y., Li, Z., Chen, X., Wang, X.: A comprehensive survey of neural architecture search: challenges and solutions. ACM Comput. Surv. **54**(4) (2021). https://doi.org/10.1145/3447582
13. Sharma, A., van Rijn, J.N., Hutter, F., Müller, A.: Hyperparameter importance for image classification by residual neural networks. In: Kralj Novak, P., Šmuc, T., Džeroski, S. (eds.) DS 2019. LNCS (LNAI), vol. 11828, pp. 112–126. Springer, Cham (2019). https://doi.org/10.1007/978-3-030-33778-0_10
14. Traoré, K.R., Camero, A., Zhu, X.X.: Landscape of neural architecture search across sensors: how much do they differ ? ISPRS Ann. Photogr. Remote Sens. Spat. Inf. Sci. **V-3-2022**, 217–224 (2022). https://doi.org/10.5194/isprs-annals-V-3-2022-217-2022, https://www.isprs-ann-photogramm-remote-sens-spatial-inf-sci.net/V-3-2022/217/2022/
15. Traoré, K.R., Camero, A., Zhu, X.X.: Fitness landscape footprint: a framework to compare neural architecture search problems (2021)

Condition-Based Maintenance Optimization Under Large Action Space with Deep Reinforcement Learning Method

Peng Bi, Yi-Ping Fang, Matthieu Roux(✉), and Anne Barros

Univ. Paris-Saclay, CentraleSupélec, LGI EA 2606, Chair Risk and Resilience of Complex Systems, Gif-sur-Yvette, France
{yiping.fang,matthieu.roux,anne.barros}@centralesupelec.fr

Abstract. Effective maintenance is essential in keeping industrial systems running and avoiding failure. Condition-based maintenance (CBM) leverages the current degradation condition of the studied object to optimize future maintenance interventions. CBM optimization problems are complex for multi-component systems, facing the issue of the curse of dimensionality brought by the increase in the number of components. Reinforcement learning provides a promising perspective to overcome the issue. In this paper, we studied CBM optimization for a multi-component system in which the components degrade subject to the gamma process independently. We considered multiple maintenance choices for individual components, leading to a large combinatorial action space. In this case, traditional deep reinforcement learning algorithms like DQN may struggle to face the inefficiency of exploration. Instead, we propose exploiting Branching Dueling Q-network (BDQ), which incorporates the action branching architecture into DQN to drastically decrease the number of estimated actions. We trained a learning agent to minimize the expected cost for a long time horizon by taking maintenance actions according to the observed exact degrading signal for each component. We compared the policy learned by the agent with some other pre-defined static policies. The numerical results demonstrate the effectiveness of the learning algorithm and its potential for application in systems with more complex structures.

Keywords: Optimal maintenance planning · Condition-based maintenance · Deep Reinforcement learning · Branching Dueling Q-Network

1 Introduction

Every component in industrial systems degrades over time. In order to maintain the regular operation of industrial equipment and reduce the loss caused by system damage, maintenance activities should take place [1]. The stochastic

B. Dorronsoro et al. (Eds.): OLA 2023, CCIS 1824, pp. 161–172, 2023.
https://doi.org/10.1007/978-3-031-34020-8_12

process is one of the most commonly studied models to describe degradation phenomena for maintenance policy [2,3]. In stochastic degradation models, the health condition of a study target is represented by a degradation metric usually constructed based on physical signals (such as the length of the crack, remaining fuel, vibration frequency, and amplitude). As soon as such degradation metric reaches a certain threshold, whether the threshold is known or not, the target will no longer keep functioning and lead to a failure state. For a system that contains multiple components, the health condition of the components will determine whether the entire system can operate successfully. In order to prevent a system-wide failure, maintenance actions on the components should be performed in a coordinated way and based on their degradation level.

Modern sensor technologies enable monitoring and assessing system state remotely and applying condition-based maintenance(CBM) policies to a degrading system whose state is observable. These policies recommend maintenance decisions using the monitored system health conditions and are often more efficient [4]. In practice, there are usually multiple maintenance choices, such as complete repair, partial repair, and replacement, of different requirements on the financial expense, human resources, and time. Often, the better the action can improve the degradation metric, the higher the cost. Therefore, CBM policy aims to find a trade-off between the cost carried by the maintenance action and the cost brought by the system failure. Most of the CBM strategies are "static", in the sense that some fixed degradation thresholds are pre-defined, and corresponding maintenance actions can only be performed if the degradation level reaches the thresholds. Those static policies may sometimes be too conservative and waste some useful life of the system. Hence a dynamic policy that can dynamically select maintenance action based on the observed degradation level is much required.

Many existing methods struggle with maintenance optimization on large-scale systems (with many components), facing the curse of dimensionality and history. Reinforcement learning (RL), and especially deep reinforcement learning (DRL), has shown promising performance in the application of optimization and control problems of dynamic and huge-dimensionality nature [5]. The RL paradigm can theoretically mitigate the curse of dimensionality associated with state spaces, both under model-free approaches that do not exploit prior offline environment information or under model-based approaches [6].

Solving maintenance optimization problems under the RL umbrella is studied in a few works. In [7], authors apply RL to schedule condition-based maintenance of fighter aircraft. Work [8] aims to minimize the long-run expected system average cost rate of the maintenance problem for a flow line system consisting of two series machines with an intermediate finite buffer in between. In [2], they developed a new dynamic maintenance policy with RL for multi-component systems with individually repairable components. [9] address their problem of minimizing the expected sum of two conflicting objective functions: the average inventory level and the average number of back-orders through Q-value-based RL algorithms. [10] proposed an RL framework for a real-time control process to improve manufacturing machines' production performance in a small factory.

Traditional RL algorithms, like Q-value-based algorithms, perform well on some small-medium systems while struggling when facing large or complex systems containing large state and action space. In [3], they considered the exact level of degradation as the maintenance cost for the duration of the maintenance contract. Work [12] proposed a CBM model based on a customized DRL algorithm DDQN(Double-DQN) for multi-component systems with independent risks. In their work, both stochastic and economic dependencies among the components are considered. Preventive maintenance policy on production lines using DDQN is applied in [13], and in [14], they applied an actor-critic DRL algorithm. They developed a new selective maintenance optimization for multi-state systems that can execute multiple consecutive missions over a finite horizon.

Consider that there are different kinds of maintenance action needs. The action space for a multi-component system will have a combinatorial increase. A few works of research studied the problem of applying reinforcement learning to a maintenance optimization with a large state and action space [11,12] and [5] developed a multi-agent reinforcement learning algorithm, providing efficient life-cycle policies for large multi-component systems operating in high-dimensional.

The main contributions of this paper are 1) the formulation of the condition-based decision-making problem into a Markov Decision Process; 2) the proposition to exploit the Branching Dueling Q-Network algorithm to solve the maintenance optimization problem with combinatorial action space; 3) a comparison of the performance of the trained agents with different maintenance policies from multiple perspectives.

This paper is structured as follows. In Sect. 2, we describe the degradation model and the maintenance strategies. Section 3 introduces the reinforcement learning algorithm. Section 4 illustrates the numerical results, and in Sect. 5, we draw the conclusion and discuss the potential work in the future.

2 Problem Description

2.1 Degradation Model of Multi-component Systems

System components degrade over time. The degradation process of different components varies due to their materials, working loads, and environment. In this paper, we use the gamma process to model the degradation of the system, which is a very common degradation process studied in the condition-based maintenance literature, and we assume each component degrades independently.

Let $X(t)$ denote the degradation level of a component at time t, following the gamma process. Then, for any $t > t' > 0$, the probability density function (PDF) g of $X(t) - X(t')$ for a given component i:

$$g(x; \alpha_i(t-t'), \beta_i) = \frac{\beta_i^{\alpha_i(t-t')} x^{\alpha_i(t-t')-1} exp(-\beta_i x)}{\Gamma(\alpha_i(t-t'))} \tag{1}$$

where α, β stands for the shape and scale parameters separately. $\Gamma(\cdot)$ stands for the gamma function. Components keep on degrading over time until they meet their failure threshold (H_i).

Due to the stochastic property of the gamma distribution, the degradation process for the same type of components (having the same shape and scale parameter) varies greatly. Figure 1 shows the different degradation paths of the same component. Such stochastic property brings difficulties in finding a stationary maintenance policy. In this paper, we model the maintenance problem as a Markov decision process, and reinforcement learning algorithms are applied to find an optimal policy over time.

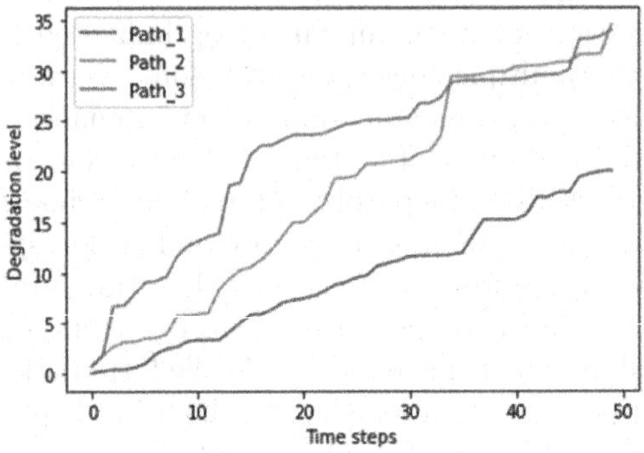

Fig. 1. Degradation paths of one component

2.2 Maintenance Strategy

Degradation level of the entire system which is composed of n components ($X_t = [X_{t,1}, X_{t,2}, ..., X_{t,n}]$) is considered fully observable. Based on the degradation level, maintenance actions can be performed at any time step t instantaneously, i.e., the durations of maintenance actions are neglected.

This paper considers five different actions for each component. Each action affects the current degradation level at different degrees and costs according to its recovery efficiency (Table 1). The better the action can recover the degradation level to a better place(i.e., smaller repair factor μ), the more expensive the cost is. "Doing nothing" means no maintenance action is performed. "Imperfect repair", "Repair" and "Imperfect replacement" can recover the current health condition to a better level, while "Replacement" will reset the degradation level to 0. Additionally, a downtime cost (c_D) will be counted when the entire system fails (Table 2). The actions performed follow the assumptions:

- Maintenance action will affect the component health condition independently.
- When a component is at its functioning state($x_i \in [0, H_i)$), all the maintenance actions are available.
- When a component is in the failure state($x_i \geq H_i$), only "Doing nothing" and "Replacement" are available.
- For each component, only one action per time step can be performed.

Table 1. Available maintenance actions

Action	Description	Repair factor (μ)	Cost (c)
0	Do nothing	1	0
1	Imperfect repair	0.6	40
2	Repair	0.5	80
3	Imperfect replacement	0.3	100
4	Replacement	0	150

Table 2. Cost of maintenance actions

Doing nothing (c_0)	Imperfect repair (c_1)	Repair (c_2)
0	40	80
Imperfect replacement (c_3)	Replacement (c_4)	Down-time cost (c_D)
100	150	200

Once performed on component i, an action $A_{t,i}(A_{t,i} \in \{0, 1, 2, 3, 4\})$ changes the degradation level from $X_{t,i}$ to $X_{t+1,i}$ following Eq. 2

$$X_{t+1,i} = \mu(a_{t,i})X_{t,i} + Y_t, \quad Y_t \sim \Gamma(\alpha_i, \beta_i) \tag{2}$$

The system-wide action A_t taken every time step is the combination of individual maintenance actions for all n components. It is denoted $A_t = [A_{t,1}, A_{t,2}, ..., A_{t,n}]$. We introduce $\delta_{t,n,i}$ as a binary variable to illustrate whether the action $a_{t,i} = n$ is performed on component i in time t. The following relation then holds $\delta_{t,0,i} + \delta_{t,1,i} + ... + \delta_{t,4,i} = 1$. Eventually, the total cost at inspection time t can be expressed as:

$$C_t = \sum_{i=1}^{n}(c_0\delta_{t,0,i} + c_1\delta_{t,1,i} + c_2\delta_{t,2,i} + c_3\delta_{t,3,i} + c_4\delta_{t,4,i}) + c_D I_t \tag{3}$$

where I_t is a binary variable that indicates the failure of the system ($I_t = 1$ when system fails and $I_t = 0$ otherwise.). The optimization goal is to find a stationary policy to minimize the expected long time cost (or maximize the reward, defined as the opposite of the cost), namely:

$$\max \quad R \tag{4}$$

$$s.t. \quad R = -\sum_{t=1}^{T}(\sum_{i=1}^{n}\sum_{p=1}^{4} c_p\delta_{t,p,i} + c_D I_t), \tag{5}$$

$$\sum_{p=0}^{4} \delta_{t,p,i} = 1, \tag{6}$$

$$\delta_{t,p,t}, I_t \in 0, 1, \tag{7}$$

where T is the total number of inspections. As t approaches $+\infty$, it becomes a infinite time horizon.

2.3 Markov Decision Process

A Markov decision process can be described as a quintuple (S, A, P, R, γ), which provides solutions for sequential decision problems under the interaction with an environment. At each time step t, the agent observes the current state $s_t \in S$, performs an action $a_t \in A$, and receives the corresponding reward $r(s_t, a_t)$ (in our case, $-C_t$). Then, the state transits to a new state $s_{t+1} \in S$ following the dynamics P of the environment. The Markov property is reflected in the fact that the next state s_{t+1} is only related to the current state s_t and the action a_t.

The performance of the agent's policy π is measured by the cumulative discounted return G_t, which represents the sum of discounted (γ is the discount factor) future rewards collected from time t to the time horizon T. Equation 8 illustrates this sum:

$$\begin{aligned} G_t &= r(s_t, a_t) + \gamma r(s_{t+1}, a_{t+1}) + ... + \gamma^{T-t} r(s_T, a_T) \\ &= \sum_{i=t}^{T} \gamma^{i-t} r(s_i, a_i) \end{aligned} \tag{8}$$

3 Reinforcement Learning Algorithms

Following the degradation model and the maintenance strategy introduced previously, we propose a dynamic model for our CBM optimization problem.

- **State space:** The degradation level for each component in the system is fully observable. The degradation condition for the n-components system is denoted by the n dimensional vector $X_t = [X_{t,1}, X_{t,2}, ..., X_{t,n}]$. The degradation level $x_{t,i}$ in each dimension is continuous, ranging from 0 to the component's failure threshold H_i, i.e., $x_{t,i} \in [0, H_i], \forall i = 1, .., n$.
- **Action space:** The action output by the agent is the combination of the "sub-actions" chosen for each component ($A_t = [A_{t,1}, A_{t,2}, ..., A_{t,n}]$). In our case, the action set for each component is the same ($A_{t,i} \in \{0, 1, 2, 3, 4\}$). Hence $A_t \in \{0, 1, 2, 3, 4\}^n$
- **Reward:** The reward at each time step is the negative cost carried by the maintenance actions, namely $R_t = -C_t$.

3.1 Dilemma of Applying Q-Learning Based Algorithm

Value-based and policy gradient methods are two main categories of today's RL algorithms. DQN [15] is often used as the baseline of the value-function-based method, while PPO [16] and DDPG [17] are two classical policy gradient methods. In this work, we chose to focus on value function-based methods.

As the number of components increases, the action space grows exponentially in size due to the combinatorial effect. The traditional Q-learning method (Tabular Q-learning) struggles with large spaces because it has to maintain an extremely large matrix. Moreover, Q-learning algorithms cannot deal with continuous state space like the model proposed in this paper. Therefore it is a wise choice to resort to DRL. Considering the standard structure of the DQN-based algorithms, the output of the neural networks is the Q-value for each action, which means the output has the same size as the action space. For a system with n components under the maintenance strategy introduced in Sect. 2.2, the size of its action spaces becomes 5^n. In this case, training a neural network with numerous parameters is challenging. Exploration is very likely to be insufficient and inefficient.

3.2 Branching Dueling Q-Network

Branching Dueling Q-Network (BDQ) is proposed in [18]. It consists in a neural architecture that distributes the representation of the actions across individual network branches, each branch representing one action dimension. The joint-action tuple that concatenates the sub-action is the output for the entire network. In practice, if k sub-actions are available for each action dimension, the DQN would need to output k^n Q-values, while the BDQ only outputs $k \times n$ Q-values, which illustrates why the BDQ seems to scale better for large and combinatorial action spaces. A feature-sharing model was added before the branching in order to let the network encode a latent representation of the input and feed more information to the branches for coordination (Fig. 2).

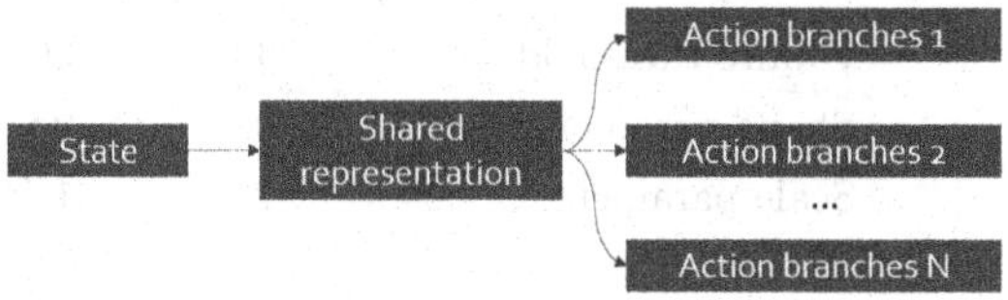

Fig. 2. Basic structure of Branching architecture

As mentioned in [18], BDQ can be thought of as an adaption of the dueling network into the action branching architecture. [19] proposed Dueling architecture instead of learning the Q value directly, but training the two branches of the last few layers of the neural network to output the state value function (V) and the advantage function (where A is defined as $A(s,a) = Q(s,a) - V(s)$). The advantage function A represents the difference of taking different actions at the same state (s); modeling them separately allows the agent to better handle states less associated with actions.

Figure 3 illustrates the detailed structure of applying BDQ on a n-component system. The BDQ maintains n branches, only leading to a linear increase in the number of outputs with increasing action dimensionality.

4 Numerical Results

We considered two cases (a three-component series system and a nine-component series system) to demonstrate the performance of the proposed BDQ on multi-component systems. In addition, a DDQN algorithm is applied for comparison.

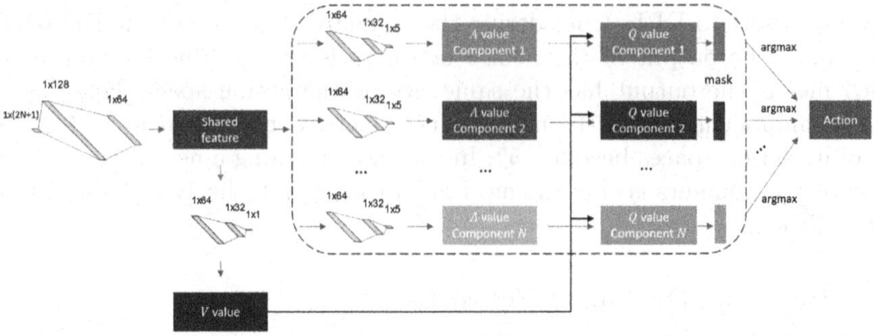

Fig. 3. BDQ structure for a N components system

The numerical results of the trained agent and the comparison to other policies are presented in this section. Table 3 presents the degradation parameters for each component.

Table 3. Components Parameters

Parameters	Description	Comp. 1	Comp. 2	Comp. 3
H_i	Failure threshold	30	33	26
α_i	Shape parameter	0.3	0.2	0.6
β_i	Scale parameter	1.2	1.1	1.1

To compare the performance of different policies, we ran simulations for each policy from a brand new system (i.e., no degradation for each component) for 50×10000 time steps. We made sure to run simulations on a sufficiently long time horizon so that it could lead to an accurate estimation of the performance of each policy. Moreover, it has the advantage to evaluate the different policies without giving too much importance to the initial state, chosen here to be an as-good-as-new state. The indicator we use to demonstrate the performance of the policy is the mean value for every 50 time steps. Besides, two static maintenance policies are used for comparison, including:

- **Corrective Maintenance(CM):** Only do replacement (action 4) as soon as one component fails.
- **Static Condition-based Maintenance(CBM):** An expert policy is given by prior knowledge. When the component state level is below $0.8H_i$, no repair action is taken (action 0); when the component state level is in $[0.8H_i, H_i]$, apply repair (action 2); otherwise, when the component fails, replacement (action 4) is triggered.

4.1 Three-Component Series System

Hyperparameters for training the BDQ and DDQN on the three-component series system can be found in Table 4. To monitor the training process, we display the cost per episode and the average cost per 100 episodes shown in blue to demonstrate the converging trend (Fig. 4(a), Fig. 5(a)).

Table 4. BDQ hyper-parameters

Parameters	Description	Value (BDQ)	Value (DQN)
lr	learning rate	1×10^{-4}	2×10^{-3}
n_s	number of episode	5×10^3	5×10^4
T	length of one episode	50	50
γ	discount factor	0.9	0.9
ϵ_0	ϵ-greedy: Initial ϵ for BDQ	0.9	–
$\Delta\epsilon_0$	ϵ_0's decay factor	0.995	–
ϵ_1	ϵ-greedy: static ϵ for DQN	–	0.01
C	frequency to update the target network	1000	2000
Bu	buffer size	5000	5000
Ba	batch size	32	32
m_s	the minimal size of begin training	1000	1000

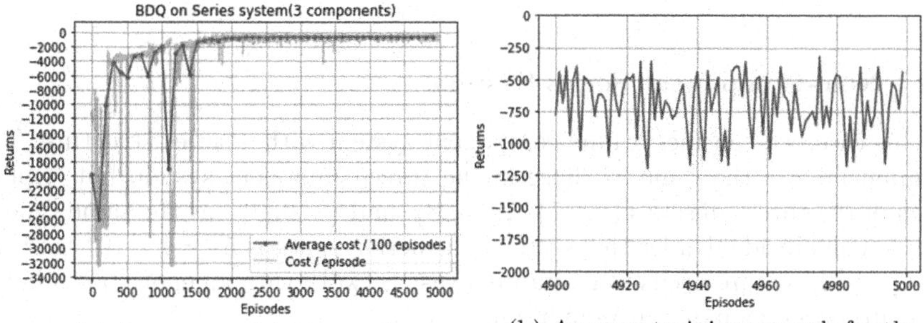

(a) Average training rewards.

(b) Average training rewards for the last 100 episodes.

Fig. 4. Average rewards during training (BDQ on series system with 3 components)

Figure 4 and Fig. 5 show that the DDQN algorithm converges faster than the BDQ algorithm while less stable. We tested the trained BDQ and DDQN agent on the proposed degradation model for 5×10^6 time steps. For every 50 time steps, we collect the total cost accumulated in the interval. Table 5 reports the simulation results (average cost, standard deviation, the 95% confidence interval, and the total time spent during simulation). From the table, we can see that for this simple series system with three components, the performance of the BDQ algorithm can reach the same level as DDQN, and both learning agents outperform the pre-defined static policy on average.

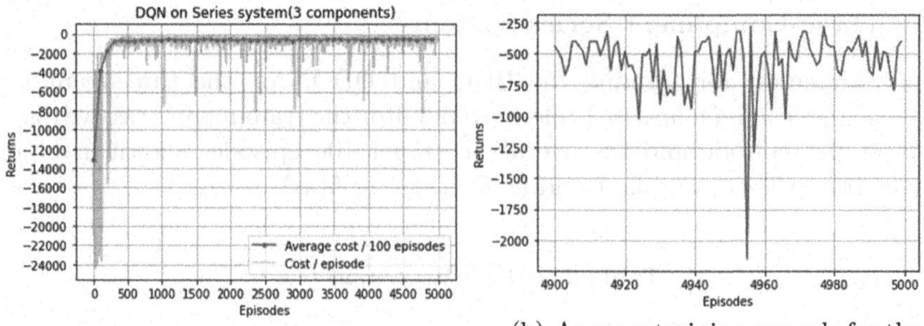

(a) Average training rewards.

(b) Average training rewards for the last 100 episodes.

Fig. 5. Average rewards during training (DDQN on series system with 3 components)

Table 5. Maintenance cost under different strategies on Series system (3 components)

Policy	C_{avg}	std_C	95%CI	t_{sim}
BDQN	−631.87	212.47	±4.16	4 m 31 s
DDQN	−603.13	202.97	±3.98	1 m 6 s
CM	−1411.56	284.27	±5.57	13.7 s
CBM	−853.49	113.32	±2.22	13.4 s

4.2 Nine-Component Series System

In this case, we consider a nine-component system with a series structure. The components are the same as those in the three-component system but tripled. Each of the three subsets {1, 2, 7}, {2, 5, 8}, and {3, 6, 9} has the same type of components inside. Under this system, the action space is $5^9 \approx 1 \times 10^6$, and the DDQN architecture needs to maintain a vast network with an enormous number of neurons and cannot converge in our training.

For BDQ, we double the size of each hidden layer in our BDQ structure to train on the nine-component system; the n_s is extended to 2×10^4 episodes while other hyper-parameters remained the same as in Table 3. Figure 6 displays the average cost per 100 episodes, where the variance is higher than training a three-component system. Table 6 shows the good performance of the policy produced by BDQ agent.

Table 6. Maintenance cost under different strategies on Series system (9 components)

Policy	C_{avg}	std_C	95%CI	t_{sim}
BDQN	−1576.00	640.32	±12.55	7 m 6 s
CM	−4054.73	474.98	±9.31	30.2 s
CBM	−2559.77	199.13	±3.90	29.9 s

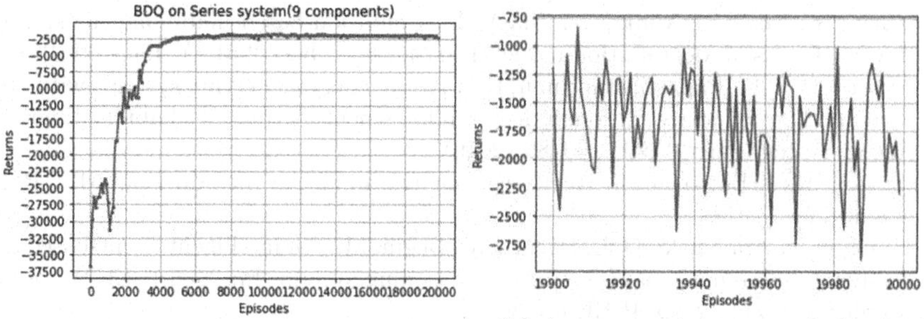

(a) Average training rewards.

(b) Average training rewards for the last 100 episodes.

Fig. 6. Average rewards during training (BDQ on series system with 9 components)

5 Conclusion and Future Work

For multi-component systems with independent repairable components, if we consider multiple action choices, the maintenance decision for the whole system will be the action combination of each component. Classic DRL algorithms like DQN are inefficient in dealing with problems with a large, multi-dimensional action space. Hence, we resort to BDQ, which enables the agent to learn the policy in a combinatorial action space efficiently.

The numerical results show that the BDQ algorithm performs well for the studied CBM optimization problem. It can perform the same as DDQN in a small series system of three components and outperform the static policies on average. When the number of components increases (nine-component), finding some certain threshold to design a static policy becomes difficult. Q-learning-based algorithms face the problem of maintaining a massive network with $5^9 \approx 10^6$ neurons in the output layer, whereas the BDQ showed good performance in training efficiency and convergence on the nine-components series system.

More work can be done to test the performance of BDQ on a multi-component system. 1) The size of the system could be further expanded. 2) The system structure could be more complex to see the potential of BDQ to recognize the system's structure. 3) Maintenance resources limit could be considered.

References

1. Peng, Y., Dong, M., Zuo, M.J.: Current status of machine prognostics in condition-based maintenance: a review. Int. J. Adv. Manuf. Technol. **50**, 297–313 (2010)
2. Yousefi, N., Tsianikas, S., Coit, D.W.: Reinforcement learning for dynamic condition-based maintenance of a system with individually repairable components. Qual. Eng. **32**(3), 388–408 (2020)
3. Yousefi, N., Tsianikas, S., Coit, D.W.: Dynamic maintenance model for a repairable multi-component system using deep reinforcement learning. Qual. Eng. **34**(1), 16–35 (2022)

4. Chen, N., Ye, Z.S., Xiang, Y., Zhang, L.: Condition-based maintenance using the inverse Gaussian degradation model. Eur. J. Oper. Res. **243**(1), 190–199 (2015)
5. Andriotis, C.P., Papakonstantinou, K.G.: Managing engineering systems with large state and action spaces through deep reinforcement learning. Reliability Engineering & System Safety **191**, 106483 (2019)
6. Sutton, R.S., Barto, A.G.: Reinforcement Learning: An Introduction. MIT Press, Cambridge (2018)
7. Tang, L., Kacprzynski, G.J., Bock, J.R., Begin, M.: An intelligent agent-based self-evolving maintenance and operations reasoning system. In: 2006 IEEE Aerospace Conference, pp. 12-pp. IEEE (2006)
8. Wang, X., Wang, H., Qi, C.: Multi-agent reinforcement learning based maintenance policy for a resource constrained flow line system. J. Intell. Manuf. **27**, 325–333 (2016)
9. Xanthopoulos, A.S., Kiatipis, A., Koulouriotis, D.E., Stieger, S.: Reinforcement learning-based and parametric production-maintenance control policies for a deteriorating manufacturing system. IEEE Access **6**, 576–588 (2017)
10. Shiue, Y.R., Lee, K.C., Su, C.T.: Real-time scheduling for a smart factory using a reinforcement learning approach. Comput. Industr. Eng. **125**, 604–614 (2018)
11. Zhang, P., Zhu, X., Xie, M.: A model-based reinforcement learning approach for maintenance optimization of degrading systems in a large state space. Comput. Industr. Eng. **161**, 107622 (2021)
12. Zhang, N., Si, W.: Deep reinforcement learning for condition-based maintenance planning of multi-component systems under dependent competing risks. Reliabil. Eng. Syst. Saf. **203**, 107094 (2020)
13. Huang, J., Chang, Q., Arinez, J.: Deep reinforcement learning based preventive maintenance policy for serial production lines. Expert Syst. Appl. **160**, 113701 (2020)
14. Liu, Y., Chen, Y., Jiang, T.: Dynamic selective maintenance optimization for multi-state systems over a finite horizon: a deep reinforcement learning approach. Eur. J. Oper. Res. **283**(1), 166–181 (2020)
15. Mnih, V., et al.: Playing Atari with deep reinforcement learning. arXiv preprint arXiv:1312.5602 (2013)
16. Schulman, J., Wolski, F., Dhariwal, P., Radford, A., Klimov, O.: Proximal policy optimization algorithms. arXiv preprint arXiv:1707.06347 (2017)
17. Lillicrap, T.P., Hunt, J.J., Pritzel, et al.: Continuous control with deep reinforcement learning. arXiv preprint arXiv:1509.02971 (2015)
18. Tavakoli, A., Pardo, F., Kormushev, P.: Action branching architectures for deep reinforcement learning. In: Proceedings of the AAAI Conference on Artificial Intelligence, vol. 32, no. 1 (2018)
19. Wang, Z., Schaul, T., Hessel, M., Hasselt, H., Lanctot, M., Freitas, N.: Dueling network architectures for deep reinforcement learning. In: International Conference on Machine Learning, pp. 1995–2003. PMLR (2016)

Learning Methods to Enhance Optimization Tools

An Application of Machine Learning Tools to Predict the Number of Solutions for a Minimum Cardinality Set Covering Problem

Brooks Emerick(✉), Myung Soon Song, Yun Lu, and Francis Vasko

Kutztown University, Kutztown, PA 19530, USA
bemerick@kutztown.edu

Abstract. The minimum cardinality set covering problem (MCSCP) is an NP-hard combinatorial optimization problem in which a set must be covered by a minimum number of subsets selected from a specified collection of subsets of the given set. It is well documented in the literature that the MCSCP has numerous, varied, and important industrial applications. For some of these applications, it would be useful to know if there are alternative optimums and the qualitative number of alternative optimums. In this article, both classification trees and neural networks are employed to qualitatively (small, medium, or large) predict the number of optimal solutions to a MCSCP. Results show that both model types have an accuracy in the low to mid 80%, with the neural network slightly outperforming the classification tree. Sensitivity and positive predictive value (PPV) are used to describe more detailed information.

Keywords: minimum cardinality set covering problem · alternative solutions · machine learning · classification tree · neural networks

1 Introduction and Literature Review

The minimum cardinality set covering problem (MCSCP), also called the unicost set covering problem, has numerous and varied industrial applications. Although it is NP-hard [4], recent advances in integer programming software [1,5] has made it possible to obtain optimal solutions for industrial problems formulated as MCSCPs. Additionally, there are industrial applications for which knowledge of alternative optimums would be very useful in practice. Three such examples from the steel industry are optimal ingot mold selection [8], metallurgical grade assignment [9], and product size consolidation [7,10]. The product size consolidation application has many and varied applications outside the steel industry.

The only work in the literature that tries to predict qualitatively the number of alternative optimums for the MCSCP is by Emerick et al. [2]. Being a preliminary study, Emerick et al. only considered MCSCPs with constraint matrices of 20% density and only analyzed the problem using classification trees.

B. Dorronsoro et al. (Eds.): OLA 2023, CCIS 1824, pp. 175–185, 2023.
https://doi.org/10.1007/978-3-031-34020-8_13

The main contributions of this article are twofold: (1) we analyze an extensive set of MCSCPs for five different densities (10%, 15%, 20%, 25%, and 30%), (2) in addition to using classification trees, we use neural networks to try to predict qualitatively the number of alternative optimums for MCSCPs and analyze the results of these two approaches.

This paper is organized as follows: we present a mathematical formulation of the MCSCP in Sect. 2 follow by our methodology in Sect. 3, which includes the construction of a large set of unique MCSCPs with varying density. We present a brief statistical summary of our representative sample and perform a principal component analysis and a correlation analysis on our set of decision variables. In Sect. 4, we train a classification tree on a subset of the sample, validate the tree on a validation subset of MCSCPs, and report the results of the classification tree. In Sect. 5, we train and validate a neural network and report the results. We summarize our findings in Sect. 6 and conclude the paper with a discussion and future work in Sect. 7.

2 Mathematical Formulation

Let $A = [a_{ij}]$ be an $m \times n$ matrix, where the entries of A are ones and zeros. The index i is the constraint index, with m total constraints; and the index j is the variable index, with n total variables. Let p be the density of ones in the matrix and assume the row and column sum of the matrix are at least one. We seek the solution to the minimum cardinality set covering problem (MCSCP), which is formulated as follows: Let $x = [x_i]$ be an $n \times 1$ vector of ones and zeros only (x is a bit string), then

$$\text{Minimize:} \quad \sum_{j=1}^{n} x_j \tag{1}$$

$$\text{Subject to:} \quad \sum_{j=1}^{n} a_{ij}x_j \geq 1 \qquad \text{for } i = 1, 2, \ldots, m \tag{2}$$

For any matrix, A, that satisfies the above constraints, there is at least one optimal solution, x, to the MCSCP. In fact, there may be alternative optima that have the same minimum value as given in Eq. (1). Like the approach detailed in Emerick et al. [2], we seek to construct predictive models using machine learning that determines, with some degree of confidence, the qualitative number of alternative solutions for any given matrix A. Emerick et al. [2] fixed the dimensions to 10×20 and density to $p = 20\%$. We generate matrices of this size so that we can determine the number of alternative optima in a reasonable amount of time at relatively low computational cost. We focus this article on the predictive nature of a classification tree and a neural network on the number of alternative solutions to the MCSCP.

3 Research Methodology

Classification trees and neural network models are efficient tools used for input-output mapping. We construct classification trees using the built-in *fitctree* function in MATLAB from the Statistical and Machine Learning Toolbox [6]. Using a set of pre-determined classification and regression tree (CART) predictor variables, this algorithm selects the optimal split predictor variable at each node that maximizes the Gini diversity index (GDI) over all other predictors' possible splits. To construct a neural network mapping from our input variables to the qualitative number of alternative solutions, we use the built-in function *feedforwardnet* in MATLAB's Statistical and Machine Learning Toolbox. This algorithm produces a network with 10 hidden layers, with the initial layer connected directly to the input variables and the output layer connected to the number of alternative optima [6]. We construct and compare the performance of each machine learning tool.

In order to construct a classification tree and a neural network to predict the number of alternative solutions to a MCSCP, we study the characteristics of a large sample of 10×20 constraint matrices with varying density. We only use 10×20 constraint matrices because using larger matrices would require a prohibitive amount of time to generate all optimums for each matrix in the training set. We create five sets of matrices, each with an identifying density of $p = 10\%, 15\%, 20\%, 25\%$, and 30%. We implement an algorithm in MATLAB that generates 50,000 unique, random matrices assuming that each row and each column is nonempty. For example, a matrix of density $p = 10\%$ will contain 20 nonzero entries, but since every column must contain a one, there will be exactly one nonzero entry in each column. Our MATLAB algorithm sorts each matrix using the sorting technique described by Emerick et al. [2]. This ensures that every generated MCSCP is unique. For more details on generating unique matrices, see [2].

A matrix with $p = 10\%$ density presents a simple case. Here, the MCSCP has exactly one nonzero entry in each column. Any MCSCP solved with a constraint matrix of this form has a minimum cardinality of 10 because there is only a single one in every column; therefore, it will take exactly m columns to cover each row. Since we are letting $m = 10$ in our study, the minimum cardinality of such an MCSCP is 10. Furthermore, in the scenario where there are exactly two ones in every row, there are two columns available to cover a single row. Hence, there are 2^{10} alternative optima. This is the maximum number of alternative optima in the $p = 10\%$ case. In contrast, if there is a single one in 9 rows, and 11 ones in a single row, the number of alternative optima is exactly 11. This scenario gives the minimum number of alternative optima in the $p = 10\%$ case. In both scenarios described above and in every case when $p = 10\%$, the number of alternative optima is exactly equal to the product of the row sum. However, the number of alternative solutions for any MCSCP of this size, with $p \neq 10\%$, cannot easily be determined by simply examining the form of the matrix A. Therefore, we seek a method to generate an accurate number of alternative solutions to any

Table 1. Summary statistics for the number of alternative optimal solutions of the global data set.

Mean	Std. Dev.	Five Number Summary					Outlier Threshold	Proportion in U_{min}	Proportion in U_{max}
		Min	Q_L	M	Q_U	*Max*			
65.50	109.28	1	4	12	55	1024	131.5	10.50%	19.39%

MCSCP, with a 10×20 constraint matrix. To this end, we study a global data set of 250,000 unique, random matrices spanning five density categories.

The summary statistics for the number of alternative solutions for all 250,000 unique matrices are given in Table 1. The maximum number of solutions is 1024, which is the case for the $p = 10\%$ density scenario with two nonzero entries in each row. The minimum value is 1 i.e., the problem has a unique solution. The sets U_{min} and U_{max} are defined as the set of all MCSCPs with exactly one solution and the set of all MCSCPs that have a very large number of alternative solutions as determined by the $1.5IQR$ rule for determining outliers, respectively. The $1.5IQR$ is a commonly used threshold for identifying outliers in a dataset. It is calculated by adding one and a half times the interquartile range (IQR) to the upper quartile of the dataset. Approximately 10.50% fall in the U_{min} category while 19.39% fall within U_{max}, which is defined by an outlier threshold of 131.50. Further, the mean and standard deviation are 65.50 and 109.28, respectively.

A frequency distribution of the data shows a heavy right skew. This fact will dictate our qualitative definition of small or large number of alternative solutions to any MCSCP.

We wish to construct a classification tree and a neural network to predict the number of alternative optima for any MCSCP. To this end, we consider 31 decision variables, two of them being the minimum cardinality, m, and the density, p. Other variables are associated with row and column statistics as well as eigenvalues and the general placement of ones in the sorted matrix A. A complete list of all variables is included in Table 2.

We first perform a principal component analysis (PCA) on our variable set [3]. In the initial stage of the PCA, x_8 is removed because it turns out to be a constant. When the threshold of 80% in cumulative proportion is used to select proper decision variables, all variables except x_8, x_{14}, and x_{25} are included to the first three principal components, which contribute most (see Fig. 1). A variable is excluded if the magnitude of the eigenvectors contributing to the first three principal components is less than a threshold of 0.2.

We also perform a correlation analysis by using a correlogram from Minitab on the set of all 30 variables from the PCA above. We find that the variables x_{16}, x_{21}, and x_{22} can be eliminated because they are perfectly linear with other variables (a correlation of $|r| = 1$). We also identify all variables that have a correlation of $|r| > 0.95$ and those with a correlation of $|r| > 0.90$ with at least one other variable. Appropriate variables are removed in each stage of this correlation analysis. These variables are identified in Table 2.

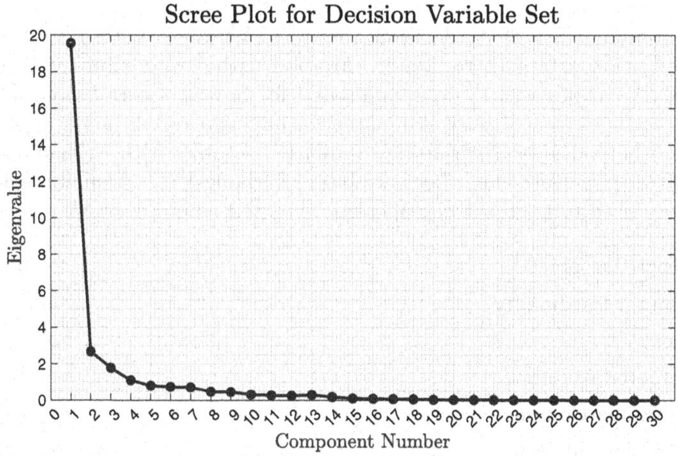

Fig. 1. Scree plot illustrating the contribution of each principal component in terms of eigenvalue. The three most dominant principal components, in terms of eigenvalues, capture 80% of the variation.

After all eliminations of variables are performed, we are left with 14 variables that are relatively less-correlated and contribute to the principal components of the data set.

4 Classification Trees

We consider two versions of a 3-output classification tree generated using the built-in function known as *fitctree* in the Statistical and Machine Learning Toolbox of MATLAB. We generate the tree using a random selection of 200,000 of the 250,000 matrices as a training set. We note that the global set consists of 50,000 matrices selected from each density, *p*. We do not select an equal number of matrices from each density for the training set. Table 2 shows the list of the 31 original decision variables. All variables that were eliminated from the principal component analysis and correlation analysis are appropriately highlighted. We proceed with 14 decision variables.

For the 3-output classification tree, the categories are defined as small (S), medium (M), and large (L). The cutoff values for each category are based on specific percentiles of the global data set. A small number of alternative optima is considered to be less than the 33rd percentile of 6 solutions. Therefore, if the MCSCP has 5 or less alternative solutions, the classification tree will identify that problem as having a small number of alternative optima. A large number of alternative solutions is defined as more than 120 solutions, which is the 80th percentile. This threshold value was chosen because the data are strongly skewed to the right. A medium number of alternative solutions is between 6 and 120 solutions.

Table 2. Complete list of 31 decision variables. Variables with † were eliminated because they are constant (1 variable). Variables with ‡ were eliminated using PCA (2 variables). Variables with * were eliminated for having a correlation of $|r| = 1$ (3 variables). Variables with ** were eliminated for having a correlation of $|r| > 0.95$ with at least one other variable (7 variables). Similarly, variables with *** had a correlation of $|r| > 0.90$ with at least one other variable (4 variables). The remaining 14 decision variables were used to train the classification tree and neural network.

Decision Variable	Notation
Minimum Cardinality	m
Density	p
Dominated Columns	x_1
Dominating Columns	x_2
Single Element Rows	x_3^{***}
Single Element Columns	x_4^{**}
Maximum Column Sum of Single Element Row	x_5
Isolated Points	x_6^{***}
Duplicate Columns	x_7^{**}
Rows with Maximum Row Sum	$x_8^{\dagger}$
Proportion of Nonzero Elements in AA^T	x_9^{**}
Maximum Eigenvalue of AA^T	x_{10}^{**}
Maximum Element of AA^T	x_{11}
Elements in the Northwest Quadrant of Sorted A	x_{12}
Elements in the Northeast Quadrant of Sorted A	x_{13}^{***}
Elements in the Southwest Quadrant of Sorted A	$x_{14}^{\ddagger}$
Elements in the Southeast Quadrant of Sorted A	x_{15}
Mean of Row Sum	x_{16}^{*}
Standard Deviation of Row Sum	x_{17}
Median of Row Sum	x_{18}^{**}
Interquartile Range of Row Sum	x_{19}
Minimum of Row Sum	x_{20}
Maximum of Row Sum	x_{21}^{*}
Mean of Column Sum	x_{22}^{*}
Standard Deviation of Column Sum	x_{23}^{**}
Median of Column Sum	x_{24}^{***}
Interquartile Range of Column Sum	$x_{25}^{\ddagger}$
Minimum of Column Sum	x_{26}
Maximum of Column Sum	x_{27}^{**}
Product of Row Sum	x_{28}
Product of Column Sum	x_{29}

We train the classification tree on a random selection of 200,000 unique matrices from the original global set of 250,000 MCSCPs. We consider the performance of the full classification tree as created by MATLAB and the pruned tree with only 10 branches. The results of the two trees are presented in Table 3.

We let S_a denote a MCSCP that actually had a small number of alternative solutions, and S_p be the number of MCSCP's that were predicted by the model to have a small number of optimal solutions. Similar definitions hold for the other sizes. We compute the sensitivities (e.g. $P(S_p \mid S_a)$) and the positive predictive value (PPV) (e.g. $P(S_a \mid S_p)$) for each output in each model. From the confusion matrices, we can see that the pruned classification tree performs slightly better than the full tree on the validation set. Indeed, the overall accuracy of the pruned tree was 83.63% as compared to the full tree at 82.54%. This fact suggests that using fewer branches and fewer decision variables is a benefit.

The results of Table 3 present the performance of each model at a glance. We have constructed a classification tree using a single training and validation set. To get a better understanding of the performance of the model, we perform a Monte Carlo Cross Validation (MCCV) with 100 simulations. In this way, we can gain a more accurate depiction of the performance by considering the mean of the accuracies. Because of high computational cost, we only do a performance analysis on the pruned tree. We construct a pruned tree using 100 training sets, each set chosen randomly without replacement with each of size 200,000 MCSCPs. The validation sets are the remaining 50,000 MCSCPs for each of the 100 trials. The average accuracy of the 10-node, pruned classification tree is 83.84% with a standard deviation of .21%.

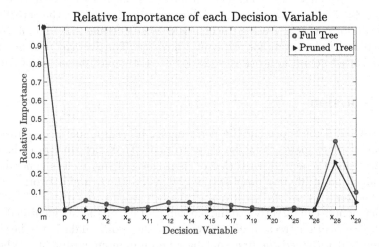

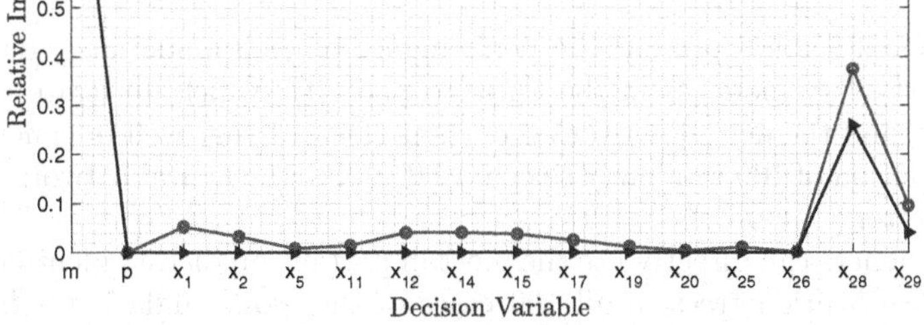

Fig. 2. A graphical depiction of the relative importance of each decision variable for the full (red) and pruned (blue) classification tree. We see that m, x_{28}, and x_{29} are the top three deciders for branch points in each tree. (Color figure online)

For each classification tree, we find that the most important predictors are minimum cardinality (m), product of the row sum (x_{28}), and product of the column sum (x_{29}). This fact is depicted in Fig. 2. We see that for the pruned tree, only these three variables are used to determine branch points. In the full tree, variables other than these three important predictors are used after the tenth branch point. We also note that density (p) and minimum of the column sum (x_{26}) have essentially no predictive power in the full model. This is in direct contrast to Emerick et al. [2] whose study found that the number of single element rows (x_3) and proportion of nonzero elements in AA^T (x_9) were the most important variables after minimum cardinality.

5 Neural Network

We train a neural network using the built-in function *feedforwardnet* in MATLAB on the exact same set of MCSCPs used to train the classification trees. We use the default number of hidden layers, in this case 10. The output of the neural network is numerical and so we reclassify the output of the model into the predetermined bin sizes for S, M, and L. In this way, we can appropriately compare the performance of each model. The training process was completed after 492 epochs and 6 validation checks.

The performance of the model is given in Table 3. Here, we can see that the neural network, at a glance, does slightly better at an accuracy of 84.82%. Similar to the tree models, the neural network is best at predicting a larger number of optima. The sensitivity of the neural network model for predicting a small number of alternative solutions seems to be significantly higher than that of the tree models. Similarly to the cross validation methods used for the classification tree, we perform a MCCV using the identical set of 100 randomly chosen, without replacement, training sets for the neural network. The average accuracy of the cross-validated neural network is 84.28% with a standard deviation of .79%.

6 Discussion

The results of Table 3 present the performance of each model at a glance. The accuracy of every model is in the lower to mid 80's, which illustrate that both the CART and the neural network models performed very well in general. After cross-validation of 100 randomly selected MCSCPs, the neural network performs slightly better.

Practitioners can directly use the most important predictors identified by the pruned classification trees in all 100 cases of the cross-validation. These trees consistently determined the first split based on minimum cardinality. When the minimum cardinality (m) is greater than 7, the MCSCP is likely to have a large number of solutions (i.e., greater than 120 alternative optima). For this branch alone, the PPV given by $P(L_a \mid L_p)$ is .9624, which is impressive for a validation set of this size. Furthermore, in this set of 100 random MCSCPs, having a minimum cardinality less than 4 will yield a small number of solutions (i.e., less than

Table 3. Confusion matrices for the full classification tree (top), pruned classification tree (middle), and neural network (bottom) on the validation set. The sensitivities, predicted values, and overall accuracy are provided for each case.

Full Tree:

		Predicted S_p	Predicted M_p	Predicted L_p	**Sensitivities:**
Actual	S_a	12090	4189	0	.7427
Actual	M_a	4271	19372	106	.8157
Actual	L_a	1	164	9807	.9835
PPV:		.7389	.8165	.9893	**.8254**

Pruned Tree:

		Predicted S_p	Predicted M_p	Predicted L_p	**Sensitivities:**
Actual	S_a	12684	3595	0	.7792
Actual	M_a	4174	19574	1	.8242
Actual	L_a	0	415	9557	.9584
PPV:		.7524	.8300	.9999	**.8363**

Neural Network:

		Predicted S_p	Predicted M_p	Predicted L_p	**Sensitivities:**
Actual	S_a	12089	4190	0	.7426
Actual	M_a	2970	20734	45	.8730
Actual	L_a	0	385	9587	.9614
PPV:		.8028	.8192	.9953	**.8482**

6 alternative optima) with a PPV of $P(S_a \mid S_p) = .8712$. Based on these findings, we can conclude that minimum cardinality is the most important predictor for determining whether there will be a small or large number of alternative optima. In addition, variables x_{28} (product of the row sum) and x_{29} (product of the column sum) are typically used to distinguish between small and medium-sized solutions in the lower branches of the full tree. Typically, the number of alternative optima increase with an increase in each of these variables. The fact that x_{28}, the product of the row sum, is an important predictor is not surprising since this value is exactly the number of alternative optima in the $p = 10\%$ case. This is in contrast to the results reported by Emerick et al. [2], who found that x_3 and x_8 were valuable for lower branches.

We consider the sensitivity for each size, i.e., $P(S_p \mid S_a)$ is the probability that the MCSCP is predicted to have a small number of alternative solutions when the model is actually in the group of small size. The sensitivity is for comparing "between models". To be precise, the sensitivity can be used to compare the CARTs and the neural network in each output (small, medium, large) because it shares the same "ballpark" (the actual count of output) in all models when calculated. In each model, the sensitivity is increasing in the order of small, medium, and large (see Table 3).

We also consider the positive predictive value (PPV) for each size, i.e., $P(S_a \mid S_p)$ is the probability that the MCSCP actually has a small number of alternative solutions when the model predicted it to be small. The PPV is for comparing "within a model". To be more specific, the PPV can be used compare each output if you fix the model of interest. In each model, the PPV is increasing in the order of small, medium, and large (see Table 3).

To make comparisons between the classification tree and the neural network, the authors originally tried to develop a random forest, but wound up with computing resource shortage – more than 230 Gb of RAM was needed, which cannot be satisfied in a standard PC. As a result, we used a simplified version of a random forest - CART as an alternative.

7 Conclusions and Future Work

In this paper, we have considered the performance of two different machine learning tools on predicting qualitatively the number of alternative optimal solutions to any MCSCP with 10 constraints and 20 variables. We constructed each model from a reduced set of 14 decision variables and tested each model using identical training and validation sets from a global set of 250,000 random MCSCPs.

For future work, we would like to develop a method for comparing the overall performance of both models and to consider a random forest CART model. We would also like to analyze how the lower threshold of the definition of "large" number of alternative optima affects the accuracy each model. In our analysis, we use the 80th percentile or 120 or more alternative solutions to define a large number of solutions. How does changing this definition influence the predictive nature of each model? The answer to this question may provide further insight into the number of alternative optima of any given MCSCP.

References

1. Bixby, R.E.: A brief history of linear and mixed-integer programming computation. Doc. Math. Extra Vol. ISMP 107–121 (2012)
2. Emerick, B., Lu, Y., Vasko, F.: Using machine learning to predict the number of alternative solutions to a minimum cardinality set covering problem. Int. J. Industr. Optim. **2**(1), 1–16 (2021)
3. Jolliffe, I.T., Cadima, J.: Principal component analysis: a review and recent developments. Philos. Trans. Roy. Soc. A: Math. Phys. Eng. Sci. **374**(2065) (2016)
4. Karp, R.M.: Reducibility among combinatorial problems. In: Miller, R.E., Thatcher, J.W. (eds) Complexity of Computer Computations, Plenum, NY, pp. 85–103 (1972)
5. Koch, T., Berthold, T., Pedersen, J., Vanaret, C.: Progress in mathematical programming solvers from 2001 to 2020. EURO J. Comput. Optim. **10**, 1–17 (2022)
6. MATLAB: Statistics and Machine Learning Toolbox Users Guide. R2020a, 1 Apple Hill Dr, Natick, MA 01760-2098 (2020)

7. Newhart, D.D., Stott, K.L., Vasko, F.J.: Consolidating product sizes to minimize inventory levels for a multi-stage production and distribution system. J. Oper. Res. Soc. **44**, 637–644 (1993)
8. Vasko, F.J., Wolf, F.E., Stott, K.L.: Optimal selection of ingot sizes via set covering. Oper. Res. **35**, 346–353 (1987)
9. Vasko, F.J., Wolf, F.E., Stott, K.L.: A set covering approach to metallurgical grade assignment. Eur. J. Oper. Res. **38**, 27–34 (1989)
10. Vasko, F.J., Wolf, F.E., Stott, K.L., Ehrsam, O.: Bethlehem Steel combines cutting stock and set covering to enhance customer service. Math. Comput. Modell. **16**(1), 9–17 (1992)

Adaptative Local Search for a Pickup and Delivery Problem Applied to Large Parcel Distribution

Matthieu Fagot[1,2(✉)], Laure Brisoux Devendeville[1], and Corinne Lucet[1]

[1] Laboratoire MIS (UR 4290), Université de Picardie Jules Verne, 33 Rue Saint-Leu, 80039 Amiens Cedex 1, France
{m.fagot,laure.devendeville,corinne.lucet}@u-picardie.fr
[2] Smile Pickup, 80440 Boves, France
mfagot@smilepickup.com

Abstract. This paper introduces an Adaptive Large Neighborhood Search algorithm that uses an epsilon-greedy movement selection strategy to solve a pickup and delivery problem for Smile Pickup, a real-life business. The algorithm also takes into account multiple time windows, heterogeneous fleets, and multiple depots as additional constraints. The algorithm utilises two diversification processes: a simulated annealing technique to update the current solution, and an epsilon-greedy strategy to balance between exploration and exploitation for the selection of neighbourhoods. We evaluated the algorithm's performance using our own benchmark PickOptBench and Li & Lim benchmarks, and found that it shows great promise in solving Smile Pickup's problem. Moreover, combining both the epsilon-greedy and simulated annealing restart strategies resulted in a 1% improvement in ALNS performance on both benchmarks. We also discovered that the algorithm found more than 70% of the best-known solutions for 4 out of the 6 classes of instances in the Li & Lim benchmark.

Keywords: Metaheuristic · Pickup and Delivery Problem · Time Windows · Reinforcement Learning

1 Introduction

In 2020 alone, the number of parcels shipped in France has increased by 12.4% to reach 1.5 billion[1]. This phenomenon can be seen all over the world and has been reinforced by the recent sanitary crisis and successive lock downs. This growth has seen the explosion of a business: pickup points. More environmentally-friendly than home deliveries and often more practical, these pickup points are

[1] https://www.data.gouv.fr/fr/datasets/observatoire-du-courrier-et-du-colis/.

CIFRE n° 2021/0599 between Smile Pickup and MIS Laboratory.

B. Dorronsoro et al. (Eds.): OLA 2023, CCIS 1824, pp. 186–199, 2023.
https://doi.org/10.1007/978-3-031-34020-8_14

usually held by convenience stores close to customers. Although this works perfectly for small parcels, bigger ones such as furniture are usually not accepted either because of size, weight or space.

Smile Pickup is a young business which manages a group of pickup points dedicated to big parcels. With such an activity comes a logistical challenge for shipping parcels from stores to pickup points using local transport solutions. This sets up a vehicle routing problem with specific constraints that need to be taken into account. The objective is to give our partners customers the choice to be delivered in one of our pickup points accommodated for receiving oversized parcels. The store then packs the order for our fleet of vehicles to deliver to the pickup point chosen by the customer as soon as possible.

In this paper we describe the particular Vehicle Routing Problem (VRP) faced by Smile Pickup: the *Smile Pickup Problem* (SPP) which combines different well known vehicle routing problems. SPP is part of the class of paired vehicle routing problems with pick up and delivery [1]. As SPP is an extension of the classic VRP and Pickup and Delivery Problem (PDP) problem [2], it can be classified as an NP-hard problem. The core of SPP is similar to the Pickup and Delivery Problem with Time Window (PDPTW) described by Li and Lim [3] with which they provide a benchmark. Regarding the time windows, Smile Pickup needs to be able to ensure great flexibility. For this purpose, we added the possibility of multiple time windows. This constraint was first proposed by DeJong et al. [4] to take into account customer brakes on home deliveries. More recently, Belhaiza [5] and Ferreira [6] have both proposed variable neighbourhood search heuristics to solve the vehicle routing problem with multiple time windows. Additional constraints such as multiple depots and an heterogeneous fleet will also be considered in our problem as described in Salhi and al. [7]. If we look at pickup and delivery problems, we can refer to the Multi-depot Dial-A-Ride Problem with Heterogeneous Vehicles (M-DARP-HV) by Braekers [8] or more recently by Detti [9]. Dial a ride problems are close to PDP problems except that they transport people and not goods. Braekers [8] gives an exact method while Detti's article [9] gives a detailed integer linear program, a tabu search algorithm and multiple variable neighbourhood searches all tested on real life instances. Our problem differs from M-DARP-HV which incorporates constraints related to the quality of service provided to patients such as a bound on the patient travel time. This makes it challenging to compare both problems.

The main features of SPP which distinguish our problem from those encountered in the literature are: stores and pickup points can be loading places and unloading places for parcels at the same time - moreover, parcels can share their origins and destinations. This prevents us from using classical exploration methods of the solution space.

The remainder of this paper is organised as follows. Section 2 gives a detailed description of Smile Pickup's vehicle routing Problem (SPP). Section 3 presents our $ALNS^{\epsilon}_{SA}$ algorithm. We detail movements and its two diversification processes: an acceptance criterion based on simulated annealing and an epsilon greedy exploration strategy. Results follow in Sect. 4 and conclusion in Sect. 5.

2 Problem Description

In this section, a detailed description of the problem faced by Smile Pickup is given. First, the problem's data is presented before describing the chosen representation of the solutions and describing the objective. The example in Fig. 1 will be used throughout this article to illustrate the problem.

2.1 Data

The problem faced by Smile Pickup spans over a total of $H \in \mathbb{N}$ consecutive days $J = \{1, ..., H\}$ during which parcels C need to be transported between places Ω using vehicles V.

places	0, 1	2	3	4	5	6
time windows	[6, 22] [0, 0]	[8, 10] [14, 16]	[8, 13] [0, 0]	[10, 16] [0, 0]	[8, 13] [0, 0]	[10, 12] [14, 18]
types	depot	store	store	pickup point	pickup point	pickup point

parcels	c_0	c_1	c_2	c_3	c_4	c_5
s_r	1	1	1	1	1	1
o_r	2	3	2	3	2	6
d_r	4	5	5	6	6	3
f_r^c (min)	5	7	1	1	5	7
f_r^d (min)	5	4	1	1	5	4
day_r	1	1	1	1	1	1

vehicles	Pu_v	K_v	o_v	τ_v
v_0	800	2	0	0
v_1	800	2	1	1

Legend for places:
▲ Depots
■ Stores
● Pickup Points

Fig. 1. Example of the data of an instance

Places. The set of places Ω is divided in three subsets: depots D where vehicles start and end their day, stores E and pickup points P which exchange parcels. A travel distance d_{ij} and duration m_{ij} is associated to each arc between places $(i, j) \in \Omega^2$. For day $\tau \in J$, a set of time windows $\{[e_{ik}^{\tau}, f_{ik}^{\tau}] | k \in \{1, ..., k_{\max}\}\}$ is assigned to each place $i \in \Omega$. $k_{\max}$ is the maximum number of time windows per place. Unused time windows are set to $[0, 0]$. When a vehicle visits a place, it must load and unload its parcels during one of the associated time windows. The vehicle can arrive early at a place even though it will have to wait until a time window opens before starting loading and unloading. For depots, the time windows model the opening hours during which vehicle may depart and return.

Parcels. A parcel $c_r \in C$ of length s_r is made available in a store or a pickup point $o_r \in E \cup P$ and needs to be delivered to it's destination $d_r \in E \cup P$. parcel c_r can be picked up and delivered from day $day_r \in J$ but can also be stored in o_r and serviced on a later day $\tau \geq day_r$ for a penalty cost p_r^{τ}. Solutions do not need to deliver all parcels but each undelivered parcel will cost P_{NL}. Furthermore, the time needed to load (resp. unload) a parcel c_r is f_r^c (resp. f_r^d).

Vehicles. For each day $\tau \in J$, a set of vehicles V_τ is available ($V = \bigcup_{\tau \in J} V_\tau$). Each vehicle $v \in V$ starts at a depot o_v and returns at the same depot at the end of the day. Vehicle $v \in V$ is given a usage cost Pu_v and a capacity K_v. The sum of the length of the parcels in a vehicle v must not exceed K_v at any time during the tour. Finally, we will call $\tau_v \in J$ the day vehicle v is associated to. Hence $v \in V_{\tau_v}$

In Fig. 1, an example of data is given. The instance spans on a single day $J = 1$ using 2 vehicles $V = V_1 = \{v_0, v_1\}$, 6 parcels $C = \{c_0, c_1, ..., c_5\}$ and 7 places including depots $D = \{0, 1\}$, stores $E = \{2, 3\}$ and pickup points $P = \{4, 5, 6\}$. As for example, parcel c_5 of length $s_5 = 1$ needs to be picked up at point 6 and delivered at store 3. Loading will take $f_5^c = 7$ min while unloading only takes $f_5^d = 4$ min. If vehicle v_0 is used, it starts from depot 0 and costs $Pu_0 = 800$. The vehicle would need to load at point 6 either between time slots 10 and 12 or 14 and 18. Unloading in 3 must take place between 8 and 13.

2.2 Solution Representation

A solution S is represented by a set of tours. A tour is assigned a single vehicle. We will consider a vehicle to be equivalent to a tour and use the same notation v. A tour $v \in V$ is an ordered list of triplets $<t_0^v, t_1^v, ..., t_k^v>$. Triplet i of tour v is such that $t_i^v = (p_i^v, C_i^{v,+}, C_i^{v,-})$ with $p_i^v \in \Omega$, $C_i^{v,+} \subseteq C$ the set of parcels loaded at p_i^v by v and $C_i^{v,-} \subseteq C$ the set of parcels unloaded at p_i^v by v. During servicing of a place, unloading will always be performed before loading. We notice that $\forall v \in S$, $p_0^v \in D$, $p_k^v \in D$, $p_0^v = p_k^v$, $C_0^{v,+} = C_0^{v,-} = \emptyset$ and $C_k^{v,+} = C_k^{v,-} = \emptyset$.

We also introduce the following notations: C^u the set of undelivered parcels, C^d the set of delivered parcels, V^+ the set of used vehicles and the function $d : C \to J$ indicating the day $d(c_r) \in J$ parcel $c_r \in C^d$ is delivered. For simplicity, we also introduce $dist(v)$ the distance travelled by vehicle $v \in V^+$.

Furthermore, a solution S is feasible if capacity constraints are respected and if for every tour, there exists at least one schedule for the associated vehicle to be able to respect the time windows of every place visited by the tour.

Using this notation we can represent the solution in Fig. 2 as follows:

$$v_0 = <(0, \emptyset, \emptyset), (2, \{c_4\}, \emptyset), (6, \emptyset, \{c_4\}), (0, \emptyset, \emptyset)>$$
$$v_1 = <(1, \emptyset, \emptyset), (3, \{c_1\}, \emptyset), (5, \emptyset, \{c_1\}), (1, \emptyset, \emptyset)>$$

In the representation in Fig. 2, the sets $C_i^{v,+}$ and $C_i^{v,-}$ are listed under place p_i. If we look at Fig. 3 presenting a solution containing all parcels, the first vehicle loads c_5 when visiting pickup point 6 for the first time. It then goes to store 3 where c_5 is unloaded and c_3 is loaded. After having picked up parcel c_4 in store 3, both c_4 and c_3 are unloaded when visiting pickup point 6 once again. The vehicle finishes his journey by coming back to depot 0.

2.3 Solution Evaluation

The objective of SPP is to find the solution S which minimises the following criteria: the cost of vehicles used, the number of undelivered parcels, the total

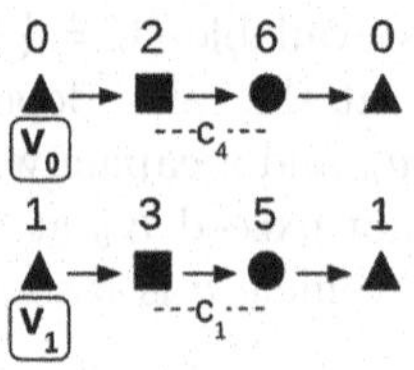

Fig. 2. An initial solution.

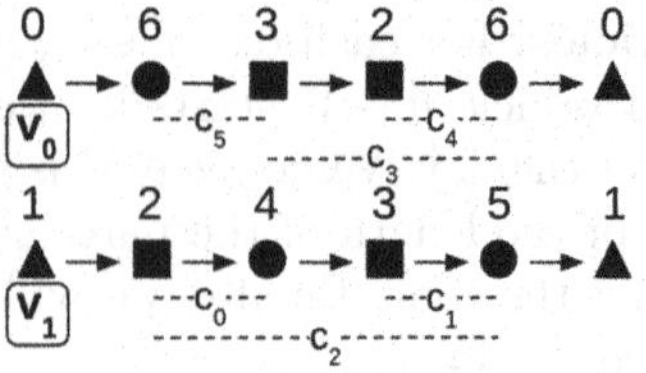

Fig. 3. An optimal solution.

distance travelled and the storage penalties. To normalise and prioritise these criteria, they are assigned weights $\alpha_{\text{veh}}, \alpha_{\text{nl}}, \alpha_{\text{dist}}$ and α_{pen}. The fitness function is shown in Eq. 1.

$$f(S) = \alpha_{\text{veh}} \sum_{v \in V^+} PU_v + \alpha_{\text{nl}} \, |C^u| + \alpha_{\text{dist}} \sum_{v \in V^+} \text{dist}(v) + \alpha_{\text{pen}} \sum_{c_r \in C^d} p_r^{d(c_r)} \quad (1)$$

3 An Adaptative Large Neighborhood Search: $\text{ALNS}^{\epsilon}_{\text{SA}}$

This section outlines the various steps involved in the proposed local search algorithm $\text{ALNS}^{\epsilon}_{\text{SA}}$ which aims at solving the Smile Pickup problem. Such a method operates by exploring the solution space by generating a set of neighbours from the current solution S and choosing one of them as the new current solution. Neighbourhoods are generated by movements dedicated to the specific problem considered. The process is iterated and the best solution S_B found is returned. After testing different local search algorithms as the Variable Neighbourhood Search (VNS) and the Large Neighbourhood Search (LNS), we selected the Adaptative Large Neighbourhood Search (ALNS) [10], an extension of the LNS in which algorithms learn how to move in the search space more efficiently. The method we use, iterates over a deterioration phase and a reconstruction phase. Neighbours are generated from specific movements of two different types: deteriorating movements and constructive movements which we will describe in Sect. 3.2.

3.1 General ALNS Algorithm

Adaptative Large Neighbourhood Search algorithm dictates the general strategy used to decide which neighbouring solution to move to at each iteration. A general scheme is presented in Algorithm 1. The algorithm starts by generating an initial solution using a greedy algorithm before iterating over the following four steps until the time limit is exceeded.

The first step is the deterioration phase (lines 5 to 7). One of the three deteriorating movements (see Sect. 3.2) is chosen and applied to the current

Algorithm 1 general scheme of ALNS algorithm

1: Nb_iterations $\leftarrow$ 0
2: S $\leftarrow$ greedy()
3: **while** stopping criteria **do**
4: Nb_iterations++
5: $S_P \leftarrow S$
6: $\mathcal{N}_s \leftarrow$ choose a deteriorating movement using distribution W_S
7: Apply $\mathcal{N}_s$ to S_P
8: **while** stopping insertion criteria **do**
9: $\mathcal{N}_s \leftarrow$ choose a constructive movement using distribution W_I
10: Apply $\mathcal{N}_s$ to S_P
11: **end while**
12: Update current solution S according to S_P.
13: Rewards (π_S^m, π_I^m) and counts (θ_S^i, θ_I^i) movements used in iteration.
14: **if** Nb_iterations mod $\Delta = 0$ **then**
15: Update movement weights W_S^m and W_I^m according to rewards and movement counts.
16: **end if**
17: **end while**

solution S_P. Each deteriorating movement m is given a weight W_S^m. Weights are used to select movements.

The second step is the constructive phase (lines 8 to 11) which rebuilds the solution by inserting unassigned parcels by performing constructive movements (see Sect. 3.2). These movements are selected and applied based on their weights W_I^m. While deteriorating movements are applied once, constructive movements are applied iteratively. The process stops when E_{max} successive movements fail to produce a feasible solution.

In the deteriorating phase as in the constructive phase, two different strategies are tested. The first one is the classical roulette wheel selection which is performed following the distribution W_S (resp. W_I) to choose the movement to apply. The second one is the epsilon greedy strategy detailed in Sect. 3.4.

The next step decides if this newly built solution is worthy enough to become the new current solution for the next iteration (line 12). It also classifies the performance of the iteration based on the solution produced. This classification will be used in the weight adjustment step. This step is detailed on Sect. 3.3.

The final step (lines 13 to 18) adapts the weights as follows: each movement applied successfully is rewarded based on the classification given in the previous step. π_S^m (resp. π_I^m) counts the rewards earned by the deteriorating (resp. constructive) movement m while θ_S^m (resp. θ_I^m) counts the number of times it was successfully applied. Each Δ iterations, the weight distributions are corrected using the formula in Eq. 2 with P_S the sum of $\frac{\pi_S^m}{\theta_S^m}$ over the set of deteriorating movements. The weights W_I^m are updated in the same way as W_S^m.

$$W_S^m \leftarrow \rho \frac{\pi_S^m}{\theta_S^m P_S} + (1-\rho) W_S^m \tag{2}$$

3.2 Movements

In the following section, the six movements created to move from neighbour to neighbour in the solution space are detailed.

Parcel Insertion Movement: *pim*
Movement *pim* first randomly selects a parcel $r \in C^u$ and tries to insert it in a tour. Candidate tours are classified in four sets ξ_j, $j = 1, \dots, 4$. The first one, ξ_1 contains tours that visit both origin o_r and destination d_r of c_r in the right order. Next, ξ_2 contains tours visiting origin o_r and ξ_3 contains those that visit only the destination d_r. Finally, ξ_4 contains tours that visit neither o_r nor d_r. Tours where d_r precedes o_r are included in ξ_2.

Let ξ_k be the first non-empty set, then operator *pim* will randomly pick out in ξ_k a tour v. Depending on ξ_k, the following processes are applied to v:

- case $\xi_k = \xi_1$. Choose a triplet $t_i^v \in v$ and $t_j^v \in v$ such that $p_i^v = o_r$, $p_j^v = d_r$ and $i < j$. Add c_r to $C_i^{v,+}$ and to $C_j^{v,-}$.
- case $\xi_k = \xi_2$. Choose triplet $t_i^v \in v$ with $p_i^v = o_r$. Add c_r to $C_i^{v,+}$. Choose t_j^v with $j \geq i$. Insert the triplet $(d_r, \emptyset, \{r\})$ in tour v between t_j^v and t_{j+1}^v.
- case $\xi_k = \xi_3$. Choose triplet $t_j^v \in v$ with $p_j^v = d_r$. Add c_r to $C_j^{v,-}$. Choose t_i^v with $i < j$. Insert triplet $(o_r, \{c_r\}, \emptyset)$ in tour v between t_i^v and t_{i+1}^v.
- case $\xi_k = \xi_4$. Choose a triplet t_i^v. Insert the triplet $(o_r, \{c_r\}, \emptyset)$ in tour v between t_i^v and t_{i+1}^v. Choose a second triplet t_j^v, $i \leq j$. Insert the triplet $(d_r, \emptyset, \{c_r\})$ in tour v between t_j^v and t_{j+1}^v.

When c_r is added to $C_i^{v,+}$, if capacity constraints are violated, then movement *pim* is rejected. In the same manner, when triplets $t_i^v = (o_r, \{c_r\}, \emptyset)$ and/or $t_j^v = (d_r, \emptyset, \{c_r\})$ are inserted in tour v, if time window constraints are violated, the movement is rejected.

Consider now the initial solution in Fig. 2 and $c_0 \in C^u$ the parcel to insert ($o_0 = 2$ and $d_0 = 4$). Then the four sets are: $\xi_1 = \emptyset$, $\xi_2 = \{v_1\}$, $\xi_3 = \emptyset$ and $\xi_4 = \{v_2\}$. We add c_0 to $C_2^{v_1,+}$ and insert the triplet $t_4^{v_1} = (d_{c_0}, \emptyset, \{c_0\})$ in tour v_1. The resulting tour is illustrated in Fig. 4(b).

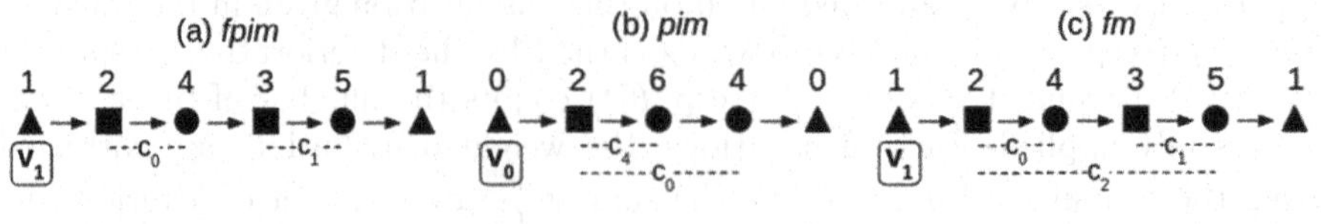

Fig. 4. Examples of solutions after the application of constructive movements.

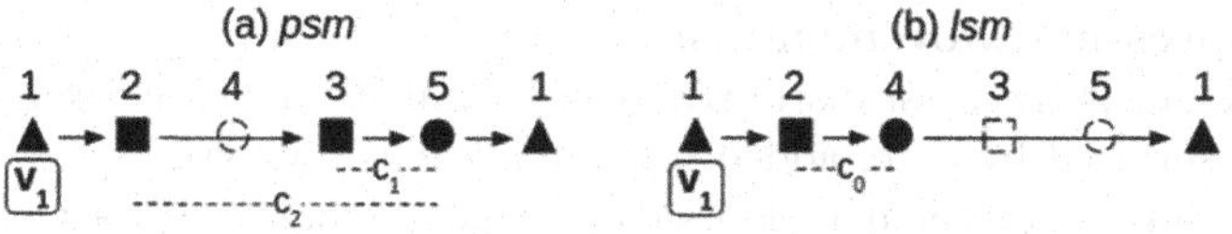

Fig. 5. Examples of solutions after the application of deteriorating movements.

Forced Parcel Insertion Movement: *fpim*
Movement *fpim* starts by randomly selecting a parcel $c_r \in C^u$ and a tour $v \in V$. Then two triplets t_i^v and t_j^v $(0 \leq i \leq j < k)$ are randomly selected. Next, triplets $(o_r, \{c_r\}, \emptyset)$ and $(d_r, \emptyset, \{c_r\})$ are respectively inserted after t_i^v and t_j^v in the right order. Capacity and time window constraints are checked on v. If unsuccessful, *fpim* is rejected. Figure 4(a) presents a possible outcome of the application of *fpim* on the solution of Fig. 2. c_0 is added to v_1 by inserting both $(2, \{c_0\}, \emptyset)$ and $(4, \emptyset, \{c_0\})$ after $t_0^{v_1}$ in the right order.

Fill Movement: *fm*
Movement *fm* randomly selects a vehicle $v = <t_0^v, ..., t_k^v>$ and one of its triplets $t_i^v = (p_i^v, C_i^{v,+}, C_i^{v,-})$ with $0 < i < k$. For every triplet $t_j^v = (p_j^v, C_j^{v,+}, C_j^{v,-})$ such that $0 < j < i$, let's define $C_{fm}^u = \{c_r \in C^u \mid o_r = p_j^v \text{ and } d_r = p_i^v\}$. Parcels of C_{fm}^u are added in a random order to both $C_j^{v,+}$ and $C_i^{v,-}$ as long as capacity and time window constraints are not broken. Ditto with triplets t_j^v where $i < j < k$, parcels of $C_{fm}^u = \{c_r \in C^u \mid o_r = p_i^v \text{ and } d_r = p_j^v\}$ are inserted in $C_i^{v,+}$ and $C_j^{v,-}$. Figure 4(c) gives an example with the application of *fm* on the solution in Fig. 4(a) where $C^u = \{c_2, c_3, c_5\}$. v_1 is selected as well as $t_1^{v_1}$. The only triplet $t_j^{v_1}$ $(1 < j \leq 4)$ where $C_{fm}^u \neq \emptyset$ is $t_4^{v_1}$: $C_{fm}^u = \{c_2\}$. Hence, c_2 is added in $C_1^{v_1,+}$ and $C_4^{v_1,-}$. No more parcel can be added because the tour is full with capacity $K_{v_1} = 2$.

Parcel Suppression Movement: *psm*
Movement *psm* starts by randomly selecting a parcel $c_r \in C^d$. Let be $t_i^v = (p_i^v, C_i^{v,+}, C_i^{v,-})$ and $t_j^v = (p_j^v, C_j^{v,+}, C_j^{v,-})$ such that $c_r \in C_i^{v,+}$ and $c_r \in C_j^{v,-}$. Movement *psm* modifies both triplets by removing c_r from $C_i^{v,+}$ and $C_j^{v,-}$. c_r is added to C^u. *psm* finishes by removing unused places $(C_i^{v,-} = C_i^{v,+} = \emptyset)$ and unused tours. Figure 5(a) shows the result of the application of *psm* on parcel c_0 of the solution in Fig. 4(c). c_0 is added to C^u and removed from $t_1^{v_1}$ and $t_2^{v_1}$. Because $C_2^{v_1,-} = C_2^{v_1,+}$ are empty, $t_2^{v_1}$ is removed from v_1.

Place Suppression Movement: *lsm*
Movement *lsm* starts by randomly selecting a tour $v \in V^+$ and one of its triplet $t_i^v = (p_i^v, C_i^{v,+}, C_i^{v,-})$. All parcels from $C_i^{v,+}$ and $C_i^{v,-}$ are removed from tour v. *psm* finishes by removing unused places $(C_i^{v,-} = C_i^{v,+} = \emptyset)$ and unused tours. Figure 5(b) shows the outcome of applying *lsm* to tour v_1 of the solution in Fig. 4(c). $t_4^{v_1} = (5, \emptyset, \{c_1, c_2\})$ is removed and $\{c_1, c_2\}$ are added to C^u. Triplet $t_3^{v_1}$ is now empty and hence removed from v.

Tour Suppression Movement: *tsm*
This movement is the most disruptive one. A tour $v \in V^+$ is chosen at random. All parcels serviced by v are added to C^u while v is removed.

A short term tabu memory is used to avoid cycling over a set of solutions. Parcels are set tabu when removed from a tour. Afterwards, the parcel can not be re-inserted in the same tour for δ iterations.

3.3 Solution Updating and Classification

After a neighbour S_P of S has been chosen by the first two steps of an iteration, we update the current solution S. Four cases are possible: (a) $f(S_P) < f(S_B)$ where S_B is the best solution visited so far, (b) $f(S_P) < f(S)$, (c) S_P satisfies the acceptance criterion and (d) S_P does not satisfy the acceptance criterion. When cases (a), (b) or (c) occur, S is replaced by S_P. Furthermore, when case (a) is met, S_B is updated with S_P. These cases are also used to choose the reward σ_i ($i \in \{1, \ldots, 4\}$) according to the quality of the solution in order to update the weights of the movements W_S and W_I [10].

The acceptance criterion is used to accept some solutions (case (c)) which degrade the fitness function and hence avoid getting stuck in local optima. Simulated annealing was chosen to manage the acceptance criterion (see Sect. 3.4).

3.4 Exploitation, Exploration and Learning Strategies

In this subsection, strategies to improve the performance of the ALNS are discussed including a simulated annealing strategy which accepts degraded solutions as well as an epsilon greedy movement selection strategy.

Simulated Annealing (SA). Simulated annealing is used as an acceptance criterion to allow degradation of the solutions. Indeed, in case (c) (see Sect. 3.3) the solution S_P is accepted with probability (see Eq. 3) where T is the temperature and S the current solution.

$$p(S_P) = e^{-\frac{f(S_P)-f(S)}{T}} \tag{3}$$

The temperature controls the range of solutions to be accepted with high probability. When T is high, worse solutions have higher chance of passing while when T is lower, only solutions with close fitness scores have a real chance of going through.

We tested two scenarios. The first one is the classical SA process where the temperature decreases progressively using a multiplicative coefficient γ_{SA}. The second one proceeds with restarts when the solution is not improved for $R^{\text{SA}}_{\text{step}}$ iterations.

Epsilon Greedy. The epsilon greedy strategy is used to balance between exploration and exploitation during movement selection. This new strategy takes in consideration the weight distributions W_I and W_S. The movement with the heaviest weight is chosen with probability $(1 - \epsilon)$ while with probability ϵ the roulette wheel is used with weight distributions W_I and W_S to select a movement. During the execution of the algorithm, ϵ is slowly decreased by a constant multiplicative coefficient γ_ϵ and reinitialised if, for $R^{\epsilon}_{\text{step}}$ iterations, there is no improvement since the last restart. The algorithm including SA with restart and epsilon-greedy strategy is named $\text{ALNS}^{\epsilon}_{\text{SA}}$.

4 Computational Experiments

In this section, we describe the experiments we performed to test our algorithms. A description of the instances is first given before talking about parameter tuning. A comparison of the performance of our algorithms on both benchmarks follows before comparing roulette wheel and epsilon greedy movement selection strategies. Finally, we tested our $\text{ALNS}^{\epsilon}_{\text{SA}}$ on Li & Lim benchmark [3] and compared with best reported solutions[2].

4.1 Instances

Our algorithms were tested on two different sets of instances: Li & Lim benchmark instances and our own instances. Li and Lim's instances are dedicated to the PDPTW. This benchmark was the closest we found to our problem. A simplification of SPP is necessary to be able to compare. The horizon is set to $H = 1$ and places have exactly one time window. Parcels do not share their place of origin and destination meaning the number of stores and pickup points equal the number of parcels. Finally, we adjust the weights α_i of the criteria in the fitness function in Eq. 1 in order to deliver all parcels and prioritise the number of vehicles used before minimising the travelled distance.

We also generated a benchmark PickOptBench to fully test our problem. These 135 instances where generated to be as close as possible to the reality faced by Smile Pickup. This was achieved by analysing the distribution of parcels across working days J and places Ω. Time windows and vehicles data where chosen based on existing ones. Here are the main characteristics for our set of instances: $H = 3$, $|D| \in \{1, 2\}$, $|E| \in \{1, 2, 4\}$, $|P| \in \{5, 10, 20, 40\}$, $k_{\text{max}} = 3$, $|C| \in \{60, 120, 240, 480, 960\}$, $|V| \in [2, 46]$ and $K_v \in [20, 30]$ (see Sect. 2).

The experimentations were conducted on an *Intel Xeon CPU E5-2680 v4* for a maximum execution time of 30 min each. For each algorithms, instances of both benchmarks were run a total of 10 times. Fitness function coefficient (see Sect. 1) used are: $\alpha_{\text{veh}} = 1000$, $\alpha_{\text{nl}} = 500$, $\alpha_{\text{dist}} = 1$, $\alpha_{\text{pen}} = 1$.

4.2 Parameter Tuning

To tune our algorithms and the different strategies implemented, we used the Irace software [11]. Tuning was performed on PickOptBench instances. We proceeded step by step and started by tuning the ALNS with simulated annealing without restart. After having fixed those parameters, we added other strategies one by one and tuned them separately. Final parameters are presented in Table 1.

2 Benchmark available on http://www.sintef.no/pdptw.

Table 1. Best configurations after tuning with irace.

algorithm	parameters tuned	configuration
ALNS SA	$<\rho, \sigma_1, \sigma_2, \sigma_3, \sigma_4, \gamma_{SA}, \delta>$	$<3, 29, 10, 1, 1, 895, 7428>$
restart	$<T_{restart}, R^{sa}_{step}>$	$<6164, 9598>$
epsilon greedy	$<\gamma_\epsilon, R^{\epsilon}_{step}>$	$<4.5 \cdot 10^5, 10^5>$

4.3 ALNS and Simulated Annealing Contributions

We first compare the following algorithms: VNS, LNS and ALNS with and without SA and $\text{ALNS}^{\epsilon}_{\text{SA}}$ on the PickOptBench and Li & Lim's benchmark. VNS_{SA} (resp. LNS_{SA}, ALNS_{SA}) is the VNS (resp. LNS, ALNS) algorithm with a simulated annealing acceptance criterion.

Table 2. Algorithm comparison on both benchmarks.

PickOptBench

algorithms	VNS	VNS_{SA}	LNS	LNS_{SA}	ALNS	ALNS_{SA}	$\text{ALNS}^{\epsilon}_{\text{SA}}$
average	12455	10088	11312	9984	12842	10088	10050
stdev	570	172	124	109	268	188	182
best average	19.26	20.00	30.37	41.48	10.37	30.37	36.30
best instance	24.67	17.63	29.33	28.44	12.89	33.26	36.52

Li & Lim Benchmark

algorithms	VNS	VNS_{SA}	LNS	LNS_{SA}	ALNS	ALNS_{SA}	$\text{ALNS}^{\epsilon}_{\text{SA}}$
average	9343	14165	17167	11109	16345	8824	8676
stdev	533	612	735	527	438	236	217
best average	0.00	0.00	0.00	0.00	0.00	26.79	46.43
best instance	9.11	0.00	0.00	0.71	0.00	26.25	31.61

For each benchmark and algorithm, Table 2 provides the average and standard deviation of the average scores obtained from 10 runs on each instance. It also computes the best instance and best average proportions, where the former indicates the percentage of instances that achieved the best known fitness score, and the latter indicates the percentage of instances where the average fitness score over 10 runs is the best known average. Table 2 contains multiple pieces of information. Firstly, it is apparent that SA significantly enhances the average fitness, except for the VNS algorithm. Moreover, the best performing algorithms are the ALNS with simulated annealing with $\text{ALNS}^{\epsilon}_{\text{SA}}$ being the most effective among all. The only algorithm that outperforms them is LNS_{SA} but only on the PickOptbench. In fact it does extremely poorly on Li & Lim's benchmark and has a lesser proportion of best instance on both benchmarks. Finally, the high null proportions observed on the first five algorithms on the Li & Lim benchmark may indicate that PickOptBench lacks sufficient discriminatory power to evaluate fully $ALNS_{SA}^{\epsilon}$.

4.4 Combining ϵ-Greedy and Simulated Annealing Restart Strategies

In this study, we compare the impact of using the roulette wheel and epsilon greedy strategies on movement choice for the $ALNS_{SA}$. We also investigate the benefits of including a restart of the SA when updating the current solution. The results of our experiments on both benchmarks are presented in Table 3. The table includes columns for the best, worst, and average fitness of the solutions returned during the 10 runs of each instance. The column "Best avg" shows the proportion of best averages on the 191 instances compared to the best known averages. The table indicates that the four strategies tested do not significantly affect the best, average or worst fitness scores. However, the epsilon greedy strategy with restart for $ALNS_{SA}^{\epsilon}$ produced the most best averages. To determine if one strategy outperformes the others, we used a paired t-test as a criteria. The starting hypothesis assumed that the two models had the same distribution of results. Only epsilon greedy with and without restart had a p-value greater than 0.005 and showed no evidence to reject the hypothesis. This shows the restart strategy did not significantly improved the epsilon greedy $ALNS_{SA}^{\epsilon}$ algorithm.

Table 3. Comparison of the different strategies on both benchmarks.

$ALNS_{SA}$		roulette wheel				epsilon greedy			
		best	average	worst	best avg	best	average	worst	best avg
SA	no restart	9491	9718	10007	21.99%	9416	9660	9897	28.27%
	restart	9439	9695	9973	27.23%	9419	9655	9905	**37.17%**

4.5 Comparison on Li and Lim Benchmark

The comparison of $ALNS_{SA}^{\epsilon}$ with the best known results found for Li & Lim benchmark [3] are presented in this section. This benchmark is composed of 56 instances organised in six classes LC1, LC2, LR1, LR2, LRC1 and LRC2, for which results are a pair of values: the number of vehicles used and the total distance travelled. $ALNS_{SA}^{\epsilon}$ finds the best known solutions on more than 70% for LC1, LC2, LR1 and LRC1. Nevertheless, only 10% are reached for LR2 and LRC2. Table 4 illustrates a small part of these results, for classes LC1 and LRC2.

Table 4. Comparison with best solutions on classes LC1 and LRC2.

LC1	best known		$\mathrm{ALNS}_{\mathrm{SA}}^{\epsilon}$		LRC2	best known		$\mathrm{ALNS}_{\mathrm{SA}}^{\epsilon}$	
	$\|V^+\|$	distance	$\|V^+\|$	distance		$\|V^+\|$	distance	$\|V^+\|$	distance
lc101	10	824.94	10	824.94	lrc201	4	1406.94	5	1497.47
lc102	10	824.94	10	824.94	lrc202	3	1374.27	5	1544.84
lc103	9	1035.35	10	826.44	lrc203	3	1089.07	4	1092.13
lc104	9	860.01	9	860.01	**lrc204**	3	818.66	3	818.66
lc105	10	824.94	10	824.94	lrc205	3	1302.2	5	1363.63
lc106	10	824.94	10	824.94	lrc206	3	1159.03	4	1210.00
lc107	10	824.94	10	824.94	lrc207	3	1062.05	4	1138.55
lc108	10	826.44	10	826.44	lrc208	3	852.76	4	937.57

5 Conclusion

This paper presented an $\mathrm{ALNS}_{\mathrm{SA}}^{\epsilon}$ algorithm for a pickup and delivery problem applied to a real life case for Smile Pickup business (SPP). Additional constraints considered are multiple time windows, heterogeneous fleet and multiple depots. $\mathrm{ALNS}_{\mathrm{SA}}^{\epsilon}$ is an Adaptive Learning Neighbourhood Search algorithm, combining two diversification processes. The first one is based on a simulated annealing technique dedicated to updating the current solution. The second one is an epsilon greedy strategy used to balance between exploration and exploitation during the generation of neighbourhoods. $\mathrm{ALNS}_{\mathrm{SA}}^{\epsilon}$ was tested on PickOptBench and Li&Lim benchmarks. Experimentation results show that such an approach is very promising for solving SPP. In addition, many levers exist to improve the performance of $\mathrm{ALNS}_{\mathrm{SA}}^{\epsilon}$. For example, we plan to improve suppression movements by integrating more relevant selection criteria than the random selection of deleted items.

References

1. Parragh, S.N., Doerner, K.F., Hartl, R.F.: A survey on pickup and delivery models part II: transportation between pickup and delivery locations. J. Betriebswirtschaft **58**, 81–117 (2006)
2. Berbeglia, G., Cordeau, J.F., Gribkovskaia, I., Laporte, G.: Static pickup and delivery problems: a classification scheme and survey. TOP **15**, 1–31 (2007)
3. Li, H., Lim, A.: A metaheuristic for the pickup and delivery problem with time windows. In: Proceedings 13th IEEE International Conference on Tools with Artificial Intelligence, ICTAI 2001, pp. 160–167 (2001)
4. de Jong, C., Kant, G., Van Vlient, A.: On finding minimal route duration in the vehicle routing problem with multiple time windows. Manuscript, Department of Computer Science, Utrecht University, Holland (1996)
5. Belhaiza, S., Hansen, P., Laporte, G.: A hybrid variable neighborhood tabu search heuristic for the vehicle routing problem with multiple time windows. Comput. Oper. Res. **52**, 269–281 (2014)

6. Ferreira, H.S., Bogue, E.T., Noronha, T.F., Belhaiza, S., Prins, C.: Variable neighborhood search for vehicle routing problem with multiple time windows. Electron. Notes Discret. Math. **66**, 207–214 (2018)
7. Salhi, S., Imran, A., Wassan, N.A.: The multi-depot vehicle routing problem with heterogeneous vehicle fleet: formulation and a variable neighborhood search implementation. Comput. Oper. Res. **52**, 315–325 (2014)
8. Braekers, K., Caris, A., Janssens, G.K.: Exact and meta-heuristic approach for a general heterogeneous dial-a-ride problem with multiple depots. Transp. Res. Part B: Methodol. **67**, 166–186 (2014)
9. Detti, P., Papalini, F., de Lara, G.Z.M.: A multi-depot dial-a-ride problem with heterogeneous vehicles and compatibility constraints in healthcare. Omega **70**, 1–14 (2017)
10. Ropke, S., Pisinger, D.: An adaptive large neighborhood search heuristic for the pickup and delivery problem with time windows. Transp. Sci. **40**, 455–472 (2006)
11. López-Ibáñez, M., Dubois-Lacoste, J., Cáceres, L.P., Birattari, M., Stützle, T.: The irace package: iterated racing for automatic algorithm configuration. Oper. Res. Perspect. **3**, 43–58 (2016)

GRAPH Reinforcement Learning for Operator Selection in the ALNS Metaheuristic

Syu-Ning Johnn[1(✉)], Victor-Alexandru Darvariu[2,3], Julia Handl[4], and Joerg Kalcsics[1]

[1] University of Edinburgh, Edinburgh, UK
shunee.johnn@sms.ed.ac.uk, joerg.kalcsics@ed.ac.uk
[2] University College London, London, UK
v.darvariu@cs.ucl.ac.uk
[3] The Alan Turing Institute, London, UK
[4] University of Manchester, Manchester, UK
Julia.Handl@manchester.ac.uk

Abstract. ALNS is a popular metaheuristic with renowned efficiency in solving combinatorial optimisation problems. However, despite 16 years of intensive research into ALNS, whether the embedded adaptive layer can efficiently select operators to improve the incumbent remains an open question. In this work, we formulate the choice of operators as a Markov Decision Process, and propose a practical approach based on Deep Reinforcement Learning and Graph Neural Networks. The results show that our proposed method achieves better performance than the classic ALNS adaptive layer due to the choice of operator being conditioned on the current solution. We also discuss important considerations such as the size of the operator portfolio and the impact of the choice of operator scales. Notably, our approach can also save significant time and labour costs for handcrafting problem-specific operator portfolios.

Keywords: Adaptive Large Neighbourhood Search · Markov Decision Process · Deep Reinforcement Learning · Graph Neural Networks

1 Introduction

Adaptive large neighbourhood search (ALNS) is a metaheuristic introduced by Ropke and Pisinger [18] to solve combinatorial optimisation problems (COPs) that iteratively deconstructs and reconstructs a part of the solution in the search for more promising solutions. This "relax-and-reoptimise" process is executed via a pair of destroy and repair heuristics called operators. Based on the principle of Shaw's large neighbourhood search (LNS) [21], ALNS contains multiple operators and an adaptive layer that iteratively selects and applies different operator pairs from a predefined operator portfolio. This is typically an embedded Roulette Wheel (RW) algorithm that selects operators in a probabilistic fashion.

B. Dorronsoro et al. (Eds.): OLA 2023, CCIS 1824, pp. 200–212, 2023.
https://doi.org/10.1007/978-3-031-34020-8_15

ALNS is renowned for its efficiency in finding good-quality solutions within reasonable computational time. However, despite the wide use of ALNS for solving various COPs, the ways in which each ALNS component contributes to its general performance is not well understood. A recent ALNS state-of-the-art review [12] indicated that only 2 out of 252 papers go beyond the straightforward implementation and concentrate on component-based analysis, including [19] which focuses on the selection of the ALNS acceptance criterion, and [25] on the effectiveness of the ALNS adaptive layer for operator selection.

We summarise two main deficiencies that exist in the current ALNS framework. Firstly, studies have shown that the adaptive layer has limited capability to dynamically select the best operators, despite being engineered to do so. Turkeš et al. [25] reported a mere 0.14% average improvement brought by the adaptive layer from the analysis of 25 ALNS implementations, indicating a need for a more efficient operator selection mechanism that reflects the contribution of individual operators accurately. Secondly, operator portfolio design for a particular problem can require considerable manual evaluation [12]. The choice of portfolio size is also delicate: too few operators might not enable the search to visit unexplored neighbourhoods, but a plethora of operators can introduce noise to the adaptive layer. To mitigate these deficiencies, we make the following contributions:

- We formulate the choice of a sequence of operators as a Markov Decision Process (MDP), in which an agent receives a reward proportional to the improvement in the solution. We draw a correspondence between value-based Reinforcement Learning (RL) methods used to solve MDPs, such as Q-learning, and the classic RW update used in ALNS. A key insight is that RL estimates are conditioned on the current solution, while RW updates are independent of it, which indicates the potential to learn a stronger operator selector through the RL framework;
- We propose a practical approach based on Deep RL for learning to select operators. Furthermore, we highlight the potential of Graph Neural Networks (GNNs) for generalizing to larger problem instances than seen during training;
- We carry out an extensive evaluation that includes a large selection of representative operators from the literature. Our results demonstrate that the proposed approach performs significantly better than the RW mechanism. We also analyse the impact of important practical considerations such as portfolio sizes and destroy operator scales on the optimality of the solutions.

2 Literature Review

In the last decade, training Machine Learning (ML) methods to solve highly complex COPs has become increasingly prominent [3], especially for the Vehicle Routing Problem (VRP) and its variants [1]. Several pioneering studies applied RL to directly construct solutions for routing-related problems. Bello et al. [2] used policy gradient algorithms to tackle the Travelling Salesperson Problem (TSP). Nazari et al. [14] proposed an end-to-end framework that outputs solutions directly from the routing-based problem instances. Moreover, Kool et al.

[10] proposed a construction heuristic that consists of an attention-based decoder trained with RL to regressively build solutions for the TSP and its variants.

ML can also be applied in many cases to enhance existing solution approaches, especially in the field of metaheuristics [24]. The reader is referred to the work of Karimi-Mamaghan et al. [8] for a comprehensive review on the integration of ML and metaheuristics to tackle COPs.

Several recent studies focused on integrating ML with classic LNS, which can be viewed as a simplified version of the ALNS metaheuristic without the adaptive layer for operator selection. As the first paper on this topic, Hottung and Tierney [7] proposed 2 generalised random-based destroy operators and a single repair operator with automated learning based on a deep neural network with an attention mechanism. Their work was the first to consider the application of RL to LNS for solving a VRP, and achieved solutions of better quality than classic optimisation approaches. Nevertheless, their proposed learning mechanism only focuses on repairing incomplete solutions during the repair phase. In another work, Falkner et al. [6] integrated a pre-trained neural construction heuristic as the repair operator in the LNS framework to solve the VRP with time windows. The destroy procedures remain handcrafted and are classified into 2 groups without any learning involved. Moreover, Oberweger et al. [15] enhanced the LNS framework with an ML-guided destroy operator to solve a staff rostering problem. For the reconstruction phase, the authors developed a mixed-integer linear program as a repair method. Lastly, Syed et al. [23] proposed a neural network in an LNS setting to solve a vehicle ride-hailing problem. However, it uses supervised learning, which requires a large training dataset and, furthermore, can only perform as well as the algorithm used for its generation.

A very recent concurrent work by Reijnen et al. [17] also applies Deep RL to improve ALNS operator selection. It considers a state space that only uses information about the search status (such as the search step), ignoring information about the current solution. In contrast to this, the design of our approach focuses on isolating the problem of operator selection from the search process, and proposing a learning mechanism that is conditioned on the decision space characteristics of the current solution. Furthermore, a fixed operator portfolio consisting of 4 destroy and 3 repair operators is used in their evaluation. In contrast, we propose a more robust operator selection system compatible with various operator portfolios of different sizes and train the system independently prior to integration with ALNS. Our approach also proposes the use of GNNs for scaling to large instances.

3 Methodology

3.1 Classic ALNS Algorithm

In ALNS [18], an initial solution is relaxed and re-optimised through iteratively employing a pair comprising a destroy operator $o_i^- \in \mathcal{D}$ and a repair operator $o_i^+ \in \mathcal{R}$ to form the new incumbent. The destroy scale d, which is randomly drawn or set as a hyper-parameter, describes the proportion of the solution that is destructed and reconstructed. In ALNS, the search can be divided into sequential segments, during which an initial score $\psi_i = 0$ is assigned to each operator (indexed by i) at the beginning and is increased by δ each time a new incumbent is formed using an operator pair that includes i. Depending on the incumbent quality, the score is increased by δ_1 if the newly-found incumbent is a global best solution, δ_2 for a local best one, and δ_3 for an accepted yet worse local solution, where $\delta_1 > \delta_2 > \delta_3$. At the end of each segment, the cumulated score for each operator i and the number of times N_i it was selected are used to compute a weight w_i that estimates the operator's capability to find promising solutions. As shown in Eq. (1), for each operator employed in the current segment K, its weight for the next segment $K+1$ is updated using a weighted average of the historical weight $w_{i,K}$ and its average performance in segment K.

$$w_{i,K+1} = \begin{cases} (1-\alpha_{\text{RW}}) \cdot w_{i,K} + \alpha_{\text{RW}} \cdot \frac{\psi_i}{N_i} & \text{if } \psi_i > 0, \\ w_{i,K} & \text{if } \psi_i = 0, \end{cases} \tag{1}$$

For each iteration within the segment, a pair of operators is selected using the RW selection algorithm with probabilities $w^-_{i,K} / \sum_{j \in \mathcal{D}} w^-_{j,K}$ and $w^+_{i,K} / \sum_{j \in \mathcal{R}} w^+_{j,K}$, where $w^{-/+}_{i,K}$ is the weight associated with each operator i in any given segment K. Initially, all operators are assigned the same score and therefore have the same selection probability. Once a new solution is formed, an ALNS acceptance mechanism, typically used in Simulated Annealing (SA), determines whether the newly-formed solution is accepted as the new incumbent to start the next iteration. The probabilistic acceptance mechanism helps to diversify the search and reduce the chance of becoming trapped in a non-promising local neighbourhood. The process continues until certain stopping criteria are met.

3.2 Operator Selection as a Markov Decision Process

Blueprint of our Approach. Our learning-based approach to improve the operator selection in ALNS consists, at a high level, of the following two steps. Firstly, we aim to isolate operator choice from the considerations of the SA process in ALNS, which introduces additional noise for navigating the solution space that may obscure the operators' contributions. To achieve this, we formulate operator selection for the COP as a standalone Markov Decision Process (MDP), in which an agent is given a limited budget of operators, and must learn to select those that lead to the best solutions. Secondly, the learned model is integrated into the ALNS loop and used to select operators in the SA process.

MDP Fundamentals. An MDP is a tuple $(\mathcal{S}, \mathcal{A}, P, R)$. In each state $s \in \mathcal{S}$, the agent selects an action $a \in \mathcal{A}(s)$ out of a set of valid actions, receiving a reward r according to a reward function $R(s, a)$. Afterwards, the agent transitions to a new state s' that depends on $P(s'|s, a)$, which is the transition function that governs the environment dynamics. Interactions happen in episodes, each of which is a finite sequence of (s, a, r, s') pairs, until a terminal state is reached. Actions are selected by the agent through the *policy* $\pi(a|s)$ that completely specifies its behaviour. The state-action value function $Q(s,a)$ is the expected reward the agent receives by picking action a at a given state s, then following π.

MDP Formulation. We are given an undirected graph $G = (V, E)$ defined by the given COP and a feature matrix $\mathbf{X}$ in which each row contains information about the node such as coordinates, demand, and distance. We formulate the MDP as below. A visualisation of an episode is shown in Fig. 1.

- *States* $\mathcal{S}$: each state S_t is a tuple $(G, \mathbf{X}, J_t, C_t, \varphi, b_t)$, wherein the graph G and feature matrix $\mathbf{X}$ remain static. J_t is the set of *tours* that start and end at the depot, forming the solution at time t. The removal list $C_t = V \backslash J_t$ holds all the d nodes temporarily removed from the solution. φ indicates the *phase*: whether a destroy or repair operator is eligible to be applied. Finally, b_t indicates the operator pair budget available to the agent.
- *Actions* $\mathcal{A}$ involve the selection of an operator o_t, with those available defined as $\mathcal{D}$ if $\varphi = 1$ (i.e., we are in the destroy phase), and $\mathcal{R}$ otherwise.
- *Transitions* P apply the selected operator o_t to the current solution. Applying a destroy operator removes d nodes from J_t and places them in the removal list C_t. Using a repair operator reinserts the nodes from C_t into J_t, leaving C_t empty and the solution J_t complete, and decreases the operator pair budget by 1. Transitions are stochastic due to the inherent randomness of the operators.
- *Rewards* R are provided once the operator budget is exhausted and the improvement in solution quality can be assessed via an objective function F. Concretely, $R(S_t, A_t) = F(S_t) - F(S_0)$ if $b_t = 0$, and 0 otherwise.

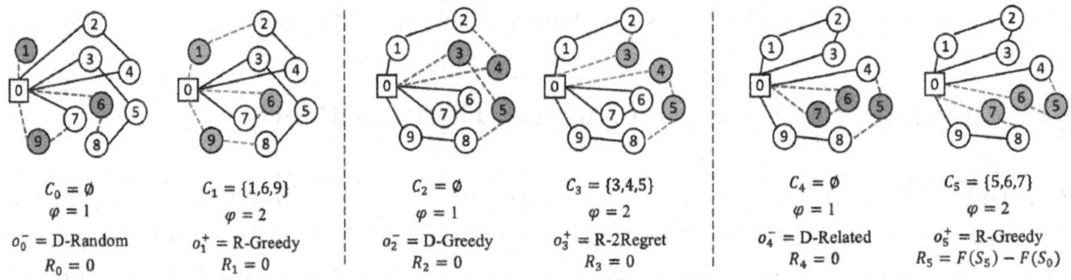

Fig. 1. Illustration of an MDP episode with budget $b = 3$ and destroy scale $d = 3$. The action spaces contain 3 destroy operators $\mathcal{D} = \{\text{Random, Greedy, Related}\}$ and 2 repairs $\mathcal{R} = \{\text{Greedy, 2Regret}\}$. The agent begins at state S_0 with $C_0 = \emptyset$ and routes $J_0 = \{[1], [2,4], [3,5,8,6], [7,9]\}$, selecting operators $o_0^- =$ Random and $o_1^+ =$ Greedy to reach S_2. The episode continues until the budget is exhausted and the terminal state S_5 with routes $J_5 = \{[1,2,3], [4,5,6], [7,8,9]\}$ is reached. Finally, it receives a reward proportional to the improvement in solution quality.

3.3 Learning an Operator Selection Policy

Q-Learning and Relationship to Roulette Wheel Update. Q-learning [27] is a model-free RL approach for solving MDPs that relies on estimating the state-action value function $Q(s, a)$, from which a policy π can be derived by acting greedily with respect to it. The agent's interactions with the environment generate (s, a, r, s') tuples, and its estimates are updated according to the rule

$$Q(s,a) \leftarrow (1 - \alpha_{\text{RL}}) \cdot Q(s,a) + \alpha_{\text{RL}} \cdot \left(r + \gamma \cdot \max_{a' \in \mathcal{A}(s')} Q(s', a') \right) \tag{2}$$

where α_{RL} is the learning rate, and γ trades immediate versus long-term rewards. Written in this form, comparing the Q-learning update in Eq. (2) and the classic RW update in Eq. (1), we notice that both use a weighted factor to balance two terms representing the historical and current estimates of performance. The key difference is that the Q-learning update is conditioned on the state and hence captures more information that may be used to select a relevant operator, while the RW update simply averages the gains of the operators irrespective of the context in which they were applied. Therefore, RW can be interpreted as a very rough approximation of the Q-learning update and, intuitively, using information about the state can allow us to obtain operator selection policies that perform at least as well. This means that Q-learning requires higher sample complexity. However, this was not an issue in practice, as we found a relatively low number of training steps suffices to reach a good policy.

Function Approximation and Graph Neural Networks. In problems with large state spaces, neural networks are commonly used to perform *function approximation* of the $Q(s, a)$ function. This helps to generalize between states that, while not being identical, share common characteristics and hence may lead to similar future rewards. The Deep Q-Network (DQN) algorithm [13], which uses this principle together with replay buffers and target networks, has been used for successfully approaching a variety of decision-making tasks.

In this work, we consider two possible neural network architectures. Firstly, we use a Multi-Layer Perceptron (MLP) formed of layers that apply a linear transformation of the inputs followed by a non-linear activation function. Despite their simplicity, MLPs are known to be universal function approximators. Secondly, we consider Graph Neural Network (GNN) architectures [20], that are explicitly designed to operate on graph-structured data. Such architectures compute an embedding for each node in the graph by iteratively aggregating the features of neighbouring nodes, resulting in node embeddings that encode both structural and feature-based information. A desirable characteristic of many GNN architectures is that their parametrization can be independent of the size of the input graph. Hence, they enable learning an approximation of the state-action value function on small instances and applying it directly on large instances – an appealing approach for COPs [3].

Integrating the Model with ALNS. As mentioned above, the resulting learned policy acts greedily with respect to the learned state-action value function, always choosing the action with the highest expected cumulative reward. This might prove problematic once integrated within ALNS, given that, in principle, greediness may cause the search to become trapped in local optima. To instead obtain a *probabilistic* policy, we use a softmax function as shown in Eq. (3), in which the temperature τ allows adjusting the level of greediness of the policy. Specifically, probabilities are uniform when $\tau \rightarrow \infty$, whereas the action with the highest expected reward has probability approaching 1 when $\tau \rightarrow 0$.

$$\pi_\tau(a|s) = \frac{\exp(Q(s,a)/\tau)}{\sum_{a' \in \mathcal{A}(s)} \exp(Q(s,a')/\tau)} \tag{3}$$

3.4 Operators for ALNS

In the literature, operators are carefully tailored to fit different problem structures and features. Despite the large variety of operator designs, the mechanisms behind them are surprisingly similar to the first version of ALNS [18]. We conducted a thorough analysis of operators in the literature and have identified the following 3 classes: *random-based* destroy that randomly removes d nodes according to specific availability criteria, *greedy-based* destroy that removes the top-ranking d nodes with respect to a particular measure, and *related-based* destroy as an extension of Shaw's destroy [21] that removes the most similar d nodes according to a certain proximity value. Variations can include perturbations or using problem-specific features including distance, time, cost, workload, demand level, inventory level, removal gain, historical information, etc.

Barring a few *random-based* operators, almost all current repair operator designs are related to *greedy-based* mechanisms that insert each node at the position with the smallest cost. Variations can include a pre-sorting that changes the order of node insertions according to certain criteria, including global minimum insertion or smallest regret value. Others can have a noise factor that perturbs the insertion cost values, or use restrictions based on historical information.

4 Experiments

4.1 Experimental Setup

Problem Settings. In this work, we consider the Capacitated Vehicle Routing Problem (CVRP) with a single depot, a group of customer nodes and a number of homogeneous vehicles each visiting an individual group of customer nodes. The capacity restriction applies to the total carrying load of vehicles. Each customer node can only be visited once. We use the R, C and RC instances (random, clustered, and mixed random-clustered nodes) of the Solomon dataset [22] each containing a depot and 100 customers. We assign the vehicle capacity to be 200, and adjust it proportionally if scaling down the instance to fewer customers.

For the portfolio design, we identified 12 popular destroy operators from the ALNS literature that span the representative categories described in Sect. 3.4: the random-based variations *random node destroy* [18] and *random route destroy* [4], the greedy-based variations *worst-node removal* [18], *neighbourhood removal* [4] and *greedy route destroy* [9], and the related-based variations *proximity destroy* [4], *cluster destroy* [16], *node neighbourhood destroy* [4], *zone destroy* [5], *route neighbourhood destroy* [5], *pair destroy* [11] and *historical node-pair removal* [16]. The repair operator portfolio is comparatively smaller. We include the group of classic *greedy repair* [18] and *k-regret repair* [16] for $k = 2$.

Operator Selection Approaches. The proposed DQN agent is compared to the following approaches. As a baseline, we consider a uniform Random sampling (RAN) of operators. We also compare against the classic RW (CRW), which can only be used within ALNS since it requires information about the SA outcomes and search progress. To make the RW mechanism applicable in the MDP setting, we make the following adaptations to obtain a method we call Learned RW (LRW). Firstly, in Eq. (1), we replace the manually-defined operator scores ψ computed from the discretised δ with the continuous objective value F. We also adjust the reward feedback frequency from every operator pair in RW to every episode in the MDP. Preliminary experimental results suggested that the performance difference between the LRW and CRW is within 2% when applied in ALNS without any prior training.

Training and Evaluation Methodology. For each instance, we generate 3 distinct sets $\mathcal{J}^{\text{train}}, \mathcal{J}^{\text{validate}}, \mathcal{J}^{\text{test}}$ of 128 randomly initialized tours each. $\mathcal{J}^{\text{train}}$ is used by DQN and LRW for model training. $\mathcal{J}^{\text{validate}}$ is used for hyperparameter tuning and model selection. Finally, $\mathcal{J}^{\text{test}}$ is used to perform the final evaluation and obtain the reported results. There are two evaluation "modes": MDP-compatible agents can be evaluated in a standalone fashion given an operator budget (CRW is excluded), while all operators (including CRW) can be evaluated on the end ALNS task. Training and evaluation is repeated across 10 random seeds for all agents, which are used to compute confidence intervals.

DQN Architectures and Inputs. For the DQN, we consider MLP and GNN representations. The MLP has 256 units in the first hidden layer, with the subsequent layers having half the size. As a GNN, we opt for the GAT [26], which allows for flexible aggregation of neighbour features. We use 3 layers and a dimension of node embeddings equal to 32. Both use a learning rate of $\alpha_{\text{RL}} = 0.0005$ and are trained for $15 \cdot 10^3$ and $25 \cdot 10^3$ steps respectively. The DQN exploration rate ϵ is linearly decayed from 1 to 0.1 in the first 10% of steps, then remains fixed. The replay buffer size is equal to 20% of the number of steps. To obtain the inputs, we construct vectors $\tilde{\mathbf{x}}_t^i$ that concatenate the static instance-specific features $\mathbf{x}^i$ with time-dependant relevant information such as whether the node i is routed in a tour and the number of tours in J_t. For the MLP, we stack the vectors in a matrix $\widetilde{\mathbf{X}}_t$ as inputs, while for the GNN the node features are provided directly. Unless otherwise stated, we use a softmax temperature $\tau = 0.01$.

4.2 Experimental Results

Evaluating Agents within MDP Framework. In this experiment, we compare the cumulative rewards gained by the DQN with an MLP representation, LRW and RAN agents on the test set $\mathcal{J}^{\text{test}}$ after undergoing training. To make

Table 1. MDP evaluation results: cumulative rewards gained by the DQN, RAN and LRW agents with destroy portfolios $\mathcal{D}$ of different sizes. Higher is better.

$\|\mathcal{D}\|$	C-instance			R-instance			RC-instance		
	DQN	RAN	LRW	DQN	RAN	LRW	DQN	RAN	LRW
2	232.2 ± 2.9	**252.2 ± 3.6**	252.1 ± 4.5	216.3 ± 5	**222.9 ± 3.7**	222.8 ± 4.4	240.2 ± 7.5	**259.4 ± 3.4**	258.9 ± 5.6
3	228.6 ± 9.0	221.7 ± 3.5	**245.9 ± 6.0**	212.9 ± 5.6	208.4 ± 4.7	**215.1 ± 3.8**	**236.1 ± 6.1**	224.0 ± 3.4	230.8 ± 7.6
4	232.5 ± 5.1	221.2 ± 6.4	**240.3 ± 5.8**	**220.2 ± 3.1**	206.9 ± 6.0	216.2 ± 4.1	**241.2 ± 5.8**	222.4 ± 5.1	239.5 ± 4.8
5	**328.8 ± 2.4**	258.5 ± 5.5	293.3 ± 8.0	**330.9 ± 4.2**	246.8 ± 3.7	273.9 ± 6.1	**331.1 ± 2.9**	260.9 ± 4.1	272.5 ± 7.6
6	**329.9 ± 4.2**	231.7 ± 5.8	284.9 ± 8.9	**329.5 ± 3.3**	217.5 ± 5.5	272.5 ± 14.9	**329.5 ± 2.6**	239.5 ± 5.4	261.6 ± 5.1
7	**328.5 ± 2.8**	246.7 ± 4.4	282.4 ± 11.0	**330.5 ± 3.8**	236.1 ± 4.8	253.7 ± 8.1	**331.7 ± 2.9**	247.6 ± 3.8	264.6 ± 6.4
8	**329.5 ± 3.9**	235.6 ± 5.2	281.6 ± 10.6	**330.9 ± 3.6**	220.4 ± 2.0	264.5 ± 7.0	**333.7 ± 3.3**	243.7 ± 4.5	254.7 ± 4.8
9	**330.7 ± 3.1**	225.8 ± 5.7	274.6 ± 12.2	**328.9 ± 4.6**	212.6 ± 3.8	260.3 ± 5.7	**332.4 ± 3.7**	226.7 ± 7.0	250.3 ± 8.4
10	**330.2 ± 4.5**	224.4 ± 4.7	276.2 ± 9.3	**330.3 ± 3.3**	206.7 ± 4.5	258.6 ± 7.0	**331.0 ± 2.6**	224.3 ± 5.8	252.6 ± 15.3
11	**330.3 ± 2.9**	222.0 ± 4.8	275.9 ± 6.4	**327.2 ± 8.0**	210.4 ± 4.6	259.1 ± 9.0	**326.1 ± 15.2**	223.3 ± 6.6	245.7 ± 8.3
12	**361.8 ± 0.2**	246.9 ± 5.5	323.7 ± 7.0	**404.2 ± 0.6**	246.8 ± 7.0	313.1 ± 17.7	**354.9 ± 4.1**	258.5 ± 4.1	283.9 ± 13.0
mean	**305.7 ± 3.7**	235.2 ± 5.0	275.5 ± 8.2	**305.6 ± 4.6**	221.4 ± 4.6	255.4 ± 8.0	**308.0 ± 5.2**	239.1 ± 4.8	255.9 ± 7.9

the training and evaluation processes less computationally intensive, we use the first 20 customer nodes and the depot from the Solomon R, C and RC instances. We define operator portfolios of different sizes ranging from 2 to 12 by sequentially adding the 12 destroy operators introduced above, together with the 2 repair operators. The destroy scale is fixed as $d = 4$ and the operator pair budget is $b = 10$, yielding MDP episodes of length 20.

Table 1 shows that the DQN agent is able to outperform competing methods as the size of $\mathcal{D}$ grows. When the destroy portfolio is smaller than 3, the DQN agent performs slightly worse due to the limited action space in which the impact of the selected actions is difficult to distinguish from chance. The DQN agent also yields smaller confidence intervals and hence a steadier performance. As expected, the RAN agent fails to show a clear increase in rewards as the portfolio size grows. The LRW agent, although showing a certain improvement, performs significantly worse than the DQN. Two performance jumps in the DQN and LRW agents were observed: from size 4 to 5 and 11 to 12 for all 3 instances, the reason for which is the inclusion of a more efficient operator in the portfolio that suits the behaviour of a greedy-based agent.

ALNS Evaluation. Using the same experimental setup as above, we apply the operator selection approaches within ALNS with a fixed number of iterations. As shown in Table 2, the DQN agent yields the lowest objective values (best results) when used to perform operator selection in ALNS for portfolios larger than 5. Interestingly, LRW is able to perform substantially better than CRW due to having undergone training on a different dataset of solutions prior to being applied. Instead, the performance of CRW is indistinguishable from RAN in the setting across all the 3 instances.

Scaling to Larger Instances with GNN. In this experiment, we train the DQN with a GNN representation and the LRW on instances of size 20, then evaluate them in an MDP setting on instances of size up to 100. The operator budget is kept the same while the destroy scale is increased proportionally to the size, i.e., $d = n/5$. We use the largest destroy portfolio with $|\mathcal{D}| = 12$. As shown

Table 2. Evaluating operator selection approaches in ALNS with destroy portfolios $\mathcal{D}$ of different sizes. Values represent the average and best objective values found within a fixed number of iterations, using each approach to select operators. Lower is better.

C-inst	$\|\mathcal{D}\|$	2	3	4	5	6	7	8	9	10	11	12	mean
DQN	*avg*	316.28	314.96	311.18	**216.33**	**213.03**	**216.34**	**215.12**	**215.75**	**214.19**	**213.09**	**188.11**	**239.49**
	min	244.64	245.28	240.32	**154.06**	**149.29**	**153.05**	**151.87**	**153.04**	**150.7**	**149.99**	**146.71**	**176.27**
LRW	*avg*	**293.84**	**299.56**	**302.66**	248.31	259.06	261.17	266.59	264.65	265.62	264.78	224.49	268.25
	min	**209.8**	**211.62**	**207.6**	172.97	180.96	178.07	187.71	183.88	182.07	185.3	160.95	187.36
RAN	*avg*	293.5	313.98	315.05	284.31	299.86	294.57	299.64	303.58	312	305.74	286.08	300.76
	min	209.95	219.79	212.33	194.6	206.28	199.23	203.63	208.35	212.2	213.38	197.54	207.03
CRW	*avg*	293.94	314.56	314.28	285.53	299.51	295.77	302.67	307.64	311.32	308.53	286.75	301.86
	min	210.27	222.46	213.42	197.82	208.25	198.37	206.6	210.7	213.69	213.38	194.63	208.14
R-inst	$\|\mathcal{D}\|$	2	3	4	5	6	7	8	9	10	11	12	mean
DQN	*avg*	330.37	334.09	331.24	**217.25**	**215.27**	**215.41**	**215.89**	**216.17**	**215.91**	**221.71**	**159.29**	**242.96**
	min	273.81	266.66	272.16	**144.97**	**143.56**	**144.11**	**144.19**	**144.58**	**144.49**	**153.64**	**108.07**	**176.39**
LRW	*avg*	**322.22**	**329.14**	**327.18**	269.97	265.26	286.74	281.67	286.63	284.94	274.17	238.46	287.85
	min	**246.19**	**254.83**	**243.38**	187.33	183.51	199.16	196.51	201.83	196.82	189.54	157.33	205.13
RAN	*avg*	323.25	337.03	337.68	298.76	317.84	310.1	321.11	321.76	327.19	321.3	294.45	319.13
	min	253.64	257.11	250.05	211.7	235.07	218.61	234.45	231.79	234.31	234.38	206.54	233.42
CRW	*avg*	322.71	333.59	336.81	298.38	320.5	308.7	318.42	321.34	327.17	324.96	296.05	318.97
	min	252.39	255.69	250.28	211.47	235.15	215.55	230.06	232.49	235.87	233.28	208.18	232.76
RC-inst	$\|\mathcal{D}\|$	2	3	4	5	6	7	8	9	10	11	12	mean
DQN	*avg*	306.06	311.02	313.78	**216.84**	**218.75**	**218.16**	**217.66**	**217.48**	**218.12**	**224.56**	**201.81**	**242.2**
	min	228.16	235.04	240.93	**151.89**	**154.74**	**153.71**	**152.85**	**152.06**	**153.79**	**161.32**	**159.09**	**176.69**
LRW	*avg*	**294.94**	**314.93**	**310.02**	278.18	282.01	286.05	292.6	293.97	293.44	292.15	261.08	290.85
	min	**205.05**	**216.86**	**209.63**	187.78	189.92	190.45	197.61	196.75	197.62	196.21	172.7	196.42
RAN	*avg*	297.2	319.87	317.7	288.95	298.41	294.26	300.61	308.9	315.42	305.53	283.71	302.78
	min	207.17	222.85	209.95	193.42	198.33	195.88	200.04	207.6	210.9	204.25	185.29	203.24
CRW	*avg*	**294.94**	319.07	320.76	289.51	299.99	296.09	299.98	310.67	315.4	309.69	285.73	303.8
	min	**205.05**	221.27	213.01	195.39	200.23	196.03	202.8	207.05	211.47	205.59	186.24	204.01

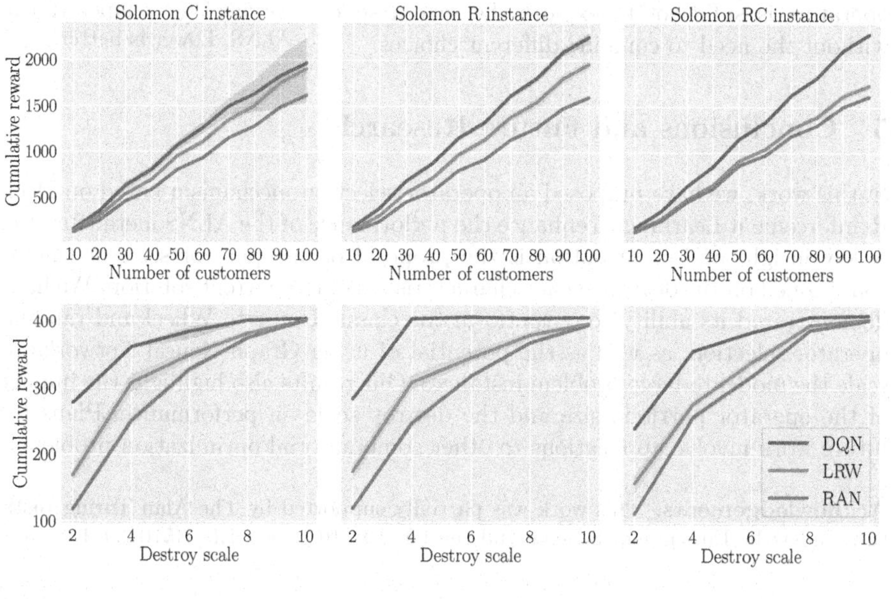

Fig. 2. Top: cumulative rewards for the DQN, LRW and RAN agents with GNN representation. Bottom: performance as a function of destroy scales. Higher is better.

in the top half of Fig. 2, the DQN+GNN agent outperforms the other methods, suggesting the strong generalization of the learned operator selection policies. A larger confidence interval is observed for the C instance, due to 1 model seed that generalizes poorly on $\mathcal{J}^{\text{test}}$ despite good performance on $\mathcal{J}^{\text{validate}}$.

Impact of Destroy Scale. Furthermore, we analyse the impact of the destroy scale on the agents' performances in the MDP setting, with a smaller scale implying the removal and reinsertion of a smaller proportion of nodes. We vary the destroy scale $d \in [2, 4, 6, 8, 10]$ with destroy portfolio $|\mathcal{D}| = 12$ on 20 nodes. Results are shown in the bottom half of Fig. 2. The gap between the DQN and other methods is largest for the smallest scale, suggesting that a careful selection of operators to remove the most expensive nodes contributes more significantly to better solution quality. In contrast, a larger destroy scale requires building up the solution from the ground, stressing the operators' reconstruction ability rather than the operator selection policy. When increasing the destroy scale, the cumulative rewards gained by different agents all converge to a similar level.

Impact of DQN Temperature. As discussed in Sect. 3.3, the temperature parameter τ controls the greediness of the resulting policy. Figure 3 shows the minimum and mean F obtained with ALNS as a function of $\tau \in \{10^{-2}, 10^{-1}, 10^{0}, 10^{1}, 10^{2}\}$, averaged over the 3 instance sets. Even though a probabilistic policy may be desirable in some ALNS scenarios, we find that performance generally degrades as the temperature increases. This suggests that, in the settings tested, the inherent stochasticity of the operators is sufficient to explore the search space without the need to combine different choices.

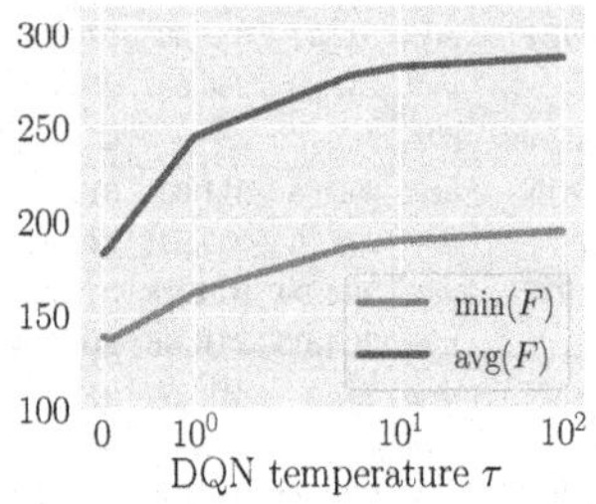

Fig. 3. Values of F when varying DQN temperature in ALNS. Lower is better.

5 Conclusions and Future Research

In this work, we have proposed an operator selection mechanism based on Deep Reinforcement Learning to enhance the performance of the ALNS metaheuristic. A key insight and contribution is the proposal of an operator selector that is conditioned on the decision space characteristics of the current solution. We have demonstrated its ability to outperform the classic Roulette Wheel and random operator selection, as well as the potential of using Graph Neural Networks to scale the model to large problem instances. Our results also highlight the impact of the operator portfolio size and the destroy scale on performance. Plans for future work involve applications to other combinatorial optimization problems.

Acknowledgements. This work was partially supported by The Alan Turing Institute under the Enrichment Scheme and the UK EPSRC grant EP/N510129/1.

References

1. Bai, R., et al.: Analytics and machine learning in vehicle routing research. Int. J. Prod. Res. **61**(1), 4–30 (2023)
2. Bello, I., Pham, H., Le, Q.V., Norouzi, M., Bengio, S.: Neural combinatorial optimization with reinforcement learning. In: ICLR Workshops (2016)
3. Bengio, Y., Lodi, A., Prouvost, A.: Machine learning for combinatorial optimization: a methodological tour d'horizon. Eur. J. Oper. Res. **290**(2), 405–421 (2021)
4. Demir, E., Bektaş, T., Laporte, G.: An adaptive large neighborhood search heuristic for the pollution-routing problem. Eur. J. Oper. Res. **223**(2), 346–359 (2012)
5. Emeç, U., Çatay, B., Bozkaya, B.: An adaptive large neighborhood search for an e-grocery delivery routing problem. Comput. Oper. Res. **69**, 109–125 (2016)
6. Falkner, J.K., Thyssens, D., Schmidt-Thieme, L.: Large neighborhood search based on neural construction heuristics. arXiv:2205.00772 (2022)
7. Hottung, A., Tierney, K.: Neural large neighborhood search for the capacitated vehicle routing problem. In: ECAI (2020)
8. Karimi-Mamaghan, M., Mohammadi, M., Meyer, P., Karimi-Mamaghan, A.M., Talbi, E.G.: Machine learning at the service of meta-heuristics for solving combinatorial optimization problems: a state-of-the-art. Eur. J. Oper. Res. **296**(2), 393–422 (2022)
9. Keskin, M., Çatay, B.: Partial recharge strategies for the electric vehicle routing problem with time windows. Transp. Res. Part C Emerg. **65**, 111–127 (2016)
10. Kool, W., Van Hoof, H., Welling, M.: Attention, learn to solve routing problems! In: ICLR (2018)
11. Mancini, S.: A real-life multi depot multi period vehicle routing problem with a heterogeneous fleet: formulation and adaptive large neighborhood search based matheuristic. Transp. Res. Part C Emerg. **70**, 100–112 (2016)
12. Mara, S.T.W., Norcahyo, R., Jodiawan, P., Lusiantoro, L., Rifai, A.P.: A survey of adaptive large neighborhood search algorithms and applications. Comput. Oper. Res. **146**, 105903 (2022)
13. Mnih, V., Kavukcuoglu, K., Silver, D., Rusu, A.A., et al.: Human-level control through deep reinforcement learning. Nature **518**(7540), 529–533 (2015)
14. Nazari, M., Oroojlooy, A., Snyder, L., Takác, M.: Reinforcement learning for solving the vehicle routing problem. In: NeurIPS (2018)
15. Oberweger, F., Raidl, G., Rönnberg, E., Huber, M.: A learning large neighborhood search for the staff rerostering problem. In: Schaus, P. (ed.) CPAIOR 2022. LNCS, vol. 13292, pp. 300–317. Springer, Cham (2022). https://doi.org/10.1007/978-3-031-08011-1_20
16. Pisinger, D., Ropke, S.: A general heuristic for vehicle routing problems. Comput. Oper. Res. **34**(8), 2403–2435 (2007)
17. Reijnen, R., Zhang, Y., Lau, H.C., Bukhsh, Z.: Operator selection in adaptive large neighborhood search using deep reinforcement learning. arXiv:2211.00759 (2022)
18. Ropke, S., Pisinger, D.: An adaptive large neighborhood search heuristic for the pickup and delivery problem with time windows. Transp. Sci. **40**(4), 455–472 (2006)
19. Santini, A., Ropke, S., Hvattum, L.M.: A comparison of acceptance criteria for the adaptive large neighbourhood search metaheuristic. J. Heurist. **24**(5), 783–815 (2018). https://doi.org/10.1007/s10732-018-9377-x
20. Scarselli, F., Gori, M., Tsoi, A.C., Hagenbuchner, M., Monfardini, G.: The graph neural network model. IEEE Trans. Neural Netw. **20**(1), 61–80 (2008)

21. Shaw, P.: Using constraint programming and local search methods to solve vehicle routing problems. In: Maher, M., Puget, J.-F. (eds.) CP 1998. LNCS, vol. 1520, pp. 417–431. Springer, Heidelberg (1998). https://doi.org/10.1007/3-540-49481-2_30
22. Solomon, M.M.: Algorithms for the vehicle routing and scheduling problems with time window constraints. Oper. Res. **35**(2), 254–265 (1987)
23. Syed, A.A., Akhnoukh, K., Kaltenhaeuser, B., Bogenberger, K.: Neural network based large neighborhood search algorithm for ride hailing services. In: Moura Oliveira, P., Novais, P., Reis, L.P. (eds.) EPIA 2019. LNCS (LNAI), vol. 11804, pp. 584–595. Springer, Cham (2019). https://doi.org/10.1007/978-3-030-30241-2_49
24. Talbi, E.G.: Machine learning into metaheuristics: a survey and taxonomy. ACM Comput. Surv. (CSUR) **54**(6), 1–32 (2021)
25. Turkeš, R., Sörensen, K., Hvattum, L.M.: Meta-analysis of metaheuristics: quantifying the effect of adaptiveness in adaptive large neighborhood search. Eur. J. Oper. Res. **292**(2), 423–442 (2021)
26. Veličković, P., Cucurull, G., Casanova, A., Romero, A., Lio, P., Bengio, Y.: Graph attention networks. In: ICLR (2018)
27. Watkins, C., Dayan, P.: Q-learning. Mach. Learn. **8**(3–4), 279–292 (1992)

Multi-objective Optimization of Adhesive Bonding Process in Constrained and Noisy Settings

Alejandro Morales-Hernández[1,2](✉), Inneke Van Nieuwenhuyse[1,2], Sebastian Rojas Gonzalez[1,2,3], Jeroen Jordens[4], Maarten Witters[4], and Bart Van Doninck[4]

[1] Flanders Make@UHasselt, Faculty of Science, Hasselt University, Diepenbeek 3590, Belgium
{alejandro.moraleshernandez,inneke.vannieuwenhuyse, sebastian.rojasgonzalez}@uhasselt.be
[2] Data Science Institute, Hasselt University, Diepenbeek 3590, Belgium
[3] Surrogate Modeling Lab, Ghent University, Ghent 9000, Belgium
sebastian.rojasgonzalez@ugent.be
[4] CoDesignS, Flanders Make, Lommel 3920, Belgium
{jeroen.jordens,maarten.witters,bart.vandoninck}@flandersmake.be

Abstract. Finding the optimal process parameters for an adhesive bonding process is challenging: the optimization is inherently multi-objective (aiming to maximize break strength while minimizing cost), constrained (the process should not result in any visual damage to the materials, and stress tests should not result in adhesive failures), and uncertain (measuring the same process parameters several times lead to different break strength). Real-life physical experiments in the lab are expensive to perform (∼6 h of experimentation and subsequent production costs); traditional evolutionary approaches are then ill-suited to solve the problem, due to the prohibitive amount of experiments required for evaluation. In this research, we successfully applied specific machine learning techniques (Gaussian Process Regression and Logistic Regression) to emulate the objective and constraint functions based on a *limited* amount of experimental data. The techniques are embedded in a Bayesian optimization algorithm, which succeeds in detecting Pareto-optimal process settings in a highly efficient way (i.e., requiring a limited number of experiments).

Keywords: multi-objective optimization · constrained optimization · machine learning · adhesive bonding

1 Introduction

Adhesive bonding is the engineering process of joining two surfaces together by a non-metallic substance [5]. This process occurs frequently in many engineering

B. Dorronsoro et al. (Eds.): OLA 2023, CCIS 1824, pp. 213–223, 2023.
https://doi.org/10.1007/978-3-031-34020-8_16

design contexts, such as the automotive industry [8] and aeronautics [9]. It is a complex process, in which several physical and chemical processes occur simultaneously [18], with outcomes that are influenced by many factors (e.g., environmental conditions, material specifications, and specific process settings). Process optimization is therefore traditionally performed by experts, based on acquired knowledge and extensive experimental campaigns [7]. Physical experiments are required in reality to detect the optimal settings for each specific adhesive process. These tend to be costly in terms of time and manual labor. Data from one industrial bonding process cannot be used to optimize another process, as not only materials and adhesives may differ but also production process specifications. Moreover, the experimental approach may easily yield suboptimal results with respect to other relevant performance metrics, such as production costs.

Evolutionary multi-objective algorithms are applicable to black-box optimization problems, and have proven to be effective derivative-free optimizers [24]. However, evolutionary algorithms require many function evaluations, which make them ill-suited for optimizing design problems that require *expensive* (often experimental) data (in terms of time or costs involved). Even when an emulator or surrogate model is used to mitigate this issue [3,4], the search process in this type of algorithm remains largely random, with convergence speeds that are sensitive to the choice of user-defined (and often problem-specific) parameters. Moreover, most machine learning techniques that might be used as emulators (including neural networks) require lots of data to be trained, which again causes a problem when function evaluations are expensive. Finally, the measurement of the objectives is affected by noise and this is often neglected during the optimization, leading to overly optimistic solutions.

In this article, we illustrate the power of Bayesian optimization (BO) approaches for the multi-objective optimization of a novel adhesive bonding process. The main contributions of this paper include:

- The use of a Gaussian Process Regression (GPR) surrogate that explicitly accounts for the *heterogeneous noise* existing in the measurements of the break strength.
- The use of an *acquisition function* to sequentially (one-by-one) select new process parameter configurations to be evaluated. This acquisition function uses information from a GPR model and a Logistic Regression classification model (to predict the feasibility of bonding process configurations).
- Results for a "*cheap simulation model*" show that the BO method is able to obtain *better* configurations than NSGA-II-based algorithms (w.r.t. average hypervolume and IGD+), with a small number of expensive evaluations required. Since real experimentation of a single process configuration can take up to 6 h, our approach is particularly relevant for settings where the analyst can only afford a very limited number of observations. As the model is general, it may constitute a powerful tool also in other constrained engineering design settings.

The remainder of this article is organized as follows. Section 2 discusses the main concepts in multi-objective optimization and introduces the bonding process problem under study. Section 3 details the main concepts of BO, and the proposed algorithms. Section 4 discusses the design of experiments, while Sect. 5 compares the results with evolutionary algorithms. Section 6 summarizes the findings and highlights some future research directions.

2 Multi-objective Adhesive Bonding Process Problem

The plasma treatment phase of the bonding process of two PolyPhenylene Sulfide (PPS) substrates (using Araldite 2011 adhesive) is our focus. The plasma treatment chemically modifies the top surface layer of the PPS substrate so that the surface energy increases, which impacts the adhesion strength (i.e., the strength of the connection between the adhesive and the substrate). In this process, the adhesion strength is very sensitive to the configuration of six parameters (see Sect. 4) that need to be specified properly.

Using lab experiments, stress tests can be performed to check the outcomes of samples that have been treated with any particular plasma parameter configuration: the lap shear strength of the sample (MPa), the failure mode (adhesive, cohesive or substrate failure), the production cost of the sample (in euros), and the potential occurrence of visual damage (the substrates will burn when heated above their maximum allowable temperature during plasma treatment).

The goal of the optimization is to set the plasma process parameters in such a way that (1) the tensile strength (TS) is maximized, (2) the production cost (PC) is minimized, and (3) adhesive failures and visual damage are avoided. Equation 1 defines this optimization problem:

$$\begin{aligned} \min \quad & [-TS(\mathbf{x}), PC(\mathbf{x})] \\ \text{s.t.} \quad & 0.5 - Pf(\mathbf{x}) \leq 0 \end{aligned} \tag{1}$$

where the notation $Pf(x)$ refers to the probability that a process configuration $\mathbf{x}$ is feasible (classified as such using the most common output observed over the replications). As the performance evaluation is expensive, the optimization algorithm should be able to detect (nearly) Pareto-optimal solutions within a small number of experiments required. We just cannot afford to collect large amounts of experimental data.

3 Bayesian Optimization: Main Concepts and Proposed Algorithm

Gaussian Process Regression (GPR) (also referred to as *kriging*, [21]) is commonly used to model an (unknown) target function. The function value prediction at an unsampled point $\mathbf{x}$ is obtained through the conditional probability $P(f(\mathbf{x})|\mathbf{X}, \mathbf{Y})$ that represents how likely the response $f(\mathbf{x})$ is, given that we observed the target function at n input locations $\mathbf{x}^{(i)}, i = 1, \ldots, n$ (contained in matrix $\mathbf{X}$), yielding function values $\mathbf{y}^{(i)}, i = 1, \ldots, n$ (contained in matrix $\mathbf{Y}$)

that may or may not be affected by noise. Ankenman et al. [1] provide a GPR model (referred to as *stochastic kriging*) that takes into account the heterogeneous noise observed in the data, and models the observed response value in the r-th replication at design point $\mathbf{x}^{(i)}$ as:

$$y_r(\mathbf{x}^{(i)}) = m(\mathbf{x}^{(i)}) + M(\mathbf{x}^{(i)}) + \epsilon_r(\mathbf{x}^{(i)}) \quad (2)$$

where $m(\cdot)$ represents the mean of the process, $M(\cdot)$ is a realization of a Gaussian random field with mean zero (also referred to as the *extrinsic uncertainty* [1]), and $\epsilon_r(\cdot)$ is the *intrinsic uncertainty* observed in replication r. Popular choices for $m(\cdot)$ are known linear or nonlinear functions of $\mathbf{x}$, an unknown constant to be estimated, or zero. $M(\cdot)$ can be seen as a function, randomly sampled from a space of functions that, by assumption, exhibit spatial correlation according to a covariance function (also referred to as *kernel*).

Figure 1 describes the steps in our Bayesian multi-objective optimization approach. The algorithm starts by evaluating an initial set of points through a Latin hypercube sample (Step 1). Simulation replications are used to estimate the objective values at these points (Step 2). In the BO literature, it is common to set the number of initial design points equal to $k = 10d$ (with d the number of dimensions of the input space; see [14]), though also smaller design sizes have been advocated [20]. The augmented Tchebycheff scalarization function [13] is applied to the objectives (Step a) to transform the problem into a single-objective optimization problem. A different set of weights is chosen on each iteration to find solutions across the entire Pareto front. Then, a metamodel is trained in Step 4 using the scalarized objective. As a metamodel, we use GPR to handle the existing uncertainty in the break strength, which variance depends on the process configuration (we thus have heterogeneous noise).

Simultaneously, a Logistic Regression (LR) classification model [22] is trained (Step 3) to determine whether a process configuration is feasible (meaning that it will not entail visual damage or adhesive failure). The probability $P(y = 1|\mathbf{x})$, with which the class "Feassible" is predicted, is used as $Pf(\mathbf{x})$ in the constraint defined by Eq. 1. The LR classifier is trained with the most common output

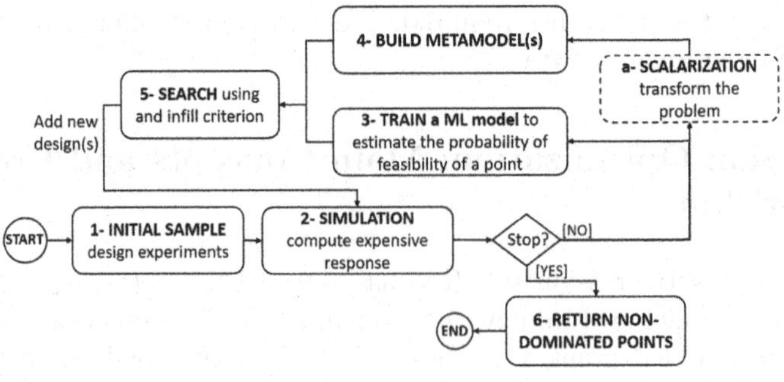

Fig. 1. Multi-objective optimization of an adhesive bonding process

observed over the replications of a given input configuration (binary class output, with 1 if the configuration is feasible and 0 otherwise).

The accuracy and fit of a GPR model can be drastically affected by random noise and just using a stochastic GPR model may not have a fair effect on the traditional EI criterion [16] to sample candidate points. As a direct consequence of the existing heterogeneous noise, the stochastic GPR may suggest (by using the EI criterion) the current best design point several times to try next, at the expense of exploring other promising regions of the design space. We propose to use the *Modified Expected Improvement* (MEI) [16] instead of the well-known EI, for having shown promising results in the optimization of problems affected by heterogeneous noise [12,17]. The estimated improvement in the objective at an arbitrary configuration $\mathbf{x}$ is:

$$\mathrm{MEI}(\mathbf{x}) = \left(\widehat{Z}(\mathbf{x}_{\min}) - \widehat{Z}(\mathbf{x})\right) \Phi\left(\frac{\widehat{Z}(\mathbf{x}_{\min}) - \widehat{Z}(\mathbf{x})}{\widehat{s}(\mathbf{x})}\right) + \widehat{s}(\mathbf{x})\phi\left(\frac{\widehat{Z}(\mathbf{x}_{\min}) - \widehat{Z}(\mathbf{x})}{\widehat{s}(\mathbf{x})}\right) \tag{3}$$

where $\widehat{Z}(\mathbf{x}_{\min})$ is the stochastic kriging prediction at $\mathbf{x}_{\min}$ (i.e. the point having the lowest sample mean for the scalarized objective among all feasible points already sampled), $\phi(\cdot)$ and $\Phi(\cdot)$ are the standard normal density and standard normal distribution function respectively, $\widehat{s}(\mathbf{x})$ is the (deterministic) ordinary kriging standard deviation, and $\widehat{Z}(\mathbf{x})$ is the stochastic kriging prediction (see [10] for a detailed mathematical formulation of this estimator).

Lastly, we propose to use the combination of the probability of feasibility (predicted by the LR model) and MEI as the acquisition function used in Step 5 to suggest a new process configuration. We refer to this acquisition function as *Constrained Modified Expected Improvement (CMEI)*:

$$CMEI(\mathbf{x}) = MEI(\mathbf{x}) * P(y = 1|\mathbf{x}) \tag{4}$$

In this work, we use the *Particle Swarm Optimization* (PSO) metaheuristic to find the infill point that maximizes CMEI (i.e., the fitness function of this inner optimization). Our choice is motivated by the good performance and low computational time observed in other studies with high-dimensional search space [23]. With PSO, the position of the particle represents the values of each variable to optimize. At the end of the search performed in Step 5, the particle representing the global best solution is evaluated with the expensive objectives, and its information is used to update the parameters of both ML models (the GPR model and the LR model). The algorithm continues searching for new infill points until the computational budget is depleted.

4 Design of Numerical Experiments

A Matlab process simulator was provided by the Joining & Materials Lab[1] to test the proposed optimization approach outside the lab environment since the

[1] https://www.flandersmake.be.

real experimentation is very expensive to perform (besides the experimental cost, testing a single process configuration can take up to 6 h). This simulator predicts the lap shear strength of the sample (MPa), failure mode (adhesive, substrate, or cohesive failure), sample production (in euros), and visual quality outcome (OK or not OK) based on the process parameters discussed above. It is *not* meant to be a perfect digital twin of the true process (such cheap digital twin does not exist), but rather a tool for the *relative* comparison of the algorithms' performance, under different conditions, at almost zero cost. Table 1 shows the range of each process parameter considered in the optimization problem.

One of the factors that introduce noise in the measurements is the so-called *contact angle*[2]: this re reflects the extent to which the adhesive can maintain good contact with the material. We use $\gamma = 30\%$ as a realistic value for the *standard deviation of the contact angle.*

Table 1. Range of the bonding process parameters (input variables)

ID	Variable	Min	Max
v1	Pre-processing	Yes or No	
v2	Power setting (W)	300	500
v3	Torch speed (mm/s)	5	250
v4	Distance between the torch and the sample (cm)	0.2	2
v5	Number of passes	1	50
v6	Time between plasma treatment and glue application (min)	1	120

Table 2. Summary of the parameters for both optimization approaches

Setting	MO-GP	cNSGA-II	GP-cNSGA-II
Size of initial design	LHS: $N = 20$		
Crossover probability	-	$c_p = 0.9$	
Mutation probability	-	$m_p = 0.1$	
Replications	$r = 5$		
Iterations/Generations	100	5	
Acquisition function	CMEI	-	EI
Acquisition function optimization	PSO*	-	
Kernel	Gaussian	-	Gaussian

We benchmark the performance of our proposal against two adaptations of the popular NSGA-II: 1) a constrained version to handle feasibility constraints (cNSGA-II) [6], and 2) an adaptation of cNSGA-II to use the surrogate prediction to generate new populations (GP-cNSGA-II) [15]. Given the experimental setting in Table 2, each algorithm evaluates exactly 120 process configurations in an expensive way, with 5 replications per configuration (i.e., 600 expensive evaluations in total). The BO algorithms start with an initial design of 20 configurations, and 100 new ones (infill points) will be obtained during the optimization. Evolutionary algorithms use a population of 20 configurations, where the

[2] Other noise factors not controlled in the simulator are not further discussed.

initial population coincides with the initial set used by BO algorithms. Then, 5 populations are generated (i.e., 5 generations) by applying genetic operators. As common in the literature, the fitness of the configuration outcomes is evaluated in the evolutionary approaches based on the sample means over a number of replications (note that, by doing so, both NSGA-II-based algorithms implicitly ignore the fact that this sample mean is in itself noisy and, hence, uncertain). The MO-GP approach takes into account both the sample mean and the sample variance though. While a total budget of 600 evaluations may seem high, it allows us to study the progress the algorithms would have obtained at lower budgets, and resemble real experimentation in the laboratory.

We evaluate the quality of the resulting fronts using the hypervolume (HV) indicator (reference point with production cost = 3, break strength = 4) [2], applied to the sample means. In addition, the modified Inverted Generational Distance (IGD+) [11] is used to quantify the distance between the Pareto front obtained by the algorithms and an ideal front (resulting from the Halton experiment, see Fig. 2). As the front obtained by the algorithms may depend on the initial design, we performed 50 macro-replications: each macro-replication starts with a different initial design, on which the algorithms then start their calculations.

5 Results

Figure 2 shows the mean responses of the simulator on a Halton sample of 60 000 process configurations assuming the contact angle could be perfectly controlled (i.e., noiseless with $\gamma = 0\%$). Interestingly, the feasible solutions seem to be clustered in areas with high break strength; moreover, the use of pre-processing seems to merely lead to a cost increase, while the resulting gains in break strength are very scarce.

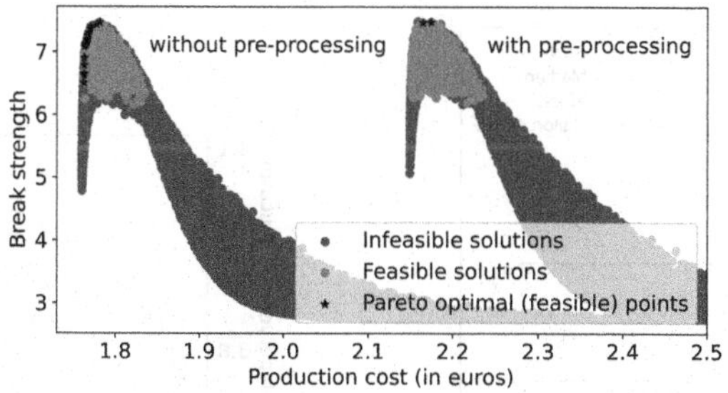

Fig. 2. Sample mean of break strength versus production cost, estimated by the simulator for 60 000 random process configurations ($\gamma = 0\%$)

Figure 3 (Left) shows the final *best* Pareto front obtained by MO-GP over 50 macro-replications (along with the *median* and *worst* fronts), for $\gamma = 30\%$. Clearly, the Pareto front obtained by MO-GP is very close to the ideal Pareto front estimated by means of the Halton set exploration. MO-GP also leads to a *faster* increase in HV, in terms of the number of expensive evaluations performed, than evolutionary approaches. This is evident from the evolution of the *average* hypervolume (across macro-replications) observed in Fig. 3 (Right). MO-GP thus is able to obtain better quality results for the Pareto front than cNSGA-II and GP-cNSGA-II, particularly at very limited budgets (after 60 expensive evaluations for instance). This makes the algorithm better suited than evolutionary algorithms in settings with expensive evaluations. As it was observed in [4], the inclusion of a surrogate in NSGA-II improves the performance of the evolutionary algorithm, but the small evaluation budget still limits its performance. The improvement in MO-GP can also be observed by comparing the average HV (MO-GP: 4.0829; cNSGA-II: 4.0204, GP-cNSGA-II: 4.0173) and IGD+ (MO-GP: 0.0585; cNSGA-II: 0.0711; GP-cNSGA-II: 0.0739). Overall, MO-GP yields Pareto fronts that are on average closer to the ideal front. Lastly, Wilcoxon's rank sum test [19] shows a significant p_value ($\alpha = 0.05$) for differences between MO-GP and the other algorithms, both in hypervolume (GP-cNSGA-II: 0, MO-DGP: 1.0e−05, cNSGA-II: 0) and IGD+ (GP-cNSGA-II: 0.012, MO-DGP: 1.8e−04, cNSGA-II: 0.0144).

Figure 3 (Right) shows a dashed line (MO-DGP), which represents the results of training MO-GP neglecting the presence of noise in the break stress. As shown, the HV curve under realistic conditions ($\gamma = 30\%$) is over the (deterministic) MO-GP after 40 expensive evaluations (20 new process configurations in addition to the starting set), which illustrates that neglecting the presence of noise in realistic settings can lead to *poor* optimal configurations.

Further analysis reveals that ±71% of the Pareto-optimal solutions put forward by GP-cNSGA-II use preprocessing (i.e., they are located in the bump

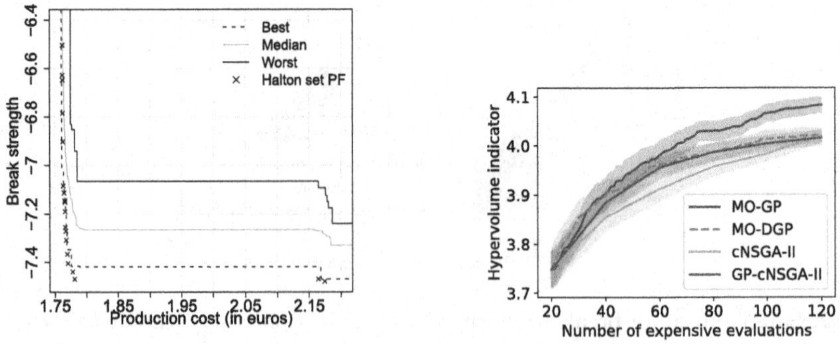

Fig. 3. Optimization results. (Left) Best, median and worst Pareto front obtained for MO-GP. (Right) The evolution of the hypervolume indicator (average of 50 macro-replications)

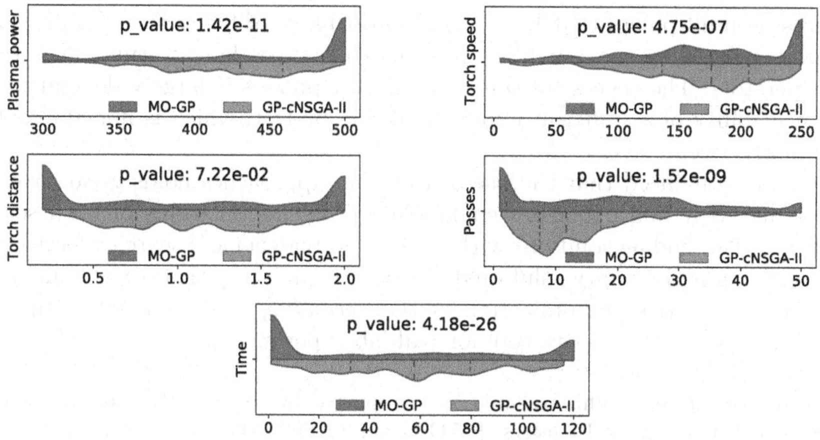

Fig. 4. Probability distributions of the Pareto-optimal input values obtained by MO-GP and GP-cNSGA-II, across 50 macro-replications. The dashed lines represent the 25%, 50% (median), and 75% percentiles of the observed distributions. *p_values* show significant differences between the configurations suggested by the algorithms.

at the right-hand side of Fig. 2), showing that the algorithm tends to "miss" solutions that do *not* require preprocessing (left-hand bump in Fig. 2). For MO-GP, ±65% of the solutions used pre-processing. Figure 4 analyzes the differences in the Pareto-optimal process configurations for the other input variables. Also here, the results show that our algorithm succeeds in finding solutions that are spread *across* the different ranges of all input variables, and the p_value of the Wilcoxon rank test indicates significant differences amongst the optimal configurations suggested by the two best algorithms (MO-GP and GP-cNSGA-II), implying that they were found on different regions of the search space.

6 Conclusions

In this paper, we apply a constrained BO algorithm to solve a bi-objective problem related to the adhesive bonding process of materials (maximizing break strength while minimizing production costs). The proposed Bayesian approach is shown to clearly outperform NSGA-II-based algorithms, which are commonly used in engineering design when solving general multi-objective, constrained problems. The difference lies in the way the experimental design is guided throughout the search: the Bayesian approach selects infill points based on an (explainable) infill criterion, which is related to the expected merit of the new infill point for optimization.

The superiority of the Bayesian approach is particularly evident in settings where the objectives are noisy: the GPR model used in the Bayesian approach accounts for this noise, whereas NSGA-II-based algorithms rely on the (noisy) sample means as information to guide the search. Moreover, the classification model embedded in the Bayesian approach ensures that the search focuses on infill points that have a high probability of being feasible. In evolutionary approaches, the search is guided by (black box, hard-to-tune) evolutionary operators. The success of this evolutionary process is largely dependent on the availability of a sufficient experimentation budget, which is not always the case in practice.

We are convinced that the use of Bayesian approaches holds great promise in solving noisy and expensive engineering problems, in terms of both search efficiency (i.e., finding solutions within a limited budget) and search effectiveness (i.e., yielding high-quality solutions). Future research will focus on the inclusion of a third objective (minimization of the *debonding* break strength) and the deployment of an interactive tool for real lab experiments.

Acknowledgements. This research was supported by the FLAIR Program and by the Research Foundation Flanders (FWO Grant 1216021N).

References

1. Ankenman, B., Nelson, B.L., Staum, J.: Stochastic kriging for simulation metamodeling. Oper. Res. **58**(2), 371–382 (2010)
2. Auger, A., Bader, J., Brockhoff, D., Zitzler, E.: Hypervolume-based multiobjective optimization: theoretical foundations and practical implications. Theoret. Comput. Sci. **425**, 75–103 (2012). https://doi.org/10.1016/j.tcs.2011.03.012
3. Briffoteaux, G., Gobert, M., Ragonnet, R., Gmys, J., Mezmaz, M., Melab, N., Tuyttens, D.: Parallel surrogate-assisted optimization: batched Bayesian neural network-assisted GA versus q-EGO. Swarm Evol. Comput. **57**, 100717 (2020)
4. Briffoteaux, G., Ragonnet, R., Tomenko, P., Mezmaz, M., Melab, N., Tuyttens, D.: Comparing parallel surrogate-based and surrogate-free multi-objective optimization of COVID-19 vaccines allocation. In: Dorronsoro, B., Pavone, M., Nakib, A., Talbi, E.G. (eds.) OLA 2022. CCIS, pp. 201–212. Springer, Cham (2022)
5. Brockmann, W., Geiß, P.L., Klingen, J., Schröder, K.B.: Adhesive Bonding: Materials Applications and Technology. Wiley, Hoboken (2008)
6. Brownlee, A.E., Wright, J.A.: Constrained, mixed-integer and multi-objective optimisation of building designs by NSGA-II with fitness approximation. Appl. Soft Comput. **33**, 114–126 (2015). https://doi.org/10.1016/j.asoc.2015.04.010
7. Budhe, S., Banea, M., De Barros, S., Da Silva, L.: An updated review of adhesively bonded joints in composite materials. Int. J. Adhes. Adhes. **72**, 30–42 (2017)
8. Cavezza, F., Boehm, M., Terryn, H., Hauffman, T.: A review on adhesively bonded aluminium joints in the automotive industry. Metals **10**(6), 730 (2020)
9. Correia, S., Anes, V., Reis, L.: Effect of surface treatment on adhesively bonded aluminium-aluminium joints regarding aeronautical structures. Eng. Fail. Anal. **84**, 34–45 (2018)
10. Forrester, A., Sobester, A., Keane, A.: Engineering Design via Surrogate Modelling (A Practical Guide), 1st edn. John Wiley and Sons, West Sussex, UK (2008)
11. Ishibuchi, H., Masuda, H., Tanigaki, Y., Nojima, Y.: Modified distance calculation in generational distance and inverted generational distance. In: International Conference on Evolutionary Multi-criterion Optimization, pp. 110–125 (2015)
12. Jalali, H., Van Nieuwenhuyse, I., Picheny, V.: Comparison of kriging-based algorithms for simulation optimization with heterogeneous noise. Eur. J. Oper. Res. **261**(1), 279–301 (2017)

13. Knowles, J.: ParEGO: a hybrid algorithm with on-line landscape approximation for expensive multiobjective optimization problems. IEEE Trans. Evol. Comput. **10**(1), 50–66 (2006)
14. Loeppky, J., Sacks, J., Welch, W.: Choosing the sample size of a computer experiment: a practical guide. Technometrics **51**(4), 366–376 (2009)
15. Qin, S., Sun, C., Jin, Y., Zhang, G.: Bayesian approaches to surrogate-assisted evolutionary multi-objective optimization: A comparative study. In: 2019 IEEE Symposium Series on Computational Intelligence (SSCI), pp. 2074–2080 (2019)
16. Quan, N., Yin, J., Ng, S.H., Lee, L.H.: Simulation optimization via kriging: a sequential search using expected improvement with computing budget constraints. IIE Trans. **45**(7), 763–780 (2013)
17. Rojas Gonzalez, S., Jalali, H., Van Nieuwenhuyse, I.: A multiobjective stochastic simulation optimization algorithm. Eur. J. Oper. Res. **284**(1), 212–226 (2020). https://doi.org/10.1016/j.ejor.2019.12.014
18. da Silva, L., Ochsner, A., Adams, R., Spelt, J.: Handbook of Adhesion Technology. Springer, Heidelberg (2011). https://doi.org/10.1007/978-3-642-01169-6
19. Steel, R.G.D., Torrie, J.H., et al.: Principles and Procedures of Statistics, a Biometrical Approach. No. Ed. 2, McGraw-Hill Kogakusha, Ltd. (1980)
20. Tao, T., Zhao, G., Ren, S.: An efficient kriging-based constrained optimization algorithm by global and local sampling in feasible region. J. Mech. Des. **142**(5), 051401 (2020)
21. Williams, C.K., Rasmussen, C.E.: Gaussian Processes for Machine Learning, vol. 2. MIT Press, Cambridge (2006)
22. Witten, I.H., Frank, E., Hall, M.A.: Data Mining: Practical Machine Learning Tools and Techniques, 3rd edn. Morgan Kaufmann, Burlington (2011)
23. Yarat, S., Senan, S., Orman, Z.: A comparative study on PSO with other metaheuristic methods. In: Mercangöz, B.A. (ed.) Applying Particle Swarm Optimization. ISORMS, vol. 306, pp. 49–72. Springer, Cham (2021). https://doi.org/10.1007/978-3-030-70281-6_4
24. Zhou, A., Qu, B.Y., Li, H., Zhao, S.Z., Suganthan, P.N., Zhang, Q.: Multiobjective evolutionary algorithms: a survey of the state of the art. Swarm Evol. Comput. **1**(1), 32–49 (2011)

Evaluating Surrogate Models for Robot Swarm Simulations

Daniel H. Stolfi[1(✉)] and Grégoire Danoy[1,2]

[1] Interdisciplinary Centre for Security, Reliability and Trust (SnT), University of Luxembourg, Esch-sur-Alzette, Luxembourg
{daniel.stolfi,gregoire.danoy}@uni.lu
[2] FSTM/DCS, University of Luxembourg, Esch-sur-Alzette, Luxembourg

Abstract. Realistic robotic simulations are computationally demanding, especially when considering large swarms of autonomous robots. This makes the optimisation of such systems intractable, thus limiting the instances' and swarms' size. In this article we study the viability of using surrogate models based on Gaussian processes, Artificial Neural Networks, and simplified simulations, as predictors of the robots' behaviour, when performing formations around a central point of interest. We have trained the predictors and tested them in terms of accuracy and execution time. Our findings show that they can be used as an alternative way of calculating fitness values for swarm configurations which can be used in optimisation processes, increasing the number evaluations and reducing execution times and computing cluster budget.

Keywords: predictors · surrogate models · machine learning · evolutionary algorithm · swarm robotics · robot formation

1 Introduction

A swarm of robots is a group of robots that show a collective behaviour, which is usually achieved from their iterations, with the objective of performing some specific tasks. One of these tasks is robot formation where the swarm members are arranged in a specific shape. These types of problems usually present unknown initial positions of the swarm members, as well as the need of path planning from these positions to the final locations. Having predefined final positions also presents a challenging adaptability to real situations, e.g. asteroid observation or escorting a rogue drone out of a restricted area, especially when there are collisions, communication losses, or robot failures.

The simulation of swarm of robots in a 3D space is related to high computing resources (and time) to achieve high levels of accuracy. For example, using a multi-physics robot simulator, e.g. ARGoS [12] to simulate a formation problem would demand from seconds to minutes, depending on the number of robots modelled [19]. If we take into account that many evaluations are needed when we face an optimisation problem and that each evaluation requires a simulation,

B. Dorronsoro et al. (Eds.): OLA 2023, CCIS 1824, pp. 224–235, 2023.
https://doi.org/10.1007/978-3-031-34020-8_17

experimentation rapidly become unaffordable. Consequently, an alternative technique, e.g. surrogate models, is desired to successfully complete such studies.

Surrogate models mimic (with different levels of accuracy) the behaviour of a complex system by giving an approximate outcome in a realistic execution time, avoiding the use of exhaustive simulations. They have been used in many applications such as modelling circuits and systems [23], wildfire forecasting [4], predicting noise emission and aerodynamic performance of propellers [13], sustainable building design [22], groundwater modelling [2], etc. We study in this article the viability of using surrogate models based on Gaussian processes, Artificial Neural Networks, and simplified simulations, as predictors of the robots' behaviour when arranging in a formation.

Our Distributed Formation Algorithm^3 (DFA^3) [19] consists of a range and bearing based approach where the robots in the swarm self-organise to arrange in a final desired formation, surrounding a central object in a sphere-like shape. There is no global coordinate system nor a different, intelligent node in the swarm. The robots just make their own local decisions based on local information following a pre-calculated optimal parameters. Since the optimisation of these robot parameters requires very accurate and time demanding simulations combined with a meta-heuristic [19], we propose the study of seven surrogate models that can be used to improve the evaluation times, predicting simulation outcomes without losing accuracy.

The rest of this paper is organised as follows. In the next section, we review the state of the art related to our proposal. In Sect. 3 our robot formation system is described and in Sect. 4, ARGoS simulations and the seven proposed surrogate models are discussed. The experimental results are in Sect. 5 and finally, Sect. 6 brings conclusion and future work.

2 Literature Review

In this section we discuss some recent research works related to robot simulations using surrogate models. The interested reader can see [1] for a review in terms of computational time, accuracy and problem size.

A quadcopter control is presented in [10]. The authors propose several machine learning techniques such as time series, Gaussian processes and neural networks, to calculate optimum control gains for a specific mission to overcome environmental uncertainties. These predictors are used in an optimisation process and tested using simulations. Having observed a better exploration of the design space, the obtained results showed performance improvements when compared to nominal control gains. In our present work, we analyse some of the predictors proposed in this article to parameterise a swarm of robots.

A surrogate approach using Kriging method is presented in [24] to optimise the design of the delta wing and the canard wing of a tube-fan hybrid Unmanned Aerial Vehicle (UAV). A multi-objective genetic algorithm is proposed to maximise lift while minimising energy consumption. The calculated solutions were validated using computational fluid dynamics simulations. We also use gaussian process regression (Kriging) among others as predictors, although applied to a different problem.

A parameter of the Rössler chaotic system to improve coverage of the CACOC (Chaotic Ant Colony Optimisation for Coverage) algorithm is tuned in [15] by using a surrogate-based method. The parameter space is efficiently explored using Bayesian optimisation avoiding using costly simulations. The authors' results showed that this method permitted to explore efficiently a bifurcation diagram by-passing periodic regions, providing two groups of points with excellent results in terms of coverage for the swarm. We have also analysed Gaussian processes as well as other methods to calculate an accurate surrogate model for our problem.

A mathematical-computational model for the control and navigation of robots is proposed in [9]. The authors combined a 2D cellular automata, Tabu search, ant colonies, and greedy approaches for selecting elitist cells, with a genetic algorithm to optimize the parameters for the two proposed surrogate models. The objective was maximising area coverage by using a pheromone-based approach. In addition, the validation of the models was done using the Webots simulator and E-Puck robots. In the present article we evaluate seven surrogate models for the formation problem in a 3D space.

In this article we propose seven surrogate models for the robot formation problem, train them and analyse their results in terms of accuracy and execution times. We aim to use these models for the optimisation of the robot swarm parameters in future works, achieving stable formations around a central point of interest.

3 Robot Formation

Our Distributed Formation Algorithm^3 (DFA^3) [19] was designed to arrange robots at the vertices of an imaginary polygon surrounding a central point of interest (Fig. 1). Achieving stable robot formations frequently involves addressing different constraints such as limited communication range, absence of absolute positions, and uncertain initial conditions. Each robot in the swarm receives a beacon signal emitted by the other robots and uses it to calculate its relative orientation and distance to the rest of swarm members. Since this system can be used not only on Earth but also in space, no localisation system, such as GPS, is used by the robots (UAVs, satellites, probes, etc.). Additionally, the robots do not have a predefined final position in the formation. The central object is tracked using its own radio signal in our experiments, although other methods can be used such as images from cameras, LIDAR (LIght Detection And Ranging) data, etc.

We have initially tested our formation algorithm in a 2D environment using E-Puck2 robots [18] and then proposed an extension of the algorithm to deal with 3D formations using UAVs [19]. A meta-heuristic was needed, e.g. a Hybrid Genetic Algorithm, to calculate the optimal parameters for the formation since it is a very complex problem requiring the use of realistic simulations. However, we have observed that the optimisation process was taking too long when the number of robots is high (720 h for 30 runs), limiting the size of the swarms.

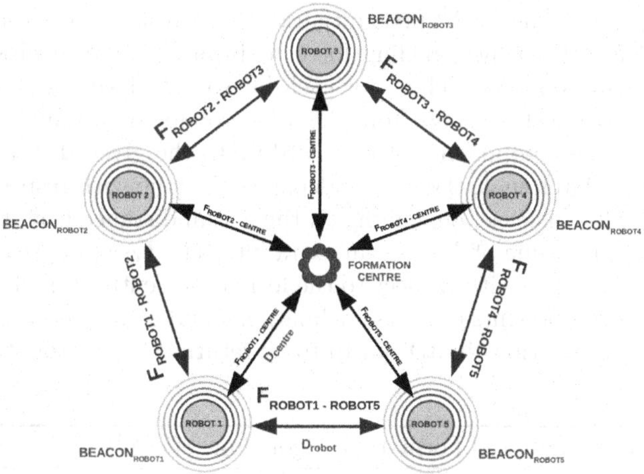

Fig. 1. Five robots in formation surrounding a central object.

Consequently, we propose the study of surrogate models [5] to speed up the evaluation of the formation parameters, allowing not only having more robots in the swarm but also assessing more accurate optimisations by increasing the number of evaluations.

3.1 Problem Definition

The formation problem is defined by $P = (\mathcal{G}, co, S, C)$, where the distance graph is given by $\mathcal{G} = (V, E, D)$, where $V = \{ROBOT_1, \ldots, ROBOT_N\}$ represents the robots in the swarm, $E = \{(i, j) \in V \times V\}$ represents the edges of the graph indicating the swarm connectivity, and $D = \{d(i, j), \forall (i, j) \in E\}$ represents the distances between robots (D_{ROBOT}). Furthermore, co stands for the central object, the distances between the robots and the central object are given by $S = \{d(co, u), u \in V\}$, and the problem's constraint is given by $C = \forall d(co, j) \in S, d(co, j) = D_{CENTRE}$, where D_{CENTRE} is the desired distance to the formation centre (radius).

There are four parameters for the swarm used to achieve stable formations: a distance threshold $D_{THRESHOLD}$ to control the attracting/repelling movement between robots, the minimum distance D_{MIN} to the centre, the intensity of the attracting/repelling force F_{CENTRE}, with respect to the central object, and the moving speed $SPEED$ of each robot.

3.2 Distributed Formation Algorithm³ (DFA³)

The pseudocode of our DFA³ [19] is detailed in Algorithm 1. Each robot executes the same algorithm using the optimal parameters and the same predefined formation radius, i.e. the desired distance to the rogue drone D_{CENTRE}, which is a constant value. The DFA^3 first initialises the vector r where the calculation of the resulting attracting/repelling force to/from the central object plus the other robots will be stored. Then, for each beacon received, the range and the vertical and horizontal bearings from the other robots are obtained and used to calculate the three components of r, according to the given distance threshold $D_{THRESHOLD}$. After that, the same calculation is done with respect to the formation centre. In this case, depending on the actual distance from the robots to the centre and the value of D_{MIN}, an extra intensity F_{CENTRE} can be applied as a repelling force (ω) with respect to the formation centre. Finally, having calculated the 3D components of r, the inclination θ and azimuth ϕ are obtained as the new moving direction (in 3D space) to be returned to the robot's controller.

Algorithm 1. Distributed Formation Algorithm^3 (DFA^3).

```
function DFA³(D_CENTRE, D_THRESHOLD, D_MIN, F_CENTRE)
    r_x ← 0, r_y ← 0, r_z ← 0
    for robot ∈ BEACONS do
        range, vBearing, hBearing ← RangeAndBearing(robot)            ▷ Other Robots
        r_x ← r_x + (range − D_THRESHOLD) × cos(hBearing) × sin(vBearing)
        r_y ← r_y + (range − D_THRESHOLD) × sin(hBearing) × sin(vBearing)
        r_z ← r_z + (range − D_THRESHOLD) × cos(vBearing)
    end for
    range, vBearing, hBearing ← RangeAndBearing(centre)                ▷ Centre
    ω ← 1.0
    if range < D_MIN then
        ω ← F_CENTRE                                                   ▷ Force intensity
    end if
    r_x ← r_x + ω(range − D_CENTRE) × cos(hBearing) × sin(vBearing)
    r_y ← r_y + ω(range − D_CENTRE) × sin(hBearing) × sin(vBearing)
    r_z ← r_z + ω(range − D_CENTRE) × cos(vBearing)
    θ ← arctan(r_y / r_x), φ ← arccos(r_z / sqrt(r_x² + r_y² + r_z²))  ▷ Next moving direction (angles)
    return θ, φ                                                        ▷ Inclination and azimuth
end function
```

4 Realistic Simulations vs. Surrogate Models

4.1 ARGoS Simulations

The proposed formation scenarios were modelled in ARGoS [12], a multi-physics robot simulator which can efficiently simulate large-scale swarms of robots of any kind. In our study, we have used the model of the Spiri UAVs [17] using the ARGoS' Range and Bearing communication model (Fig. 2) to simulate robot communications. Each UAV is only aware of the relative distance and angles of the other robots, calculated from the received beacon signals. The experimental area is a cube of 30 × 30 × 30 m and the distance to the centre of the formation was set to three metres, according to the simulation area dimensions.

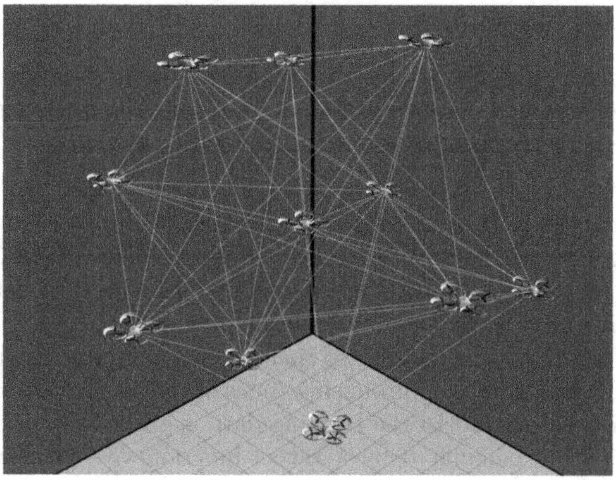

Fig. 2. Robot formation in the ARGoS simulator.

The DFA[3] was implemented in each robot controller and was parameterised using the aforementioned formation parameters, i.e. $D_{THRESHOLD}$, D_{MIN}, F_{CENTRE}, $SPEED$. A stable formation depends on the values of these parameters, requiring an optimisation process taking into account the distance to the centre (D_{CENTRE}) and the number of robots. As these simulations are very computationally demanding, an alternative technique, e.g. surrogate models, is needed when there are many robots in the swarm. In the next section we propose an evaluation function to compute the correctness of the formation depending on its parameters.

4.2 Formation Fitness

We have proposed in [19] the fitness function $F(\vec{x})$ (Eq. 1) to evaluate the formation in terms of shape, distance to the centre, and how equally spaced are the robots (avoiding forming local clusters). The terms in $F(\vec{x})$ are the minimum error ($Em(\vec{x})$) and maximum error ($EM(\vec{x})$), both calculated from the distance of every robot in the swarm to the centre, with respect to the desired distance D_{CENTRE}. The last term ($D(\vec{x})$) is meant to evaluate how spread are the robots throughout the surface (sphere). These terms are to be minimised by using an optimisation algorithm. Thus, the lower the value of $F(\vec{x})$ the better.

$$F(\vec{x}) = Em_j(\vec{x}) + EM_j(\vec{x}) + D_j(\vec{x}) \tag{1}$$

$$Em(\vec{x}) = |\min \delta(i, centre) - D_{CENTRE}|, \quad i \in \{1 \dots N\} \tag{2}$$

$$EM(\vec{x}) = |\max \delta(i, centre) - D_{CENTRE}|, \quad i \in \{1 \dots N\} \tag{3}$$

$$D(\vec{x}) = |2.0 \times D_{CENTRE} - \min \delta(l, m)|, \quad \forall l, m \in \{1 \dots N\}, l \neq m \tag{4}$$

4.3 Surrogate Models

We propose seven surrogate models to replace the costly simulations by predictions. Five are based on Gaussian processes, the sixth is using an artificial neural network, and the last one is based on a fast cellular simulator.

Gaussian Processes (GPs). Bayesian optimisation aims to solve black-box problems by generating surrogate models of the problems using Gaussian processes (GPs) [16]. GPs are both interpolators and smoothers of data and can be used as effective predictors when the solutions' landscape ($F(\vec{x})$ in our study) is a smooth function of the parameter space. It calculates a distribution of the objective function by sampling promising zones of the solution space. The Gaussian distribution associated to the training data is given by a mean vector and a covariance matrix, calculated by a kernel function. We propose testing five different kernel functions: *gp_lin* (linear), *gp_sexp* (squared exponential), *gp_nn* (neural network), *gp_m32* (Matérn $\nu = 3/2$), and *gp_m52* (Matérn $\nu = 5/2$), provided by the R package "gplite" [11]. We have set up 1000 maximum iterations and 100 restarts for training each of these predictors.

Artificial Neural Network (ANN). Neural networks have been used in numerous machine learning research works in the last years. We propose an artificial neural network with four neurons as inputs corresponding to our problem's variables, one linear output neuron, and five neurons in the hidden layer (experimentally chosen taking into account the required training time). We have used resilient backpropagation (RPROP) with weight backtracking [14] during the training process, which performs a direct adaptation of the weight step based on local gradient information. RPROP has the advantage that for many problems no choice of parameters is needed to obtain optimal convergence times. We have used the R package "neuralnet" [7] to implement this predictor and performed 100 repetitions to select the best calculated network (minimum error).

Cellular Simulator (C-Sim). Finally, we propose a cellular simulator to implement a simplified model of the robot simulation without using inertial real physics. We kept the same arena dimensions as in the ARGoS simulations as well as the robot's starting positions. We have defined four constants, $\{\omega_1, \omega_2, \omega_3, \omega_4\}$ to calibrate this model as part of its training, reducing the error between the fitness values obtained and the reference values provided by ARGoS. An optimisation of the four ω_i values is then to be conducted using a genetic algorithm (GA) implemented by using the jMetalPy package [3], to fit the C-Sim model to the realistic ARGoS simulations. Binary Tournament [8] was used as selection operator, Uniform Crossover [20] as recombination operator ($P_c = 0.9$), Integer Polynomial Mutation [6] as mutation operator ($P_m = 1/L$), while an elitist replacement was used to update the algorithm population after each generation. This generational GA has a population of $\lambda = 100$ individuals and will perform 10000 evaluations per run.

5 Experimental Results

In this section we first present the training of the seven predictors followed by the testing phase where we address the accuracy of each surrogate model and the corresponding improvement in evaluation times.

5.1 Training

We have calculated a training set initially consisting of 300 fitness values corresponding to different formation parameters randomly chosen to sample the parameter space. When there is a robot collision (which can happen when the swarm is misconfigured), ARGoS simulations stop and the calculated fitness value turns into a penalisation value. This represents discontinuities in the fitness function which unnecessarily complicate the training process. For this first study, we decided to treat these values as outliers and keep them out of the training process. We propose the Mean Square Error (MSE) as a metric to evaluate the predictors' accuracy. It is calculated as shown in Eq. 5, where n is the number of data points, Y_i are the observed values (from ARGoS), and $\widehat{Y}_i$ are the estimated values (from predictors).

$$MSE = \frac{1}{n}\sum_{i=1}^{n}(Y_i - \widehat{Y}_i)^2 \tag{5}$$

All in all, we present in Table 1 the results of the training of the seven predictors using surrogate models. We can see that the number of observations is lower when the number of robots is higher as collisions are more likely to happen. GP using neural networks as kernel function (*gp_nn*) has shown the most accurate results in terms of MSE. Predictions from ANN have not been so accurate and C-Sim, despite of being a simulation based model, has obtained the worst results. Note that GP using a linear kernel has not converged for swarms of five robots during the training process.

Table 1. MSE values for the training process of predictors based on GP's, ANN, and C-Sim. Note that *gp_lin* has not converged for five robots.

# robots	# Obs	gp_lin	gp_sexp	gp_nn	gp_m32	gp_m52	ann	c-sim
3	238	13.030	6.138	**5.081**	5.408	5.408	5.835	12.788
5	192	⋆11.665	4.059	**3.643**	3.720	3.898	4.966	16.630
10	118	7.542	2.236	**0.995**	1.759	1.926	2.525	13.546

Moreover, Fig. 3 shows the boxplots representing the distribution of the measured error values for each predictor and swarm. It can be seen that despite outliers, the surrogate models based on GP present the most accurate values (except by the linear kernels). Finally, in Table 2 the elapsed training times for each predictor are presented. It can be seen that GP models are quite fast compared to

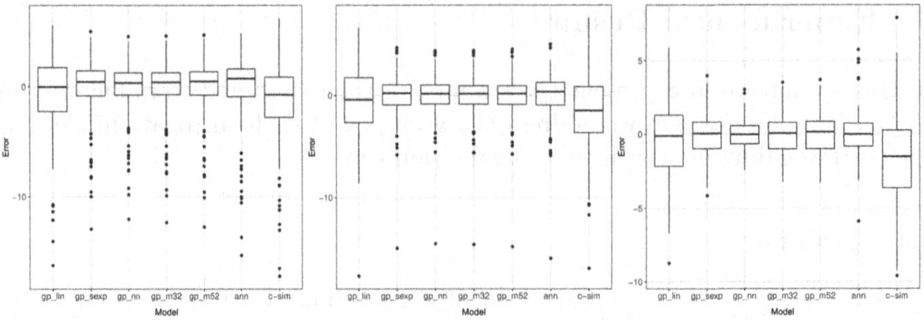

(a) 3 robots (b) 5 robots (c) 10 robots

Fig. 3. Boxplots showing the error distribution between the ARGoS simulation fitness values and the predictors' ones.

Table 2. Elapsed training times in seconds for the seven predictors. Note that C-Sim includes 30 parallel runs of the optimisation GA.

# robots	gp_lin	gp_sexp	gp_nn	gp_m32	gp_m52	ann	c-sim
3	**0.457**	1.324	2.203	1.071	1.060	2674	2859
5	5.622	**0.558**	0.754	0.876	0.955	1428	2365
10	**0.099**	0.190	0.402	0.318	0.354	330	1758

ANN and C-Sim. Taking this into account, plus their accuracy during the training stage, GP models look promising as surrogate models for the simulations of robot formations. In the next section we test all the calculated predictors on a number of unseen swarm configurations to address their accuracy beyond the training set.

5.2 Testing

The testing dataset consists in 3000 new fitness values corresponding to formations developed by swarms of three, five, and ten robots. Table 3 shows the MSE values calculated using the testing dataset compared with the predictions. Again, in concordance with the training stage, the outliers corresponding to robot collisions have been removed. It can be seen that all GPs predictors (except *gp_lin*) have performed well, being *gp_nn* the most promising one ($MSE = 6.199$ for three robots, $MSE = 5.146$ for five robots, and $MSE = 3.966$ for ten robots), despite Matérn being slightly better for swarms of ten robots ($MSE = 3.841$). In congruence with the observed during training, our proposed ANN predictors are not good enough to compete with GPs and the cellular simulator (C-Sim) is not accurate whatsoever ($MSE > 11$ for all the cases studied). Figure 4 shows the distribution of the results where we can note the accuracy of *gp_sexp*, *gp_nn*, *gp_m32*, and *gp_m52*, compared with the rest of the predictors.

Table 3. MSE values for the predictions done using the GP's, ANN, and C-Sim models.

# robots	# Obs.	gp_lin	gp_sexp	gp_nn	gp_m32	gp_m52	ann	c-sim
3	2220	12.564	6.596	**6.199**	6.279	6.399	8.330	11.318
5	1853	13.483	5.471	**5.146**	5.231	5.314	7.160	15.723
10	1096	11.680	3.977	3.966	**3.841**	3.872	5.319	16.076

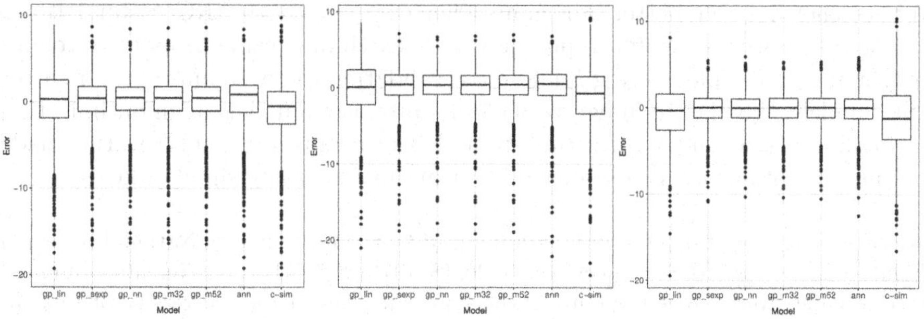

(a) 3 robots (b) 5 robots (c) 10 robots

Fig. 4. Error distribution between the predictions and the ARGoS' fitness values.

Table 4. Average computing times in seconds for the seven predictors compared with the corresponding ARGoS simulations.

# robots	ARGoS	gp_lin	gp_sexp	gp_nn	gp_m32	gp_m52	ann	c-sim
3	6.609	0.171	0.171	0.170	0.170	0.171	0.106	**0.004**
5	10.126	0.170	0.169	0.170	0.170	0.170	0.106	**0.004**
10	36.410	0.170	0.171	0.169	0.169	0.170	0.106	**0.007**

We have also studied the time needed for calculating the fitness for a given configuration of a robot swarm, using the different surrogate models, and comparing them to simulations using ARGoS. Table 4 shows the different average execution times calculated from 300 evaluations per swarm. We can see that the cellular simulator is by far the fastest model (milliseconds), although its predictions are not accurate as was aforementioned. The rest of predictors present calculation times under two tenths of seconds which represents a huge improvement compared with full ARGoS simulations which go from about 7 to more than 35 s, depending on the swarm size.

6 Conclusions

In this paper we have addressed the training and testing of seven surrogate models to be used as predictors of the formation accuracy (fitness) for swarms of three, five, and ten robots. We have defined the formation problem and the predictors based on Gaussian Processes (GPs), Artificial Neural Networks (ANN), and a simplified cellular simulator. Then, we have trained them using a dataset calculated from real ARGoS simulations and tested on unseen configurations in terms of accuracy (Median Square Error) and computation times. Our results show that GPs predictors have achieved the best results in accuracy and that they were very competitive in terms of execution times. These are good news as we expect that the GPs models will scale appropriately with the number of robots, showing even better time gains when compared with ARGoS simulations.

As future works we aim to pursue this research line, using dropout to reduce overfitting and implementing k-fold cross validation to try to improve our surrogate models. We plan to integrate the more promising models in an optimisation algorithm, e.g. a genetic algorithm, to address the optimisation of swarms made of many robots which is not affordable using just accurate simulations.

Acknowledgements. This work is supported by the Luxembourg National Research Fund (FNR) – ADARS Project, ref. C20/IS/14762457. The experiments presented in this paper were carried out using the HPC facilities of the University of Luxembourg [21]– see https://hpc.uni.lu.

References

1. Alizadeh, R., Allen, J.K., Mistree, F.: Managing computational complexity using surrogate models: a critical review. Res. Eng. Des. **31**(3), 275–298 (2020). https://doi.org/10.1007/s00163-020-00336-7
2. Asher, M.J., Croke, B.F.W., Jakeman, A.J., Peeters, L.J.M.: A review of surrogate models and their application to groundwater modeling. Water Resour. Res. **51**(8), 5957–5973 (2015). https://doi.org/10.1002/2015WR016967
3. Benítez-Hidalgo, A., Nebro, A.J., García-Nieto, J., Oregi, I., Ser, J.D.: jMetalPy: a python framework for multi-objective optimization with metaheuristics. Swarm Evolut. Comput. 100598 (2019). https://doi.org/10.1016/j.swevo.2019.100598
4. Cheng, S., Prentice, I.C., Huang, Y., Jin, Y., Guo, Y.K., Arcucci, R.: Data-driven surrogate model with latent data assimilation: Application to wildfire forecasting. J. Comput. Phys. **464**, 111302 (2022). https://doi.org/10.1016/j.jcp.2022.111302
5. Cozad, A., Sahinidis, N.V., Miller, D.C.: Learning surrogate models for simulation-based optimization. AIChE J. **60**(6), 2211–2227 (2014). https://doi.org/10.1002/aic.14418
6. Deb, K.: Multi-Objective Optimization Using Evolutionary Algorithms. John Wiley & Sons Inc., USA (2001)
7. Fritsch, S., Guenther, F., Wright, M.N., Suling, M., Mueller, S.M.: neuralnet: training of neural networks (2022). https://CRAN.R-project.org/package=neuralnet. Accessed 8 Dec 2022
8. Goldberg, D.E., Deb, K.: A comparative analysis of selection schemes used in genetic algorithms. Found. Genet. Algorithms **1**, 69–93 (1991). https://doi.org/10.1016/B978-0-08-050684-5.50008-2
9. Lopes, H.J., Lima, D.A.: Evolutionary Tabu inverted ant cellular automata with elitist inertia for swarm robotics as surrogate method in surveillance task using e-puck architecture. Robot. Auton. Syst. **144**, 103840 (2021). https://doi.org/10.1016/j.robot.2021.103840

10. do Nascimento, R.G., Fricke, K., Viana, F.: Quadcopter control optimization through machine learning. In: AIAA Scitech 2020 Forum. No. 0 in AIAA SciTech Forum, American Institute of Aeronautics and Astronautics, January 2020. https://doi.org/10.2514/6.2020-1148
11. Piironen, J.: gplite: General purpose gaussian process modelling (2022). https://cran.r-project.org/package=gplite. Accessed 8 Dec 2022
12. Pinciroli, C., et al.: ARGoS: a modular, parallel, multi-engine simulator for multi-robot systems. Swarm Intell. **6**(4), 271–295 (2012). https://doi.org/10.1007/s11721-012-0072-5
13. Poggi, C., et al.: Surrogate models for predicting noise emission and aerodynamic performance of propellers. Aerosp. Sci. Technol. **125**, 107016 (2022). https://doi.org/10.1016/j.ast.2021.107016, sI: DICUAM 2021
14. Riedmiller, M., Braun, H.: A direct adaptive method for faster backpropagation learning: the rprop algorithm. In: IEEE International Conference on Neural Networks, vol. 1, pp. 586–591 (1993). https://doi.org/10.1109/ICNN.1993.298623
15. Rosalie, M., Kieffer, E., Brust, M.R., Danoy, G., Bouvry, P.: Bayesian optimisation to select Rössler system parameters used in chaotic ant colony optimisation for coverage. J. Comput. Sci. **41**, 101047 (2020). https://doi.org/10.1016/j.jocs.2019.101047
16. Seeger, M.: Gaussian processes for machine learning. Int. J. Neural Syst.**14**(02), 69–106 (2004). https://doi.org/10.1142/s0129065704001899
17. Spiri Robotics: Spiri mu (2022). https://spirirobotics.com/. Accessed 26 Sep 2022
18. Stolfi, D.H., Danoy, G.: Optimising autonomous robot swarm parameters for stable formation design. In: Proceedings of the Genetic and Evolutionary Computation Conference, pp. 1281–1289. GECCO 2022, Association for Computing Machinery, New York, NY, USA (2022). https://doi.org/10.1145/3512290.3528709
19. Stolfi, D.H., Danoy, G.: An evolutionary algorithm to Optimise a distributed UAV swarm formation system. Appl. Sci. **12**(20) (2022). https://doi.org/10.3390/app122010218
20. Syswerda, G.: Uniform crossover in genetic algorithms. In: Proceedings of the 3rd International Conference on Genetic Algorithms, pp. 2–9. Morgan Kaufmann Publishers Inc. (1989)
21. Varrette, S., Cartiaux, H., Peter, S., Kieffer, E., Valette, T., Olloh, A.: Management of an academic HPC & research computing facility: the ULHPC experience 2.0. In: Proceedings of the 6th ACM High Performance Computing and Cluster Technologies Conference (HPCCT 2022). Association for Computing Machinery (ACM), Fuzhou, China, July 2022
22. Westermann, P., Evins, R.: Surrogate modelling for sustainable building design - a review. Energy Build. **198**, 170–186 (2019). https://doi.org/10.1016/j.enbuild.2019.05.057
23. Yelten, M.B., Zhu, T., Koziel, S., Franzon, P.D., Steer, M.B.: Demystifying surrogate modeling for circuits and systems. IEEE Circuits Syst. Mag. **12**(1), 45–63 (2012). https://doi.org/10.1109/MCAS.2011.2181095
24. Yue, H., Medromi, H., Ding, H., Bassir, D.: A novel hybrid drone for multi-propose aerial transportation and its conceptual optimization based on surrogate approach. J. Phys. Conf. Ser. **1972**(1), 012103 (2021). https://doi.org/10.1088/1742-6596/1972/1/012103

Interactive Job Scheduling with Partially Known Personnel Availabilities

Johannes Varga[1](✉), Günther R. Raidl[1], Elina Rönnberg[2], and Tobias Rodemann[3]

[1] Institute of Logic and Computation, TU Wien, Vienna, Austria
{jvarga,raidl}@ac.tuwien.ac.at

[2] Department of Mathematics, Linköping University, Linköping, Sweden
elina.ronnberg@liu.se

[3] Honda Research Institute Europe, Offenbach, Germany
tobias.rodemann@honda-ri.de

Abstract. When solving a job scheduling problem that involves humans, the times in which they are available must be taken into account. For practical acceptance of a scheduling tool, it is further crucial that the interaction with the humans is kept simple and to a minimum. Requiring users to fully specify their availability times is typically not reasonable. We consider a scenario in which initially users only suggest single starting times for their jobs and an optimized schedule shall then be found within a small number of interaction rounds. In each round users may only be suggested a small set of alternative time intervals, which are accepted or rejected. To make the best out of these limited interaction possibilities, we propose an approach that utilizes integer linear programming and a theoretically derived probability calculation for the users' availabilities based on a Markov model. Educated suggestions of alternative time intervals for performing jobs are determined from these acceptance probabilities as well as the optimization's current state. The approach is experimentally evaluated and compared to diverse baselines. Results show that an initial schedule can be quickly improved over few interaction rounds, and the final schedule may come close to the solution of the full-knowledge case despite the limited interaction.

Keywords: Job scheduling · human machine interaction · preference learning · integer linear programming

1 Introduction

We consider a class of job scheduling problems in which the personnel of a company is involved as a bottleneck resource. The central aim is to schedule jobs of employees in an interactive way that works from the humans' perspectives as simple, stress-free, and with low cognitive effort—while at the same time a

J. Varga acknowledges the financial support from Honda Research Institute Europe.

B. Dorronsoro et al. (Eds.): OLA 2023, CCIS 1824, pp. 236–247, 2023.
https://doi.org/10.1007/978-3-031-34020-8_18

cost function is minimized. In the simplest, and from the users' perspective most convenient case, each user just suggests a starting time for each of her or his jobs. As the jobs also require further shared resources, this directly obtained schedule will rarely be feasible nor are cost-aspects considered. Ideally, we would have full knowledge of all the users' times at which they would be available for performing their jobs, in which case we could solve an optimization problem in one shot. In many practical scenarios, however, it is impossible or far too inconvenient to request such complete information. We therefore start with the users' initial suggestions and perform a small number of simple *interaction rounds* to get more freedom for finding better schedules. In each such round the solution approach is allowed to suggest each user a small number of additional time intervals for scheduling her or his jobs. The users are then supposed to indicate their acceptance or rejection of these intervals. Hereby, we intentionally avoid that users are requested to specify larger amounts of additional availability intervals on their own. With the increased knowledge on the users' availabilities, the optimization can aim at improving the solution in each round.

The main challenge we address in this work is to, in each round, come up with *meaningful queries* for further time intervals to perform jobs in. Queried time intervals are most meaningful when (a) they would allow the optimization to obtain a better schedule and (b) the users are likely to accept them. For example, very large time intervals may aid the optimization the most, but as they are rather unlikely to be accepted by the users, they are usually not that meaningful. To consider (b), the likelihood that users accept queried time intervals, in some reasonable way, we need to exploit at least some stochastic assumptions on the users' unknown availabilities. Ideally, we would have precise user-specific stochastic models available, for example derived from historic availability data. Here we assume that such information is not available and instead build upon just a simple stochastic model represented by a two-stage Markov process. In essence, we only assume to know average probabilities of users to be available/unavailable in a timestep under the condition that the user is known to be available/unavailable in the directly preceding timestep.

The overall scenario can also be seen as *active learning*, as the solution approach queries the users to learn more information, which is further exploited in the optimization. Our main contributions are (a) to propose this general interactive scheduling setting, (b) to narrow it down to a specific *Interactive Job Scheduling Problem* (IJSP) to make concrete computational investigations on, (c) an Integer Linear Programming (ILP) model as optimization core for solving the IJSP, (d) an exact and computationally efficient calculation of the probabilities for users to accept potentially queried time intervals based on the two-stage Markov process and the already known availability information from the users, and (e) to propose a heuristic solution approach for the IJSP that utilizes this probability calculation. In an experimental evaluation, this solution approach is compared to a greedy baseline approach as well as to solving the full-knowledge case. Results show that already with a very moderate amount of interaction and the simplistic assumptions of the two-stage Markov model, schedules may be obtained that come close to those of solving the full-knowledge case.

In the IJSP, we assume that each job is associated with and requires one specific user and one of a set of available machines. On each machine, only one job can be performed at any time in a non-preemptive manner. As planning horizon we consider several days and time is discretized. Jobs have individual but machine-independent durations. Scheduling a job induces costs, for example for used electricity, and we consider these costs to be time-dependent. For example, when electricity is bought on the spot market, (expected) electricity costs may change significantly over time. For avoiding to have to deal with infeasible schedules, we allow that jobs remain unscheduled at additional penalty costs. The objective is to find a feasible schedule of minimum total cost.

The core of this problem, if neglecting the users, can be described in the common three-field notation for scheduling problems as $\mathrm{P}m||$ TEC, where $\mathrm{P}m$ refers to the m machines and that job durations do not depend on the machines, and where the objective is to minimize the Total Energy Costs (TEC). The similar problem $\mathrm{P}m||C_{\max}$,TEC, which additionally takes the makespan into account for the objective, has been considered in the literature. Solving approaches for it include a Mixed Integer Linear Program (MILP) [2,10], a problem specific heuristic and a genetic algorithm [10], as well as a greedy heuristic and local search [2]. Also similar is the scheduling problem $\mathrm{R}m||$TEC where jobs have in general different processing times on different machines. For it, Ding et al. [5] proposed a MILP and a Dantzig-Wolfe decomposition. The MILP was further improved by Cheng et al. [4] and by Saberi-Aliabad et al. [8].

In interactive optimization approaches, most works only consider a single user who guides the optimization process. For instance, Saha et al. [9] develop approaches based on evolutionary algorithms that cooperate with human designers to find aesthetic, aerodynamic, and structurally efficient designs for automotives, and Aghaei-Pour et al. [1] consider a multiobjective optimization problem where the human interactively specifies preferences on the solution, which are also considered within evolutionary algorithms. Interactive optimization with multiple users is less common. For instance, Jatschka et al. [6] consider a MILP-based cooperative optimization approach that interacts with many users to learn an objective function for distributing service points in mobility applications.

We perform active learning on the availability times of the users. This has also been done in the domain of calendar scheduling. There, a calendar scheduling agent assists the user in arranging meetings with others and to do so it learns the user's preferences over time. Existing approaches use decision trees [7], the weighted-majority algorithm or the Winnow algorithm [3] for the learning task.

The next section formalizes our IJSP and introduces the ILP used as optimization core. Section 3 presents our solution approaches: a greedy baseline method and the advanced heuristic that makes use of estimated acceptance probabilities for time interval suggestions. The calculation of acceptance probabilities based on a two-stage Markov model is subsequently detailed in Sect. 4. Section 5 shows experimental results, and Sect. 6 concludes this work.

2 Interactive Job Scheduling Problem

The IJSP is formally introduced as follows. Let the planning period be given by $t^{\text{max-day}}$ days, each with $t^{\max}$ uniform timesteps, and let $T = \{t \mid t = (t^{\text{day}}, t^{\text{time}}),\ t^{\text{day}} = 1, \dots, t^{\text{max-day}},\ t^{\text{time}} = 1, \dots, t^{\max}\}$ be a set of pairs where each pair refers to a specific timestep at a specific day. To refer to a time interval within a day and the corresponding set of timesteps, we use the notation $[t_1, t_2] = \{(t_1^{\text{day}}, t_1^{\text{time}}), \dots, (t_2^{\text{day}}, t_2^{\text{time}})\}$ for $t_1, t_2 \in T \mid t_1^{\text{day}} = t_2^{\text{day}},\ t_1^{\text{time}} \le t_2^{\text{time}}$, and adding a scalar Δ to a tuple $t \in T$ is defined as $t + \Delta = (t^{\text{day}}, t^{\text{time}} + \Delta)$.

Denote the set of users by U and let the set of jobs of user $u \in U$ be J_u. Let each job $j \in J_u$ have a duration $d_j \in \{1, \dots, t^{\max}\}$ and use the notation $T_j[t] = [t, t + d_j - 1]$ to refer to the subset of timesteps where job j is performed if started at timestep t. Furthermore, the possible starting times of job $j \in J$ are restricted to the set $T_j^{\text{job}} = \bigcup_{t^{\text{day}}=1}^{t^{\text{max-day}}} \{(t^{\text{day}}, 1), \dots, (t^{\text{day}}, t^{\max} - d_j + 1)\}$, because of the job duration. Denote the set of all jobs by $J = \bigcup_{u \in U} J_u$, and let $n = |J|$. To perform a job, two resources are needed: the availability of the user associated with the job and a machine. Denote the set of machines by M.

Using machine $i \in M$ in timestep $t \in T$ induces time-dependent cost $c_{it} \ge 0$, e.g., for electricity depending on expected spot market prices. For a job to be feasibly scheduled, it needs to be given non-preemptive access to its user and a machine for the complete duration of the job. If a job $j \in J$ cannot be feasibly scheduled, this induces cost $q_j \ge 0$, e.g., for over-time or extra personnel. We assume that the cost for leaving a job unscheduled is always higher than the highest cost of scheduling it, i.e., $q_j \ge d_j \max_{i \in M,\, t \in T} c_{it}$, $j \in J$.

The dynamic and interactive aspect of our problem is represented by $\mathcal{T} = (\mathcal{T}_j)_{j \in J}$ where $\mathcal{T}_j \subseteq T_j^{\text{job}}$ are the timesteps in which job j may start in when considering the respective user's currently known availabilities. More details on $\mathcal{T}$ are addressed later.

Assuming for now $\mathcal{T}$ is given and fixed, we aim at finding a feasible schedule of minimum cost. This can be expressed by the following ILP, in which the binary decision variables x_{jit} indicate if job $j \in J$ is scheduled on machine $i \in M$ to start with timestep $t \in \mathcal{T}_j$, or not.

$$\text{ILP}(\mathcal{T}) \quad \min \sum_{j \in J} \sum_{i \in M} \sum_{t \in \mathcal{T}_j} \sum_{t' \in T_j[t]} c_{it'} x_{jit} + \sum_{j \in J} q_j \left(1 - \sum_{i \in M} \sum_{t \in \mathcal{T}_j} x_{jit}\right) \tag{1}$$

$$\text{s.t.} \quad \sum_{i \in M} \sum_{t \in \mathcal{T}_j} x_{jit} \le 1 \qquad j \in J \tag{2}$$

$$\sum_{j \in J} \sum_{t \in \mathcal{T}_j \mid t' \in T_j[t]} x_{jit} \le 1 \qquad i \in M,\ t' \in T \tag{3}$$

$$\sum_{j \in J_u} \sum_{i \in M} \sum_{t \in \mathcal{T}_j \mid t' \in T_j[t]} x_{jit} \le 1 \qquad u \in U,\ t' \in T \tag{4}$$

$$x_{jit} \in \{0, 1\} \qquad j \in J,\ i \in M,\ t \in \mathcal{T}_j \tag{5}$$

The first and second term of the objective function (1) correspond to the total cost for machine usage and unscheduled jobs, respectively. Constraints (2) ensure that each job is scheduled at most once, constraints (3) limit the number of scheduled jobs per machine and timestep to one, and constraints (4) limit the number of jobs per user and timestep to one.

As indicated, this model can be solved for different sets $\mathcal{T}$ that reflect the user availability information in the current stage of the decision-making. As an important characteristic of the problem is that the user availability is not assumed to be fully known, we introduce the following notation for the currently available information. Let $T_u^{\text{avail}} \subseteq T$ be a subset of timesteps where user $u \in U$ has confirmed to be available. Feasible start times for each job $j \in J_u$ can then be derived as $T_j^{\text{feas}} = \{t \in T_j^{\text{job}} \mid T_j[t] \subseteq T_u^{\text{avail}}\}$. Further, let $T_j^{\text{infeas}} \subseteq T$ refer to time steps where job $j \in J$ is not allowed to start since the user is known to be unavailable in at least one time step in $T_j[t]$, $t \in T_j^{\text{infeas}}$.

Based on these confirmed availabilities and unavailabilities, it is possible to solve the model ILP($\mathcal{T}$) for two extreme cases. For $\mathcal{T} = (T_j^{\text{feas}})_{j \in J}$, only the timesteps that the respective users have so far confirmed to be available are included, and thus the solution to ILP$((T_j^{\text{feas}})_{j \in J})$ is feasible for the IJSP and in general provides a pessimistic bound. For $\mathcal{T} = (T_j^{\text{job}} \setminus T_j^{\text{infeas}})_{j \in J}$, all timesteps except those where the users are already known to be not available are included, and the solution to ILP$((T_j^{\text{job}} \setminus T_j^{\text{infeas}})_{j \in J})$ provides an optimistic bound; but the corresponding schedule may not be feasible with respect to user availability.

The interactive aspect of the problem is that users can be queried concerning their availabilities. A query is represented by a pair $(u, [t, t'])$ specifying a user $u \in U$ and a time interval from $t \in T$ to $t' \in T$. If the user is available in the full interval of the query, this information is directly included in the sets T_u^{avail} and T_j^{feas}. If the user is unavailable in at least one timestep of the interval, the interval is rejected and included in the set I_u^{rej}. In such update, I_u^{rej} is made sure not to contain any interval that is a superinterval of another interval, as such superintervals are redundant. The interaction with the users is made in a number of rounds, and before each new round an updated ILP($\mathcal{T}$) can be solved. Let the number of rounds be denoted by $B \in \mathbb{N}_{>0}$, and let the allowed number of queries in each round be $b \in \mathbb{N}_{>0}$. In each round, a user may be queried multiple times. The choice of queries to make in a round is critical for the outcome of the scheduling, and our strategy for this is described in the next section.

3 Solving Approaches

The challenge in each round is to find a set of queries that are likely to be accepted and reduce the objective value as much as possible if accepted. We consider only queries that are reasonable in the following sense. They concern the scheduling of jobs outside the users' already known availabilities, and we do not want to have more than one query for a user for the same day. Denote with $T_j^{\text{query}} = T_j^{\text{job}} \setminus T_j^{\text{infeas}} \setminus T_j^{\text{feas}}$ all starting times of job j that would require a confirmed user query. Most beneficial queries—if accepted—can then be determined by solving the model ILP$((T_j^{\text{query}} \cup T_j^{\text{feas}})_{j \in J})$ with the additional constraints

$$\sum_{j\in J}\sum_{i\in M}\sum_{t\in T_j^{\text{query}}} x_{jit} \leq b \tag{6}$$

$$\sum_{j\in J_u}\sum_{i\in M}\sum_{\bar{t}\in T_j^{\text{query}}|\bar{t}^{\text{day}}=t^{\text{day}}} x_{ji\bar{t}} \leq 1 \qquad u\in U, t^{\text{day}}\in\{1,\ldots,t^{\text{max-day}}\} \tag{7}$$

where the former limits the total number of user queries to b and the latter prevents multiple queries for the same user on the same day. Having obtained a solution x, each value of one of a variable x_{jit} for $u \in U$, $j \in J_u$, $i \in M$ and $t \in T_j^{\text{job}} \setminus T_j^{\text{infeas}} \setminus T_j^{\text{feas}}$ results in a query $[t, t+d_j]$ for user u. We refer to this approach to determine user queries by GREEDY.

This approach can possibly be improved by assuming that the user availabilities behave according to some model that yields an acceptance probability for each query. To exploit such probabilities, we remove the starting times from T_j^{query} whose associated queries have probabilities below a given threshold $0 \leq p^{\text{lim}} \leq 1$, i.e., which we do not consider promising. Queries are again obtained by solving ILP$((T_j^{\text{query}} \cup T_j^{\text{feas}})_{j\in J})$ with constraints (6) and (7), but now with these reduced sets T_j^{query}. As model for the acceptance probabilities, the next section proposes one based on a two-state Markov process, and consequently, we refer to this advanced model-based solution approach by MARKOV(p^{lim}).

4 Probability Calculation for Two-State Markov Process

Consider a single user $u \in U$ and a single day $t^{\text{day}} \in \{1,\ldots,t^{\text{max-day}}\}$. For better readability we refer to the timesteps of this day in the following by $T_{t^{\text{day}}} = \{1,\ldots,t^{\max}\}$. Assume that the average duration of the periods when a user is available, and the average duration of the unavailable-periods are known. When we want to exploit just this minimal information, it is natural to model a user's availabilities by a simple two-state Markov process. The two states of this process are 0 and 1, representing that the user either is unavailable in the current timestep or available, respectively. Moreover, let us introduce the additional artificial timesteps 0 and $t^{\max}+1$ before the start of the day and after the end of the day. In both of these timesteps, the user is not available and therefore the corresponding state is 0. Proceeding from one timestep to the next, we associate probabilities ρ_{00}, ρ_{01}, ρ_{10}, and ρ_{11} for staying in state 0, transitioning from 0 to 1, transitioning from 1 to 0, and staying in state 1, respectively. Naturally, $\rho_{00} = 1-\rho_{01}$ and $\rho_{11} = 1-\rho_{10}$ must hold. This Markov process is depicted in Fig. 1a. The transition probabilities are computed based on the fact that the expected number of steps the Markov process stays in state 1 is $1/\rho_{10}$ and $1/\rho_{01}$ for state 0. In this section we only consider one user, and for the sake of simplicity we omit the index regarding this user.

Given the current set of known availability times T^{avail} and the set of so far rejected time intervals I^{rej}, we now want to determine the probability that the user is available in some given time interval $[\tau, \tau']$, $1 \leq \tau \leq \tau' \leq t^{\max}$. For this

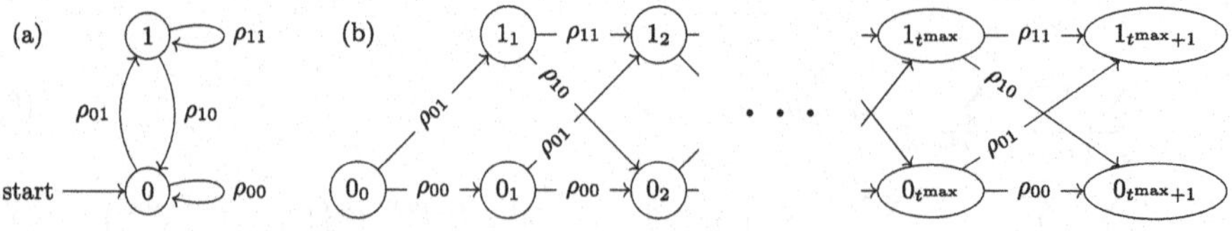

Fig. 1. (a) Two-state Markov process and (b) corresponding unrolled state graph.

purpose we unroll the Markov process into a state graph over all timesteps from 0 to $t^{\max} + 1$ as follows and illustrated in Fig. 1b.

As the user is supposed to be not available outside of $T_{t^{\text{day}}}$, the initial state at the beginning of the day is represented by the single node 0_0. Then, we have nodes 0_t and 1_t for each timestep $t \in T_{t^{\text{day}}}$, indicating the availability or non-availability of the user in timestep t. We also add node $0_{t^{\max}+1}$ and for now $1_{t^{\max}+1}$ to allow a correct modeling of the transition to the time after the considered time horizon by the two-state Markov process. All nodes of two successive timesteps are connected with arcs corresponding to the state transitions of the Markov process, and they are weighted with the respective transition probabilities ρ_{00}, ρ_{01}, ρ_{10}, and ρ_{11}.

Ignoring known user availabilities T^{avail} and rejected time intervals I^{rej} for now, this state graph has been constructed in such a way that each path from node 0_0 to either node $0_{t^{\max}+1}$ or $1_{t^{\max}+1}$, which we call terminal nodes, corresponds to exactly one outcome of the Markov process over $t^{\max} + 1$ timesteps, and each possible outcome of the Markov process has an individual corresponding path. We refer by the *probability* of a path to the product of the path's arc weights, and with the probability of a set of paths to the sum of the paths' probabilities. The probability of all paths from node 0_0 to any of the terminal nodes is then one as this covers all possible outcomes of the Markov process.

Next, we consider the already known availability times T^{avail} of the user by removing all nodes 0_t for T^{avail} with their incident arcs. This effectively reduces the set of possible paths, and thus represented Markov process outcomes, to those where state 1 is achieved in all timesteps from T^{avail}. Moreover, we also remove node $1_{t^{\max}+1}$ with its ingoing arcs in order to model that the user is unavailable after the last actual timestep $t^{\max}$.

To modify the graph w.r.t. the intervals in which the user is known to be available was straightforward since all timesteps of such intervals must have state 1. A time interval rejected by the user requires more care since it implies only that for *at least* one timestep in the interval – but not necessary all – the Markov process is in state 0. Only a rejected time interval $[t, t] \in I^{\text{rej}}$ of length one can thus be handled directly by removing node 1_t with its incident arcs as the Markov process has to be in state 0 in this timestep. For a longer rejected interval $[t_1, t_2] \in I^{\text{rej}}$ we ensure that only paths are kept in the graph where the Markov process achieves state 0 at least once within this interval. More specifically, observe that if the Markov process is in timestep $t \in [t_1, t_2]$ and state 0 has not been obtained in timesteps $[t_1, t]$ yet, then there has to

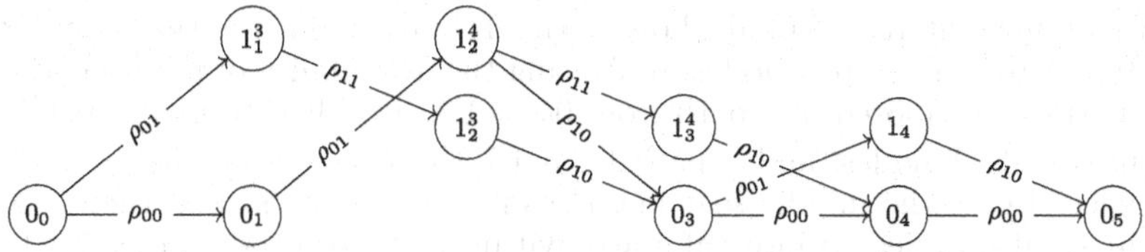

Fig. 2. The state graph for $t^{\max} = 4$, $T^{\text{avail}} = \{2\}$, and $I_u^{\text{rej}} = \{(1,3),(2,4)\}$.

follow at least one timestep $t' \in [t+1, t_2]$ in which state 0 is achieved. To model this aspect, we add further nodes $1_t^{t_2}$ for $t \in [t_1, t_2 - 1]$, $[t_1, t_2] \in I^{\text{rej}}$ to our graph. Former arcs $(0_t, 1_{t+1})$ and $(1_t, 1_{t+1})$, $t \in T_{t^{\text{day}}} \cup \{0\}$, are now replaced by arcs $(0_t, 1_{t+1}^{t_2})$ and $(1_t, 1_{t+1}^{t_2})$, respectively if there is a rejected time interval $[t_1, t_2] \in I^{\text{rej}}$ starting in the next timestep $t_1 = t+1$ and ending in timestep t_2. Note that there can be at most one interval in $[t_1, t_2] \in I^{\text{rej}}$ that starts at timestep t_1 since I^{rej} has been guaranteed not to contain a proper subinterval of $[t_1, t_2]$. Each new node $1_t^{t_2}$ further has an outgoing arc to node 0_{t+1} if this node still exists, corresponding to the transition to state 0. Moreover, there is an outgoing arc from each node $1_t^{t_2}$ to node $1_{t+1}^{t_2}$ as long as $t + 1 < t_2$ for the case of staying in state 1. Due to the absence of an arc from node $1_{t_2-1}^{t_2}$ to some successor node in which state 1 is kept, it is effectively enforced that state 0 is reached at least once within the rejected time interval $[t_1, t_2]$. Remaining nodes without ingoing arcs except 0_0 and their outgoing arcs are pruned as they do not play an active further role. An example of such a final state graph is shown in Fig. 2.

Now, we want to utilize this graph to derive the probability that the considered user is available in a given time interval $[\tau, \tau']$. The key observation to do this efficiently is that each path from node 0_0 to a node v passes through exactly one predecessor of v. Therefore the total probability $p_{0_0,v}^{\text{path}}$ of all paths from 0_0 to v, denoted by $\text{Paths}(0_0, v)$, can be computed recursively as

$$\begin{aligned} p_{0_0,v}^{\text{path}} &= \sum_{P \in \text{Paths}(0_0,v)} \prod_{(u,u') \in P} \rho(u,u') \\ &= \sum_{u \in N^-(v)} \sum_{P \in \text{Paths}(0_0,u)} \left(\prod_{(u,u') \in P} \rho(u,u') \right) \cdot \rho(u,v) = \sum_{u \in N^-(v)} p_{0_0,u}^{\text{path}} \rho(u,v), \end{aligned} \tag{8}$$

where P denotes one specific 0_0–v path represented by the corresponding set of arcs and $N^-(v)$ is the set of predecessors of node v. Denoting the set of successors of node v by $N^+(v)$, the probabilities $p_{v,0_{t^{\max}+1}}^{\text{path}}$ of all paths from a node v to node $0_{t^{\max}+1}$ can be correspondingly computed recursively by

$$p_{v,0_{t^{\max}+1}}^{\text{path}} = \sum_{w \in N^+(v)} p_{w,0_{t^{\max}+1}}^{\text{path}} \rho(v,w). \tag{9}$$

We are now interested in all those paths from 0_0 to $0_{t^{\max}+1}$ that stay for the timesteps τ to τ' in state 1 nodes, indicating the availability of the user. Each of these paths is composed of a path from 0_0 to $1_\tau^{t_2}$, a path P from $1_\tau^{t_2}$ to $1_{\tau'}^{t_2}$ that only uses state 1 nodes, and a path from $1_{\tau'}^{t_2}$ to $0_{t^{\max}+1}$ for some $t_2 \geq \tau' + 1$. As a special case the middle segment P can also start in 1_τ and then it either ends in $1_{\tau'}$ if no rejected interval starts within $[\tau, \tau']$ or otherwise in $1_{\tau'}^{t_2}$ for an appropriate $t_2 \geq \tau' + 1$. There are only a few possibilities for the middle segment P and the probability of all paths that stay in state 1 nodes for the timesteps from τ to τ' can be computed with a sum over these possibilities. For us, the conditional probability in respect to all paths in the graph, i.e., those respecting T^{avail} and I^{rej} and ending in $0_{t^{\max}+1}$, is of main interest, which is

$$p^{\text{avail}}([\tau, \tau'] \mid T^{\text{avail}}, I^{\text{rej}}, 0_{t^{\max}+1}) = \frac{\sum_{P \in \text{1-Paths}(\tau,\tau')} p^{\text{path}}_{0_0, P_\tau} \cdot \rho_{11}^{\tau'-\tau} \cdot p^{\text{path}}_{P_{\tau'}, 0_{t^{\max}+1}}}{p^{\text{path}}_{0, t^{\max}+1}}, \tag{10}$$

where the sum is taken over all middle segments 1-Paths(τ, τ'), and P_τ and $P_{\tau'}$ are the first and last nodes of a middle segment P, respectively. The denominator is the probability of all paths from 0_0 to $0_{t^{\max}+1}$, and the nominator the probability of only those paths that stay in state 1 nodes in timesteps τ to τ'.

5 Experimental Evaluation

We implemented the approaches in Julia 1.8.3, using the solver Gurobi 10.0 (https://www.gurobi.com) and the package JuMP as interface to Gurobi. As real world instances were not available to us we created artificial benchmark instances and used them to compare the approaches with each other. Each test run was performed on a single core of an AMD EPYC 7402 and Gurobi was given a timelimit of 15 min for each ILP, which always led to final gaps below 5%.

5.1 Instance Generation

We consider a time horizon of $t^{\text{max-day}} = 5$ days, each starting at 6am and ending at 10pm, with a time granularity of 15 min per timestep. Random time intervals are determined by a function rand_interval$(\mu^{\text{start}}, \sigma^{\text{start}}, \mu^{\text{dur}}, \sigma^{\text{dur}})$ that first draws a random value from a normal distribution with mean μ^{start} and standard deviation σ^{start} and rounds it to the closest timestep in T, which is then the start of the time interval. The duration of the interval is then determined by drawing another random value from a normal distribution with mean μ^{dur} and standard deviation σ^{dur}, rounding it to the closest positive integer. Should the interval exceed $t^{\max}$, it is capped at this last timestep of our time horizon.

For each user $u \in U$ a set of timesteps $T_u^{\text{avail}*}$ at which she or he is, in total, available is determined for each day independently as follows. With a probability of 90%, the user is assumed to be available in rand_interval(9 am, 1 h, 4 h, 1 h)

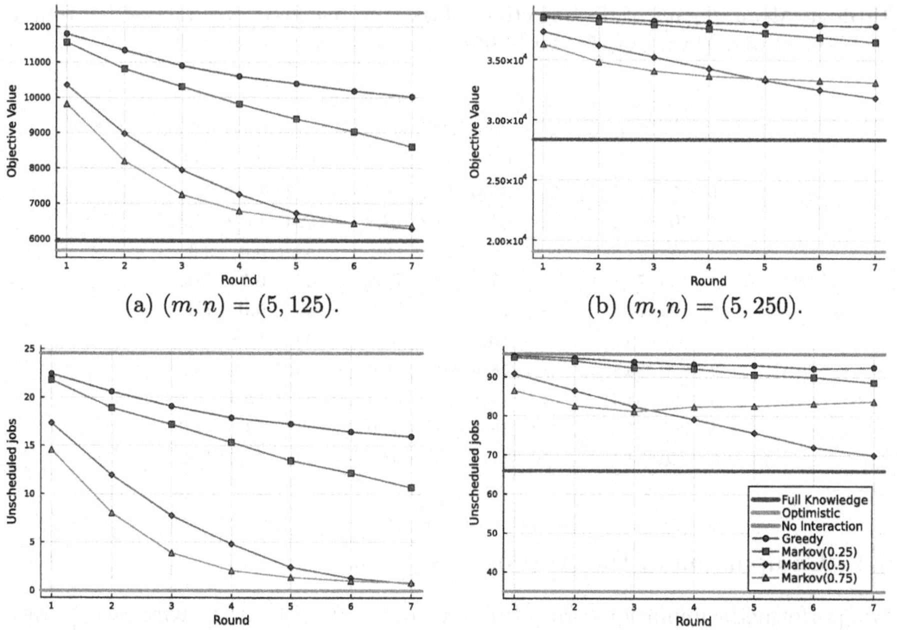

(a) $(m, n) = (5, 125)$. (b) $(m, n) = (5, 250)$.

(c) $(m, n) = (5, 125)$. (d) $(m, n) = (5, 250)$.

Fig. 3. Development of the objective value (a) and (b) respectively number of unscheduled jobs (c) and (d) for two different instance sizes.

and, again with a probability of 90%, the user is assumed to be available in rand_interval(1 pm, 1 h, 5 h, 1 h). If the two intervals overlap the union is taken.

For each job $j \in J_u$ of a user $u \in U$ the duration d_j is chosen uniformly at random from 30 min to 4 h. Moreover, a starting time t_j is selected at random so that the job can in principle be scheduled within $T_u^{\text{avail}*}$. The initially provided set of availabilities for user $u \in U$ is then $T_u^{\text{avail}} = \bigcup_{j \in J_u} T_j[t_j]$.

We generate 50 instances for $m \in \{1, 2, 3, 4, 5\}$ machines and either 25 or 50 jobs per machine $n \in \{25m, 50m\}$. When considering the generated user availabilities, each machine can execute roughly 30 jobs on average, thus for $n = 25m$ usually it is possible to schedule all jobs, while for $n = 50m$ this is not the case. Each user has five jobs, thus there are either $5m$ or $10m$ users. We allow $|U|$ user queries in each round, for a total of seven rounds.

The costs are based on the real-world spot market prices c_t^{kWh} for electricity in Germany from week 26 in 2022 from https://energy-charts.info. We use as cost $c_{i,t} = 15\,\text{min} \cdot P_i c_t^{\text{kWh}}$, where the electric power P_i is assumed to differ among the machines $i \in M$ and is thus chosen uniformly at random from [50 kW, 150 kW]. The cost q_j for not scheduling a job $j \in J$ is set to 40 Euro $\cdot\, d_j$, which is roughly twice the cost of scheduling the job in the most expensive timesteps.

Table 1. Mean %-gaps of the objective values after five and seven interaction rounds for GREEDY and MARKOV(p^{lim}) with different limits p^{lim}.

m	n	Round 5				Round 7			
		GREEDY	MARKOV(p^{lim})			GREEDY	MARKOV(p^{lim})		
			0.25	0.5	0.75		0.25	0.5	0.75
1	25	77.7	54.5	**39.6**	40.0	70.7	36.9	**18.0**	24.6
2	50	95.2	74.5	27.0	**23.8**	87.0	52.6	**11.9**	13.3
3	75	77.5	62.3	18.2	**15.0**	69.1	48.8	**8.4**	9.3
4	100	79.5	64.3	15.7	**12.1**	71.8	47.9	**6.7**	7.9
5	125	77.8	60.4	13.4	**10.7**	71.5	46.7	**6.0**	7.3
1	50	40.2	33.9	**19.4**	22.6	37.1	29.5	**12.8**	21.7
2	100	36.8	31.9	**18.6**	19.2	35.6	29.1	**13.0**	18.2
3	150	34.4	31.6	**17.8**	18.0	33.4	28.7	**12.1**	17.0
4	200	35.0	32.2	**18.3**	18.8	34.0	29.2	**12.6**	17.6
5	250	34.4	31.8	**17.7**	18.3	33.8	29.1	**12.4**	17.2

5.2 Comparison of the Approaches

We performed simulations for GREEDY and MARKOV(p^{lim}) with acceptance probability thresholds $p^{\text{lim}} \in \{0.25, 0.5, 0.75\}$ on all benchmark instance. After each round we determine the best schedule that is feasible for the information collected up to this round. Figure 3 shows the development of the mean objective value and mean number of unscheduled jobs, respectively, over the rounds. Values are aggregated over the 50 instances with $m = 5$ machines and $n = 125$ respectively $n = 250$ jobs. Furthermore, we determine the best feasible schedule with the information that is available before the first round ("No Interaction"), the best schedule when ignoring user availabilities ("Optimistic") and the best schedule with full knowledge about the users' availabilities ("Full Knowledge") and show these as horizontal lines in the figures. Table 1 additionally shows the mean optimality gaps of the objective values from GREEDY and MARKOV(p^{lim}) in respect to "Full Knowledge" after five and seven rounds in percent.

We observe that MARKOV(0.5) and MARKOV(0.75) quickly converge towards the best possible schedule. For $n = 125$, the original objective values without interaction could almost be halved after already five rounds, while for $n = 250$, 18% and 15% of the original costs could be saved after seven rounds. Moreover, for $n = 125$, the final optimality gaps of these two approaches are by a factor of more than nine better than the final gap of GREEDY. In contrast $p^{\text{lim}} = 0.25$ leads to much slower convergence with an improvement over GREEDY of only roughly 35%. Remarkably, MARKOV(0.75) performs best in the first rounds, while MARKOV(0.5) catches up later on and performs best in the end. The reason is that the two-state Markov process has the steady state between 0.5 and 0.75 and therefore MARKOV(0.75) does not query days it knows nothing about while MARKOV(0.5) does; while it takes more iterations to get enough information about these days, this information provides more flexibility in scheduling the jobs.

6 Conclusions

We considered a job scheduling problem in which humans are involved as resource and where their availabilities can only be partially revealed in a small number of interaction rounds, within which few time interval queries can be made. The proposed solution approach calculates probabilities for users to accept suggested time intervals based on a two-state Markov process. An ILP is used as optimization core and to select time intervals for the next round of queries, aiming for sufficiently high probabilities of acceptance and a maximum cost reduction. Experiments on artificial test instances show that an initial solution quickly improves over the interaction rounds and may soon get close to a solution of the full-knowledge case, despite the very restricted interaction and the simple assumptions of the two-state Markov process. In future work it would be interesting to replace the proposed probability computation by a machine learning model trained on historic user availability data. Moreover, alternative ways to consider the estimated acceptance probabilities of user queries in the optimization core should be investigated.

References

1. Aghaei-Pour, P., Rodemann, T., Hakanen, J., Miettinen, K.: Surrogate assisted interactive multiobjective optimization in energy system design of buildings. Optim. Eng. **23**(1), 303–327 (2022)
2. Anghinolfi, D., Paolucci, M., Ronco, R.: A bi-objective heuristic approach for green identical parallel machine scheduling. Eur. J. Oper. Res. **289**(2), 416–434 (2021)
3. Blum, A.: Empirical support for winnow and weighted-majority algorithms: results on a calendar scheduling domain. Mach. Learn. **26**(1), 5–23 (1997)
4. Cheng, J., Chu, F., Zhou, M.: An improved model for parallel machine scheduling under time-of-use electricity price. IEEE Trans. Autom. Sci. Eng. **15**(2), 896–899 (2018)
5. Ding, J.Y., Song, S., Zhang, R., Chiong, R., Wu, C.: Parallel machine scheduling under time-of-use electricity prices: new models and optimization approaches. IEEE Trans. Autom. Sci. Eng. **13**(2), 1138–1154 (2016)
6. Jatschka, T., Raidl, G.R., Rodemann, T.: A general cooperative optimization approach for distributing service points in mobility applications. Algorithms **14**(8), 232 (2021). https://www.mdpi.com/1999-4893/14/8/232
7. Mitchell, T.M., Caruana, R., Freitag, D., McDermott, J., Zabowski, D., et al.: Experience with a learning personal assistant. Commun. ACM **37**(7), 80–91 (1994)
8. Saberi-Aliabad, H., Reisi-Nafchi, M., Moslehi, G.: Energy-efficient scheduling in an unrelated parallel-machine environment under time-of-use electricity tariffs. J. Clean. Prod. **249**, 119393 (2020)
9. Saha, S., Minku, L.L., Yao, X., Sendhoff, B., Menzel, S.: Exploiting linear interpolation of variational autoencoders for satisfying preferences in evolutionary design optimization. In: 2021 IEEE Congress on Evolutionary Computation, pp. 1767–1776 (2021)
10. Wang, S., Wang, X., Yu, J., Ma, S., Liu, M.: Bi-objective identical parallel machine scheduling to minimize total energy consumption and makespan. J. Clean. Prod. **193**, 424–440 (2018)

Multi-armed Bandit-Based Metaheuristic Operator Selection: The Pendulum Algorithm Binarization Case

Pablo Ábrego-Calderón[1](✉), Broderick Crawford[1], Ricardo Soto[1], Eduardo Rodriguez-Tello[2], Felipe Cisternas-Caneo[1], Eric Monfroy[3], and Giovanni Giachetti[4]

[1] Pontificia Universidad Católica de Valparaíso, Valparaíso, Chile
{pablo.abrego.c,felipe.cisternas.c}@mail.pucv.cl,
{broderick.crawford,ricardo.soto}@pucv.cl

[2] Cinvestav, Unidad Tamaulipas, Km. 5.5 Carretera Victoria - Soto La Marina, 87130 Victoria, Tamaulipas, Mexico
ertello@cinvestav.mx

[3] Université d' Angers, LERIA, Angers, France
eric.monfroy@univ-angers.fr

[4] Universidad Andres Bello, Santiago, Chile
giovanni.giachetti@unab.cl

Abstract. Multi-armed bandit (MAB) is a well-known reinforcement learning algorithm that has shown outstanding performance for recommendation systems and other areas. On the other hand, metaheuristic algorithms have gained much popularity due to their great performance in solving complex problems with endless search spaces. Pendulum Search Algorithm (PSA) is a recently created metaheuristic inspired by the harmonic motion of a pendulum. Its main limitation is to solve combinatorial optimization problems, characterized by using variables in the discrete domain. To overcome this limitation, we propose to use a two-step binarization technique, which offers a large number of possible options that we call scheme. For this, we use MAB as an algorithm that learns and recommends a binarization schemes during the execution of the iterations (online). With the experiments carried out, we show that it delivers better results in solving the Set Covering problem than using a fixed binarization scheme.

Keywords: Pendulum Search Algorithm · Multi-Armed Bandit · Set Covering Problem · Reinforcement Learning · Binarization Schemes

1 Introduction

In recent years, combinatorial optimization problems have become more frequent and complex. Often these problems must be solved in a reasonable time and with good results. Metaheuristics are general-purpose algorithms thet can

B. Dorronsoro et al. (Eds.): OLA 2023, CCIS 1824, pp. 248–259, 2023.
https://doi.org/10.1007/978-3-031-34020-8_19

to solve different optimization problems with minor modifications [17]. The process of optimization consists of two main phases: exploration and exploitation. In the exploration phase, the algorithm explores the search space for founding promising areas where good results can be expected. In the exploitation phase, the optimization process is intensified by expecting a better solution [20]. This process allows high-quality solutions to be found in a short times.

In case the problem to be solved is combinatorial, these general-porpuse algorithms must go through a binarization process, which allows for keeping the value of variables in the discrete domain [18]. For this, a two-step technique is used, which depending on its configuration, can get better or worse results.

Moreover, the development of machine learning in computer science has made significant contributions to various areas of engineering, and operations research was no exception. The use of these techniques across multiple optimization processes has catapulted the performance of the previously described algorithms. That is why for this work, an another reinforcement learning algorithm will be implemented to improve the performance in the selection of binarization schemes of metaheuristic algorithms in order to obtain better results in the optimization of combinatorial problems [13]. For this, the Multi-Armed Bandit (MAB) algorithm will be developed, binarizing the Pendulum Search Algorithm (PSA). The proposed solution is assessed using the well-known Set Covering Problem (SCP) as a test case.

This paper continues with the explanation of the problem to be solved (SCP) in Sect. 2, then in Sect. 3 the metaheuristic to be used is explained, to continue with the explanation of the Multi-Armed Bandit algorithm in Sect. 4. Sections 5 and 6 define the proposal and analyze the experimental results. Finally, the conclusions are presented in Sect. 7.

2 Set Covering Problem

The Set Covering Problem is a classic NP-Hard combinatorial optimization problem. It consists of finding a subset of elements that satisfies a set of constraints at the lowest possible cost [8].

$$\text{Minimize} \quad Z = \sum_{j=1}^{N} c_j x_j \tag{1}$$

$$\text{s.t.} \quad \sum_{j=1}^{N} a_{i,j} \cdot x_j \geq 1 \qquad \forall i \in \{1, \ldots, M\} \tag{2}$$

$$x_j \in \{0, 1\} \quad \forall j \in \{1, \ldots, N\} \tag{3}$$

where c_j corresponds to the cost of column j and vector x is the decision variable. M is the number of constraint and N the number of variables of the solution. $a_{i,j}$ is a binary value that indicate if the constraint i is covered by the column j.

Equation 2 tells us that all constraints must be covered by at least one column and Eq. 3 tells us the domain of the decision variables.

The complete mathematical model of the Set Covering Problem is explained in more detail in [11]. This problem formulation has inspired the modeling of different real-world problems such as emergency humanitarian logistics [3], disaster management system [14], dynamic vehicle routing problem [21], among others.

The tremendous practical applicability of this optimization problem has motivated our research to develop new algorithms with good performance for solving it.

3 Pendulum Search Algorithm

Pendulum Search Algorithm is a population-based metaheuristic recently created by Nor Azlina and Kamarulzaman [1] to solve continuous optimization problems. This metaheuristic was born in response to the premature convergence problems of the Sine Cosine Algorithm (SCA). The authors mimic the harmonic motion of the pendulum to improve premature convergence. Unlike SCA, the harmonic motion of the pendulum decreases with an exponential function. This exponential function would enhance the exploration and exploitation balance [2].

The search agents are initialized randomly and their position is updated using Eq. 4.

$$X_{i,j}^{t} = X_{i,j}^{t} + pend_{i,j}^{t} \cdot (Best_j - X_{i,j}^{t}) \tag{4}$$

where $X_{i,j}^{t}$ is the position of the i-th solution in the j-th dimension in t-th iteration, $Best_j$ it is the j-th dimension of the global best solution, and $pend_{i,j}^{t}$ is a parameter which is calculated for the i-th solution in the j-th dimension in t-th iteration (Eq. 5).

$$pend_{i,j}^{t} = 2 \cdot e^{(-t/tmax)} \cdot cos(2 \cdot \pi \cdot rand) \tag{5}$$

where t is the current iteration, $tmax$ is the maximum number of iterations and $rand$ is a uniform random number between [0,1]. The pseudo-code of PSA is shown in Algorithm 1.

The promising PSA performances previously demonstrated by the authors are unusable when we try to solve binary combinatorial problems, because PSA was initially designed for solving continuous optimization problems. To enable PSA to tackle binary combinatorial problems it is necessary to transform the solutions from the continuous domain to a binary one [6,7].

The Two-Step Technique is the most widely used mechanism in the literature for binarizing continuous solutions. First, transfer functions are applied to leave the continuous solution in the range [0, 1]; then, a binarization operator is applied with the transferred number. We call each combination of these functions a binarization scheme. For more details on Two-Step Technique please refer to [6,7].

Algorithm 1. Pendulum Search Algorithm

Input: The population $X = \{X_1, X_2, ..., X_i\}$
Output: The updated population $X' = \{X'_1, X'_2, ..., X'_i\}$ and *Best*

Initialize random population X
Evaluate the objective function of each individual in the population X
Identify the best individual in the population (*Best*)
for *iteration* (t) **do**
  for *solution* (i) **do**
    for *dimension* (j) **do**
      Update $pend^t_{i.j}$ by Eq. (5)
      Update the position of $X^t_{i,j}$ using Eq. (4)
    end for
  end for
  Evaluate the objective function of each individual in the population X
  Update *Best*
end for
Return the updated population X' where *Best* is the best result

4 Multi-armed Bandit

This technique is a reinforcement learning algorithm, where learning consists of an agent performing actions that generate a stimulus from its environment. This signal is called a reward, and the objective is to maximize its value with the following actions. For more details on reinforcement learning, refer to [19].

The metaphor of this technique puts us in a casino, where we have a series of slot machines, of which the probability of success is unknown, nor is the reward that will be obtained in each game. In this way, we must test each of them until we know their behavior and use a strategy that allows us to maximize the gain at the end of the experiment. Another common objective in this kind of experiment is to minimize the difference between what is obtained and the maximum value expected (known as the regret value) [5,15]. Based on this, the exploration-exploitation dilemma is presented, where we must negotiate between activating the arm that has the best rewards, and trying the others to find possible better rewards [4]. This trade-off is essential to obtain good results.

4.1 Action Selection: UCB

One of the most common techniques to select actions in reinforcement learning algorithms is the greedy criterion which chooses the action that has delivered the best reward. For that, we will calculate each action's average reward in each iteration to make the decision.

We will define the estimated value of action a at iteration t as $Q_t(a)$.

$$Q_t(a) \doteq \frac{\sum_{i=1}^{t-1} R_i \cdot \mathbb{1}_{A_i=a}}{\sum_{i=1}^{t-1} \mathbb{1}_{A_i=a}} \tag{6}$$

where R_i is the reward obtained in iteration i, A_i is the arm selected in iteration i and $\mathbb{1}_{predicate}$ defines a variable that will be 1 only if the *predicate* is true and 0 otherwise. If the fraction's denominator equals zero, then a default value for $Q_t(a)$ will be selected, such as 0 or another value depending on where we want the optimal value to start. We also define the greedy action of the t iteration as $A_t \doteq argmax_a Q_t(a)$ where $argmax_a$ corresponds to the action a that maximizes the value of Q_t.

As mentioned above, MAB has an extensive exploration and exploitation factor, so a correct balance between both characteristics is essential to achieve good results. This trade-off is reflected in the different equations that model the selection of actions in each iteration. One of the most used is the so-called Upper Confidence Bound, which uses the average reward and a parameter to control exploration and exploitation.

$$A_t \doteq argmax_a \left[Q_t(a) + c\sqrt{\frac{\ln t}{N_t(a)}} \right] \tag{7}$$

where $N_t(a)$ corresponds to the number of times that arm a has been actuated up to time t (the denominator in Eq. (6)), $\ln t$ is the natural logarithm of t, and the constant $c > 0$ controls the degree of exploration that the algorithm will have. The higher the c value, the more importance it will be the factor that follows it.

5 Binary MAB-Pendulum Search Algorithm

To improve the proposal presented in [1], the MAB algorithm will be used, which will learn to select the best binarization schemes to use in the binary version of PSA (BPSA) metaheuristic.

In [12], the importance of binarization schemes for the search result, solving SCP, is analyzed. The authors conclude that it directly affects and give recommendations of schemes to use in certain instances.

As the selected scheme is so important [13], in this paper it is proposed to use Multi-armed Bandit as a reinforcement learning algorithm, recognized for its use in recommendation systems [10], to implement a dynamic binarization model in the metaheuristic BPSA, solving SCP. In this way, our actions pull would be made up of the 40 different combinations between the transfer functions, shown in Table 1 and the binarization rules, shown in Table 2. In other words, each of the arms of MAB would be represented by each tuple of the actions pull.

Table 1. S-shape and V-shape transfer functions

Name	Function
S1	$T(d_w^j) = \frac{1}{1 + e^{-2d_w^j}}$
S2	$T(d_w^j) = \frac{1}{1 + e^{-d_w^j}}$
S3	$T(d_w^j) = \frac{1}{1 + e^{\frac{-d_w^j}{2}}}$
S4	$T(d_w^j) = \frac{1}{1 + e^{\frac{-d_w^j}{3}}}$
V1	$T(d_w^j) = \left\lvert erf\left(\frac{\sqrt{\pi}}{2} d_w^j\right)\right\rvert = \left\lvert \frac{\sqrt{2}}{\pi} \int_0^{\frac{\sqrt{\pi}}{2}x} e^{-t^2} dt \right\rvert$
V2	$T(d_w^j) = \left\lvert tanh\left(d_w^j\right)\right\rvert$
V3	$T(d_w^j) = \left\lvert \frac{d_w^j}{\sqrt{1+\left(d_w^i\right)^2}} \right\rvert$
V4	$T(d_w^j) = \left\lvert \frac{2}{\pi} arctan\left(\frac{\pi}{2} d_w^j\right)\right\rvert$

An important factor for the success of any reinforcement learning algorithm is the modeling of the reward that will be obtained with each action performed. For this investigation, the way in which the reward will be calculated is with its percentage of fitness improvement delivered by the metaheuristic in each iteration, expressed as follows:

$$r = 100 \cdot \left[\frac{f_{old} - f_{new}}{f_{old}}\right] \tag{8}$$

where f_{old} is the best fitness of the last iteration, and f_{new} is the best fitness of the current iteration.

As mentioned in Sect. 4, there are many ways to choose which arm to use in each iteration, but based on the literature [9,16,22], one that has given good results is the call UCB, described in the Sect. 4.1.

6 Experimentation

This section will discuss the experimentation process that was carried out, with its respective results.

As mentioned in [12], one of the best binarization schemes is the combination of the V4 transfer function and the Elitist binarization rule. This is why we will compare the result of the BPSA execution, configured with V4-Elitist in all its iterations, versus the MAB-BPSA proposal, binarizing in each iteration with the technique provided by the MAB algorithm.

Regarding the development, first of all, the instances of the SCP problem were captured from OR-library, from which instances 41, 51, 61, a1, b1, c1, nre1

Table 2. Binarization Rules

Name	Functions
D_1: Standard	$X^j_{new} = \begin{cases} 1 \ if \ rand \leq T\left(d^j_w\right) \\ 0 \quad otherwise \end{cases}$
D_2: Complement	$X^j_{new} = \begin{cases} Complement\left(X^j_w\right) \ if \ rand \leq T\left(d^j_w\right) \\ 0 \quad otherwise \end{cases}$
D_3: Static	$X^j_{new} = \begin{cases} 0 \quad if \ T\left(d^j_w\right) \leq \alpha \\ X^j_w \ if \ \alpha < T\left(d^j_w\right) \leq \frac{1}{2}(1+\alpha) \\ 1 \quad if \ T\left(d^j_w\right) \geq \frac{1}{2}(1+\alpha) \end{cases}$
D_4: Elitist	$X^j_{new} = \begin{cases} X^j_{Best} \ if \ rand < T\left(d^j_w\right) \\ 0 \quad otherwise \end{cases}$
D_5: Elitist Roulette	$X^j_{new} = \begin{cases} P[X^j_w = \delta^j] = \frac{f(\delta)}{\sum_{\delta \in X} f(\delta)} \ if \ \alpha < T\left(d^j_w\right) \\ P[X^j_w = 0] = 1 \quad otherwise \end{cases}$

and nrf1 were used. In this way we have a good representation of the different families of instances.

Second, the BPSA and MAB algorithms, along with the different binarization techniques, were developed in the Python v3.9 with the NumPy library to optimize matrix calculations.

Finally, the experiments were carried out on a Macbook Air computer with an Apple M1 processor, 7-core CPU and 8GB of RAM. Each instance was executed 31 times independently, enough quantity to have a confidence idea of the behavior of the algorithm in each one of the instances.

Regarding the configuration of parameters, a population size of 40 individuals and 500 iterations were used for both tested approaches, 31 independent runs were performed and the value for the value c of Eq. 7 is $\sqrt{2}$, based on the suggestion of [9].

6.1 Experimental Results

The results obtained are shown in Table 3. This table has ten columns, where the first one represents the instance of the Set Covering Problem evaluated, the second one represents the global optimum of instances and the next four columns are repeated for each algorithm executed. The first of these shows the best fitness obtained among the thirty-one independent runs, the second of these shows the average across the thirty-one independent runs, the third one of these shows the standard deviation of the thirty-one independent runs, and the fourth one shows the Relative Percentage Deviation (RPD) between global optimum and the best fitness obtained among the thirty-one independent runs. RPD is defined as follows:

$$RPD = \frac{Z - Z_{opt}}{Z_{opt}} \times 100 \tag{9}$$

where Z corresponds to the best value found and Z_{opt} the global optimal value that is expected to be reached.

Table 3. Comparison of fitness between fixed and dynamic binarization.

Inst.	Opt	V4-ELIT				MAB			
		Best	Avg-fit	Std-dev-fit	RPD	Best	Avg-fit	Std-dev-fit	RPD
41	429	**433**	**433**	**0**	**0.932**	**433**	**433**	**0**	**0.932**
51	253	267	267	0	5.534	**257**	**266.452**	**2.173**	**1.581**
61	138	141	142.774	1.91	2.174	**141**	**141**	**0**	**2.174**
a1	253	257	257.065	0.25	1.581	**257**	**257**	**0**	**1.581**
b1	69	69	69.161	0.374	0	**69**	**69.097**	**0.301**	**0**
c1	227	**230**	**232.387**	**1.202**	**1.322**	231	232.645	0.798	1.762
d1	60	60	60.355	0.661	0	**60**	**60.871**	**0.619**	**0**
nre1	29	**29**	**29**	**0**	**0**	**29**	**29**	**0**	**0**
nrf1	14	**14**	**14**	**0**	**0**	**14**	**14**	**0**	**0**
Avg	163.556	166.667	167.194	0.489	1.283	**165.667**	**167.007**	**0.432**	**0.892**

The following criteria were used to determine the best algorithm:

* **Best fitness obtained and RPD:** This allows us to see what our best result was and how far it is from the global optimum.
* **Standard Deviation:** A low standard deviation indicates the results obtained with the thirty-one independent runs were close.
* **Average between thirty-one independent:** An average close to the optimal value indicates the results obtained with the thirty-one independent runs the algorithm performed well.

With this in mind, our proposed BPSA with MAB as binarization schemes selector won in 5 out of 9 instances, tied in 3 out of 9 instances, and only lost in one instance.

Figure 1 shows the convergence plots of the best execution of each algorithm run. The X-axis shows the iterations and the Y-axis shows the best fitness obtained during the process.

In these two figures, we can see the BPSA with MAB has a slower convergence compared to BPSA with a fixed binarization scheme.

Thus, we can demonstrate that using MAB as a selector of binarization schemes helps to balance the exploration and exploitation of BPSA and to find better solutions.

On the other hand, in Fig. 2 we can see 3 horizontal bar graphs showing the number of times each actions was selected. The one on the left shows how they had been selected in the first 50 iterations of the run, then at 200 iterations, and on the right at 500 iterations. In this way we can analyze the behavior that MAB algorithm had throughout the execution. From Fig. 2, which shows the average selection of instance 51, it stands out that in the first iterations, where the algorithm has a greater exploratory component, the 4 actions that are most selected have as a binarization rule (D_2 at Table 2) the complement function, while in advanced iterations the actions with the Elitist or Elitist Roulete function are selected more (D_4 and D_5 at Table 2). This result makes sense with what is described in [12].

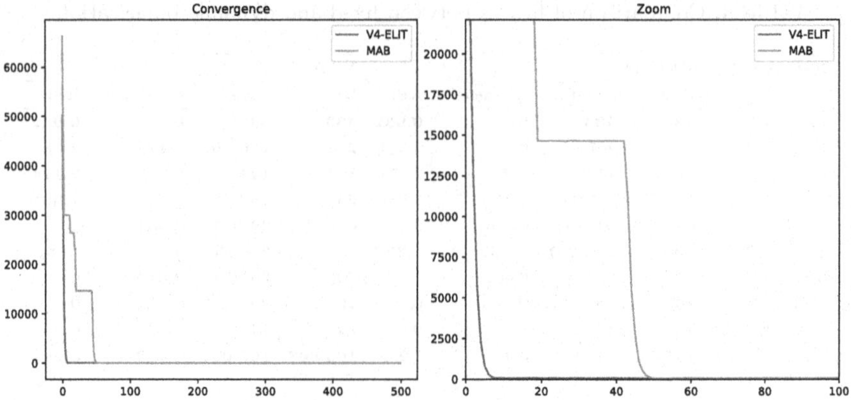

Fig. 1. Fitness for each iteration (convergence) and Zoom for instance d1

Table 4. Comparison of time between fixed and dynamic binarization.

Instance	V4-ELIT		MAB	
	Avg-time (s)	Std-dev-time (s)	Avg-time (s)	Std-dev-time (s)
41	**69.429**	4.15	71.727	9.763
51	88.634	4.35	89.497	13.175
61	42.355	3.303	**39.683**	5.878
a1	204.238	5.852	**193.114**	17.502
b1	**113.087**	4.817	123.087	17.13
c1	408.542	10.897	**383.78**	41.616
d1	218.491	6.603	**199.662**	20.153
nre1	**257.875**	7.011	277.048	34.237
nrf1	**148.318**	5.306	174.743	21.352
average	**172.330**	5.810	172.482	20.090

As mentioned above, the use of metaheuristics lies in their efficiency in delivering good results in reasonable times. The Table 4 shows the execution times of each proposal, where the first one represents the instance of the Set Covering Problem evaluated and the next two columns are repeated for each algorithm executed. The first-one of these shows the average time in seconds across the thirty-one independent runs, and the second one of these shows the standard deviation of the thirty-one independent runs.

As can be seen in this table, both proposals have very similar implementation times. This indicates that there is not a large computational increase when incorporating a machine learning technique such as MAB.

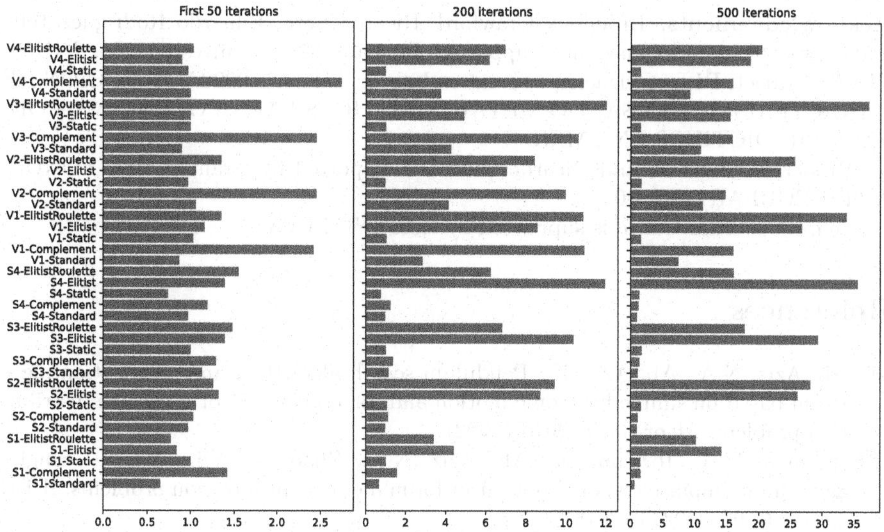

Fig. 2. Average number of selections on instance 51

7 Conclusion

The ease of accessing large computational capacities at reduced costs has enabled the use of machine learning techniques such as Multi Armed Bandit. The research on hybrid algorithms between metaheuristics and machine learning with the aim of improving the search process is increasing every year and the present work is an example of that. In particular, Multi Armed Bandit was successfully incorporated into Pendulum Search Algorithm where it was used to dynamically and intelligently select binarization schemes.

Preliminary results indicate that our proposed hybrid performs better when compared to Pendulum Search Algorithm using a fixed binarization scheme (V4-Elitist). Better results were obtained by improving the balance of diversification and intensification in the Pendulum Search Algorithm search process. During the diversification process, Multi Armed Bandit determined that the most exploratory binarization schemes are those that include the Complement binarization rule. In contrast, for the intensification process, Multi Armed Bandit determined that the most exploitative binarization schemes are those that include the Elitist or Elitist Roulette binarization rule.

Regarding computation times, the results indicate that there is no great increase in computation times when comparing both proposals. This dismisses that incorporating machine learning techniques to metaheuristic algorithms increases the computational time.

As future work, Multi Armed Bandit could be used to select other operators specific to metaheuristics or use this work and apply it to another optimization problem. On the Multi Armed Bandit side, other action selection techniques could be studied or another action reward function could be applied.

Acknowledgements. Broderick Crawford, Ricardo Soto, Eduardo Rodriguez-Tello and Felipe Cisternas-Caneo are supported by Dirección de Investigación, VINCI-PUCV; Project: DI Investigación Asociativa Interdisciplinaria 2022 "SELECCIÓN DE CARACTERÍSTICAS USANDO METAHEURÍSTICAS PARA POTENCIAR MODELOS PREDICTIVOS EN SALUD".

Broderick Crawford and Ricardo Soto are supported by Grant ANID/ FONDECYT/REGULAR/1210810.

Felipe Cisternas-Caneo is supported by Beca INF-PUCV.

References

1. Ab. Aziz, N.A., Ab. Aziz, K.: Pendulum search algorithm: an optimization algorithm based on simple harmonic motion and its application for a vaccine distribution problem. Algorithms **15**(6) (2022)
2. Rahman, T.A., Ibrahim, Z., Ab. Aziz, N.A., Zhao, S., Aziz, N.H.A.: Single-agent finite impulse response optimizer for numerical optimization problems. IEEE Access **6**, 9358–9374 (2018)
3. Alizadeh, R., Nishi, T.: Hybrid set covering and dynamic modular covering location problem: Application to an emergency humanitarian logistics problem. Appl. Sci. **10**(20), 7110 (2020)
4. Audibert, J.-Y., Munos, R., Szepesvári, C.: Exploration-exploitation tradeoff using variance estimates in multi-armed bandits. Theoret. Comput. Sci. **410**(19), 1876–1902 (2009)
5. Auer, P., Cesa-Bianchi, N., Fischer, P.: Finite-time analysis of the multiarmed bandit problem. Mach. Learn. **47**(2), 235–256 (2002)
6. Becerra-Rozas, M., et al.: Continuous metaheuristics for binary optimization problems: an updated systematic literature review. Mathematics **11**(1), 129 (2022)
7. Crawford, B., Soto, R., Astorga, G., García, J., Castro, C., Paredes, F.: Putting continuous metaheuristics to work in binary search spaces. Complexity **2017** (2017)
8. Crawford, B., Soto, R., Monfroy, E., Astorga, G., García, J., Cortes, E.: A meta-optimization approach for covering problems in facility location. In: Figueroa-García, J.C., López-Santana, E.R., Villa-Ramírez, J.L., Ferro-Escobar, R. (eds.) WEA 2017. CCIS, vol. 742, pp. 565–578. Springer, Cham (2017). https://doi.org/10.1007/978-3-319-66963-2_50
9. DaCosta, L., Fialho, A., Schoenauer, M., Sebag, M.: Adaptive operator selection with dynamic multi-armed bandits. In: Proceedings of the 10th Annual Conference on Genetic and Evolutionary Computation, pp. 913–920 (2008)
10. Elena, G., Milos, K., Eugene, I.: Survey of multiarmed bandit algorithms applied to recommendation systems. Int. J. Open Inf. Technol. **9**(4), 12–27 (2021)
11. Lanza-Gutierrez, J.M., Caballe, N.C., Crawford, B., Soto, R., Gomez-Pulido, J.A., Paredes, F.: Exploring further advantages in an alternative formulation for the set covering problem. Math. Probl. Eng. **2020** (2020)
12. Lanza-Gutierrez, J.M., Crawford, B., Soto, R., Berrios, N., Gomez-Pulido, J.A., Paredes, F.: Analyzing the effects of binarization techniques when solving the set covering problem through swarm optimization. Expert Syst. Appl. **70**, 67–82 (2017)
13. Lemus-Romani, J., et al.: A novel learning-based binarization scheme selector for swarm algorithms solving combinatorial problems. Mathematics **9**(22), 2887 (2021)
14. Mandal, S., Patra, N., Pal, M.: Covering problem on fuzzy graphs and its application in disaster management system. Soft. Comput. **25**(4), 2545–2557 (2021)

15. Patil, V., Ghalme, G., Nair, V., Narahari, Y.: Achieving fairness in the stochastic multi-armed bandit problem. In: AAAI, pp. 5379–5386 (2020)
16. Rodriguez-Tello, E., Narvaez-Teran, V., Lardeux, F.: Dynamic multi-armed bandit algorithm for the cyclic bandwidth sum problem. IEEE Access **7**, 40258–40270 (2019)
17. Song, H., Triguero, I., Özcan, E.: A review on the self and dual interactions between machine learning and optimisation. Progr. Artif. Intell. **8**(2), 143–165 (2019). https://doi.org/10.1007/s13748-019-00185-z
18. Soto, R., et al.: A reactive population approach on the dolphin echolocation algorithm for solving cell manufacturing systems. Mathematics **8**(9), 1389 (2020)
19. Sutton, R.S., Barto, A.G.: Reinforcement Learning: An Introduction, 2nd edn. The MIT Press, Cambridge (2018)
20. Talbi, E.-G.: Metaheuristics: From Design to Implementation. Wiley, Hoboken (2009)
21. Xiang, X., Qiu, J., Xiao, J., Zhang, X.: Demand coverage diversity based ant colony optimization for dynamic vehicle routing problems. Eng. Appl. Artif. Intell. **91**, 103582 (2020)
22. Fialho, Á., Da Costa, L., Schoenauer, M., Sebag, M.: Dynamic multi-armed bandits and extreme value-based rewards for adaptive operator selection in evolutionary algorithms. LION **3**, 176–190 (2009)

15. Patil, V., Ghalme, G., Nair, V., Narahari, Y.: Achieving fairness in the stochastic multi-armed bandit problem. In: AAAI, pp. 5379–5386 (2020)
16. Rodriguez-Tello, E., Narvaez-Teran, V., Lardeux, F.: Dynamic multi-armed bandit algorithm for the cyclic bandwidth sum problem. IEEE Access 7, 40258–40270 (2019)
17. Song, H., Triguero, I., Özcan, E.: A review on the self and dual interactions between machine learning and optimisation. Progr. Artif. Intell. 8(2), 143–165 (2019). https://doi.org/10.1007/s13748-019-00185-z
18. Sun, B., [illegible]: [illegible] population approach [illegible] the dolphin echolocation [illegible] for solar cell [illegible] tuning. [illegible] Mathematics 8(?), 1389 (2020)
19. Sutton, R.S., Barto, A.G.: Reinforcement Learning: An Introduction, 2nd edn. The MIT Press, Cambridge (2018)
20. Talbi, E.-G.: Metaheuristics: From Design to Implementation. Wiley, Hoboken (2009)
21. Wang, X., [illegible], L., Ma, [illegible], Xie, X.: [illegible] diversity based [illegible] [illegible] problems. [illegible] Appl. Artif. Intell. 91, 103582 (2020)
22. [illegible], F., [illegible]: [illegible] adaptive operator selection [illegible] (2021)

Optimization Applied to Learning Methods

Binary Black Widow with Hill Climbing Algorithm for Feature Selection

Ahmed Al-saedi and Abdul-Rahman Mawlood-Yunis(✉)

Physics and Computer Science Department, Wilfrid Laurier University, Waterloo, ON N2L3C5, Canada
alsa0290@mylaurier.ca, amawloodyunis@wlu.ca

Abstract. Feature Selection (FS) is a pre-processing step in most big data processing applications. Its purpose is to remove inconsequential and redundant features from data to determine a final set of data properties that best describe the data as a whole. The FS process is an NP-hard problem. It tries to determine the optimal subset, i.e., produces all conceivable solutions to acquire only the best. In the last few years, metaheuristic algorithms (MAs) have been coined as an ideal solution for FS problems, particularly in high-dimensional data cases. This work is an extension of our previous effort in finding an effective solution to FS problems by applying a recently developed metaheuristic algorithm called the Black Widow Optimization (BWO) algorithm. We combine our previous algorithm, the Binary Black Widow Algorithm (BBWO), with a Hill-Climbing Algorithm to solve the slow convergence problem of the BBWO. The newly developed algorithm, BBWO-HCA, is tested using 28 UCI datasets and compared with six well-regarded algorithms in the domain. The test results show that the BBWO-HCA outperforms our previous BBWO solution and almost all comparable solutions tested.

Keywords: Feature Selection · Evolutionary Algorithm · Metaheuristic Algorithm · Classification · Machine Learning · Data Mining

1 Introduction

It has become a challenge for researchers and developers to cope with the explosive growth of available data, the dimensions of which are expanding daily. Feature selection (FS) is a pre-processing step in most big data processing and machine learning applications, particularly in data mining applications. It is used to remove noisy and inconsequential features to determine a subset of features that best represent and portray the data, thus, boosting the quality of the data obtained.

Classical search approaches, such as random search and complete search have been used to solve FS problems [17]. While these methods ensure the optimal

Supported by Wilfrid Laurier University research fund.

B. Dorronsoro et al. (Eds.): OLA 2023, CCIS 1824, pp. 263–276, 2023.
https://doi.org/10.1007/978-3-031-34020-8_20

solution for small datasets, their execution is impractical for large datasets. FS is an NP-hard problem, it tries to determine the optimal subset. For example, if a dataset contains n features, then 2^n solutions must be formulated and assessed, i.e., the problem complexity is $O(\infty)$. It also requires an enormous amount of computational power and an excessive amount of time.

In the last few years, metaheuristic algorithms (MAs) have been identified as an ideal solution to FS problems, particularly in cases involving high-dimensional data. Researchers used the Simulated Annealing algorithm, Ant Colony Optimization algorithm, Particle Swarm Optimization algorithm, Genetic Algorithm, etc. to solve FS problems and have obtained valuable results. For example, see [1,4,5,8,11,12,18].

MAs are the most appropriate alternative method for addressing the limitations of lengthy, far-reaching searches that entail high computational cost. Despite some desirable results, however, most MAs are impeded by the limitations imposed by a *local optimum* and a disproportion between the *explorative* and *exploitative* scope of the algorithm. Moreover, each dataset has a different number of features, and no single method is the most appropriate for FS problems, i.e., one can still find room for improvements. These observations motivated this work to look for means to overcome the limitations described and develop a novel FS solution.

We selected a recent algorithm, the Black Widow Optimization algorithm (BWO) [9], to study FS problems due to its success in optimizing engineering design problems. BWO is a nature-inspired algorithm that mimics the black widow's life cycle. It is inspired by the singular mating behaviour exhibited by the black widow spider, a process that includes an exclusive stage called *cannibalism.* The BWO approach is designed to deliver rapid convergence and to avoid local optima, and, because BWO maintains equilibrium between the exploration and exploitation stages [9], a property that most MAs applied to FS problems are lacking [14], the BWO is particularly appropriate for solving several kinds of optimization problems that involve a number of local optima.

This work is an extension of our previous effort in trying to find an effective solution to FS problems by applying the BWO algorithm. In our previous effort, we modified the BWO algorithm to solve feature selection problems and developed the Binary Black Widow Optimization (BBWO) algorithm [3].

Despite the competitive results of the BBWO, its performance can be further improved by enhancing the *slow convergence* caused by the use of a population of solutions and a lack of *local exploitation.* In this article, we describe an improved version of the BBWO. We combined the BBWO with the Hill-Climbing Algorithm (HCA). The newly developed algorithm, the BBWO-HCA, is tested using 28 UCI datasets and compared with six well-regarded algorithms in the domain. The algorithms are Binary particle swarm optimization (BPSO) [10], Binary multi-verse optimization algorithm (BMWO) [2], Binary grey wolf optimizer algorithm (BGWO) [1,7], Binary moth-flame optimization algorithm (BMFO) [19], Binary whale optimization algorithm (BWOA) [11], and Binary bat algorithm (BBAT) [13].

The rest of this article is organized as follows: In Sect. 2, the proposed algorithm is presented. In Sect. 3, the experiment setup, test results, and result discussion are presented. Finally, in Sect. 4, the research is concluded and some future works are identified.

2 BBWO-HCA Feature Selection Algorithm

The results presented in [3] show that the BBWO produces impressive results and, in some cases, is competitive with the best-known algorithms. The results also reveal that the BBWO performance can be further improved by enhancing the *slow convergence* due to the use of a population of solutions and a lack of *local exploitation*. The BBWO-HCA aims to increase the *exploitation* process of the BBWO by incorporating it with a local metaheuristic algorithm based on the *Hill-Climbing Algorithm* (HCA). HCA is a well-known local search algorithm. It has been tested on various problems and has shown to be an effective and efficient method that can produce sound results [6]. The BBWO-HCA's main steps, *selection*, *procreation*, and *mutation*, are described in the subsections below.

2.1 Solutions Representation

In the BBWO-HCA, each solution represents a single black widow. All possible solutions to all FS problems are envisioned in terms of the attributes of the black widow spider. In programming terms, this is equivalent to saying each spider is represented by a class and spider attributes are class instance variables, or each spider is an array and spider attributes are array values. The spider population is modeled as an N_{var} dimensional array, i.e., an array of spider objects, and the FS problem becomes an N_{var} dimensional optimization problem.

The BBWO-HCA algorithm uses binary values to represent a population of solutions (N_{pop}). In binary representation, a solution is shown by a one-dimensional array. The length of the array varies in accordance with the feature number of the original dataset. For example, if S features are contained in the dataset, the solution length is S. The cell value in the array will be '1' or '0'. The value '1' indicates that the corresponding feature is selected, whereas '0' indicates that the feature is not selected. In general, when the number of features is N_f and the population size is $\mid N_{pop} \mid$, the array size of the problem will be $N_f \times \mid N_{pop} \mid$.

2.2 Initialization

The population of solutions offered by the BBWO-HCA is randomly generated by assigning a value of either "0" or "1" to each cell of the solution. The process begins by initializing the population size and the number of features. The algorithm then arbitrarily assigns either '0' or '1' by looping through each solution in the population. This process is repeated until all solutions in the population have been initialized.

2.3 Fitness Function and Evaluation

Each solution is evaluated according to a fitness function. The function employed is shown in Eq. 1. A similar function is used by [1,15]. The k-nearest neighbour algorithm (KNN) [16] is used in the solution evaluation, i.e., the KNN classifier determines the accuracy of the solution.

$$f = \alpha\gamma_R(D) + \beta\frac{\mid R \mid}{\mid C \mid} \tag{1}$$

In Eq. 1, $\gamma_R(D)$ represents the classification error rate of the KNN classifier, $\mid R \mid$ is the cardinality of the selected subset, $\mid C \mid$ is the total number of the original features in the dataset, and α, β are two weight parameters corresponding to the importance of classification quality and subset length, $\alpha \in [0, 1]$ and $\beta = (1 - \alpha)$. A similar approach is adapted by [7,11,19].

After initializing the population of solutions, we assign to each solution (widow) a fitness value, which represents the quality of the solution. The fitness value of each solution is calculated using the fitness function and is evaluated using the KNN classifier. This is because the BBWO-HCA is a wrapper-based FS approach.

2.4 Transformation Function

The positions of the search agents generated from the standard BWO are continuous values. This cannot be directly applied to our problem because it contradicts the binary nature of the FS on selection or non-selection (0 or 1).

The sigmoidal function in 2 and 3, which is considered a form of the transformation function, is used in our proposed method as a part of the reproduction process to convert any continuous values to binary equivalents. The performance of the transformation function has been investigated and adopted by many researchers, e.g., [1,7,19].

$$z_{s_w} = \frac{1}{1 + e^{-z_w}} \tag{2}$$

$$z_{binary} = \begin{cases} 0, & \text{if } rand < z_{s_w} \\ 1, & \text{if } rand \geq z_{s_w} \end{cases} \tag{3}$$

where each of z_{s_w} is a continuous value (feature) in the search agent for the S-shaped function, specifically in the solution w at dimension d (w = 1,. . . ,d), and is a random number drawn from the uniform distribution ∈ [0,1]. The z_{binary} value can be 0 or 1 depending on the value of *rand* compared to the values of z_{s_w}, where e is a mathematical constant known as Euler's number.

2.5 Reproduction Process

BWO is inspired by Darwin's natural selection theory, which is defined as generational descent accompanied by modification where species are subtly adjusted over time and new species arise as a result.

In the BBWO-BCA algorithm, the procreation process begins and parents (in pairs) are selected randomly to perform the procreating steps by mating to bring forth the new generation. An array known as *alpha* will be generated to complete further reproduction. Offspring c_1 and c_2 will be produced by taking α with the following equation in which w_1 and w_2 are parents.

$$\begin{cases} c_1 = \alpha \times w_1 + (1-\alpha) \times w_2 \\ c_2 = \alpha \times w_2 + (1-\alpha) \times w_1 \end{cases} \quad (4)$$

2.6 Cannibalism Process

The BWO includes an exclusive stage, *cannibalism.* Cannibalism can be classified into three kinds: *sexual cannibalism* where the husband gets eaten by the female black widow during or after mating, *sibling-cannibalism* where the weaker siblings are eaten by the stronger siblings, and *mother cannibalism* where the mother is eaten by her strongest child. The BBWO-HCA uses this concept of cannibalism and determines the weak or strong spiders by calculating and evaluating their fitness values. The best solutions (surviving spiders) from the reproduction process will be selected and stored in population two, i.e., pop_2.

2.7 Mutation Process and New Population Generation

The procedure of mutations begins by randomly selecting a number of solutions (widows) from the pop_1 which will be mutated individually. Two cells from each selected solution are randomly exchanged, and the new mutation solutions will be kept in pop_3. The new generation can finally be generated as a combination of pop_2 and pop_3, which will then be evaluated to return the optimal solution (W^*) of values bearing the N dimension.

In the BBWO-HCA, the cannibalism rate (C_R), the procreation rate (P_r), and the mutation rate (M_r) are used as parameters. The value of the (C_R) is determined by the fitness values obtained by Eq. 1, and the P_r and M_r are identical to those of the standard BWO.

2.8 HCA Steps

The algorithm uses the best solution (W^*) of the BBWO as an initial solution for the HCA. The solution is modified by selecting one feature randomly and flipping the value of that feature, i.e., if the feature value is "0" it is changed to "1" (which indicates adding one feature), and if the value is "1" it is changed to "0" (which indicates deleting one feature). If the fitness value of the modified solution is improved, it will replace the old one, otherwise, it discards the new solution.

Next, the HCA iteration counter and BBWO best solution (W^*) are updated, and the stopping criteria of the BBWO is checked. If the BBWO stopping condition is met, i.e., the max iterations are reached, the algorithm stops and returns the best solution (W^*), otherwise, a new iteration for the BBWO starts.

The pseudocode for the BBWO is shown in Fig. 1 and the additional steps involved in implementing the HCA are shown in Fig. 2. Together, they form the pseudocode for the BBWO-HCA.

3 Experiment Setup and Results

28 well-known datasets from the University of California Irvine (UCI)[1] machine learning repository have been used to investigate the performance and strength of our proposed methods. The dataset is randomly split into 80% for the training set and 20% for the test set. These rates are widely accepted data partition rates. The datasets vary in the number of features and instances. Table 1 presents a brief description of the datasets. Each row in the table represents the number of features, objects, classes, and the domain to which each of these datasets belong.

The performance of our proposed method, the BBWO-HCA, is compared with six well-respected binary FS algorithms: (BPSO [10], BMVO [2], BGWO [1,7], BMFO [19], BWOA [11], BBAT [13]) based on the two evaluation criteria, classification *accuracy* and the number of *features selected.*

To ensure an impartial comparison and a correct evaluation between our proposed method and other FS algorithms, we re-implemented the six FS algorithms using the same parameters values as illustrated in Table 2 and the same transformation function as explained in Sect. 2.4. The algorithms are run independently multiple times and the average accuracy and the average number of features selected are reported.

3.1 BBWO-HCA vs. BBWO Results and Discussion 1

Table 3 shows the comparison between our two algorithms (BBWO-HCA and BBWO algorithms) based on the two evaluation criteria (the classification accuracy and feature selected). The best *classification accuracy* and the *lower number of features* selected are highlighted in bold.

The results show that the BBWO-HCA is more efficient than the BBWO in terms of maximizing classification accuracy. The BBWO-HCA outperforms the BBWO in 15 datasets and obtains the same results in 13 datasets in terms of classification accuracy. When considering the average accuracy for all datasets, the performance of the BBWO-HCA is better than the BBWO. This is shown in Fig. 3.

[1] The datasets can be downloaded here: https://archive.ics.uci.edu/ml/datasets.php.

```
# Set the value of the parameters
population size (N_pop) = 20; number of iterations (maxIteration) = 10; number of
features (N_f) = dimension size; procreate rate (P_r) = 0.6; mutation rate (M_r) = 0.4
# Initialization process
Generating the initial population of solutions randomly (N_pop × N_f).
Each solution represents one widow, which is indicated in one-dimension vector 1×N_f
Calculate the fitness value for each solution using Eq.1
Evaluate all solutions in the population based on their fitness value and save them in
pop1
Set the best solution in the population as W*
based on P_r calculate the number of reproductions N_r
based on M_r calculate the number of mutations N_m
Define I=0
while I < maxIteration do
    # and cannibalism processes
    for i = 1 to ( N_r/ 2 ) do
        Randomly select two solutions w1, w2 as parents from pop1
        Generate two children c1, c2 using Eq. 4
        Transformation c1, c2 to binary nature using Eq.2 and 3
        Calculate the fitness value of c1, c2 using Eq.1
        Destroy the father w1 or w2 based on their fitness value (cannibalism process)
        c1 or c2 based on their fitness value (sibling cannibalism)
        Save the remaining solutions in pop2
end for
#Mutation process
    for i = 1 to N_m do
        Randomly select a solution from pop1
       Apply the mutation process on the selected solution
       Save the result (the new solution) in pop3
    end for
#Update the population
Update the population = pop2+pop3
Evaluate all solutions in the population using Eq.1
Update W* if there is a better solution
I= I+1
end while
returning the best solution W*
```

Fig. 1. The Binary Black Widow Algorithm Pseudocode for FS

In terms of minimizing the total number of features selected, the results show that the BBWO-HCA obtains better results than the BBWO in 26 datasets, the same results in one dataset, and worse results in one dataset. The results also show that on average, the BBWO-HCA is more efficient than the BBWO in this regard. This is shown in Fig. 4.

#Hill climbing algorithm
#Set the new solution = the best solution
S^*=W^*
#Set the fitness value of S^* = fitness values of W^*
$F(S^*)$= $F(W^*)$
$H = 1$
While H < maxIteration **do**
  S=S^*
  Randomly select one feature (i) in $S^*, i = 1, 2..., N_f$
  if $S_i^* = 0$, then $S_i^* = 1$; else $S_i^* = 0$
  if $F(S^*) < F(W^*)$ then $S^* = W^*$ else $S^* = S$
  H=H+1
end while
I= I+1
end while
returning the best solution =W^*

Fig. 2. BBWO with the Hill-climbing (BBWO-HCA) Algorithm

Table 1. Datasets description.

No.	Datasets	Features	Objects	Classes	Domain
1	Breastcancer	9	699	2	Medical
2	BreastEW	30	569	2	Medical
3	CongressEW	16	435	2	Politics
4	Exactly	13	1000	2	Medical
5	Exactly2	13	1000	2	Medical
6	HeartEW	13	270	5	Medical
7	IonosphereEW	34	351	2	Electronic
8	Lymphography	18	148	4	Medical
9	M-of-n	13	1000	2	Medical
10	PenglungEW	325	73	2	Medical
11	SonarEW	60	208	2	Medical
12	SpectEW	22	267	2	Medical
13	Tic-tac-toe	9	958	2	Game
14	Vote	16	300	2	Politics
15	WaveformEW	40	5000	3	Physical
16	Zoo	16	101	7	Artificial
17	Colon	2000	62	2	Medical
18	Parkinsons	22	195	2	Medical
19	Lungcancer	21	226	2	Medical
20	Leukemia	7129	72	2	Medical
21	Dermatology	34	366	6	Medical
22	Semeion	256	1593	10	Handwriting
23	Satellite	36	5100	2	Physical
24	Spambase	57	4601	2	Computer
25	Segment	19	2310	7	Images
26	Credit	20	1000	2	Business
27	KrvskpEW	36	3196	2	Game
28	Plants-100	64	1599	100	Agriculture

Table 2. BBWO-HCA Parameters

Parameter Name	Value	Parameter Name	Value
Population-size	20	No. of iterations	10
Number of independent runs	20	K (KNN classifier)	5
Dimension-size	No. of features	Number of iterations for hill climbing	20
pr(procreate rate)	0.6	mr (mutation rate)	0.4
α	0.99	β	0.01

Table 3. Compression between BBWO-HCA and BBWO

Datasets	Classification accuracy		Feature selected	
	BBWO-HCA	BBWO	BBWO-HCA	BBWO
Breastcancer	**0.98**	0.97	3.00	3.00
BreastEW	**0.95**	0.94	**4.60**	12.25
CongressEW	0.95	0.95	1.50	4.60
Exactly	**1**	0.91	**5.25**	3.75
Exactly2	0.77	0.77	**2.00**	3.65
HeartEW	**0.85**	0.84	**2.55**	3.80
IonosphereEW	**0.90**	0.88	**9.45**	13.75
Lymphography	0.85	0.85	**4.05**	6.80
M-of-n	**1**	0.95	**5.75**	7.00
PenglungEW	0.90	0.90	**100.85**	151.75
SonarEW	**0.87**	0.86	**14.75**	24.40
SpectEW	**0.82**	0.81	**6.50**	8.50
Tic-tac-toe	**0.82**	0.80	**4.05**	3.80
Vote	**0.95**	0.93	**1.55**	4.05
WaveformEW	0.88	0.88	**19.80**	20.60
Zoo	**0.92**	0.92	**4.60**	5.05
Parkinsons	0.90	0.90	**2.65**	7.70
Lungcancer	**0.92**	0.90	**4.70**	7.00
Colon	**0.89**	0.87	**888.35**	980.22
Leukemia	0.86	0.86	**3499.75**	3531.90
Dermatology	0.97	0.97	**11.25**	14.70
Semeion	**0.94**	0.93	**120.00**	130.00
Satellite	0.99	0.99	**4.40**	9.20
Spambase	0.93	0.93	**21.10**	28.60
Segment	0.96	0.96	**6.60**	7.70
Credit	0.79	0.79	**6.00**	7.22
KrvskpEW	0.97	0.95	**15.80**	19.00
Plants-100	**0.81**	0.80	**32.20**	32.50
Average	0.9050	0.8932	**171.53**	180.44
Rank	1	2	1	2

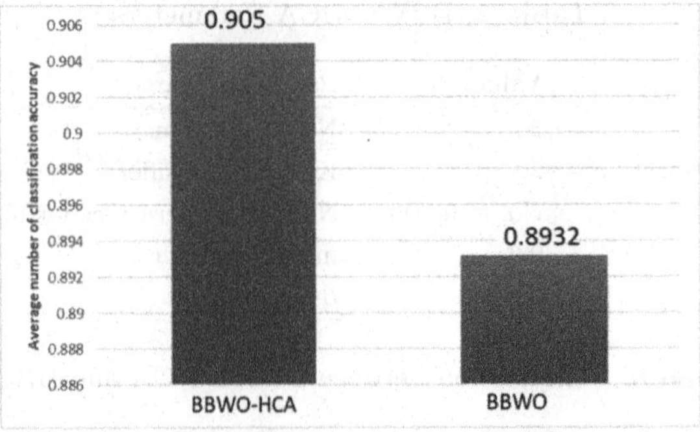

Fig. 3. Average of classification accuracy of BBWO-HCA vs. BBWO

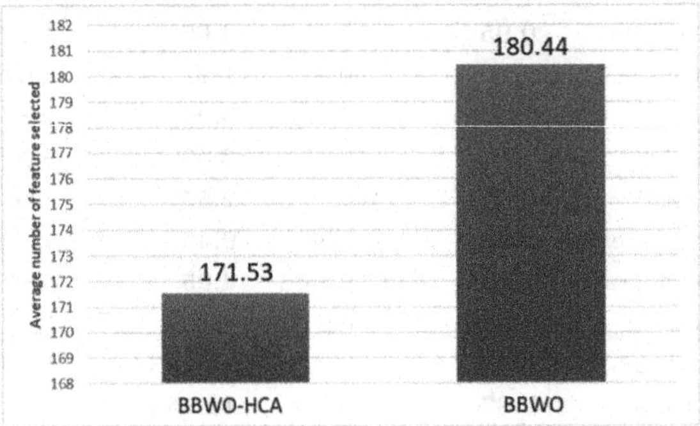

Fig. 4. Average of feature selection of BBWO-HCA vs. BBWO

3.2 BBWO-HCA vs. Six FS Algorithms, Results and Discussion 2

We compared the BBWO-HCA results with six FS algorithms (BPSO, BMVO, BGWO, BMFO, BWOA, BBAT). The results are presented in Tables 4 and 5. The test results reveal that the **BBWO-HCA** outperforms all six algorithms unless an algorithm already reached the best possible solution, in which case the BBWO-HCA results are the same as the other algorithm. For example, the **BBWO-HCA** produced better results than the **BPSO** in 26 datasets and the same results in two datasets.

In to the number of features selected, the test results reveal that the BBWO-HCA outperforms all six FS algorithms for all 28 datasets tested, see Table 5. These results are depicted pictorially in Figs. 5 and 6, and they show that the BBWO-HCA is an effective algorithm for solving FS problems.

Table 4. Comparison BBWO-HCA with all algorithms based on the classification accuracy

Datasets Name	**BBWO HCA**	BPSO	BMVO	BGWO	BMFO	BWOA	BBAT
Breastcancer	**0.98**	0.96	0.97	0.96	0.97	0.97	0.96
BreastEW	**0.95**	0.94	0.94	0.95	0.94	0.93	0.94
CongressEW	**0.95**	0.92	0.95	0.95	0.95	0.95	0.94
Exactly	**1**	0.76	0.89	0.74	0.90	0.91	0.73
Exactly2	**0.77**	0.77	0.76	0.75	0.76	0.74	0.74
HeartEW	**0.85**	0.81	0.85	0.84	0.85	0.85	0.82
IonosphereEW	**0.90**	0.86	0.88	0.88	0.88	0.88	0.88
Lymphography	**0.85**	0.82	0.84	0.82	0.85	0.83	0.81
M-of-n	**1**	0.83	0.99	0.88	0.98	0.98	0.81
PenglungEW	**0.90**	0.87	0.89	0.89	0.89	0.88	0.88
SonarEW	**0.87**	0.86	0.87	0.86	0.86	0.87	0.86
SpectEW	**0.82**	0.81	0.81	0.82	0.82	0.81	0.81
Tic-tac-toe	**0.82**	0.74	0.81	0.78	0.82	0.81	0.76
Vote	**0.95**	0.91	0.94	0.94	0.94	0.94	0.93
WaveformEW	**0.88**	0.86	0.88	0.87	0.88	0.88	0.83
Zoo	**0.92**	0.89	0.89	0.88	0.88	0.90	0.89
Parkinsons	**0.90**	0.88	0.89	0.86	0.89	0.89	0.88
Lungcancer	**0.92**	0.88	0.91	0.90	0.91	0.91	0.90
Colon	**0.89**	0.86	0.89	0.87	0.87	0.87	0.87
Leukemia	**0.86**	0.83	0.86	0.85	0.86	0.86	0.85
Dermatology	**0.97**	0.89	0.96	0.95	0.97	0.97	0.92
Semeion	**0.94**	0.92	0.93	0.92	0.93	0.93	0.92
Satellite	**0.99**	0.99	0.99	0.99	0.99	0.99	0.99
Spambase	**0.93**	0.88	0.93	0.92	0.93	0.93	0.89
Segment	**0.96**	0.94	0.96	0.96	0.96	0.96	0.94
Credit	**0.79**	0.76	0.79	0.77	0.78	0.79	0.78
KrvskpEW	**0.97**	0.90	0.96	0.95	0.97	0.97	0.87
Plants-100	**0.81**	0.78	0.79	0.78	0.80	0.79	0.77
Average	0.9050	0.8614	0.8935	0.8760	0.8939	0.8925	0.8632
Rank	1	7	3	5	2	4	6

Table 5. Comparison BBWO-HCA all algorithms based on the features selected

Datasets	BBWO-HCA	BPSO	BMVO	BGWO	BMFO	BWOA	BBAT
Breastcancer	**3.00**	3.40	4.55	5.15	4.35	4.60	3.45
BreastEW	**4.60**	11.40	10.95	13.55	13.20	12.12	13.15
CongressEW	**1.50**	5.20	4.25	5.95	5.40	4.20	5.55
Exactly	**5.25**	5.30	7.25	7.05	7.05	6.65	5.80
Exactly2	**2.00**	3.95	2.33	5.40	3.15	2.10	3.50
HeartEW	**2.55**	4.25	3.45	4.15	3.70	3.45	4.65
IonosphereEW	**9.45**	14.35	13.35	15.90	15.00	12.55	16.55
Lymphography	**4.05**	7.60	6.66	7.56	7.35	6.35	7.55
M-of-n	5.75	5.50	7.22	8.00	6.75	**7.35**	6.15
PenglungEW	**100.85**	154.80	152.35	155.20	152.45	146.35	156.85
SonarEW	**14.75**	27.05	25.55	28.05	28.95	23.85	27.00
SpectEW	**6.50**	8.95	8.22	10.15	8.20	8.75	9.85
Tic-tac-toe	**4.05**	4.20	4.55	4.55	4.41	4.05	4.28
Vote	**1.55**	5.05	4.95	6.35	5.85	4.50	6.15
WaveformEW	19.80	22.00	22.15	**21.45**	21.35	19.45	20.25
Zoo	**4.60**	5.59	6.35	6.65	6.13	5.75	6.50
Parkinsons	**2.65**	8.00	8.45	9.15	9.10	8.20	9.25
Lungcancer	**4.70**	7.25	8.66	9.35	8.90	8.05	8.95
Colon	**888.35**	961.65	963.25	965.55	962.15	943.55	963.35
Leukemia	**3499.75**	3555.82	3571.85	3535.85	3534.55	3511.35	3513.50
Dermatology	**11.25**	16.85	15.95	16.70	16.60	16.45	16.50
Semeion	**120.00**	131.85	127.00	128.6	131.60	126.80	126.70
Satellite	**4.40**	13.01	10.55	12.40	11.40	10.10	12.45
Spambase	**21.10**	29.77	26.55	30.50	26.25	26.50	27.25
Segment	**6.60**	8.72	9.25	9.90	9.95	8.90	9.60
Credit	**6.00**	8.41	7.95	8.50	8.30	7.72	8.75
KrvskpEW	**15.80**	19.73	19.81	21.30	18.56	17.92	18.45
Plants-100	**32.20**	35.55	33.15	33.80	33.32	34.15	35.50
Average	171.53	181.61	181.66	181.66	180.85	178.27	180.26
Rank	1	5	6	6	4	2	3

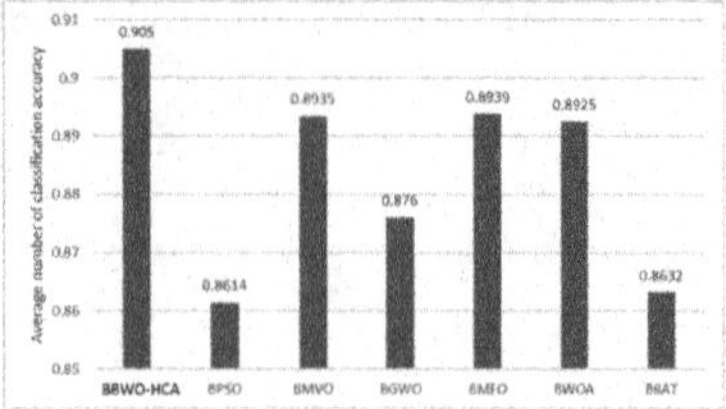

Fig. 5. Average number of classification accuracy of all algorithms

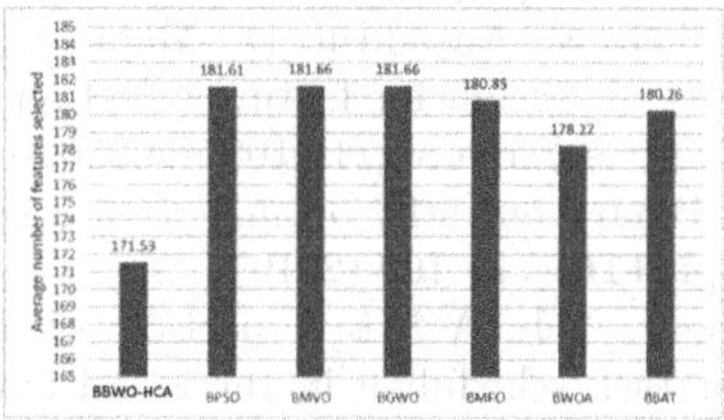

Fig. 6. Average number of features selected of all algorithms

4 Conclusion and Future Works

Recently, a novel algorithm, the Black Widow Algorithm (BWO), has been developed to solve optimization problems. BWO is derived from nature; it mimics the singular mating behaviour exhibited by the black widow spider.

Initially, we developed the BBWO algorithm based on the BWO for solving FS problems. In this work, we further improved the BBWO by combining it with the Hill-Climbing Algorithm. The newly developed algorithm, BBWO-HCA, is tested using 28 UCI datasets and compared with six well-regarded algorithms in the domain. The test results show that the BBWO-HCA outperforms the BBWO and almost all comparable algorithms for most datasets tested.

This work opened the door for further FS and optimization studies. Examples of such studies are:

- The test results of the BBWO-HCA can be further analyzed to determine the impact of the dataset size, number of features in the dataset, number of instances, etc. on the performance of the algorithm.
- Combining the BBWO with other algorithms and studying the outcomes of these new combinations are open future works.
- The BBWO and BBWO-HCA can be applied to various other areas of study to solve many other real-world optimization problems such as text mining, clustering, image processing, and routing problems.

References

1. Abdel-Basset, M., El-Shahat, D., El-henawy, I., de Albuquerque, V.H.C., Mirjalili, S.: A new fusion of grey wolf optimizer algorithm with a two-phase mutation for feature selection. Expert Syst. Appl. **139**, 112824 (2020)
2. Al-Madi, N., Faris, H., Mirjalili, S.: Binary multi-verse optimization algorithm for global optimization and discrete problems. Int. J. Mach. Learn. Cybern. **10**(12), 3445–3465 (2019). https://doi.org/10.1007/s13042-019-00931-8
3. Al-Saedi, A., Mawlood-Yunis, A.R.: Binary black widow optimization algorithm for feature selection problems. In: Simos, D.E., Rasskazova, V.A., Archetti, F., Kotsireas, I.S., Pardalos, P.M. (eds.) Learning and Intelligent Optimization (LION 2022). LNCS, vol. 13621, pp. 93–107. Springer, Cham (2022). https://doi.org/10.1007/978-3-031-24866-5_7

4. Alweshah, M., Alkhalaileh, S., Albashish, D., Mafarja, M., Bsoul, Q., Dorgham, O.: A hybrid mine blast algorithm for feature selection problems. Soft. Comput. **25**(1), 517–534 (2020). https://doi.org/10.1007/s00500-020-05164-4
5. Arora, S., Anand, P.: Binary butterfly optimization approaches for feature selection. Expert Syst. Appl. **116**, 147–160 (2019)
6. Dordaie, N., Navimipour, N.J.: A hybrid particle swarm optimization and hill climbing algorithm for task scheduling in the cloud environments. ICT Express **4**(4), 199–202 (2018)
7. Emary, E., Zawbaa, H.M., Hassanien, A.E.: Binary grey wolf optimization approaches for feature selection. Neurocomputing **172**, 371–381 (2016)
8. Hammouri, A.I., Mafarja, M., Al-Betar, M.A., Awadallah, M.A., Abu-Doush, I.: An improved dragonfly algorithm for feature selection. Knowl.-Based Syst. **203**, 106131 (2020)
9. Hayyolalam, V., Kazem, A.A.P.: Black widow optimization algorithm: a novel meta-heuristic approach for solving engineering optimization problems. Eng. Appl. Artif. Intell. **87**, 103249 (2020)
10. Mafarja, M., Jarrar, R., Ahmad, S., Abusnaina, A.A.: Feature selection using binary particle swarm optimization with time varying inertia weight strategies. In: Proceedings of the 2nd International Conference on Future Networks and Distributed Systems, pp. 1–9 (2018)
11. Mafarja, M., Mirjalili, S.: Whale optimization approaches for wrapper feature selection. Appl. Soft Comput. **62**, 441–453 (2018)
12. Mirjalili, S.: The ant lion optimizer. Adv. Eng. Softw. **83**, 80–98 (2015)
13. Mirjalili, S., Mirjalili, S.M., Yang, X.S.: Binary bat algorithm. Neural Comput. Appl. **25**(3), 663–681 (2014)
14. Mostafa, R.R., Ewees, A.A., Ghoniem, R.M., Abualigah, L., Hashim, F.A.: Boosting chameleon swarm algorithm with consumption AEO operator for global optimization and feature selection. Knowl.-Based Syst. **246**, 108743 (2022)
15. Neggaz, N., Houssein, E.H., Hussain, K.: An efficient henry gas solubility optimization for feature selection. Expert Syst. Appl. **152**, 113364 (2020)
16. Syriopoulos, P.K., Kotsiantis, S.B., Vrahatis, M.N.: Survey on KNN methods in data science. In: Simos, D.E., Rasskazova, V.A., Archetti, F., Kotsireas, I.S., Pardalos, P.M. (eds.) Learning and Intelligent Optimization (LION 2022). LNCS, vol. 13621, pp. 379–393. Springer, Cham (2022). https://doi.org/10.1007/978-3-031-24866-5_28
17. Venkatesh, B., Anuradha, J.: A review of feature selection and its methods. Cybern. Inf. Technol. **19**(1), 3–26 (2020)
18. Yang, X.S.: A new metaheuristic bat-inspired algorithm. In: Gonzáilez, J.R., Pelta, D.A., Cruz, C., Terrazas, G., Krasnogor, N. (eds.) Nature Inspired Cooperative Strategies for Optimization (NICSO 2010). Studies in Computational Intelligence, vol. 284, pp. 65–74. Springer, Heidelberg (2010). https://doi.org/10.1007/978-3-642-12538-6_6
19. Zawbaa, H.M., Emary, E., Parv, B., Sharawi, M.: Feature selection approach based on moth-flame optimization algorithm. In: 2016 IEEE Congress on Evolutionary Computation (CEC), pp. 4612–4617 (2016)

Optimization of Fuzzy C-Means with Alternating Direction Method of Multipliers

Benoit Albert(✉), Violaine Antoine, and Jonas Koko

Université Clermont Auvergne, Clermont Auvergne INP, CNRS, LIMOS, 63000 Clermont-Ferrand, France
{benoit.albert,violaine.antoine,jonas.koko}@uca.fr

Abstract. Among the clustering methods, K-Means and its variants are very popular. These methods solve at each iteration the first-order optimality conditions. However, in some cases, the function to be minimized is not convex, as for the Fuzzy C-Means version with Mahalanobis distance (FCM-GK). In this study, we apply the Alternating Directions Method of Multiplier (ADMM) to ensure a good convergence. ADMM is often applied to solve a separable convex minimization problem with linear constraints. ADMM is a decomposition/coordination method with a coordination step provided by Lagrange multipliers. By appropriately introducing auxiliary variables, this method allows the problem to be decomposed into easily solvable convex subproblems while keeping the same iterative structure. Numerical results have demonstrated the significant performance of the proposed method compared to the standard method, especially for high-dimensional data.

Keywords: Clustering · FCM · Mahalanobis distance · Optimization · ADMM

1 Introduction

Clustering is a data analysis process that consists to split n objects of the dataset into c subsets, with the idea that each group (subset) has similar objects and that the subsets are quite distinguishable from each other [15]. It allows the detection of hidden structures in data sets without prior knowledge. Several different approaches exist, the methods are distinguished by the nature of the partitions created. Among the models using centroids to represent clusters, there is a variant of K-Means called Fuzzy C-Mean (FCM) [2,3] which allows to take into account the uncertainty. This method creates a fuzzy partition that model the degree to which each object belongs to each cluster. It is still used in various fields such as bioinformatics [1] and image analysis [5,21]. The similarity between objects and centroids in the FCM algorithm is calculated with the Euclidean distance. The algorithm of Gustafson and Kessel FCM-GK [13] is an extension of FCM that adjusts an adaptive distance for each cluster. It allows us to take

B. Dorronsoro et al. (Eds.): OLA 2023, CCIS 1824, pp. 277–286, 2023.
https://doi.org/10.1007/978-3-031-34020-8_21

into account the shape of the clusters, to detect not only spherical structures but also ellipsoidal structures. Indeed, based on the Mahalanobis distance, the algorithm adapts symmetric positive definite matrices interpreted as the inverse of the fuzzy covariance matrices of clusters. FCM and GK are two non-convex optimization problems under constraints for which the standard optimization method is the alternating optimization (AO) method, an iterative method of the Gauss-Seidel type.

The Alternating Direction Method of Multiplier (ADMM) is a simple but powerful decomposition-coordination method. It decomposes the problem into subproblems, and the solutions obtained locally are coordinated by Lagrange multipliers to find a solution for the global problem. This method was introduced in the mid-1970s for the numerical approximation of non-smooth convex problems from mechanics [9,12]. This method has been used in many fields, first of all in nonlinear mechanics [8,11,12,16], also in image restoration [17], in neural networks [7], in large scale optimization [6,16], etc. A summary of ADMM applications in machine learning is available in [4]. The standard ADMM focuses on the minimization of separable (convex) functions with linear coupling constraints.

In this paper, we extend the application of ADMM to the non-convex cost function of FCM-GK. ADMM divides the FCM-GK problem into a sequence of simpler, uncoupled subproblems, through the appropriate introduction of unknown auxiliary variables. The solution formulation is close to the one obtained by alternating optimization for the original variables (centroids, distance-related membership matrices). The auxiliary variables sub-problem leads to the solution of small uncoupled linear systems. Numerical experiments on UCI machine learning data show that the proposed FCM-ADMM algorithm is robust, insensitive to random initialization, and generally creates better partitioning.

The paper is organized into four sections. Section 2 presents the GK model and the standard method optimization (AO). Then, in Sect. 3, we describe the application of the ADMM method in this context. In Sect. 4, the numerical experiments are presented. Finally, the conclusion and perspectives are given in Sect. 5.

2 FCM-GK Model

2.1 Optimisation Problem

Let the data set represented by $\boldsymbol{X} = (\boldsymbol{x}_1 \dots \boldsymbol{x}_n)$ contain n objects $\boldsymbol{x}_i \in \mathbb{R}^p$, p is the number of attributes. The objective is to group objects into c clusters $2 \leq c < n$. The variables used in the FCM-GK method are

- the matrix of membership degrees $(n \times c)$, $\boldsymbol{U} = (u_{ij})$ such that,

$$u_{ij} \in [0,1], \quad \sum_{j=1}^{c} u_{ij} = 1, \quad \sum_{i=1}^{n} u_{ij} > 0. \tag{1}$$

- the centroids of each group $\boldsymbol{\mathcal{V}} = \{\boldsymbol{v}_1, \dots, \boldsymbol{v}_c\}$, $\boldsymbol{v}_j \in \mathbb{R}^p$,

- the positive definite matrices, $\boldsymbol{\mathcal{S}} = \{\boldsymbol{S}_1, \ldots, \boldsymbol{S}_c\}$, inducing the norm of each group, $\boldsymbol{S}_j \in \mathbb{R}^{p \times p}$.

The K-Means algorithm and its variants focuses on minimizing the intra-class inertia. In FCM-GK, the unknown variables $(\boldsymbol{U}, \boldsymbol{\mathcal{V}}, \boldsymbol{\mathcal{S}})$ are determined by optimizing the following problem

$$\min_{(\boldsymbol{U},\boldsymbol{\mathcal{V}},\boldsymbol{\mathcal{S}})} J(\boldsymbol{U}, \boldsymbol{\mathcal{V}}, \boldsymbol{\mathcal{S}}) = \sum_{i=1}^{n} \sum_{j=1}^{c} u_{ij}^m \boldsymbol{q}_{ij}^\top \boldsymbol{S}_j \boldsymbol{q}_{ij}, \tag{2}$$

with the constraints, $\forall i, j \in [1, n] \times [1, c]$,

$$u_{ij} \geq 0, \sum_{j=1}^{c} u_{ij} = 1, \sum_{i=1}^{n} u_{ij} > 0, \tag{3}$$

$$\det(\boldsymbol{S}_j) = \rho_j, \quad \forall j \in [1, c] \tag{4}$$

where

$$\boldsymbol{q}_{ij} = \boldsymbol{x}_i - \boldsymbol{v}_j. \tag{5}$$

The fuzzy parameter m allows us to control the fuzziness of the partition. It's usually fixed at 2 [19]. The constraint Eq. (4) avoid trivial solution for the minimization is the solution with all $\boldsymbol{S}_j$ matrices zero. From a geometric point of view, ρ_j is the constant volume of the cluster j.

2.2 Alternating Optimization Method (AO)

The method used by Gustafson and Kessel to resolve this constrained problem is the alternating optimization method (AO) [13]. It is also used for the other versions of *k-means*, such as PFCM [20] and ECM [18]. Starting from $(\boldsymbol{U}^0, \boldsymbol{\mathcal{V}}^0, \boldsymbol{\mathcal{S}}^0)$, the method successively minimizes $\boldsymbol{U}$, $\boldsymbol{\mathcal{V}}$ and $\boldsymbol{\mathcal{S}}$ using first-order optimality conditions:

$$\boldsymbol{U}^{k+1} = \arg\min_{\boldsymbol{U} \in \mathcal{U}} J(\boldsymbol{U}, \boldsymbol{\mathcal{V}}^k, \boldsymbol{\mathcal{S}}^k), \tag{6}$$

$$\boldsymbol{\mathcal{V}}^{k+1} = \arg\min_{\boldsymbol{\mathcal{V}}} J(\boldsymbol{U}^{k+1}, \boldsymbol{\mathcal{V}}, \boldsymbol{\mathcal{S}}^k), \tag{7}$$

$$\boldsymbol{\mathcal{S}}^{k+1} = \arg\min_{\boldsymbol{\mathcal{S}} \in \mathcal{S}_1} J(\boldsymbol{U}^{k+1}, \boldsymbol{\mathcal{V}}^{k+1}, \boldsymbol{\mathcal{S}}). \tag{8}$$

With the two sets of constraints (3) and (4):

$$\mathcal{U} = \left\{ u_{ij} \geq 0, \quad \sum_{j=1}^{c} u_{ij} = 1, \quad \sum_{i=1}^{n} u_{ij} > 0 \right\},$$

$$\mathcal{S}_1 = \{\boldsymbol{S}, \quad p \times p \text{ symmetric positive matrix}, \det(\boldsymbol{S}) = 1\}.$$

Algorithm 1 describes the FCM-GK algorithm. It stops when the partition is stabilized, i.e. when the absolute error between two successive $\boldsymbol{U}$ matrices (membership degrees) is smaller than a threshold fixed at 10^{-3}. Note that for t iterations, its complexity is $O(tnc^2p)$ [10].

Algorithm 1. FCM-GK

```
Intput : c
err = 0, k = 0,
U^0 random initialization or through FCM.
while err > 10^{-3} do
    k ← k + 1
    compute 𝒱^k : v_j^k = Σ_{i=1}^n u_{ij}^{k-1} x_i / Σ_{i=1}^n u_{ij}^{k-1}, q_{ij}^k = x_i − v_j^k.
    compute 𝒮^k : Σ_j^k = Σ_{i=1}^n u_{ij}^{k-1} q_{ij}^k (q_{ij}^k)^⊤, S_j^k = det(Σ_j)^{1/p} (Σ_j^k)^{-1}.
    compute U^k : u_{ij}^k = [ Σ_{ℓ=1}^c ((q_{ij}^k)^⊤ S_j^k q_{ij}^k) / ((q_{iℓ}^k)^⊤ S_ℓ^k q_{iℓ}^k) ]^{-1}.
    err ← || U^k − U^{k-1} ||
end while
Output : U^k, 𝒱^k, 𝒮^k
```

3 Alternating Direction Method of Multipliers (ADMM)

The main idea of Alternating Direction Methods of Multipliers, introduced in the mid-1970s, is to use a decomposition/coordination process where the coordination is realized by Lagrange multipliers [8,9,12].

3.1 Augmented Lagrangien's Formulation

ADMM does not only minimize the objective function but also the associated augmented Lagrangian. Before formulating the latter, it is necessary to introduce auxiliary variables into the original problem to obtain a constrained block optimization problem. First, we write the characteristic functions of the original constraints to introduce them in the function to be minimized.

$$I_{\mathcal{U}}(\boldsymbol{U}) = \begin{cases} 0 & \text{if } \boldsymbol{U} \in \mathcal{U} \\ +\infty & \text{else} \end{cases}, \quad I_{\mathcal{S}_1}(\boldsymbol{S}_j) = \begin{cases} 0 & \text{if } \boldsymbol{S}_j \in \mathcal{S}_1 \\ +\infty & \text{else.} \end{cases}$$

and $I_{\mathcal{S}_1}(\boldsymbol{\mathcal{S}}) = \sum_j I_{\mathcal{S}_1}(\boldsymbol{S}_j)$. In addition to the auxiliary variables $\boldsymbol{\mathcal{Q}}$ (5), we introduce the variables $\boldsymbol{\mathcal{P}}$

$$\boldsymbol{p}_{ij} = u_{ij}\boldsymbol{q}_{ij} = u_{ij}(\boldsymbol{x}_i - \boldsymbol{v}_j).$$

Thus, we reformulate the cost function as (2) which becomes:

$$J(\boldsymbol{U}, \boldsymbol{\mathcal{V}}, \boldsymbol{\mathcal{S}}, \boldsymbol{\mathcal{Q}}, \boldsymbol{\mathcal{P}}) = \sum_{i=1}^{n}\sum_{j=1}^{c} \boldsymbol{p}_{ij}^{\top}\boldsymbol{S}_j\boldsymbol{p}_{ij}. \tag{9}$$

To simplify the writings we note:
$\mathbb{U} = (\boldsymbol{U}, \boldsymbol{\mathcal{V}}, \boldsymbol{\mathcal{S}})$ the set of variables of the problem and $\mathbb{Q} = (\boldsymbol{\mathcal{Q}}, \boldsymbol{\mathcal{P}})$ the set of auxiliary variables. The constrained minimization problem becomes (2)–(8)

$$\min J(\mathbb{U}, \mathbb{Q}) + I_{\mathcal{U}}(\boldsymbol{U}) + I_{\mathcal{S}_1}(\boldsymbol{\mathcal{S}}) \tag{10}$$

under constraints

$$q_{ij} = x_i - v_j, \tag{11}$$
$$p_{ij} = u_{ij} q_{ij}. \tag{12}$$

The coupling constraints are defined in such a way as to guarantee the equivalence (in terms of solution) with the original problem (2)–(8), while allowing independent optimization of variables. With (10)–(12), the function of the augmented Lagrangian is :

$$\begin{aligned} \mathscr{L}_r(\mathbb{U}, \mathbb{Q}, \mathbb{Y}) = & J(\mathbb{U}, \mathbb{Q}) + I_{\mathcal{U}}(U) + I_{\mathcal{S}_1}(\boldsymbol{\mathcal{S}}) \\ & + \sum_{i,j} \left[\boldsymbol{y}_{ij}^\top (\boldsymbol{q}_{ij} - \boldsymbol{x}_i + \boldsymbol{v}_j) + \boldsymbol{z}_{ij}^\top (\boldsymbol{p}_{ij} - u_{ij} \boldsymbol{q}_{ij}) \right] \\ & + \frac{r}{2} \sum_{i,j} \left[\| \boldsymbol{q}_{ij} - \boldsymbol{x}_i + \boldsymbol{v}_j \|^2 + \| \boldsymbol{p}_{ij} - u_{ij} \boldsymbol{q}_{ij} \|^2 \right] \end{aligned} \tag{13}$$

where $r > 0$ is the penalty term, $\| \cdot \|$ is the Euclidean norm, $\boldsymbol{y}_{ij}$ and $\boldsymbol{z}_{ij}$ are the Lagrange multipliers associated with the constraints of the auxiliary variables (11) and (12), represented by $\mathbb{Y} = (\boldsymbol{\mathcal{Y}}, \boldsymbol{\mathcal{Z}})$.

3.2 Application of ADMM

We apply the ADMM method to the augmented Lagrangian (13) by the following iterative algorithm. Starting with $\mathbb{Q}^0 : (\boldsymbol{\mathcal{Q}}^0, \boldsymbol{\mathcal{P}}^0)$ and $\mathbb{Y}^0 :(\boldsymbol{\mathcal{Y}}^0, \boldsymbol{\mathcal{Z}}^0)$, we successively compute $\mathbb{U}^k : (U^k, \boldsymbol{\mathcal{V}}^k, \boldsymbol{\mathcal{S}}^k)$, $\mathbb{Q}^k : (\boldsymbol{\mathcal{Q}}^k, \boldsymbol{\mathcal{P}}^k)$ and $\mathbb{Y}^k :(\boldsymbol{\mathcal{Y}}^k, \boldsymbol{\mathcal{Z}}^k)$ by the following procedure.

$$\mathbb{U}^{k+1} = \arg\min_{\mathbb{U}} \mathscr{L}_r(\mathbb{U}, \mathbb{Q}^k, \mathbb{Y}^k), \tag{14}$$
$$\mathbb{Q}^{k+1} = \arg\min_{\mathbb{Q}} \mathscr{L}_r(\mathbb{U}^{k+1}, \mathbb{Q}, \mathbb{Y}^k), \tag{15}$$
$$\boldsymbol{y}_{ij}^{k+1} = \boldsymbol{y}_{ij}^k + r(\boldsymbol{q}_{ij}^{k+1} - \boldsymbol{x}_i + \boldsymbol{v}_j^{k+1}), \tag{16}$$
$$\boldsymbol{z}_{ij}^{k+1} = \boldsymbol{z}_{ij}^k + r(\boldsymbol{p}_{ij}^{k+1} - u_{ij}^{k+1} \boldsymbol{q}_{ij}^{k+1}). \tag{17}$$

Note that the iterations of the ADMM method (14)–(17) admit exact updates if 1) the function is bi-convex, i.e., convex along $\mathbb{U}$ for $\mathbb{Q}$ fixed and reciprocally, and 2) if the constraints are bi-affine, i.e., affine in $\mathbb{U}$ for $\mathbb{Q}$ fixed and reciprocally [4]. In (10), $I_{\mathcal{S}_1}(\boldsymbol{\mathcal{S}})$ is non-convex because of the constraint $\det(\boldsymbol{S}_j) = 1$. To ensure the convergence of the method, it is sufficient to fix a number it_a of repetitions of the relaxation blocks (14)–(15), recommended $it_a = 5$, before updating the multipliers [11,16].

Solution of the subproblem (14) in $\mathbb{U}$

Assuming the auxiliary variables $\mathbb{Q}$ and the multipliers $\mathbb{Y}^k$ fixed, the problem (14) of the augmented Lagrangian (13) is decoupled according to each variable of $\mathbb{U}$, to be optimized separately.

$$\mathcal{V}^{k+1} = \arg\min_{\mathcal{V}} \sum_{i=1}^{n}\sum_{j=1}^{c} (\boldsymbol{y}_{ij}^{k})^\top (\boldsymbol{q}_{ij}^{k} - \boldsymbol{x}_i + \boldsymbol{v}_j) + \frac{r}{2} \parallel \boldsymbol{q}_{ij}^{k} - \boldsymbol{x}_i + \boldsymbol{v}_j \parallel^2, \quad (18)$$

$$\begin{aligned} \boldsymbol{U}^{k+1} = \arg\min_{\boldsymbol{U}} I_{\mathcal{U}}(\boldsymbol{U}) + \sum_{i=1}^{n}\sum_{j=1}^{c} (\boldsymbol{y}_{ij}^{k})^\top (\boldsymbol{p}_{ij}^{k} - u_{ij}\boldsymbol{q}_{ij}^{k}) \\ + \frac{r}{2} \parallel \boldsymbol{p}_{ij}^{k} - u_{ij}\boldsymbol{q}_{ij}^{k} \parallel^2, \end{aligned} \quad (19)$$

$$\mathcal{S}^{k+1} = \arg\min_{\mathcal{S}} \sum_{i=1}^{n}\sum_{j=1}^{c} (\boldsymbol{p}_{ij}^{k})^\top \boldsymbol{S}_j \boldsymbol{p}_{ij}^{k} + I_{\mathcal{S}_1}(\mathcal{S}). \quad (20)$$

The subproblems (18)–(20) are solved by taking the first-order optimality conditions, as for the AO method. Thus, the formulations obtained are quite close :

$$\boldsymbol{v}_j^{k+1} = \frac{1}{n}\sum_{i=1}^{n} \left(\boldsymbol{x}_i - \boldsymbol{q}_{ij}^{k} - \frac{1}{r}\boldsymbol{y}_{ij}^{k} \right), \quad (21)$$

$$u_{ij}^{k+1} = \frac{1}{r^2 \alpha_i^k \parallel \boldsymbol{q}_{ij}^{k} \parallel^2} \left[r\alpha_i^k (\boldsymbol{q}_{ij}^{k})^\top \tilde{\boldsymbol{z}}_{ij}^{k} + 1 - \sum_{\ell=1}^{c} \frac{(\boldsymbol{q}_{i\ell}^{k})^\top \tilde{\boldsymbol{z}}_{i\ell}^{k}}{\parallel \boldsymbol{q}_{i\ell}^{k} \parallel^2} \right], \quad (22)$$

$$\boldsymbol{S}_j^{k+1} = \det(\boldsymbol{\Sigma}_j^k)^{1/p} (\boldsymbol{\Sigma}_j^k)^{-1}, \quad (23)$$

with,

$$\tilde{\boldsymbol{z}}_{ij}^{k} = \boldsymbol{z}_{ij}^{k} + r\boldsymbol{p}_{ij}^{k}, \quad \alpha_i^k = \frac{1}{r}\sum_{j=1}^{c} \frac{1}{\parallel \boldsymbol{q}_{ij}^{k} \parallel^2}, \quad \boldsymbol{\Sigma}_j^k = \sum_{i=1}^{n} \boldsymbol{p}_{ij}^{k} (\boldsymbol{p}_{ij}^{k})^\top.$$

Solution of the subproblem (15) **in** $\mathbb{Q}$

Now assuming the variables $\mathbb{U}$ and the multipliers $\mathbb{Y}^k$ are fixed. The sub-problem in $\mathbb{Q}$: $(\mathcal{Q}, \mathcal{P})$ is an unconstrained optimization problem. Since $\mathcal{Q} \mapsto F(\mathbb{Q}) = \mathscr{L}_r(\mathbb{U}^{k+1}, \mathbb{Q}, \mathbb{Y}^k)$ is quadratic, the unique solution is obtained by solving the gradient equation $\nabla F(\mathbb{Q}) = 0$. A simple calculation allows to obtain the following linear system in $(\boldsymbol{q}_{ij}, \boldsymbol{p}_{ij})$.

$$r(1 + (u_{ij}^{k+1})^2)\boldsymbol{q}_{ij} - r u_{ij}^{k+1} \boldsymbol{p}_{ij} = u_{ij}^{k+1} \boldsymbol{z}_{ij}^{k} - \boldsymbol{y}_{ij}^{k} + r(\boldsymbol{x}_i - \boldsymbol{v}_j^{k+1}) \quad (24)$$

$$-r u_{ij}^{k+1} \boldsymbol{q}_{ij} + (2\boldsymbol{S}_j^{k+1} + r\mathbb{I})\boldsymbol{p}_{ij} = -\boldsymbol{z}_{ij}^{k} \quad (25)$$

It follows that at each iteration, we solve nc linear systems of size $2p$

$$\boldsymbol{A}_{ij}^{k} \begin{bmatrix} \boldsymbol{q}_{ij}, \\ \boldsymbol{p}_{ij} \end{bmatrix} = \boldsymbol{b}_{ij}^{k} \quad (26)$$

$$\boldsymbol{A}_{ij}^k = \begin{bmatrix} r(1+(u_{ij}^{k+1})^2)\mathbb{I} & -ru_{ij}^{k+1}\mathbb{I} \\ -ru_{ij}^{k+1}\mathbb{I} & 2S_j^{k+1} + r\mathbb{I} \end{bmatrix}, \boldsymbol{b}_{ij}^k = \begin{bmatrix} u_{ij}^{k+1}\boldsymbol{z}_{ij}^k - \boldsymbol{y}_{ij}^k + r(\boldsymbol{x}_i - \boldsymbol{v}_j^{k+1}) \\ -\boldsymbol{z}_{ij}^k \end{bmatrix}.$$

Algorithm

Algorithm 2 summarises the ADMM method. The stopping criterion is now the relative error on all primal and dual variables less than a threshold fixed at 10^{-3}. For t iterations, the complexity of our method is the same $O(tnc^2p)$. We initialize ADMM with random $\mathbb{U}$, same then AO, and construct all other variables $\mathbb{Q}$ (11)–(12). The Lagrange multipliers are initialized by solving the first order optimality condition (10)–(12), deriving the Lagrangian according to the variables $\mathcal{Q}, \mathcal{P}$: $\boldsymbol{z}_{ij}^0 = 2\boldsymbol{S}_j^0\boldsymbol{p}_{ij}^0, \boldsymbol{y}_{ij}^0 = u_{ij}^0\boldsymbol{z}_{ij}^0, \forall i, j$.

Algorithm 2. ADMM

Intput : Number of clusters c, penalty term r
$err = 1, k = 0,$
Random initialization or through ADMM(euclidean).
while $err > 10^{-3}$ **do**
    $k \leftarrow k+1$
    for 1 until 5 **do** *(it_a repetitions of the relaxation blocks)*
        $\mathcal{V}^k$,$\mathcal{S}^k$ and $\boldsymbol{U}^k$ respectively according to (21), (23) et (22)
        $\mathcal{Q}^k, \mathcal{P}^k$ solving the system (26)
    end for
    $\mathcal{Y}^k, \mathcal{Z}^k$ respectively according to (16) et (17)
    $err \leftarrow \parallel (\mathbb{U},\mathbb{Q})^k - (\mathbb{U},\mathbb{Q})^{k-1} \parallel / \parallel (\mathbb{U},\mathbb{Q})^k \parallel$
end while
Output : $\boldsymbol{U}^k$,$\mathcal{V}^k$,$\mathcal{S}^k$

---

4 Numerical Experiences

In this section, we studied the performance of our ADMM method for the FCM problem with Mahalanobis distance. We used Matlab (R2021). The penalty term r influences the performance of ADMM. In order to fine-tune this parameterization, we normalized all the data between $[-1, 1]$. Thus we take for r default the product of the dimensions to ensure the coordination of the variables, $r_d = 4cnp$. To find the optimal value r^*, we test several values and keep the one that converges the fastest in term of iterations.

In our study, we have fixed $m = 2$ and $\rho_j = 1, \forall j \in [1, c]$, such as [13]. We compare the three following algorithms:

- **FCM-GK**, the original method with alternating optimization on the GK model.
- ADMM_{r^*}, ADMM applied to the augmented Lagrangian (13), with the optimal penalty value r^*.
- ADMM_{r_d}, the same algorithm with the default value r_d.

In order to evaluate these different methods, we will use an external evaluation criterion that measures the similarity between two hard partitions: the clustering result and the ground truth. It is however necessary to transform the fuzzy partition into hard partition by assigning each object to the cluster with the highest membership. We used the Ajusted Rand Index (ARI) introduced by Hubert et al. [14]. The ARI value is between 0 and 1, 1 corresponding to identical partitions.

We used 11 data sets. The first five corresponds to real data from the UCI library[1]: IRIS, WINE, SEEDS, WDBC, and DRYBEAN. We also used six synthetic data sets[2]: A1, A3, DIM32, DIM64, S1, and S3. We have referenced Table 1 their characteristics, i.e. the number of classes c, objects n and attributes p, as well as the optimal penalty parameter r^*, and by default r_d. For insensitivity of the results to the initialization for every algoithms, we first ran ADMM with the Euclidean distance (ADMM$_{eu}$) with $r = 2.5$ and set a maximum number of iterations to 50 starting with random $\boldsymbol{U}^0$.

Table 1. Characteristics of data sets.

	IRIS	WINE	SEEDS	WDBC	DRYBEAN	A1	A3	DIM32	DIM64	S1	S3
c	3	3	3	2	7	20	50	16	16	15	15
n	150	178	210	569	13611	3000	7500	1024	1024	5000	5000
p	4	13	7	30	16	2	2	32	64	2	2
r^*	13	30	480	710	2.10^5	2000	1000	300	50	100	800
r_d	7200	27768	17640	136560	6.097.728	$4,8.10^4$	3.10^6	2^{21}	2^{22}	6.10^5	6.10^5

Table 2 shows that the ADMM methods perform better overall than the FCM-GK method, except for DRYBEAN, where FCM-GK is better. It seems that the larger number of individuals per class and the ratio between the number of clusters and the number of individuals explain this behavior.

Table 2. ARI score (UCI).

	IRIS	WINE	SEEDS	WDBC	DRYBEAN
FCM-GK	0.74	0.34	0.72	0.41	**0.70**
ADMM$_{r^*}$	**0.78**	0.81	0.71	**0.74**	0.32
ADMM$_{r_d}$	0.72	**0.90**	0.71	**0.74**	0.32

Table 3, corresponding to the results for the synthetic data, confirms this characteristic: the greater the number of clusters (A1, A3) the lower the ARI score. On the other hand, the greater the number of dimensions (DIM32, DIM64), the better the score.

[1] https://archive.ics.uci.edu/ml/datasets.php.
[2] https://cs.joensuu.fi/sipu/datasets/.

Although their complexity is of the same order of magnitude, the ADMM method is the fastest especially with the default penalty r_d. Table 4 lists the number of iterations for some data, measured for ten different random initializations ($ADMM_{eu}$).

Table 3. ARI score (Synthetic data).

	A1	A3	DIM032	DIM064	S1	S3
FCM-GK	**0.90**	**0.93**	0.44	0.18	**0.97**	**0.66**
$ADMM_{r^*}$	0.23	0.16	**0.57**	**0.68**	0.33	0.24
$ADMM_{r_d}$	0.20	0.16	**0.57**	**0.68**	0.33	0.26

Table 4. Number of iterations (mean ± standard deviation).

	IRIS	WINE	SEEDS	WDBC	A1	S1
$ADMM_{eu}$	30 ± 1	33 ± 3	30 ± 4	26 ± 1	10 ± 2	10 ± 0
FCM-GK	67 ± 0	113 ± 0	41 ± 0	35 ± 0	197 ± 75	92 ± 37
$ADMM_{r^*}$	35 ± 0	41 ± 0	6 ± 0	7 ± 0	4 ± 0	3 ± 0
$ADMM_{r_d}$	$\mathbf{2 \pm 0}$	$\mathbf{2 \pm 0}$	$\mathbf{4 \pm 0}$	$\mathbf{3 \pm 0}$	$\mathbf{2 \pm 0}$	$\mathbf{2 \pm 0}$

5 Conclusion

We have proposed an application of the ADMM method for the FCM clustering model with the Mahalanobis distance. The interest of this method is to divide the problem into a sequence of simpler sub-problems, easy to solve. Convergence to the same minimum, assumed to be global, is ensured. The results obtained on several data sets (real or synthetic) show good performances, in terms of ratios of well-classified samples, when the number of clusters is not too large or when the number of dimensions is significantly higher. To simplify the use of our method, we have proposed a default value for the penalty term (hyperparameter), whose convergence is assured and close to that of the optimal value. Our methods need less iterations than FCM-GK to converge and consequently are faster regarding the execution time.

The results are very encouraging. To confirm them, we wish to apply our method to a biology dataset, where many objects to be classified have a large number of attributes. To facilitate the use of our method, a formulation with an adaptive penalty is envisaged to replace the study of the optimal r. Finally, our study opens the possibility to apply the ADMM method to other clustering methods, having a non-convex objective function particularly those using alternating optimization.

References

1. Anter, A.M., Hassenian, A.E., Oliva, D.: An improved fast fuzzy c-means using crow search optimization algorithm for crop identification in agricultural. Expert Syst. Appl. **118**, 340–354 (2019)
2. Bezdek, J.C.: Fuzzy mathematics in pattern classification. Cornell University (1973)
3. Bezdek, J., Dunn, J.: Optimal fuzzy partitions: a heuristic for estimating the parameters in a mixture of normal distributions. IEEE Trans. Comput. **100**(8), 835–838 (1975)
4. Boyd, S., Parikh, N., Chu, E., Peleato, B., Eckstein, J.: Distributed optimization and statistical learning via alternating direction method of multipliers. Found. Trends Mach. Learn. **3**, 1–122 (2011)
5. Cai, W., Zhai, B., Liu, Y., Liu, R., Ning, X.: Quadratic polynomial guided fuzzy c-means and dual attention mechanism for medical image segmentation. Displays **70**, 102106 (2021)
6. Eckstein, J.: Parallel alternating direction multiplier decomposition of convex programs. J. Optim. Theory Appl. **1**, 39–62 (1994)
7. Wang, J., Yu, F., Chen, X., Zhao, L.: ADMM for efficient deep learning with global convergence. In: Proceedings of the 25th ACM SIGKDD International Conference on Knowledge Discovery and Data Mining, pp. 111–119 (2019)
8. Fortin, M., Glowinski, R.: Augmented Lagrangian Methods: Applications to the Numerical Solution of Boundary-Value Problems. North-Holland, Amsterdam (1983)
9. Gabay, D., Mercier, B.: A dual algorithm for the solution of nonlinear variational problems via finite element approximations. Comput. Math. Appl. **2**, 17–40 (1976)
10. Ghosh, S., Dubey, S.K.: Comparative analysis of k-means and fuzzy c-means algorithms. Int. J. Adv. Comput. Sci. Appl. **4**(4), 35–39 (2013)
11. Glowinski, R., Le Tallec, P.: Augmented Lagrangian and Operator-Splitting Methods in Nonlinear Mechanics. SIAM, Philadelphia (1989)
12. Glowinski, R., Marocco, A.: Sur l'approximation par éléments finis d'ordre un, et la résolution par pénalisation-dualité, d'une classe de problèmes de Dirichlet non linéaires. RAIRO-AN **9**(2), 41–76 (1975)
13. Gustafson, D., Kessel, W.: Fuzzy clustering with a fuzzy covariance matrix. In: 1978 IEEE Conference on Decision and Control Including the 17th Symposium on Adaptive Processes, pp. 761–766. IEEE (1979)
14. Hubert, L., Arabie, P.: Comparing partitions. J. Classif. **2**(1), 193–218 (1985)
15. Jain, A.K., Dubes, R.C.: Algorithms for Clustering Data. Prentice-Hall Inc., Hoboken (1988)
16. Koko, J.: Parallel Uzawa method for large-scale minimization of partially separable functions. J. Optim. Theory Appl. **158**, 172–187 (2013)
17. Koko, J., Jehan-Besson, S.: An augmented Lagrangian method for $TV_g + L^1$-norm minimization. J. Math. Imaging Vis. **38**, 182–196 (2010)
18. Masson, M.H., Denoeux, T.: ECM: an evidential version of the fuzzy c-means algorithm. Pattern Recogn. **41**(4), 1384–1397 (2008)
19. Pal, N.R., Bezdek, J.C.: On cluster validity for the fuzzy c-means model. IEEE Trans. Fuzzy Syst. **3**(3), 370–379 (1995)
20. Pal, N.R., Pal, K., Keller, J.M., Bezdek, J.C.: A possibilistic fuzzy c-means clustering algorithm. IEEE Trans. Fuzzy Syst. **13**(4), 517–530 (2005)
21. Yin, S., Li, H.: Hot region selection based on selective search and modified fuzzy c-means in remote sensing images. IEEE J. Sel. Top. Appl. Earth Obs. Remote Sens. **13**, 5862–5871 (2020)

Partial K-Means with M Outliers: Mathematical Programs and Complexity Results

Nicolas Dupin[1(✉)] and Frank Nielsen[2]

[1] Univ Angers, LERIA, SFR MATHSTIC, 49000 Angers, France
nicolas.dupin@univ-angers.fr
[2] Sony Computer Science Laboratories Inc., Tokyo, Japan
Frank.Nielsen@acm.org

Abstract. A well-known bottleneck of Min-Sum-of-Square Clustering (MSSC, the celebrated k-means problem) is to tackle the presence of outliers. In this paper, we propose a Partial clustering variant termed PMSSC which considers a fixed number of outliers to remove. We solve PMSSC by Integer Programming formulations and complexity results extending the ones from MSSC are studied. PMSSC is NP-hard in Euclidean space when the dimension or the number of clusters is greater than 2. Finally, one-dimensional cases are studied: Unweighted PMSSC is polynomial in that case and solved with a dynamic programming algorithm, extending the optimality property of MSSC with interval clustering. This result holds also for unweighted k-medoids with outliers. A weaker optimality property holds for weighted PMSSC, but NP-hardness or not remains an open question in dimension one.

Keywords: Optimization · Min-Sum-of-Square · Clustering · K-means · outliers · Integer Programming · Dynamic Programming · Complexity

1 Introduction

The K-means clustering of n d-dimensional points, also called Min Sum of Square Clustering (MSSC) in the operations research community, is one of the most famous unsupervised learning problem, and has been extensively studied in the literature. MSSC was is known to be NP hard [4] when $d > 1$ or $k > 1$. Special cases of MSSC are also NP-hard in a general Euclidean space: the problem is still NP-hard when the number of clusters is 2 [1], or in dimension 2 [15]. The case $K = 1$ is trivially polynomial. The 1-dimensional (1D) case is polynomially solvable with a Dynamic Programming (DP) algorithm [19], with a time complexity in $O(KN^2)$ where N and K are respectively the number of points and clusters. This last algorithm was improved in [9], for a complexity in $O(KN)$ time using memory space in $O(N)$. A famous iterative heuristic to solve MSSC

B. Dorronsoro et al. (Eds.): OLA 2023, CCIS 1824, pp. 287–303, 2023.
https://doi.org/10.1007/978-3-031-34020-8_22

was reported by Lloyd in [14], and a local search heuristic is proposed in [12]. Many improvements have been made since then: See [11] for a review.

A famous drawback of MSSC clustering is that it is not robust to noise nor to outliers [11]. The K-medoid problem, the discrete variant of the K-means problem addresses this weakness of MSSC by computing the cluster costs by choosing the cluster representative amongs the input points and not by calculating centroids. Although K-medoids is more robust to noise and outliers, it induces more time consuming computations than MSSC [5,10]. In this paper, we define Partial MSSC (PMSSC for short) by considering a fixed number of outliers to remove as in partial versions of facility location problems like K-centers [7] and K-median [3], and study extensions of exact algorithms of MSSC and report complexity results. Note that a K-means problem with outliers, studied in [13,20], has some similarities with PMMSC, we will precise the difference with PMMSC. To our knowledge, PMSSC is studied for the first time in this paper.

The remainder of this paper is structured as follows. In Sect. 2, we introduce the notation and formally describe the problem. In Sect. 3, Integer Programming formulations are proposed. In Sect. 4, we give first complexity results and analyze optimality properties. In Sect. 5, a polynomial DP algorithm is presented for unweighted MSSC in 1D. In Sect. 6, relations with state of the art and extension of these result are discussed. In Sect. 7, our contributions are summarized, discussing also future directions of research. To ease the readability, the proofs are gathered in an Appendix.

2 Problem Statement and Notation

Let $E = \{x_1, \dots, x_N\}$ be a set of N distinct elements of $\mathbb{R}^L$, with $L \in \mathbb{N}^*$. We note discrete intervals $[\![a, b]\!] = [a, b] \cap \mathbb{Z}$, so that we can use the notation of discrete index sets and write $E = \{x_i\}_{i \in [\![1,N]\!]}$. We define $\Pi_K(E)$, as the set of all the possible partitions of E into K subsets:

$$\Pi_K(E) = \left\{ P \subset \mathcal{P}(E) \,\middle|\, \forall p, p' \in P,\ p \cap p' = \emptyset \text{ and } \bigcup_{p \in P} = E \text{ and } \operatorname{card}(P) = K \right\}$$

MSSC is special case of K-sum clustering problems. Defining a cost function f for each subset of E to measure the dissimilarity, K-sum clustering are combinatorial optimization problems indexed by $\Pi_K(E)$, minimizing the sum of the measure f for all the K clusters partitioning E:

$$\min_{\pi \in \Pi_K(E)} \sum_{P \in \pi} f(P) \tag{1}$$

Unweighted MSSC minimizes the sum for all the K clusters of the average *squared distances* from the points of the clusters to the centroid. Denoting with d the Euclidean distance in $\mathbb{R}^L$:

$$\forall P \subset E, \quad f_{\text{UMSSC}}(P) = \min_{c \in \mathbb{R}^L} \sum_{x \in P} d(x, c)^2 = \sum_{x \in P} d\left(x, \frac{1}{|P|} \sum_{y \in P} y\right)^2 \tag{2}$$

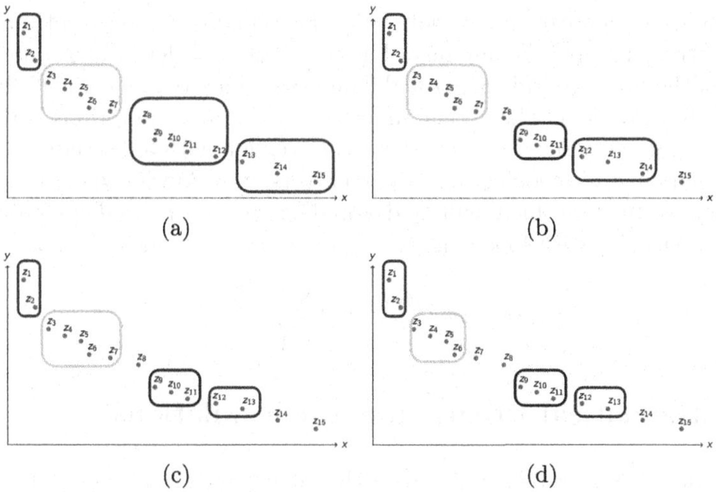

Fig. 1. MSSC clustering of a Pareto front in 4 clusters: (a) no outliers, (b) 2 outliers, (c) 3 outliers, and (d) 4 outliers.

The last equality can be proven using convexity and order one optimality conditions. In the weighted version, a weight $w_j > 0$ is associated to each point $x_j \in E$. For $x \in E$, $w(x)$ denotes the weight of point x. Weighted version of MSSC considers as dissimilarity function:

$$\forall P \subset E, \quad f_{\text{MSSC}}(P) = \min_{c \in \mathbb{R}^L} \sum_{x \in P} w(x) \times d(x,c)^2 \tag{3}$$

Unweighted cases correspond to $w_j = 1$. Analytic computation of weighted centroid holds also with convexity:

$$f_{MSSC}(P) = \sum_{x \in P} w(x) \times d\left(x, \frac{1}{\sum_{z \in P} w(z)} \sum_{y \in P} w(y) \times y\right)^2 \tag{4}$$

We consider a partial clustering extension of MSSC problem, similarly to the partial p-center and facility location problems [3,7]. A bounded number $M < N$ of the points may be considered outliers and removed in the evaluation. It is an optimal MSSC computation enumerating each subset $E' \subset E$ removing at most M points, i.e. such that $|E \setminus E'| \leqslant M$. It follows that PMSSC can be written as following combinatorial optimization problem:

$$\min_{E' \subset E : |E \setminus E'| \leqslant M} \; \min_{\pi \in \Pi_K(E')} \sum_{P \in \pi} f_{MSSC}(P) \tag{5}$$

Figure 1 shows an example of MSSC and PMSSC with $M \in \{2, 3, 4\}$.

In the "robust K-means problem" studied in [13,20], also denoted or "K-means problem with outliers", "robust" also denotes the partial variant with a defined number of outliers. It is not the usual meaning of robust optimization in the operations research community. These papers consider only the unweighted version of the problem, this paper highlights the difficulty of meaningfully formulating such a problem. The crucial difference with our assumptions is that their partial version concerns a discrete clustering version with a discrete set of possible centroids like K-medoids, not a partial version of MSSC where the centroid is continuous. Such problem will be denoted as "partial K-medoids problem", it is defined with (5) with following $f_{medoids}$ measure instead of f_{MSSC}:

$$f_{medoids}(P) = \min_{c \in P} \sum_{x \in P} d(x,c)^2 \tag{6}$$

3 Mathematical Programming Formulations

Partial MSSC can be formulated with Integer Programming formulations, extending the ones from MSSC [2,17,18].

For $n \in [\![1;N]\!]$ and $k \in [\![1;K]\!]$, we use binary variables $z_{n,k} \in \{0,1\}$ defined with $z_{n,k} = 1$ if and only if point x_n is assigned to cluster $k \in [\![1,K]\!]$. Using definition (3), the weighted centroid of cluster K is defined as a continuous variable $c_k \in \mathbb{R}_+^L$. It give rises to a first quadratic formulation:

$$\min_{z_{n,k},c_k} \sum_{k=1}^{K} \sum_{n=1}^{N} w_n d(x_n, c_k)^2 z_{n,k} \tag{7}$$

$$s.t: \quad \sum_{k=1}^{K} z_{n,k} \leqslant 1 \qquad \forall n, \tag{8}$$

$$\sum_{n'=1}^{N} \sum_{k=1}^{K} z_{n',k} \geqslant N - M \tag{9}$$

Objective function (7) holds also for (1) and (3) with $z_{n,k}$ encoding subsets $P \in \pi$. If $M = 0$, constraint (9) is equivalent to $\sum_{k=1}^{K} z_{n',k} = 1$ for each index n', point $x_{n'}$ shall be assigned to exactly one cluster. Constraints (8) impose that each point is assigned to at least one cluster. Constraint (9) aggregates that at most M points are unassigned, i.e. $\sum_{k=1}^{K} z_{n',k} = 0$ for these $x_{n'}$, and the other ones fulfill $\sum_{k=1}^{K} z_{n'',k} = 1$.

As for unpartial MSSC, last quadratic formulation is not solvable by mathematical programming solvers like `Cplex` and `Gurobi` because of non-convexity of the objective function. A compact reformulation, as for unpartial MSSC, allows such straightforward resolution. Using additional continuous variables $s_{n,k} \geqslant 0$ as the squared distance from point x_n to its cluster centroid c_k if $z_{n,k} = 1$ and 0 otherwise. It induces following convex quadratic formulation with quadratic convex constraints with a big M that can be set to $D = \max_{i,i'} d_{i,i'}^2$:

$$\min_{z_{n,k}, s_{n,k}, c_k} \sum_{k=1}^{K} \sum_{n=1}^{N} w_n s_{n,k} \tag{10}$$

$$s.t: \quad \sum_{k=1}^{K} z_{n,k} \leqslant 1 \quad \forall n, \tag{11}$$

$$s.t: \quad \sum_{n'=1}^{N} \sum_{k=1}^{K} z_{n',k} \geqslant N - M \tag{12}$$

$$s_{n,k} \geqslant d(x_n, c_k)^2 - D(1 - z_{n,k}) \quad \forall n, k, \tag{13}$$

Previous formulations have a common weakness, it induces symmetric solutions with permutations of clusters, which makes Branch & Bound tree search inefficient. As in [2] for unpartial MSSC, an extended reformulation can improve this known bottleneck. Enumerating each subset of E, $p \in \mathcal{P} = 2^E$, c_p denotes the clustering cost of p with formula (4), and we define a binary variable $z_p \in \{0,1\}$ with $z_p = 1$ if and only if subset p is chosen as a cluster. We define binaries $y_n \in \{0,1\}$ with $y_n = 1$ if and only if point x_n is chosen to be counted as outlier and not covered.

$$\text{PMSSC} = \min_{z} \sum_{p \in \mathcal{P}} c_p z_p \tag{14}$$

$$s.c: \forall n, \ \textstyle\sum_{p \in \mathcal{P}} \mathbb{1}_{n \in p} z_p \geqslant 1 - y_n \tag{15}$$

$$\textstyle\sum_n y_n \leqslant M \tag{16}$$

$$\textstyle\sum_{p \in \mathcal{P}} z_p \leqslant K \tag{17}$$

Objective function (14) is linear in the extended reformulation. Constraint (16) bounds the maximal budget of uncovered points. Constraint (17) bounds the maximal number of clusters, having more clusters decreases the objective function. Constraints (15) express that either a point x_n is uncovered when $y_n = 1$ and there is no need to select a subset which contains x_n, or one subset (at least) contains x_n. Note that $\mathbb{1}_{n \in p} z_p$ is one if and only if subset p contains point x_n. These constraints are written with inequalities, equalities are valid also to have the same optimal solutions. Inequalities are preferred for numerical stability with Column Generation (CG) algorithm.

Variables z_p, contrary to variables y_n, are of an exponential size and cannot be enumerated. CG algorithm applies to generate only a subset of z_p variables to compute the continuous (LP) relaxation of (14)–(17). We consider the Restricted Master Problem (RMP) for a subset of z_p variables in $\mathcal{P}' \subset \mathcal{P}$ of the LP relaxation, so that dual variables are defined for each constraint:

$$\begin{array}{lll} \text{RMP}(\mathcal{P}') = & \min\limits_{z \geqslant 0} \sum\limits_{p \in \mathcal{P}'} c_p z_p & \\ s.c: \forall n, & y_n + \sum_{p \in \mathcal{P}'} \mathbb{1}_{n \in p} z_p \geqslant 1 & (\pi_n) \\ & -\sum_n y_n \geqslant -M & (\lambda) \\ & -\sum_{p \in \mathcal{P}'} z_p \geqslant K & (\sigma) \end{array} \tag{18}$$

Inequalities imply that dual variables $\sigma, \lambda, \pi_n \geqslant 0$ are signed. This problem is feasible if $E \in \mathcal{P}'$ or if a trivial initial solution is given. Applying strong duality:

$$\begin{array}{lll} \mathrm{RMP}(\mathcal{P}') = & \max\limits_{\pi_n,\sigma,\lambda \geqslant 0} -K\sigma - M\lambda + \sum\limits_n \pi_n & \\ s.t: \forall p \in \mathcal{P}', -\sigma + \sum_n 1\!\!1_{n \in p} \pi_n \leqslant c_p & & (z_p) \\ \forall n, & \pi_n - \lambda \leqslant 0 & (y_n) \end{array} \tag{19}$$

Having only a subset of z_p variables, RMP is optimal if for the non generated z_p variables, we have $-\sigma + \sum_n 1\!\!1_{n \in p} \pi_n \leqslant c_p$. Otherwise, a cluster p should be added in the RMP if $-\sigma + \sum_n 1\!\!1_{n \in p} \pi_n > c_p$. It defined CG sub-problems:

$$\mathrm{SP} = \min_{p \in \mathcal{P}} c_p - \sum_n 1\!\!1_{n \in p} \pi_n \tag{20}$$

CG algorithm iterates adding subsets p such that $c_p - \sum_n 1\!\!1_{n \in p} \pi_n < -\sigma$. Once SP$\geqslant -\sigma$, the RMP is optimal for the full extended formulation.

As constraints (16) are always in the RMP, partial clustering induces the same pricing problem with [2]. Primal variables y_n influence numerical values of RMP, and thus the values of dual variables π_n, σ that are given to the pricing problem, but not the nature of sub-problems. Sub-problems SP can be solved with `Cplex` or `Gurobi`, using the same reformulation technique as in (10)–(13). Defining binaries $z_n \in \{0,1\}$ such that $z_n = 1$ iff point x_n is assigned to the current cluster, sub-problem SP is written as:

$$\mathrm{SP} = \min_{p \in \mathcal{P}} c_p - \sum_n \pi_n z_n \tag{21}$$

Considering continuous variables $c \in \mathbb{R}^d$ for the centroid of the optimal cluster, and $s_n \geqslant 0$, the squared distance from point x_n to centroid c if $z_n = 1$ and 0 otherwise. It gives rise to the following convex quadratic formulation:

$$\begin{array}{ll} \mathrm{SP} = & \min\limits_{z_n, s_n, c_d} \sum\limits_{n=1}^{N} s_n - \sum\limits_n \pi_n z_n \\ s.t: \forall n, & s_n \geqslant d(x_n, c)^2 - D(1 - z_n) \end{array} \tag{22}$$

CG algorithm can thus be implemented using `Cplex` or `Gurobi` for LP computations of RMP and for computations of SP. This gives a lower bound of the integer optimum. Integer optimality can be obtained using Branch & Price.

4 First Complexity Results, Interval Clustering Properties

PMSSC polynomially reduces to MSSC: if any instance of PMSSC (or a subset of instances) is polynomially solvable, this is the case for any corresponding instance of MSSC considering the same points and a value $M = 0$ and the same algorithm. Hence, NP-hardness results from [1,4,15] holds for PMSSC:

Theorem 1. *Following NP-hardness results holds for PMSSC:*

- *PMSSC is NP-hard for general instances.*
- *PMSSC is NP-hard in a general Euclidean space.*
- *PMSSC is NP-hard for instances with a fixed value of $K \geqslant 2$.*
- *PMSSC is NP-hard for instances with a fixed value of $L \geqslant 2$.*

After Theorem 1, it remains to study cases $K = 1$ and $L = 1$, where MSSC is polynomial. In the remainder of this paper, we suppose that $L = 1$, i.e. we consider the 1D case. Without loss of generality in 1D, we consider $d(x, y) = |x - y|$. We suppose that $E = \{x_1 < \cdots < x_N\}$, a sorting procedure running in $O(N \log N)$ time may be applied. A key element for the polynomial complexity of MSSC is the interval clustering property [16]:

Lemma 1. *Having $L = 1$ and $M = 0$, each global minimum of MSSC is only composed of clusters $C_{i,i'} = \{x_j\}_{j \in [\![i,i']\!]} = \{x \in E \mid \exists j \in [\![i,i']\!],\, x = x_j\}$.*

The question is here to extend this property for PMSSC. Considering an optimal solution of PMSSC the restriction to no-outliers points is an optimal solution of PMSSC and an interval clustering property holds:

Proposition 1. *Having $L = 1$ and an optimal solution of PMSSC induce an optimal solution of MSSC removing the outliers. In this subset of points, the optimality property of interval clustering holds.*

Proposition 1 is weaker than Lemma 1, selected points are not necessarily an interval clustering with the indexes of E. This stronger property is false in general for weighted PMSSC, one can have optimal solutions with outliers to remove inside the natural interval cluster as in the following example with $M = 1$, $L = 1$ and $K = 2$:

- $x_1 = 1$, $w_1 = 10$
- $x_2 = 2$, $w_2 = 1000$
- $x_3 = 3$, $w_2 = 1$
- $x_4 = 100$, $w_4 = 100$
- $x_5 = 101$, $w_5 = 1$

Optimal PMSSC consider x_2 as outlier, $\{x_1; x_3\}$ and $\{x_4; x_5\}$ as the two clusters. For $K = 1$, changing the example with $x_4 = 3.001$ and $x_5 = 3.002$, gives also a counter example with $K = 1$ with $\{x_1; x_3; x_4; x_5\}$ being the unique optimal solution. These counter-examples use a significant difference in the weights. In the unweighted PMSSC, interval property holds as in Lemma 1, with outliers (or holes) between the original interval clusters:

Proposition 2. *Having $L = 1$, each global minimum of unweighted PMSSC is only composed of clusters $C_{i,i'} = \{x_j\}_{j \in [\![i,i']\!]}$. In other words, the K clusters may be indexed $C_{i_1,j_1}, \ldots, C_{i_K,j_K}$ with $1 \leqslant i_1 \leqslant j_1 < i_2 \leqslant j_2 < \cdots < i_K \leqslant j_K \leqslant N$ and $\sum_{k=1}^{K} (j_k - i_k) \geqslant N - M - K$.*

As in [5], the efficient computation of cluster cost is a crucial element to compute the polynomial complexity. Cluster costs can be computed from scratch, leading to polynomial algorithm. Efficient cost computations use inductive relations for amortized computations in $O(1)$ time, extending the relations in [19]. We define for i, i' such that $1 \leqslant i \leqslant i' \leqslant N$:

- $b_{i,i'} = \sum_{k=i}^{i'} \frac{w_k}{\sum_{l=i}^{i'} w_l} x_k$ the weighted centroid of $C_{i,i'}$.
- $c_{i,i'} = \sum_{j=i}^{i'} w_j d(x_j, b_{i,i'})^2$ the weighted cost of cluster $C_{i,i'}$.
- $v_{i,i'} = \sum_{j=i}^{i'} w_j$

Proposition 3. *Following induction relations holds to compute efficiently $b_{i,i'}, v_{i,i'}$ with amortized $O(1)$ computations:*

$$v_{i,i'+1} = w_{i'+1} + v_{i,i'} \quad , \quad \forall 1 \leqslant i \leqslant i' < N \tag{23}$$

$$v_{i-1,i'} = w_{i-1} + v_{i,i'} \quad , \quad \forall 1 < i \leqslant i' \leqslant N \tag{24}$$

$$b_{i,i'+1} = \frac{w_{i'+1} x_{i'+1} + b_{i,i'} v_{i,i'}}{v_{i,i'+1}} \quad , \quad \forall 1 \leqslant i \leqslant i' < N \tag{25}$$

$$b_{i-1,i'} = \frac{w_{i-1} x_{i-1} + b_{i,i'} v_{i,i'}}{v_{i-1,i'}} \quad , \quad \forall 1 < i \leqslant i' \leqslant N \tag{26}$$

Cluster costs are then computable with amortized $O(1)$ computations:

$$c_{i,i'+1} = c_{i,i'} + w_{i'+1}(x_{i'+1} - b_{i,i'})^2 + v_{i,i'}(b_{i,i'+1} - b_{i,i'})^2 \tag{27}$$

$$c_{i-1,i'} = c_{i,i'} + w_{i-1}(x_{i-1} - b_{i,i'})^2 + v_{i,i'}(b_{i-1,i'} - b_{i,i'})^2 \tag{28}$$

Trivial relations $v_{i,i} = w_i$, $b_{i,i} = x_i$ and $c_{i,i} = 0$ are terminal cases.

Proposition 3 allows to prove Propositions 4 and 5 to compute efficiently cluster costs. Proposition 3 is also a key element to have first complexity results with $K = 1$ and $M \leqslant 1$ in Propositions 6, 7.

Proposition 4. *Cluster costs $c_{1,i}$ for all $i \in [\![1; N]\!]$ can be computed in $O(N)$ time using $O(N)$ memory space.*

Proposition 5. *For each $j \in [\![1; N]\!]$ cluster costs $c_{i,j}$ for all $i \in [\![1; j]\!]$ can be computed in $O(j)$ time using $O(j)$ memory space.*

Proposition 6. *Having $L = 1$ and $K = 1$, unweighted PMSSC is solvable in $O(N)$ time using $O(1)$ additional memory space.*

Proposition 7. *Having $L = 1$, $M = 1$ and $K = 1$, weighted PMSSC is solvable in $O(N)$ time using $O(N)$ memory space.*

5 DP Polynomial Algorithm for 1D Unweighted PMSSC

Proposition 2 allows to design a DP algorithm for unweighted PMSSC, extending the one from [19]. We define $O_{i,k,m}$ as the optimal cost of unweighted PMSSC with k clusters among points $[\![1, i]\!]$ with a budget of m outliers for all $i \in [\![1, N]\!]$, $k \in [\![1, K]\!]$ and $m \in [\![0, M]\!]$. Proposition 8 sets induction relations allowing to compute all the $O_{i,k,m}$, and in particular $O_{N,K,M}$:

Proposition 8 (Bellman equations). *Defining $O_{i,k,m}$ as the optimal cost of unweighted MSSC among points $[\![1, i]\!]$ for all $i \in [\![1, N]\!]$, $k \in [\![1, K]\!]$ and $m \in [\![0, M]\!]$, we have the following induction relations*

$$\forall i \in [\![1, N]\!], \quad O_{i,1,0} = c_{1,i} \tag{29}$$

$$\forall m \in [\![1, M]\!], \forall k \in [\![1, K]\!], \forall i \in [\![1, m+k]\!], \quad O_{i,k,m} = 0 \tag{30}$$

$$\forall m \in [\![1, M]\!], \forall i \in [\![m+2, N]\!], \quad O_{i,1,m} = \min\left(O_{i-1,1,m-1}, c_{1+m,i}\right) \tag{31}$$

$$\forall k \in [\![2, K]\!], \forall i \in [\![k+1, N]\!], \quad O_{i,k,0} = \min_{j \in [\![k,i]\!]} \left(O_{j-1,k-1,0} + c_{j,i}\right) \tag{32}$$

$\forall m \in [\![1, M]\!], \forall k \in [\![2, K]\!], \forall i \in [\![k+m+1, N]\!],$

$$O_{i,k,m} = \min\left(O_{i-1,k,m-1}, \min_{j \in [\![k+m,i]\!]} \left(O_{j-1,k-1,m} + c_{j,i}\right)\right) \tag{33}$$

Using Proposition 8, a recursive and memoized DP algorithm can be implemented to solve unweighted PMSSC in 1D. Algorithm 1 presents a sequential implementation, iterating with index i increasing. The complexity analysis of Algorithm 1 induces Theorem 2, unweighted PMSSC is polynomial in 1D.

Theorem 2. *Unweighted PMSSC is polynomially solvable in 1D, Algorithm 1 runs in $O(KN^2(1+M))$ time and use $O(KN(1+M))$ memory space to solve unweighted 1D instances of PMSSC.*

6 Discussions

6.1 Relations with State of the Art Results for 1D Instances

Considering the 1D standard MSSC with $M = 0$, the complexity of Algorithm 1 is identical with the one from [19], it is even the same DP algorithm in this sub-case written using weights. The partial clustering extension implied using a $M+1$ time bigger DP matrix, multiplying by M the time and space complexities. This had the same implication in the complexity for p-center problems [6,7]. Seeing Algorithm 1 as an extension of [19], it is a perspective to analyze if some improvement techniques for time and space complexity are valid for PMSSC.

As in [7], a question is to define a proper value of M in PMSSC. Algorithm 1 can give all the optimal $O_{N,K,m}$ for $m \leqslant M$, for a good trade-off decision. From a statistical standpoint, a given percentage of outliers may be considered.

Algorithm 1: DP algorithm for unweighted PMSSC in 1D

sort E in the increasing order
initialize $O_{i,k,m} := 0$ for all $m \in [\![0; M]\!], k \in [\![1; K-1]\!], i \in [\![k; N-K+k]\!]$
compute $c_{1,i}$ for all $i \in [\![1; N-K+1]\!]$ and store in $O_{i,1,0} := c_{1,i}$
for $i := 2$ to N
  compute and store $c_{i',i}$ for all $i' \in [\![1; i]\!]$
  compute $O_{i,k,0} := \min_{j \in [\![k,i]\!]} (O_{j-1,k-1,0} + c_{j,i})$ for all $k \in [\![2; \min(K,i)]\!]$
  for $m = 1$ to $\min(M, i-2)$
    compute $O_{i,1,m} := \min(O_{i-1,1,m-1}, c_{1+m,i})$
    for $k = 2$ to $\min(K, i-m)$
      compute $O_{i,k,m} := \min\left(O_{i-1,k,m-1}, \min_{j \in [\![k+m,i]\!]} (O_{j-1,k-1,m} + c_{j,i})\right)$
    end for
  end for
  delete the stored $c_{i',i}$ for all $i' \in [\![1; i]\!]$
end for
initialize $\mathcal{P} = \emptyset$, $\underline{i} = \overline{i} = N$, $m = M$
for $k = K$ to 1 with increment $k \leftarrow k-1$
  compute $\overline{i} := \min\{i \in [\![\underline{i} - m; \underline{i}]\!] | O_{\underline{i},k,m} := O_{\underline{i}-i,k,m-i+\underline{i}}\}$
  $m := m - \overline{i} + \underline{i}$
  compute and store $c_{i',\overline{i}}$ for all $i' \in [\![1; \overline{i}]\!]$
  find $\underline{i} \in [\![1, \overline{i}]\!]$ such that $\underline{i} := \arg\min_{j \in [\![k+m,i]\!]} \left(O_{j-1,k-1,m} + c_{j,\overline{i}}\right)$
  add $[x_{\underline{i}}, x_{\overline{i}}]$ in $\mathcal{P}$
  delete the stored $c_{i',\overline{i}}$ for all $i' \in [\![1; \overline{i}]\!]$
end for
return $O_{N,K,M}$ the optimal cost and the selected clusters $\mathcal{P}$

If we consider that 1% (resp 5%) of the original points may be outliers, it induces $M = 0,01 \times N$ (resp $M = 0,05 \times N$). In these cases, we have $M = O(N)$ and the asymptotic complexity of Algorithm 1 is in $O(KN^3)$ time and using $O(KN^2)$ memory space. If this remains polynomial, this cubic complexity becomes a bottleneck for large vales of N in practice.

In [7], partial min-sum-k radii has exactly the same complexity when $\alpha = 2$, which is quite comparable to PMSSC but considering only the extreme points of clusters with squared distances. PMSSC is more precise with a weighted sum than considering only the extreme points, having equal complexities induce to prefer partial MSSC for the application discussed in [7]. A reason is that the $O(N^2)$ time computations of cluster costs are amortized in the DP algorithm. Partial min-sum-k radii has remaining advantages over PMSSC: cases $\alpha = 1$ are solvable in $O(N \log N)$ time and the extension is more general than 1D instances and also valid in a planar Pareto Front (2D PF). It is a perspective to study PMSSC for 2D PFs, Fig. 1 shows in that case that it makes sense to consider an extended interval optimality as in [5,7].

6.2 Definition of Weighted PMSSC

Counter-example of Proposition 1 page 293 shows that considering both (diverse) weights and partial clustering as defined in (5) may not remove outliers, which was the motivating property. This has algorithmic consequences, Algorithm 1 and the optimality property are specific to unweighted cases. One can wonder the sense of weighted and partial clustering after such counter-example, and if alternative definitions exist.

Weighted MSSC can be implied by an aggregation of very similar points, the weight to the aggregated point being the number of original points aggregated in this new one. This can speed-up heuristics for MSSC algorithms. In this case, one should consider a budget of outliers M, which is weighted also by the points. Let m_n the contribution of a point x_n in the budget of outliers. (35) would be the definition of partial MSSC with budget instead of (5):

$$X = \left\{ E' \subset E : \sum_{x_n \in E \backslash E'} m_n x_n| \leqslant M \right\} \tag{34}$$

$$\min_{x \in X} \min_{\pi \in \Pi_K(x)} \sum_{P \in \pi} f(P) \tag{35}$$

(5) is a special case of (35) considering $m_n = 1$ for each $n \in [\![1;N]\!]$. Note that this extension is compatible with the developments of Sect. 3, replacing respectively constraints (9) and (16) by linear constraints (36) and (37). These new constraints are still linear, there are also compatible with the convex quadratic program and the CG algorithm for the extended formulation:

$$\sum_{n'=1}^{N} \left(1 - m_{n'} \sum_{k=1}^{K} z_{n',k} \right) \geqslant M \tag{36}$$

$$\sum_{n=1}^{N} m_n y_n \leqslant M \tag{37}$$

For the DP algorithm of Sect. 5, we have to suppose $m_n \in \mathbb{N}$. Note that it is the case with aggregation of points, fractional or decimal m_n are equivalent to this hypothesis, it is not restrictive. Bellman equations can be adapted in that goal: (30), (31) and (33) should be replaced by:

$$\forall m \in [\![1, M]\!],\ \forall k \in [\![1, K]\!],\ \forall i,\ \sum_{j=1}^{i} m_i \leqslant m \Longrightarrow O_{i,k,m} = 0 \tag{38}$$

$$\forall m \in [\![1, M]\!],\ \forall i,\ \ m_i > m \Longrightarrow O_{i,1,m} = c_{\alpha_m,i} \tag{39}$$

$$\forall m \in [\![1, M]\!],\ \forall i,\ \ m_i \leqslant m \Longrightarrow O_{i,1,m} = \min\left(O_{i-1,1,m-m_i}, c_{\alpha_m,i}\right) \tag{40}$$

where α_m is the minimal index such that $\sum_{j=1}^{\alpha_m} m_j > m$.

$$m_i \leqslant m \Longrightarrow O_{i,k,m} = \min\left(O_{i-1,k,m-m_i}, \min_{j \in [\![1,i]\!]} \left(O_{j-1,k-1,m} + c_{j,i}\right) \right) \tag{41}$$

$$m_i > m \Longrightarrow O_{i,k,m} = \min_{j \in [\![1,i]\!]} \left(O_{j-1,k-1,m} + c_{j,i}\right) \tag{42}$$

This does not change the complexity of the DP algorithm. However, we do not have necessarily the property $M < N$ anymore. In this case, DP algorithm in 1D is pseudo-polynomial.

6.3 From Exact 1D DP to DP Heuristics?

If hypotheses $L = 1$ and unweighted PMSSC are restrictive, Algorithm 1 can be used in a DP heuristic with more general hypotheses. In dimensions $L \geqslant 2$, a projection like Johnson-Lindenstrauss or linear regression in 1D, as in [10], reduces heuristically the original problem, solving it with Algorithm 1 provides a heuristic clustering solution by re-computing the cost in the original space. This may be efficient for 2D PFs, extending results from [10].

Algorithm 1 can be used with weights. For the cost computations, Propositions 4 and 5 make no difference in complexity. Algorithm 1 is not necessarily optimal in 1D in the unweighted case, it gives the best solution with interval clustering, and no outliers inside clusters. It is a primal heuristic, it furnishes feasible solutions. One can refine this heuristic considering also the possibility of having at most one outlier inside a cluster. Let $c_{i,i'}^{(0)}$ be the cost of cluster $x_i, \ldots, x_{i'}$ as previously and also $c_{i,i'}^{(1)}$ the best cost of clustering $x_i, \ldots, x_{i'}$ with one outlier inside that can be computed as in Proposition 7. The only adaptation of Bellman equations that would be required is to replace (31), (33) by:

$$\forall m \in [\![1, M]\!],\ \forall i \in [\![m+2, N]\!],\ O_{i,1,m} = \min\left(O_{i-1,1,m-1}, c_{1+m,i}^{(0)}, c_{1+m,i}^{(1)})\right) \quad (43)$$

$\forall m \in [\![1, M]\!]$, $\forall k \in [\![2, K]\!]$, $\forall i \in [\![k+m+1, N]\!]$,

$$O_{i,k,m} = \min\left(O_{i-1,k,m-1}, \min_{j \in [\![k+m,i]\!], l \in \{0,1\}} \left(O_{j-1,k-1,m-l} + c_{j,i}^{(l)}\right)\right) \quad (44)$$

Note that if case $L = 1$ and $K = 1$ is proven polynomial, one may compute in polynomial time $c_{j,i}^{(m)}$ values of optimal clustering with m outliers with points indexed in $[\![j, i]\!]$ and solve weighted PMSSC in 1D with similar Bellman equations. This is still an open question after this study.

6.4 Extension to Partial K-medoids

In this section, we consider the partial K-medoids problem with M outliers defined by (5) and (6), as in [13,20]. To our knowledge, the 1D sub-case was not studied, a minor adaptation of our results and proofs allows to prove this sub-case is polynomially solvable. Indeed, Lemma 1 holds with K-medoids as proven in [5]. Propositions 4 and 5 have their equivalent in [8], complexity of such operations being in $O(N^2)$ time instead of $O(N)$ for MSSC. Propositions 1 and 2 still hold with the same proof for K-medoids. Proposition 8 and Algorithm 1 are still valid with the same proofs, the only difference being the different computation of cluster costs. In Theorem 2 this only changes the time complexity: computing the cluster costs $c_{i,i'}$ is in $O(N^3)$ time instead of $O(N^2)$, it is not bounded by the $O(KN^2(1+M))$ time to compute the DP matrix. This results in the theorem:

Theorem 3. *Unweighted partial K-medoids problem with M outliers is polynomially solvable in 1D, 1D instances are solvable in $O(N^3 + KN^2(1+M))$ time and using $O(KN(1+M))$ memory space.*

7 Conclusions and Perspectives

To handle the problem of MSSC clusters with outliers, we introduced in this paper partial clustering variants for unweighted and weighted MSSC. This problem differs from the "robust K-means problem" (also noted "K-means problem with outliers"), which consider discrete and enumerated centroids unlike MSSC. Optimal solution of weighted PMSSC may differ from intuition of outliers: We discuss about this problem and present another similar variant. For these extensions of MSSC, mathematical programming formulations for solving exactly MSSC can be generalized. Solvers like `Gurobi` or `Cplex` can be used for a compact and an extended reformulation of the problem. NP-hardness results of these generalized MSSC problems holds. Unweighted PMSSC is polynomial in 1D and solved with a dynamic programming algorithm which relies on the optimality property of interval clustering. With small adaptations, "K-means problem with outliers" defined as the unweighted partial K-medoids problem with M outliers is also polynomial in 1D and solved with a similar algorithm. We show that a weaker optimality property holds for weighted PMSSC. The relations with similar state-of-the-art results and adaptation of the DP algorithm to DP heuristics are also discussed.

This work opens perspectives to solve this new PMSSC problem. The NP-hardness complexity of weighted PMSSC for 1D instances is still an open question. Another perspective is to extend 1D polynomial DP algorithms for PMSSC for 2D PFs, as in [5,7]. Approximation results may be studied for PMSSC also, trying to generalize results from [13,20]. Using only quick and efficient heuristics without any guarantee would be sufficient for an application to evolutionary algorithms to detect isolated points in PFs, as in [7]. Adapting local search heuristics for PMSSC is also another perspective [10]. If K-medoids variants with or without outliers are used to induce more robust clustering to noise and outliers, the use of PMSSC is promising to retain this property without having slower calculations of cluster costs with K-medoids. Finally, using PMSSC as a heuristic for K-medoids is also a promising venue for future research.

Appendix: Proofs of Intermediate Results

Proof of Lemma 1: We prove the result by induction on $K \in \mathbb{N}$. For $K = 1$, the optimal cluster is $E = \{x_j\}_{j \in [\![1,N]\!]}$. Note that $N \leqslant K$ is also a trivial case, we suppose $1 < K < N$ and the Induction Hypothesis (IH) that Lemma 1 is true for $K-1$. Let an optimal clustering partition, denoted with clusters $\mathcal{C}_1, \dots, \mathcal{C}_K$ and centroids $c_1 < \dots < c_K$, where c_i is the centroid of cluster $\mathcal{C}_i$. Strict inequalities are a consequence of Lemma 2. Necessarily, $x_N \geqslant c_K$ and $x_N \in \mathcal{C}_K$ because x_N is assigned to the closest centroid. Let $A = \{i \in [\![1,N]\!] \mid \forall k \in [\![i,N]\!], x_k \in \mathcal{C}_K\}$

and let $j = \min A$. If $j = 1$, $E = \mathcal{C}_K = \{x_j\}_{j \in [\![1,N]\!]}$, it is in contradiction with $K > 1$. $j - 1 \in A$ is in a contradiction with $j = \min A$. Hence, we suppose $j > 1$ and $j - 1 \notin A$. Necessarily $x_{j-1} \in \mathcal{C}_{K-1}$, c_{K-1} is the closest centroid among $c_1, \dots, c_{K-1}$. For each $l \in [\![1, j-2]\!]$, x_l is strictly closer from centroid c_{K-1} than from centroid c_K, then $x_l \notin \mathcal{C}_K$ and $A = [\![j, N]\!]$. On one hand, it implies that $\mathcal{C}_K = \{x_l\}_{l \in [\![j,N]\!]}$. On the other hand, the other clusters are optimal for $E' = E \backslash \mathcal{C}$ with weighted $(K-1)$-means clustering. Applying IH proves that the optimal clusters are of the shape $\mathcal{C}_{i,i'} = \{x_j\}_{j \in [\![i,i']\!]}$. □

Lemma 2. *We suppose $L = 1$ and $K < N$. Each global optimal solution of weighted MSSC indexed with clusters $\mathcal{C}_1, \dots, \mathcal{C}_K$ and centroids such that $c_1 \leqslant \dots \leqslant c_K$, where c_i is the centroid of cluster $\mathcal{C}_i$, fulfills necessarily $c_1 < \dots < c_K$.*

Proof of Lemma 2: Ad absurdum, we suppose that an optimal solution exists with centroids such that $c_{k'} = c_k$. Having $K < N$, there exist a point x_n that is not a centroid (note that points of E are distinct in the hypotheses of this paper). Merging clusters $\mathcal{C}_{k'}$ and $\mathcal{C}_k$ does not change the objective function as the centroid are the same. Removing x_n from its cluster and defining it in a singleton cluster strictly decreases the objective function, it is a strictly better solution than the optimal solution. □

Proof of Proposition 1: Let X the set of selected outliers in an optimal solution of weighted PMSSC. Ad absurdum, if there exists a strictly better solution of weighted MSSC in $E \setminus X$, adding X as outliers would imply a strictly better solution for PMSSC in E, in contradiction with the global optimality of the given optimal solution. Lemma 1 holds in $E \setminus X$. □

Proof of Proposition 2: Let X the set of selected outliers in an optimal solution of unweighted PMSSC. Ad absurdum, we suppose that there exists a cluster $\mathcal{C}$ of centroid c with $x_j = \min \mathcal{C}$, $x_{j'} = \max \mathcal{C}$ and $x_i \in X$ such that $x_j < x_i < x_{j'}$. If $c \leqslant x_i$, the objective function strictly decreases when swapping x_i and $x_{j'}$ in the cluster and outlier sets. If $c \geqslant x_i$, the objective function strictly decreases when swapping x_i and x_j in the cluster and outlier sets. This is in contradiction with the global optimality. For the end of the proof, let us count the outliers. We have: $i_1 - 1 + i_2 - 1 - j_1 + \dots + i_K - 1 - j_{K-1} + N - j_K \leqslant M$ which is equivalent to $N + \sum_{k=1}^{K} (i_k - j_k) \leqslant M + K$. □

Proof of Proposition 3: Relations (23) and (24) are trivial with the definition of $v_{i,i'}$ as a sum. Relations (25) and (26) are standard associativity relations with weighted centroids. We prove here (28), the proof of (27) is similar. $c_{i-1,i'} - w_{i-1}(x_{i-1} - b_{i-1,i'})^2 = \sum_{j=i}^{i'} w_j (x_j - b_{i-1,i'})^2$
$= \sum_{j=i}^{i'} w_j \left((x_j - b_{i,i'})^2 + (b_{i,i'} - b_{i-1,i'})^2 + 2(x_j - b_{i,i'})(b_{i,i'} - b_{i-1,i'}) \right)$
$= c_{i,i'} + (b_{i,i'} - b_{i-1,i'})^2 \sum_{j=i}^{i'} w_j + 2(b_{i,i'} - b_{i-1,i'}) \sum_{j=i}^{i'} w_j (x_j - b_{i,i'})$.
It gives the result as $\sum_{j=i}^{i'} w_j (x_j - b_{i,i'}) = \sum_{j=i}^{i'} w_j x_j - b_{i,i'} \sum_{j=i}^{i'} w_j = 0$. □

Proof of Proposition 4: We compute and store values $c_{1,i}$ with i increasing starting from $i = 1$. We initialize $v_{1,1} = w_1$, $b_{1,1} = x_1$ and $c_{1,1} = 0$ and compute values $c_{1,i+1}, b_{1,i+1}, v_{1,i+1}$ from $c_{1,i}, b_{1,i}, v_{1,i}$ using (27) (25) (23). Such computation is in $O(1)$ time, so that cluster costs $c_{1,i}$ for all $i \in [\![1;N]\!]$ are computed in $O(i)$ time. In memory, only four additional elements are required: $b_{1,i+1}, v_{1,i+1}, b_{1,i}, v_{1,i}$. The space complexity is given by the stored $c_{1,i}$ values. □

Proof of Proposition 5: Let $j \in [\![1;N]\!]$. We compute and store values $c_{i,j}$ with i decreasing starting from $i = j$. We initialize $v_{j,j} = w_j$, $b_{j,j} = x_j$ and $c_{j,j} = 0$ and compute values $c_{i-1,j}, b_{i-1,j}, v_{i-1,j}$ from $c_{i,j}, b_{i,j}, v_{i,j}$ using (28) (26) (24). Such computation is in $O(1)$ time, so that cluster costs $c_{i,j}$ for all $i \in [\![1;j]\!]$ are computed in $O(j)$ time. In memory, only only four additional elements are required $b_{i-1,j}, v_{i-1,j}, b_{i,j}, v_{i,j}$ the space complexity is in $O(i)$. □

Proof of Proposition 6: Using interval optimality for unweighted PMSSC, we compute successively $c_{1,N-M}, c_{2,N-M+1}, \ldots, c_{M+1,N}$ and store the best solution. Computing $c_{1,N-M}, b_{1,N-M}, v_{1,N-M}$ is in $O(N-M)$ time with a naive computation. Then $c_{1,N-M+1}, b_{1,N-M+1}, v_{1,N-M+1}$ are computed from $c_{1,N-M}, b_{1,N-M}, v_{1,N-M}$ in $O(1)$ time using successively (23), (25) and (27). Then $c_{2,N-M+1}, b_{2,N-M+1}, v_{2,N-M+1}$ are computed from $c_{1,N-M+1}, b_{1,N-M+1}, v_{1,N-M+1}$ in $O(1)$ time using successively (24), (26) and (28). This process is repeated M times, there are $O(N-M) + O(M)$ operations, it runs in $O(N)$ time. Spatial complexity is in $O(1)$. □

Proof of Proposition 7: We enumerate the different costs considering all the possible outliers. We compute $c_{1,N}, b_{1,N}, v_{1,N}$ in $O(N)$ time. Adapting Proposition 2, each cluster cost removing one point can be computed in $O(1)$ time. The overall time complexity is in $O(N)$. □

Proof of Proposition 8: (29) is the standard case $K = 1$. (30) is a trivial case where the optimal clusters are singletons. (31) is a recursion formula among $K = 1$ cases, either point x_i is chosen and in this case all the outliers are points x_l with $l \leqslant m$ or x_i is an outlier and it remains an optimal PMSSC with $K = 1$ and $M = m - 1$ among the $i - 1$ first points. (32) is a recursion formula among $M = 0$ cases distinguishing the cases for the composition of the last cluster. (32) are considered for MSSC in [9,19]. (33) is an extension of (32). $O_{i,k,m}$ is $O_{i-1,k,m-1}$ if point x_i is not selected. Otherwise, the cluster k is a $\mathcal{C}_{j,i}$ and the optimal cost of other clusters is $O_{j-1,k-1,m}$. □

Proof of Theorem 2: by induction, one proves that at each loop i of In Algorithm 1, the optimal values of $O_{i,k',m'}$ are computed for all k', m' using Proposition 8. Space complexity is given by the size of DP matrix $(O_{i,k,m})$, it is in $O(KN(1+M))$. Each value requires at most N elementary operations, building the DP matrix runs in $O(KN^2(1+M))$ time. The remaining of Algorithm 1 is a standard backtracking procedure for DP algorithms, running in $O(N^2)$ time, the time complexity of DP is thus in $O(KN^2(1+M))$. Lastly, unweighted PMSSC is polynomially solvable with Algorithm 1, the memory space of inputs are in $O(N)$, mostly given y the N points of E, and using $K, M \leqslant N$, the time and space complexity are respectively bounded by $O(N^4)$ and $O(N^3)$. □

References

1. Aloise, D., Deshpande, A., Hansen, P., Popat, P.: NP-hardness of Euclidean sum-of-squares clustering. Mach. Learn. **75**(2), 245–248 (2009)
2. Aloise, D., Hansen, P., Liberti, L.: An improved column generation algorithm for minimum sum-of-squares clustering. Math. Prog. **131**(1), 195–220 (2012)
3. Charikar, M., Khuller, S., Mount, D.M., Narasimhan, G.: Algorithms for facility location problems with outliers. In: SODA, vol. 1, pp. 642–651 (2001)
4. Dasgupta, S.: The hardness of k-means clustering. Department of Computer Science and Engineering, University of California, San Diego (2008)
5. Dupin, N., Nielsen, F., Talbi, E.-G.: K-medoids clustering is solvable in polynomial time for a 2d pareto front. In: Le Thi, H.A., Le, H.M., Pham Dinh, T. (eds.) WCGO 2019. AISC, vol. 991, pp. 790–799. Springer, Cham (2020). https://doi.org/10.1007/978-3-030-21803-4_79
6. Dupin, N., Nielsen, F., Talbi, E.-G.: Clustering a 2d pareto front: p-center problems are solvable in polynomial time. In: Dorronsoro, B., Ruiz, P., de la Torre, J.C., Urda, D., Talbi, E.-G. (eds.) OLA 2020. CCIS, vol. 1173, pp. 179–191. Springer, Cham (2020). https://doi.org/10.1007/978-3-030-41913-4_15
7. Dupin, N., Nielsen, F., Talbi, E.: Unified polynomial dynamic programming algorithms for p-center variants in a 2D Pareto Front. Mathematics **9**(4), 453 (2021)
8. Dupin, N., Nielsen, F., Talbi, E.G.: k-medoids and p-median clustering are solvable in polynomial time for a 2d Pareto front. arXiv preprint arXiv:1806.02098 (2018)
9. Grønlund, A., Larsen, K.G., Mathiasen, A., Nielsen, J.S., Schneider, S., Song, M.: Fast exact k-means, k-medians and Bregman divergence clustering in 1d. arXiv preprint arXiv:1701.07204 (2017)
10. Huang, J., Chen, Z., Dupin, N.: Comparing local search initialization for k-means and k-medoids clustering in a planar pareto front, a computational study. In: Dorronsoro, B., Amodeo, L., Pavone, M., Ruiz, P. (eds.) OLA 2021. CCIS, vol. 1443, pp. 14–28. Springer, Cham (2021). https://doi.org/10.1007/978-3-030-85672-4_2
11. Jain, A.: Data clustering: 50 years beyond k-means. Pattern Recogn. Lett. **31**(8), 651–666 (2010)
12. Kanungo, T., Mount, D.M., Netanyahu, N.S., Piatko, C.D., Silverman, R., Wu, A.Y.: A local search approximation algorithm for k-means clustering. In: Proceedings of the Eighteenth Annual Symposium on Computational Geometry, pp. 10–18 (2002)
13. Krishnaswamy, R., Li, S., Sandeep, S.: Constant approximation for k-median and k-means with outliers via iterative rounding. In: Proceedings of the 50th Annual ACM SIGACT Symposium on Theory of Computing, pp. 646–659 (2018)
14. Lloyd, S.: Least squares quantization in PCM. IEEE Trans. Inf. Theory **28**(2), 129–137 (1982)
15. Mahajan, M., Nimbhorkar, P., Varadarajan, K.: The planar k-means problem is NP-hard. Theoret. Comput. Sci. **442**, 13–21 (2012)
16. Nielsen, F., Nock, R.: Optimal interval clustering: application to Bregman clustering and statistical mixture learning. IEEE Signal Process. Lett. **21**(10), 1289–1292 (2014)
17. Peng, J., Wei, Y.: Approximating k-means-type clustering via semidefinite programming. SIAM J. Optim. **18**(1), 186–205 (2007)

18. Piccialli, V., Sudoso, A.M., Wiegele, A.: SOS-SDP: an exact solver for minimum sum-of-squares clustering. INFORMS J. Comput. **34**(4), 2144–2162 (2022)
19. Wang, H., Song, M.: Ckmeans. 1d. dp: optimal k-means clustering in one dimension by dynamic programming. R J. **3**(2), 29 (2011)
20. Zhang, Z., Feng, Q., Huang, J., Guo, Y., Xu, J., Wang, J.: A local search algorithm for k-means with outliers. Neurocomputing **450**, 230–241 (2021)

An Optimization Approach for Optimizing PRIM's Randomly Generated Rules Using the Genetic Algorithm

Rym Nassih(✉) and Abdelaziz Berrado

Equipe AMIPS - Ecole Mohammadia d'Ingénieurs, Mohamed V University in Rabat, Avenue Ibn Sina, BP765 Agdal, Rabat, Morocco

rymnassih@research.emi.ac.ma, berrado@emi.ac.ma

Abstract. The Patient Rule Induction Method (PRIM) is a bump hunting algorithm that generates a big number of rules in high dimensional data. Despite the high accuracy it provides, in this case it lacks of interpretability when the set of rules is big. To address it, we aim, in this paper to optimize the number of rules using Genetic Algorithm (GA) by formulating a combinatorial optimization problem to minimize the ruleset and maximize the performance of the ruleset. We applied this approach on a real-life dataset involving slope stability, one of the most important subjects in civil engineering, by choosing random feature spaces to generate the rules. We also set a performance score that balances between the confidence and the support of the rules in a ruleset and that has to be maximized to select a ruleset as a potential candidate. The results obtained show that optimizing with GA gives a more powerful set of rules that eases the interpretation. However, if the goal of the study is to detect small groups, we should minimize the performance of the ruleset by looking at the weakest groups, hence with the lowest support.

Keywords: PRIM · genetic algorithm · rules induction · combinatorial optimization

1 Introduction

Machine learning (ML) and optimization are closely tied, in fact, each time we aim at fitting an ML method on a dataset, we are solving an optimization problem. For instance, in [6] authors propose a binary swarm particle optimization [15] based on association rule where the rules are generated without specifying the minimum confidence and support requested, thus generating the best rules from the given database. Authors in [7] also propose to apply different optimization algorithm to construct the best road traffic noise's model to detect abnormal noise in the urban area. Moreover, we can use optimization in data preparation, hyperparameter tuning or model selection that can be found in the following papers [8–12]. In this paper we aim at using optimization to increase the interpretability of the results generated for one class by the Patient Rule Induction Method (PRIM).

B. Dorronsoro et al. (Eds.): OLA 2023, CCIS 1824, pp. 304–312, 2023.
https://doi.org/10.1007/978-3-031-34020-8_23

The Patient Rule Induction Method [3] is a bump hunting algorithm that is used in a supervised learning setting to find regions, in the input variables subspace, that are associated with the highest or lowest occurrence of a target label of a class variable chosen by the data analyst. Although the resulting rules/regions when taken individually are highly actionable, their high number and redundancy complicate the interpretation task which requires selecting, organizing and eliminating useless/redundant rules. This method is mostly characterized by the choice of the search space in which we aim at discovering the subgroups of interest. Even if it takes place in a supervised context, PRIM is not a classification algorithm.

This work aims at optimizing the number of rules generated by PRIM in a way to select a representative subset of rules that provide the same information as the initial set of rules. We explore achieving this goal using the Genetic Algorithm (GA) [1]. This initial research orientation, seems to reconcile at best power, generality and ease of programming. In fact, the GA is more adapted than other algorithms such as Gradient Descent or Newton method [5] when it comes to modeling and solving the problem at hand given the morphology of the rules generated by PRIM. Moreover, the rules are generated by randomly selecting iteratively different search spaces, aiming at giving to the experts' new insights in datasets and discovering other subgroups of interest.

2 Overview of PRIM and Problem Definition

2.1 The Patient Rule Induction Method

The Patient Rule Induction Method is an algorithm introduced in 1999 by Friedman and Fisher [3]. It innovates in the search of the interesting subgroups by performing a bump hunting. It consists of two phases. Given a chosen subset of the input variables subspace, the first step is the top-down peeling in which the search of the interesting subgroups is done. The method starts by peeling dimension by dimension constructing at the end a rectangle shaped group, hence the world box to refer to the output. The boxes are constructed, such that each box is delimited by the variables defining the chosen subspace. At each peel the portion of data removed can be controlled by the data analyst and is usually 5%, hence the term patient. The final box found after the peeling process may not be optimal because of past greedy suboptimal choices. Thus, the second step of PRIM is the bottom up pasting in which the boxes are expanded by iteratively enlarging their boundaries as long as the outcome's density increases. Figure 1 illustrates PRIM's process to find a box in the first phase of the top-down peeling. These two steps are repeated recursively to find other regions until stopping criteria are reached.

The outcome of PRIM is a set of boxes/regions that define rules relating the targeted label of the depended variable to the explanatory attributes in the data subspace chosen initially by the analyst. Each rule is shaped as:

$$\textit{if cond } 1, \ldots, \textit{cond } n \rightarrow \textit{target variable} = \textit{class}$$

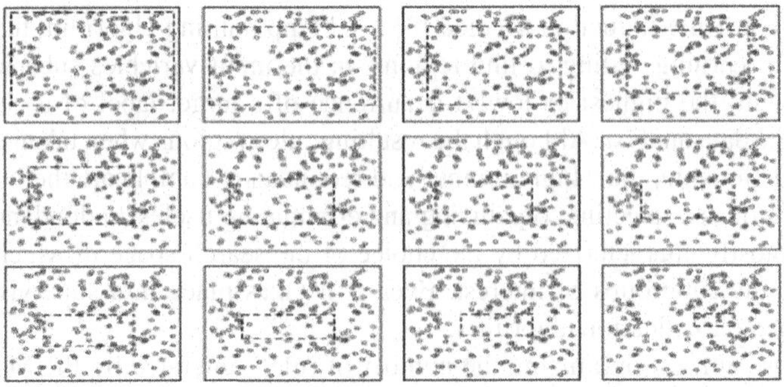

Fig. 1. Search procedure of one box

Contrary to other methods such as decision trees, PRIM has a better exploration of the dataset and can handle discrete and continuous variables. Also, the variation of the hyperparameters such as the support, the pasting and the peeling thresholds can give lot of different rules each time, which can even increase the set of possible rules useful to the user. Authors in [2] and [13] give a literature review of PRIM where they show the strength of PRIM in comparison with the most used search methods such as the decision trees and the regression. Indeed, the decision trees can peel up to 50% of the dataset in each split, loosing potential interesting rules whereas PRIM controls the peeling threshold. The one complexity with PRIM is the big interaction with the user in choosing the most optimal hyperparameter to find the best rules. Often experts test a lot of combination of support, peeling threshold and pasting threshold to find good results.

2.2 Postprocessing of PRIM's Rules

PRIM offers a number of tools to post-process [3] or inspect the result of rule induction. The most important that could be used to optimize each rule are: Analyzing variables redundancy, Interbox Dissimilarity and Relative frequency ratio plots.

After inducing the rules, we can look at ways to optimize each one of them by pruning irrelevant variables. For each variable we consider the decrease in box/region density when it is removed. The variable yielding the smallest decrease in box density is provisionally removed, and we do the same for the remaining variables.

The second tool to optimize each rule, is to look at the Interbox Dissimilarity. Indeed, the covering procedure produces a sequence of boxes that cover a subregion of the attribute space. These boxes can overlap or be disjoint, be close or far apart, depending on the nature of the target function. By inspecting the dissimilarities of pairs of boxes B_i and B_j in the sequence, we inspect the difference between the support of the smallest box $B_{i,j}$ that covers both of them, and the support of the union. It's defined as:

$$D(Bi,\ Bj) = \beta(Bi, j) - \beta\,(Bi \cup Bj)$$

with $\beta(B)$ the support of a box B, defined as the fraction of observations in the entire dataset that it covers, and $D(B_i, B_j)$ is in the interval [0,1]. This measure is used to reduce the number of boxes by assuming that if $D(B_i, B_j)$ approaches 1 then boxes are dissimilar and we can keep them both in our set of rules, and if it's under 0.3 then we remove the box because of low support.

From the perspective of interpretation, it is important to be aware of possible alternative definitions for each induced box B_k. Hence the Relative frequency ratio plots allow us to compare the relative frequency distribution of values of each input variables x_j within the box $p_j(x_j \mid x \in B_k)$ to that over the entire data sample $p_j(x_j)$.

The ratio is defined as:

$$r_{jk}(x_j) = p_j(x_j|x \in B_k)p_j(x_j).$$

A uniform distribution for $r_{jk}(x_j)$ implies that x_j is totally irrelevant to B_k because the relative frequency of its values is the same inside or outside of the box, and a highly peaked distribution for $r_{jk}(x_j)$ implies that the input x_j is highly relevant to the definition of B_k.

Although these post processing tools contribute to improving each rule induced by PRIM, they do not solve the redundancy problem as they may result in pruning important variables because of their greedy approach. Furthermore, working on each rule individually makes rules postprocessing time consuming.

As an alternative, we attempt in this work to postprocess and thereafter prune the overall set of rules by modeling this problem as an optimization problem. The genetic algorithm seems to have the potential for interpreting the rules without dealing with the optimization of each box.

3 Formulation of the Optimization Problem

Let R be the ensemble of boxes generated by PRIM in the subspace of input variables selected by the analyst. The optimization problem is to select a subset, S, of rules from the overall set of rules R with an optimum density.

Let *PerfS* represents a performance indicator of S that balances confidence and support of the ruleset generated. It should be noted that our research is exploring different variations of *PerfS* and assessing the sensitivity of the optimal subset of rules S to the definition of *PerfS*. We define *PerfS* as:

$$\boldsymbol{PerfS} = \sqrt{}(\mathbf{d} * \boldsymbol{\beta})$$

where **d** is the average density of S and **β** the average support of S.

This whole problem is formulated as a combinatorial optimization system with the objectives:

$$\begin{cases} \textbf{Maximize } \boldsymbol{PerfS} \\ \textbf{Minimize}|S| : \ |S| \text{ is the cardinality of S, i.e. the number of rules in S} \end{cases}$$

To facilitate the application of Genetic Algorithm in this problem we introduce positive weights W_G and W_S and reformulate it through the objective function f(S) as follows:

$$\boldsymbol{Maximizef}(S) = \boldsymbol{W_G.G} - \boldsymbol{W_S.}|S| \ \boldsymbol{with} \ \boldsymbol{S_CR}$$

The weights, W_G and Ws introduced are specified according to the objective to be achieved in terms of trade-offs between a number of rules and the performance of S.

In the Genetic Algorithm, every realizable solution of our problem is individually processed. The set of rules in S is represented by a string and is considered as an individual. Thus, we note the rule set S by the string $s_1s_2...s_n$ and n the total number of rules in R:

$$S = s_1 s_2 \ldots s_n$$

$$\begin{cases} s_r = 1 \text{ if the rule number r is in S} \\ s_r = 0 \text{ if the rule number r is not in S} \end{cases}$$

4 Implementation of the Genetic Algorithm

The first step of the algorithm is the genesis of the population of the rules, that is the choice of the starting devices that we are going to evolve. We choose to do a random initialization: the values of the genes are drawn randomly according to a uniform distribution **«1 if the rule is drawn 0 otherwise».** Each string contains a set of rules. Each rule, that is randomly included in each string, has a 50% chance of being chosen to belong to any string. The size N of this population is a parameter whose choice results from a compromise between calculation time and solution quality.

After defining the genesis of the population, we move to the "selection and elimination" step. We call "Generation" the population at a given moment t. Once we carry out the evaluation of the generation, a selection is made from the adaptation function. Only individuals passing the selection test can access the intermediate generation and reproduce there. In fact, the size of the intermediate generation is twice smaller (N/2 chains of rules) than the previous generation. Thus, we reduce the number of rules in our set. For the selection step we have chosen to work with the "tournament selection". In this approach, two individuals are randomly selected and they compete to access the middle generation. This step is repeated until the intermediate generation is filled (N/2 chains). It is quite possible that some individuals participate in several tournaments: if they win several times, they will be entitled to be copied several times in the middle generation, which will promote the sustainability of their genes.

Once the intermediate generation is half full, individuals are randomly divided into couples. The parents' chromosomes (parameter sets) are then copied and recombined to form two descendants with characteristics from both parents. Thus, the generation t + 1 is formed. The crossing operator promotes the exploration of the research space. It ensures the mixing of genetic material and the accumulation of favorable mutations which creates new combinations of component parameters. As a crossing method, we used the single point crossover [4].

A mutation is defined as the inversion of a bit in a chromosome. This is equivalent to modifying the value of a parameter of the device. The mutations play the role of noise and prevent the evolution from freezing. They allow to ensure a global as well as local search, according to the weight and the number of mutated bits. Moreover, they guarantee that the global optimum can be reached. On the other hand, a population that

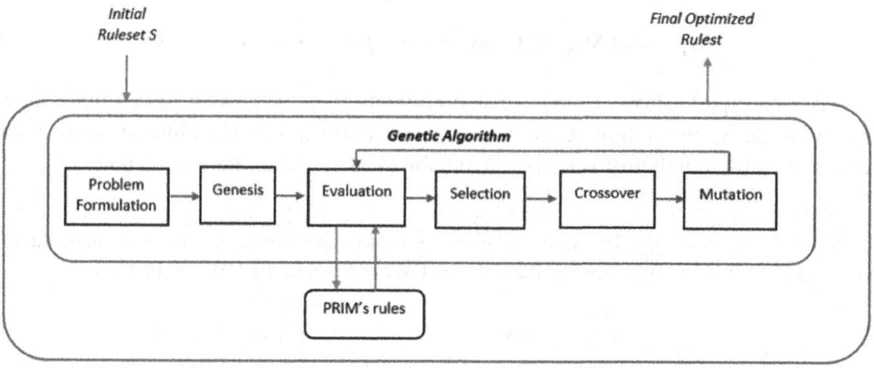

Fig. 2. Genetic Algorithm for optimizing PRIM's rules

is too small can homogenize because of stochastic errors: genes favored by chance can spread to the detriment of others. This other mechanism of evolution, which exists even in the absence of selection, is known as genetic drift. From the point of view of the device, this means that we risk to end up with devices that are not necessarily optimal. Mutations counteract this effect by constantly introducing new genes into the population.

The total number of generations, which we set as a parameter, is used as a stopping condition. If this condition is not satisfied, we return to step 2, the evaluation step. Figure 2 displays the procedure of the optimization with the genetic algorithms.

5 Experimentation and Results

5.1 Experiment

To test our concept, we use in the experiment a real-life dataset on slope stability. This problematic is considered as one of the most important in civil engineering, since an unstable slope can cause serious damages. The dataset set used can be found in [14]. It contains 168 instances with balanced data having 84 examples for the 'class = stable' and 84 for the 'class = unstable'. Six input variables are used: the bulk density (θ), the cohesion (C), the angle of internal friction (Ø), the slope angle (β), the slope height (H) and the pore water pressure ratio R_u.

In this illustration we considered the rules of both classes and we generated 10 random feature combinations among the six available for the study. We also kept to the hyperparameters selected by the expert which are a support of 0.05, and a peeling threshold of 0.07.

The random features are: (C, Φ, R_u), (θ, C, Φ, R_u), (β, R_u), (θ, C, Φ, β), (θ, Φ, β), (θ, H, R_u), $(\theta, C, \Phi, H, R_u)$, (θ, C, β), (Φ, β, H), (θ, C, H). For each one of these feature space, PRIM performs a top-down peeling and a bottom up pasting to select the good boxes.

We have chosen two pairs of values for Wc and Wr:

$$\mathbf{W_G=1\ et\ W_S=0} \quad \text{we obtain} \quad f(S)=W_G.G \tag{1}$$

$$\mathbf{W_G=10\ et\ W_S=0.1} \quad \text{we obtain} \quad f(S)=W_G.G-W_S.|S| \tag{2}$$

The first Eq. (1) allows us to have an optimization on the classification rates without taking into account the number of rules. The second Eq. (2) takes into an account the reduction of the number of rules but favors the classification rate on this number.

Table 1. Results obtained for the two classes with only generating random rules using PRIM, with applying the GA on the ruleset for <Wc = 1 Wr = 0> and for <Wc = 10 Wr = 0.1>

	Class = 'stable'		Class = 'unstable'	
	PerfS	*Number of rules in S*	*PerfS*	*Number of rules in S*
$\mathbf{W_G = 1 \quad W_S = 0}$	68.49%	9	70%	11
$\mathbf{W_G = 10 \quad W_S = 0.1}$	67%	15	68.85%	13
PRIM random rules	56.33%	38	48.93%	45

As displays in Table 1, for the class 'stable' we obtain 38 rules with a performance of 56.33% with PRIM's random generation on 10 feature combination, 9 rules and 68.49% of the performance if we only consider the performance in the equation and 67% of performance for 15 rules if we have $W_S = 0.1$.

For the class 'unstable', the initial number of rules generated by random features with PRIM is 45, which is hardly interpretable. With $W_G = 1$ and $W_S = 0$ the performance rises to 70% for a number of 11 rules and for the implementation of the GA with $W_G = 10$ and $W_S = 0.1$ the performance is 68.85% for a ruleset of 13 rules.

5.2 Results and Discussion

The results obtained are in adequacy with the goal of the algorithm and its results. PRIM aims at discovering small groups, which is called the subgroup discovery [16, 17]. Therefore, the rules obtained have a small support but a big range of confidence from 100% to 20%.

In our implementation we tested the possibility to consider GA as a pruning method for PRIM, however the use of it could lead to missing valuable rules for one's problematic. The key would be to either maximize the performance to select the strongest ruleset, hence rules, to understand the subject or minimize the performance to detect the outliers in the dataset.

To test our concept, we only took 10 random combinations of the features to variate and search space and we noticed that PRIM gives rules with a complexity between 1 to 4, it's rare to have numerous variables in the rules. This increased the interpretability and the explainability of the rules since they are easily readable and usable.

Nonetheless, for further validation we are increasing the number of random combinations and studying whether or not our solution converges to the same outputs.

6 Conclusion

In this paper we propose a novel concept involving the genetic algorithm and the Patient Rule Induction Method in a context where the search space is randomly generated. The results obtained satisfy to the goal of the experiment, however the interpretability of the results can differ according to the domain knowledge and the purpose of the implementation.

As PRIM is considered in the literature as a subgroup discovery algorithm, having a performant ruleset does not mean that the other rules should be ignored, it only allows a better reading of the results to analyze the dataset. Thus, performing the combinatorial problem to maximize and minimize at the same the performance can give a big range of the big boxes as well as the smallest ones.

As ongoing works, we are testing the approach on other datasets and analyzing whether or not imbalanced data can impact the performance of the model, also we are conducting a sensitivity analysis on the number of combination that could give the best results.

References

1. Goldberg, D.E.: Genetic Algorithms in Search, Optimization and Machine Learning. Addison-Wesley Longman Publishing Co., Inc., Boston (1989)
2. Nassih, R., Berrado, A.: Potential for PRIM based classification: a literature review. In: The Third European International Conference on Industrial Engineering and Operations Managements in Pilsen, Czech Republic, p. 7 (2019)
3. Friedman, J.H., Fisher, N.I.: Bump hunting in high-dimensional data. Stat. Comput. **9**(2), 123–143 (1999)
4. Sastry, K., Goldberg, D., Kendall, G.: Genetic algorithms. In: Burke, E.K., Kendall, G. (eds.) Search Methodologies, pp. 97–125. Springer, Boston, MA (2005). https://doi.org/10.1007/0-387-28356-0_4
5. Ruder, S.: An overview of gradient descent optimization algorithms. arXiv preprint arXiv: 1609.04747 (2016)
6. Sarath, K.N.V.D., Ravi, V.: Association rule mining using binary particle swarm optimization. Eng. Appl. Artif. Intell. **26**(8), 1832–1840 (2013)
7. Nedic, V., Cvetanovic, S., Despotovic, D., Despotovic, M., Babic, S.: Data mining with various optimization methods. Expert Syst. Appl. **41**(8), 3993–3999 (2014)
8. Minaei-Bidgoli, B., Punch, W.F.: Using genetic algorithms for data mining optimization in an educational web-based system. In: Cantú-Paz, E., et al. (eds.) GECCO 2003. LNCS, vol. 2724, pp. 2252–2263. Springer, Heidelberg (2003). https://doi.org/10.1007/3-540-45110-2_119
9. Liu, Y., Chung, Y.Y.: Mining cancer data with discrete particle swarm optimization and rule pruning. In: 2011 IEEE International Symposium on IT in Medicine and Education, vol. 2, pp. 31–34. IEEE (2011)
10. Alatas, B., Akin, E.: Multi-objective rule mining using a chaotic particle swarm optimization algorithm. Knowl.-Based Syst. **22**(6), 455–460 (2009)
11. Bottou, L., Curtis, F.E., Nocedal, J.: Optimization methods for large-scale machine learning. SIAM Rev. **60**(2), 223–311 (2018)
12. Kausar, N., Palaniappan, S., Samir, B.B., Abdullah, A., Dey, N.: Systematic analysis of applied data mining based optimization algorithms in clinical attribute extraction and classification for diagnosis of cardiac patients. In: Hassanien, A.-E., Grosan, C., Fahmy Tolba, M. (eds.)

Applications of Intelligent Optimization in Biology and Medicine. ISRL, vol. 96, pp. 217–231. Springer, Cham (2016). https://doi.org/10.1007/978-3-319-21212-8_9
13. Nassih, R., Berrado, A.: Towards a patient rule induction method-based classifier. In: 2019 1st International Conference on Smart Systems and Data Science (ICSSD), pp. 1–5. IEEE (2019)
14. Kaveh, A., Hamze-Ziabari, S.M., Bakhshpoori, T.: Soft computing-based slope stability assessment: a comparative study. Geomech. Eng. **14**(3), 257–269 (2018)
15. Kennedy, J., Eberhart, R.: Particle swarm optimization. In: Proceedings of ICNN 1995-International Conference on Neural Networks, vol. 4, pp. 1942–1948. IEEE (1995). https://doi.org/10.1109/ICNN.1995.488968
16. Herrera, F., Carmona, C.J., González, P., Del Jesus, M.J.: An overview on subgroup discovery: foundations and applications. Knowl. Inf. Syst. **29**, 495–525 (2011)
17. Atzmueller, M. Subgroup discovery. Wiley Interdiscip. Rev. Data Min. Knowl. Discov. **5**(1), 35–49 (2015)

Real-World Applications

A Fast Methodology to Find Decisively Strong Association Rules (DSR) by Mining Datasets of Security Records

Claudia Cavallaro, Vincenzo Cutello, Mario Pavone (☑), and Francesco Zito

Department of Mathematics and Computer Science, University of Catania, V.le Andrea Doria 6, 95125 Catania, Italy {claudia.cavallaro,cutello}@unict.it, mpavone@dmi.unict.it, francesco.zito@phd.unict.it

Abstract. Cybersecurity bulletins officially recognize and publicly share the vulnerabilities of Information Systems. The attacks exploit various aspects of those vulnerabilities, compromising confidentiality, integrity or availability of the data collected. We analyze a public dataset of security records so to obtain some common features and to be able to forecast future attacks. We propose an intervention based on history of attacks through data mining methods and so a more dynamic risk analysis, by concentrating on some specific classes of cyberattacks in a period of two years. We devise a fast algorithm to find strong rules which provide an estimate of the probability that these attacks will occur so to identify adequate controls and countermeasures.

Keywords: Pattern analysis \cdot Cyber security \cdot Association Rules \cdot Data Mining \cdot Anomaly detection \cdot Optimization

1 Introduction

Cyberattacks affect different sectors such as healthcare, government, financial and automotive industries. Incidents due to malware attacks impact industrial production and critical infrastructures, causing significant delays in control operations and consequent process anomalies.

Particularly for programmable cars, a compromise of the system can lead to risks to people safety, as well as to their privacy. Connected cars are targeted via Spear Phishing mechanisms which lead to the download of malicious attachments and payloads, or by Hardware Trojans which provide covert access to the onboard computer system and can disrupt communication of Controller Area Network buses. The vehicle can be affected by Ransomware attacks which encrypt user data causing operational disruptions. Via the infotainment system, the victim driver is threatened that the ignition of the car will be suspended until a ransom is paid.

[©] The Author(s), under exclusive license to Springer Nature Switzerland AG 2023 B. Dorronsoro et al. (Eds.): OLA 2023, CCIS 1824, pp. 315–326, 2023. https://doi.org/10.1007/978-3-031-34020-8_24

Public data and technical reports regarding cybersecurity provide a description of vulnerabilities and exposures discovered over time, but the records contained in the published databases are very numerous if we consider long periods of time. Furthermore, faced with two or more vulnerabilities, it is generally not possible to decide which one is more urgent to deal with and, in particular, each vulnerability can have different impacts on different systems. Software developers are often forced to work within a limited time frame and are unable to analyze all security weaknesses. So they have to focus on targeting the most serious weaknesses or the ones related to specific characteristics, such as vulnerability metrics, type of exploits and so on.

It is therefore necessary to establish a priority among all the mitigation and detection measures to be adopted, on the basis of the frequent relationships among them, such as: basic metrics of vulnerability, weaknesses, attack tactics and techniques, operating systems or architectures.

We propose in this paper to simplify the standard and computational challenging general data mining problem of finding strong association rules by concentrating the search onto the prediction of specific attack and vulnerabilities, and in doing so create an information structure which can be easily updated.

Our work is organized as follows: Sect. 2 presents public security datasets, and some research work of data mining applied to the field of cybersecurity. Section 3 explains the methodology chosen to mine frequent patterns efficiently and prioritize actions to safeguard security. In Sect. 5 the results of our analysis show the forecasting of attacks based on past records. Section 6 concludes our study and outlines some future research directions.

2 Dataset and Background

Since 1999 the MITRE Corporation collects a catalog of known cybersecurity vulnerabilities and the NIST (National Institute of Standards and Technology) assigns to each of them a severity score, based on a standard called CVSS (Common Vulnerability Scoring System), and publishes them in the National Vulnerability Database (NVD), available online[1]. CVSS estimates the severity of a vulnerability and it is used by vendors, developers, researchers, security managers in companies and public administration and security agencies that deal with the publication of bulletins.

Common vulnerability and exposures entries, CVE for short, are reported with a unique identifier which is tagged with CVE-YYYY-XXXX, where YYYY is the year the vulnerability was discovered and XXXX is a sequential integer. The CVE archive, available online[2], provides a description of the vulnerabilities included in MITRE reports. The CVE's perspective is to catalog errors after they have occurred and to investigate possible solutions. At the same time, MITRE is responsible for providing a list of CWE (Common Weakness Enumeration) to show the weaknesses in the architecture or in the code.

[1] https://nvd.nist.gov.

[2] https://cve.mitre.org.

All the catalogued CWE IDs are represented with a hierarchical tree organization, called "View 1000 - Research Concepts"[3]. The *Pillars* are the parents from whom the first branch starts, and which, then, describe common classifications of weaknesses.

CAPEC (Common Attack Pattern Enumeration and Classification), available online[4], describes and classifies the attack patterns. The MITRE ATT&CK (Adversarial Tactics, Techniques and Common Knowledge) is a framework that describes all the main procedures used by attackers to violate systems and possibly gain persistent access to them. Attack procedures include tactics, which identify the attackers ultimate goals and the main purpose of their actions. Each attack tactic contains different techniques, which are concrete actions aimed at a specific goal and specify what an attacker achieves when finished. The MITRE ATT&CK matrices, available online[5] for the Enterprise, Mobile and ICS domains report the technical-tactics of violations and persistence of fixed corporate, mobile and industrial control systems.

ENISA, the European Network and Information Security Agency, aggregates the records from the official databases mentioned above and from other resources such as the Vulnerability Database (VULDB), online[6], into a single .csv file. Each row contains information about these features: CVE ID, source database, severity level, impact score, exploitability score, attack vector, complexity, privilege, scope, confidentiality impact, integrity impact, availability impact, CWE ID, CAPEC ID, date published, attack technique ID and tactic.

In [12], ENISA presents a technical cybersecurity report about $2018-19$, but it does not provide any prediction about future attacks. In this work we analyzed its aggregate information to establish what could be the next information (within some tolerance) that could be reported in the security bulletins. The ENISA statistics do not highlight the coexistence of vulnerabilities, weaknesses and attacks in frequent tuples of the dataset, features which are co-present in its rows according to a fixed minimum frequency.

We are interested in records that have common characteristics for a fixed minimum percentage of the analyzed data (a total of about $230k$ rows).

2.1 Data Mining and Cybersecurity

We will now overview some works that link data mining to the field of cybersecurity.

In [15] the authors, starting from the Record Audit data - Snort log, identify IP numbers and probable attacks, but they do not deal with the pattern detection of vulnerability features.

Fan et al. in [7] created a dataset by adding code changes and summary for C / C ++ vulnerability to the CVE archive.

[3] https://cwe.mitre.org/data/definitions/1000.html.
[4] https://capec.mitre.org.
[5] https://attack.mitre.org/matrices.
[6] https://vuldb.com.

In [14], Murtaz et al. show that vulnerabilities can be treated like Markov Chains, and so they can predict the next vulnerability by using only the previous one.

The authors of [13] extract associations of words, used in websites for Cybersecurity, through the well-known Apriori data mining algorithm.

Dodiya et al. [6] provide statistical distributions of the NVD, such as the number of new vulnerabilities reported by year, security levels, access complexity and integrity impact.

Threat searching can involve anomaly detection on machine logs, where behavioral data analysis is automatically separated from outliers using NLP and deep learning [4]. A Big Data Platform [16] was created to centralize collection of logs and metrics from heterogeneous data sources. It can be accessed so to perform a semi-supervised anomaly detection using the results of log clustering and visualize in real time the health of services through dashboards.

Anomaly detection finds application in many domains, including Cultural Heritage [8] and Urban Informatics [5]. In particular, data mining methods are also used to forecast next destinations [3].

3 Mining Association Rules

Let us start by introducing a mathematical formalization of the problem. Let $\mathcal{D}$ be a dataset (matrix) with m rows and n columns. Each column represents a specific attribute $ID_1, ID_2, \ldots, ID_n$, and each row represents a complete set of values for the n attributes. Any attribute ID_i and any of its values v found in the rows of $\mathcal{D}$, define the element $< ID_i = v >$.

Given now any element I, the singleton $\{I\}$, also called 1-element itemset or itemset of length 1, is said to be "*infrequent*" if it is contained in a number k of rows of the dataset where $\frac{k}{m} < min\_supp$, i.e. is smaller than the fixed minimum support (we use the notation $supp(\{I\}, \mathcal{D}) < min\_supp$). The minimum support represents then a fraction or percentage value of the rows of the dataset. If $supp(\{I\}, \mathcal{D}) \geq min\_supp$ then $\{I\}$ is said to be "*frequent*". We generalize the above concept to itemsets of length h for any $1 \leq h \leq n$, as follows: an itemset of length h is a set of h elements, $\{I_1, I_2, \ldots, I_h\}$, such that

- each element I_i represents the value of an attribute, i.e. $I_i =< ID_{j_i} = a >$ for some attribute ID_{j_i} and a one of the values of ID_{j_i};
- two distinct elements I_{i_1} and I_{i_2} represent values of two different attributes.

The frequency of the itemset $\{I_1, I_2, \ldots, I_h\}$ is the number of rows of $\mathcal{D}$ which contain its values. As in the case of itemsets of length 1, the itemset is frequent if $supp(\{I_1, I_2, \ldots, I_h\}, \mathcal{D}) \geq min\_supp$, otherwise is said to be infrequent.

Since $supp(\{I_1, I_2, \ldots, I_h\}, \mathcal{D}) \leq supp(S, \mathcal{D})$ for any $S \subseteq \{I_1, I_2, \ldots, I_h\}$, it is clear that if $\{I_1, I_2, \ldots, I_h\}$ is frequent, all its subsets are also frequent. Thus, if any of its subsets is infrequent then the itemset is infrequent as well.

To clarify the above, let us consider the example of the dataset in Table 1. We have a dataset with 20 rows and 5 columns, corresponding at the attributes

ID_1, ID_2, ID_3, ID_4, ID_5. If we choose $min\_supp = 0.3$, i.e. 30% of the total number of rows (6 in our case) the following elements, or 1-itemsets, are frequent (shown with their frequencies):

$$a1, 6; a2, 6; a3, 6; b1, 7; b2, 8; c1, 6; c2, 6; d3, 9; e4, 6.$$

The itemsets $\{a1, b2\}$ and $\{b1, c2, d3, e4\}$ have frequencies 6, so they are both frequent. The itemset $\{a2, b1\}$, instead, has frequency 3 and thus it is not frequent.

Table 1. Dataset with 5 attributes and 20 rows

ID_1	ID_2	ID_3	ID_4	ID_5
a1	b2	c3	d2	e1
a2	b1	c1	d3	e6
a1	b2	c1	d2	e1
a2	b1	c2	d3	e4
a1	b2	c1	d3	e2
a2	b1	c2	d3	e4
a1	b2	c3	d2	e2
a2	b3	c1	d5	e3
a4	b3	c3	d1	e2
a2	b4	c4	d3	e5
a1	b2	c4	d2	e2
a2	b2	c5	d6	e3
a3	b1	c2	d3	e4
a3	b4	c1	d1	e3
a3	b2	c1	d2	e2
a3	b1	c2	d3	e4
a1	b2	c3	d4	e3
a3	b5	c3	d1	e1
a3	b1	c2	d3	e4
a5	b1	c2	d3	e4

3.1 Mining Datasets

There are many algorithms available in literature for mining data and produce association rules. Given that the problem is clearly computationally challenging, many of these algorithms employ heuristics (see the excellent survey [9] for a comprehensive list of heuristics approach) or population based algorithms such as genetic algoritms (see for instance [17]) or particle swarm optimization (see [1]).

We briefly mention now the two most famous algorithms to find frequent itemsets. We start with Apriori [2], the most famous and first to be used algorithm for such a purpose, along with its successor FP-Growth [11]. It first computes the support of each single item and, then, it does the same for each itemset of cardinality 2, 3 and so on. In addition, the comparison of candidates for all rows becomes more expensive as the iterations of the algorithm increase and therefore the size of the itemsets to be generated increases. The Apriori algorithm requires $l + 1$ scan of the dataset to find the longest patterns, of length l.

The second algorithm is Prefix-Span (PREFIX-projected Sequential PAtterN mining), a data mining algorithm introduced by Pei et al. [10], which is used for marketing strategies.

Both algorithms would produce the entire collection of frequent itemsets. In our working example (itemsets are shown followed by their frequencies) the following itemsets are frequent:

$$\{a1\} : 6; \{a2\} : 6; \{a3\} : 6; \{b1\} : 7; \{b2\} : 8; \{c1\} : 6; \{c2\} : 6; \{d3\} : 9; \{e4\} : 6;$$
$$\{a1, b2\} : 6; \{c2, b1\} : 6; \{d3, b1\} : 7; \{e4, b1\} : 6; \{d3, c2\} : 6; \{c2, e4\} : 6; \{d3, e4\} : 6;$$
$$\{d3, c2, b1\} : 6; \{c2, e4, b1\} : 6; \{d3, e4, b1\} : 6; \{d3, c2, e4\} : 6; \{d3, c2, e4, b1\} : 6$$

3.2 Association Rules and Confidence

The concept or Association Rules $A \Rightarrow B$ was presented in [2] along with its related confidence value *Confidence* $(A \Rightarrow B)$, which represents, for instance, in market basket analysis the probability of buying a set of objects B, called consequent, given the purchase of a set of objects A, called antecedent, within the same transaction. More formally, given the probability distribution which generated the rows in the dataset, *Confidence* $(A \Rightarrow B) = P(B|A)$.

To generate an association rule $A \Rightarrow B$, where A and B are itemsets, we will take the support of $A \cup B$, and divide it by the support of A, thus computing, among the rows in the Dataset which contain A, the percentage of rows which contain also B.

If the itemsets satisfy two fixed parameters, that are the *min_supp* and also the minimum value of Confidence c (see below), the predictions are called Strong Rules. So, formally we have

Definition 1. *Given a dataset* $\mathcal{D}$ *and given two fixed parameters,* $0 \leq min\_supp \leq 1$ *and the minimum value of Confidence* $0 \leq c \leq 1$*, and given two disjoint itemsets* A, B *such that* $supp(A \cup B, \mathcal{D}) \geq min\_supp$*, the association rule* $A \Rightarrow B$ *is strong if* $\frac{supp(A \cup B, \mathcal{D})}{supp(A, \mathcal{D})} \geq c$.

In our work, we set as *Minimum Confidence* value $c = 75\%$ to get only itemsets that have a higher (or equal) confidence and also a support that exceeds or equals the Minimum Support chosen (30%).

When searching for strong rules we pay particular attention to maximal frequent itemsets, i.e. itemsets which are frequent but such that by adding one more element would no longer be frequent. Thus, given a maximal frequent itemset

M and any itemset B such that $B \cap M = \emptyset$, we know that the association rule $M \Rightarrow B$, will not be strong, since $M \cup B$ is not frequent.

Going back to the example of Table 1, we have two maximal itemsets of cardinality greater than 1, namely $\{b1, c2, d3, e4\}$ and $\{a1, b2\}$. The Association Rule $\{d3\} \Rightarrow \{b1, c2, e4\}$ is not Strong because its confidence value is equal to 66.6%. Instead, the Association Rules $\{e4\} \Rightarrow \{b1, c2, d3\}$, $\{c2\} \Rightarrow \{b1, d3, e4\}$, and $\{b1\} \Rightarrow \{c2, d3, e4\}$ are all strong and, in particular, the first two have 100% confidence value while the last one 86%.

Table 2 shows the 15 strong association rules. In particular, rules 6, 7, 8, 12, 13, 14 are a consequence of the fact that rule 3 is strong. Same reasoning could be applied to the other rules which are a consequence of rules 4 and 5.

Table 2. Strong Rules for the maximal itemset of the example in Table 1

Rule n.	Antecedent	Consequent	Antecedent support	Itemset support	Confidence
1	$\{a1\}$	$\{b2\}$	6	6	100%
2	$\{b2\}$	$\{a1\}$	8	6	75%
3	$\{b1\}$	$\{e4, d3, c2\}$	7	6	86%
4	$\{c2\}$	$\{e4, b1, d3\}$	6	6	100%
5	$\{e4\}$	$\{c2, b1, d3\}$	6	6	100%
6	$\{b1, c2\}$	$\{e4, d3\}$	6	6	100%
7	$\{b1, d3\}$	$\{e4, c2\}$	7	6	86%
8	$\{b1, e4\}$	$\{c2, d3\}$	6	6	100%
9	$\{c2, d3\}$	$\{e4, b1\}$	6	6	100%
10	$\{c2, e4\}$	$\{b1, d3\}$	6	6	100%
11	$\{d3, e4\}$	$\{c2, b1\}$	6	6	100%
12	$\{b1, d3, e4\}$	$\{c2\}$	6	6	100%
13	$\{b1, c2, e4\}$	$\{d3\}$	6	6	100%
14	$\{b1, c2, d3\}$	$\{e4\}$	6	6	100%
15	$\{c2, d3, e4\}$	$\{b1\}$	6	6	100%

4 Mining Security Datasets for Decisively Strong Rules

In a field such as security, we are more interested in association rules where the antecedent is a set of events and the consequent is a specific type of attack. Same kind of reasoning may be applied in the medical field, where we are interested in diagnosing the likely disease given a list of symptoms.

So, we are considering the case that B contains a single element, i.e. $B = \{id_j\}$ and $A = \{id_1, id_2, ..., id_i\}$ is an itemset with i elements non containing id_j. We have formally

$$\text{Confidence}(\{id_1, \ldots, id_i\} \Rightarrow \{idj\}) = P(B|A) = \frac{supp(\{id_1, \ldots, id_i, id_j\}, \mathcal{D})}{supp(\{id_1, id_2, ..., id_i\}, \mathcal{D})} \quad (1)$$

with $\{id_1, \ldots, id_i\} \cap \{id_j\} = \emptyset$.

In other words, we would like to be able to infer which attack technique is likely being used, so to apply proper countermeasures. Obviously, such an ability is particularly important if the attack is not very common, i.e. the probability of such an attack, though frequent, is not likely or very likely.

Equation 1 gives us the probability that, given some specific attribute values $id_1, id_2, ..., id_i$ for weaknesses and vulnerabilities that occur as frequent itemsets, they will appear together with attack tactics and techniques id_j as maximal frequent itemsets. The greater the confidence the greater the reliability in forecasting a certain type of attack, and therefore priority will be given to defensive actions related to it.

In view of the above, let us define then, among the attributes in the dataset $\mathcal{D}$ a specific attribute target T.

We introduce now the following definition, by recalling that any event whose probability is not higher than 0.5 is typically called *unlikely*.

Definition 2. *Given two fixed parameters, min_supp and the minimum value of Confidence c, and given an itemset A and a single value $I \notin A$ such that $supp(A \cup \{I\}, \mathcal{D}) \geq min\_supp$, the association rule $A \to \{I\}$ is a Decisively Strong Rule (DSR for short), if $\{I\}$ is frequent, i.e. $min\_supp \leq supp(\{I\}, \mathcal{D})$ but unlikely, i.e. $min\_supp \leq supp(\{I\}, \mathcal{D}) \leq 0.5$ and $\frac{supp(A \cup \{I\}, \mathcal{D})}{supp(A), \mathcal{D}} \geq c$.*

Our goal is to find all the decisively strong rules given the attribute target T, i.e. association rules $A \to \{t_i\}$ where t_i is a frequent (at least 30%) but unlikely value of the attribute target T.

Since both A and $\{t_i\}$ are frequent, i.e. their supports are both at least 30% of the rows of the dataset, it follows that if m are the rows of $\mathcal{D}$, since t_i is unlikely, $supp(\{t_i\}, \mathcal{D}) = \alpha \cdot m$ with $0.3 \leq \alpha \leq 0.5$, while $supp(A \cup \{T_i\}, \mathcal{D}) = \beta \cdot m$ with $0.3 \leq \beta \leq \alpha$ then

$$\frac{supp(A \cup \{t_i\}, \mathcal{D})}{supp(\{t_i\}, \mathcal{D})} = \frac{\beta}{\alpha}$$

from which it follows that $supp(A \cup \{t_i\}, \mathcal{D}) = \frac{\beta}{\alpha} supp(\{t_i\}, \mathcal{D})$. Thus, we need to mine the sub-dataset where t_i occurs for itemsets with a minimum support of $\frac{\beta}{\alpha}$.

We notice that since $\alpha \leq 0.5$ and $\beta \geq 0.3$ we have

$$\frac{\beta}{\alpha} \geq \frac{0.3}{0.5} = 0.6$$

For instance, let us consider Table 1 and suppose our target is the value $b1$ of $ID2$. The sub-table containing the value $b1$ is shown in Table 3. Since the support of $\{b1\}$ is $\frac{7}{20} < 0.5$ we need to look for itemsets with support at least $\frac{3}{10} \cdot \frac{20}{7} = \frac{6}{7}$, and we find, as expected, just $\{c2, d3, e4\}$.

Table 3. Dataset for target $b1$

ID_1	ID_2	ID_3	ID_4	ID_5
$a2$	$b1$	c1	$d3$	e6
$a2$	$b1$	c2	$d3$	e4
$a2$	$b1$	c2	$d3$	e4
$a3$	$b1$	c2	$d3$	e4
$a3$	$b1$	c2	$d3$	e4
$a3$	$b1$	c2	$d3$	e4
$a5$	$b1$	c2	$d3$	e4

The algorithm, called *DSR*, formally described in the pseudocode 1, takes as input the dataset $\mathcal{D}$, the minimum support value *min_supp*, the confidence value c, a specific target attribute T and a frequent value t_i for T.

To explain how *DSR* works, we will use the following notations:

- t_i will denote the singleton $\{T = t_i\}$
- $\mathcal{D}(t_i)$ denotes the projections of the dataset $\mathcal{D}$ on the value t_i for T, i.e. the dataset obtained eliminating all the rows where $T \neq t_i$.
- $supp(A, \mathcal{D}(t_i))$ the support of the itemset A in the dataset $\mathcal{D}(t_i)$ while $supp(A, \mathcal{D})$ is the support of the itemset A in the whole dataset $\mathcal{D}$.

Let us suppose that $F = \{ID_i = x_i\}$ is the collection of frequent elements all of length 1, therefore for each element $x \in F$, we have $supp(\{x\}, \mathcal{D}) \geq min\_supp$. Let also $F_T = \{t_1, \ldots, t_h\}$ be the set of the frequent values of target attribute T. Thus, for each $t_i \in F_T$ we have $supp(\{t_i\}, \mathcal{D}) \geq min\_supp$.

Our goal is to find all subsets $F' \subseteq F$, such that $F' \Rightarrow t_i$ is a DSR for some t_i frequent value of T. So,

$$supp(F', \mathcal{D}(t_i)) \geq \frac{min\_supp}{supp(\{t_i\}, \mathcal{D})} \geq 50\% \quad \textbf{Searching condition}$$

$$\frac{supp(F' \cup \{t_i\}, \mathcal{D})}{supp(F', \mathcal{D})} > c \quad \textbf{Pruning condition}$$

DSR uses, as a subroutine, any fast algorithm to find maximal frequent itemsets but on possibly quite small sub-datasets. For our tests, we used Apriori.

Algorithm 1. Pseudo-code of DSR.

```
procedure DSR(D, min_supp, c, T)
    min_supp = 0.3,   c = 0.75
    Compute F set of frequent elements,   F_T set of frequent values of T;
    for each attribute t_i ∈ F_T do
        supp(t_i, D) = α_i
        F'(t_i) = ∅
        for each x ∈ F do
            if supp(x, D(t_i)) > min_supp / α_i then
                add x to F'(t_i)
            end if
            Use General Algorithm to find max. freq. itemsets from F'(t_i) in D_{t_i}
            for each maximal frequent set A do
                if supp(A ∪ {t_i}, D) > c · supp(A, D) then
                    output DSR: A ⇒ t_i
                end if
            end for
        end for
    end for
end procedure
```

5 Results

In order to predict future threats we divided the ENISA dataset into the set of vulnerabilities and exposures published up to December 31st of 2018 (training set) and the set of CVEs available for the first half of 2019 (testing set) for comparisons with the obtained prediction. We add a new feature column in the original ENISA dataset, and so processed the Pillars, as attribute targets, instead of the single CWE ID because they group the weaknesses in a more generic way and consequently the mitigation of the data predicted could be addressed on a wider range.

We set the minimum confidence to 75% and *min_supp* to 30% and searched for decisively strong rules of the form $\{id_1, \ldots, id_k\} \Rightarrow \{attack\_technique\_id\}$. For year 2018 we found just one attack technique with support between 0.3 and 0.5, namely $T1027$ (*Obfuscated Files or Information*) with support value 39.99% and another attack technique $T1148$ (*Impair Defenses: Impair Command History Logging*), whose support value is 66.25% therefore higher than 0.3 but not unlikely according to our definition.

For the target value $T1027$, we obtained 24 DSR but only two with an antecedent which are maximals, the following:

- A = {CVSS_Complexity = Low,CVSS_Scope = Unchanged, CWE Pillar = Improper neutralization, CAPEC = Leverage Alternate Encoding, Attack Tactic = Defense Evasion}
- B = {CVSS_attack = Network, CWE Pillar = Improper neutralization, CAPEC = Leverage Alternate Encoding, Attack Tactic = Defense Evasion}

So, the two DSR found are $A \Rightarrow \{T1027\}$ and $B \Rightarrow \{T1027\}$.

The total number of all strong rules (with min_supp=0.3 and min_conf=0.75) that we could have obtained with traditional data mining algorithm would have been 1073, so our procedure is way faster and it avoids many useless generation.

To test the accuracy of the found rules, we extracted the frequent itemsets of the testing set (the first semester of 2019) that contain the same $T1027$ attack_technique. By comparing the obtained prediction of the 2 DSR rules of 2018 with the restricted frequent itemsets of 2019, we obtained a perfect matching.

To justify, experimentally, our choice of considering just target values with support not higher than 50%, we use as an example the attack technique $T1148$ (*Impair Defenses: Impair Command History Logging*) which has support 0.66.

From the sub-datasets containing the value $T1148$ we searched for frequent itemsets with support $(0.3/0.66) = 0.45$. We found 303 frequent itemsets but only 5 maximal:

1. {CVSS_severity = HIGH, CVSS_scope = Unchanged, CAPEC = Subverting Environment Variable Values, Attack Tactic = Defense Evasion}
2. {CVSS_complexity = Low, CVSS_scope = Unchanged, CVSS_availability = None, CAPEC = Subverting Environment Variable Values, Attack Tactic = Defense Evasion}
3. {CVSS_complexity = Low, CVSS_scope = Unchanged, CWE Pillar = Improper Neutralization, CAPEC = Subverting Environment Variable Values, Tactic = Defense Evasion}
4. {CVSS_complexity = Low, CVSS_scope = Unchanged, CVSS_confidentiality = High, CVSS_integrity = High, CAPEC = Subverting Environment Variable Values, Attack Tactic = Defense Evasion}
5. {CVSS_attack = Network, CVSS_complexity = Low, CVSS_priveleges = None, CVSS_scope = Unchanged, CAPEC = Subverting Environment Variable Values, Attack Tactic = Defense Evasion}

The above 5 maximal frequent itemsets are the antecedent to 5 DSR with minimum confidence 0.75 and consequent $T1148$. After extracting the frequent itemsets of the first semester of 2019 which contain $T1148$ we found that only 2 maximal itemsets out of 5 are also found for 2019. It follows that in this case the accuracy is only 40%.

6 Conclusion

In this work, we addressed the general data mining problem of finding strong association rules so to predict specific attacks and discover unknown vulnerabilities. We proposed a framework which takes into account frequent but not very likely attacks and proposed a fast way to compute strong association rules which turn out to be highly accurate. Our data-driven approach to deal with potential attacks in order of priority, could in future research be extended by experimentally setting the parameters of minimal support, confidence and likelihood of target values. Keeping into account past and recent work using population based methodologies [1,17] and heuristics [9] a possible future works could involve population-based metaheuristics for the choice of such parameters.

References

1. Agrawal, M., Mishra, M., Kushwah, S.P.S.: Association rules optimization using improved PSO algorithm. In: 2015 International Conference on Communication Networks (ICCN). IEEE (2015)
2. Agrawal, R., Imieliński, T., Swami, A.: Mining association rules between sets of items in large databases. ACM SIGMOD Rec. **22**(2), 207–216 (1993)
3. Cavallaro, C., Verga, G., Tramontana, E., Muscato, O.: Suggesting just enough (Un)crowded routes and destinations. In: CEUR Workshop Proceedings, vol. 2706, pp. 237–251 (2020)
4. Cavallaro, C., Ronchieri, E.: Identifying anomaly detection patterns from log files: a dynamic approach. In: Gervasi, O., et al. (eds.) ICCSA 2021. LNCS, vol. 12950, pp. 517–532. Springer, Cham (2021). https://doi.org/10.1007/978-3-030-86960-1_36
5. Cavallaro, C., Vizzari, G.: A novel spatial–temporal analysis approach to pedestrian groups detection. Procedia Comput. Sci. **207**, 2364–2373 (2022)
6. Dodiya, B., Singh, U.K., Gupta, V.: Trend analysis of the CVE classes across CVSS metrics. Int. J. Comput. Appl. **183**(33), 23–30 (2021)
7. Fan, J., Li, Y., Wang, S., Nguyen, T.N.: A C/C++ code vulnerability dataset with code changes and CVE summaries. In: Proceedings of the 17th International Conference on Mining Software Repositories. ACM (2020)
8. Fouladvand, S., Osareh, A., Shadgar, B., Pavone, M., Sharafi, S.: DENSA: an effective negative selection algorithm with flexible boundaries for self-space and dynamic number of detectors. Eng. Appl. Artif. Intell. **62**, 359–372 (2017)
9. Ghafari, S.M., Tjortjis, C.: A survey on association rules mining using heuristics. WIREs Data Min. Knowl. Discov. **9**(4), e1307 (2019)
10. Han, J., et al.: PrefixSpan: mining sequential patterns efficiently by prefix-projected pattern growth. In: Proceedings of the 17th International Conference on Data Engineering, pp. 215–224. IEEE (2001)
11. Han, J., Pei, J., Yin, Y., Mao, R.: Mining frequent patterns without candidate generation: a frequent-pattern tree approach. Data Min. Knowl. Discov. **8**(1), 53–87 (2004)
12. Katos, V., et al.: State of vulnerabilities 2018/2019 : analysis of events in the life of vulnerabilities. European Network and Information Security Agency (2020). for Cybersecurity, E.U.A.
13. Li, Z., Li, X., Tang, R., Zhang, L.: Apriori algorithm for the data mining of global cyberspace security issues for human participatory based on association rules. Front. Psychol. **11**, 582480 (2021)
14. Murtaza, S.S., Khreich, W., Hamou-Lhadj, A., Bener, A.B.: Mining trends and patterns of software vulnerabilities. J. Syst. Softw. **117**, 218–228 (2016)
15. Saboori, E., Parsazad, S., Sanatkhani, Y.: Automatic firewall rules generator for anomaly detection systems with Apriori algorithm. In: 2010 3rd International Conference on Advanced Computer Theory and Engineering (ICACTE). IEEE (2010)
16. Tisbeni, S.R., et al.: A big data platform for heterogeneous data collection and analysis in large-scale data centers. In: Proceedings of International Symposium on Grids & Clouds 2021 — PoS (ISGC2021). Sissa Medialab (2021)
17. Yan, X., Zhang, C., Zhang, S.: Genetic algorithm-based strategy for identifying association rules without specifying actual minimum support. Expert Syst. Appl. **36**(2, Part 2), 3066–3076 (2009)

Characterization and Categorization of Software Programs on X86 Architectures

Javier Jareño[1], Juan Carlos de la Torre[1], and Bernabé Dorronsoro[1,2](✉)

[1] Departamento de Ingeniería Informática, Escuela Superior de Ingeniería, Universidad de Cádiz, Cádiz, Spain
{javier.jareno,bernabe.dorronsoro,juan.detorre}@uca.es

[2] School of Computer Science, Faculty of Engineering, The University of Sydney, Camperdown, Australia
bernabe.dorronsorodiaz@sydney.edu.au

Abstract. The rapid technological growth in the computer industry has brought a wide variety of computer architectures and available programs with it. This makes the important problem of optimizing software performance increasingly complex. The reason is that automatic processing is required in order to consider in the decisions both the characteristics of the program itself and those of the system where it will be executed, in order to obtain remarkable results. With the aim of advancing in the knowledge on automatic and adaptive optimization of computer software, this paper presents a novel system for characterizing and clustering programs on x86 architectures according to their intrinsic characteristics, extracted with the Intel Software Development Emulator and `perf` tools. As a case of study, a subset of the programs collected in EEMBC, a benchmark suite for analyzing the performance of computing devices, has been selected. The results show how 70 different programs can be grouped together into 11 clusters wherein they share similar features.

Keywords: Software characterization · Software features extraction · Clustering · x86 Architecture · Intel Software Development Emulator · `perf`

1 Introduction

Due to the huge number of computer architectures (hardware), programming languages, and computer programs (software) available, the characterization of software takes special relevance as far as optimization techniques are concerned, since achieving a perfect generic optimization system for any program and architecture is an unmanageable task. However, the possibility of categorizing programs according to their features would open the door towards the automatic optimization of programs, adapted to their own characteristics and to the target hardware. In this work, we propose the characterization of software programs by categorizing them into groups that share program features.

B. Dorronsoro et al. (Eds.): OLA 2023, CCIS 1824, pp. 327–340, 2023.
https://doi.org/10.1007/978-3-031-34020-8_25

The characterization of a benchmark (or test software) lies in the extraction of a predefined series of program features. They must be sufficiently representative to be able to distinguish two different programs from each other, so their selection requires a previous study. In addition, the features will be strongly related to the architecture on which the software runs, so it is consistent to restrict the software characterization to the specific desired architecture.

The dependence of a characterization on an architecture occurs because different hardware architectures have different physical features, such as cache sizes, ALUs, RAM, hard disk, parallelism systems, instruction sets, or buses, among others. This means that the set of features that can be extracted from the same software in different contexts may vary.

The characterization of benchmark programs is a very important step for their subsequent categorization. The aim is to obtain representative features of a program that sufficiently distinguish it from others with different behavior. The categorization of benchmarks makes it possible to create different subsets of programs, depending on the objective of the study to be carried out on them. In the future, this will allow optimization systems to specialize in dealing with sets of programs with similar characteristics, instead of the inefficient current approach that applies a generic optimization for any type of program. Such achievement will be the cornerstone of a new generation of smart compilers.

For the study on characterization and categorization of software, it is essential to have a diverse set of programs that are the object of study. This set is known as a benchmark suite, and it must be complete, accurate, and consistent to serve as a relative measure of the performance and characterization of a program. Therefore, it is important to understand what type of computational load it generates on the processors, as well as how it was done, since two programs can generate the same computational load in terms of percentage of CPU used and at the same time have a completely different characterization.

In the present work, the benchmark suite considered for our characterization and categorization study is a subset of the one offered by EEMBC [7], consisting of 70 benchmarks. This is comprised of different suites that in turn contain a series of related benchmarks. They are all written in C, and the different suites define sets of benchmarks that perform tasks referred to a specific topic. The existing suites are: `automotive`, `consumer`, `digital_entertainment`, `embench`, `multibench`, `networking`, `networking_V2`, `office_automation` and `telecommunications`.

The purpose of each one can be deduced from its name, however, they perform simulations. That is to say, when evaluating network operations in one of the benchmarks, the real operations are not actually performed, meaning that they could be executed with the network card being disabled. Instead, disk writes/reads are made, collecting the data that is supposed to be sent and received through the network card. Therefore, using the benchmark suite names as a grouping feature is not representative, and it is required to properly characterize each of the benchmarks.

It should be noted that the methodology designed in the present study can be used with any benchmark suite written in C under x86 architecture, a study has not been carried out on the range of application in terms of programs with dependencies on modules written in other languages, or dynamic libraries, and they are a matter for future works. Additionally, older x86 architectures with possible outdated instruction sets are not considered, because they might include certain deprecated assembly instructions that the program emulator used is not able to account for.

The main contribution of this work is the proposal of a novel methodology for the effective classification and categorization of software programs on x86 architectures. For that, an automatic software feature extraction system is designed, based on the use of Intel Software Development Emulator (SDE) and `perf` tools. Specifically, a study of the measurements is carried out to ensure their correct use and interpretation, as well as a sensitivity study to select which features are representative enough to differentiate the benchmarks from each other. As a second contribution, a software program clustering system is proposed based on the extracted features. For this purpose, an agglomerative connective method based on the elbow method [6] will be used to define the number of clusters.

The document is organized as follows: Sect. 2 discusses the current state of the art on software feature extraction and its categorization into groups by similarity. Section 3 defines the software feature extraction method we propose, as well as the tools used for that, and Sect. 4 summarizes the experiments performed, mentioning their setup, as well as a discussion of the results obtained. Finally, Sect. 5 discusses the conclusions obtained from the study and the main lines of future work identified after this work.

2 State of the Art

Benchmark characterization and data clustering are frequently studied problems in the literature, usually independently addressed. However, we can find several works dealing with the classification and characterization of software.

Poovey et al. [11] study the characterization of software on the same benchmark used in the present study (EEMBC), in addition to offering a manual grouping based on the results obtained. For this purpose, software characteristics are divided into the following categories: cache references and misses for different block sizes, distribution of instruction types (%ALU, %Memory, %Jump and %Others), inter-process communication, and jump prediction failure.

Hoste et al. [8] propose a microarchitecture-independent characterization system to study the inherent properties of benchmarks. For that, authors focus on the use of features concerning dependency distance distribution to registers, jump prediction, instruction set, data flow for a given size, and memory access patterns.

Joshi et al. [9] study a microarchitecture-independent characterization and clustering system. They establish the set of features extracted from the benchmarks, as well as the tools used (SCOPE and STATISTICA). For clustering, they perform a dimensionality reduction of the features with Principal Component Analysis (PCA) and a subsequent categorization with k-means and hierarchical clustering. The Bayesian Information Criterion method is used to define the number of clusters [10].

Conte et al. [2] focus the characterization on memory instructions, since the other types of instructions are architecture instruction set dependent. They define that one of the most important aspects to take into account when studying a benchmark is the cache reference miss rate, as this implies a considerable workload.

In comparison to the previous studies, the feature extraction of the present study diverges from the existing ones in the use of features intrinsic to the benchmarks within a fixed x86 architecture, because the aim is to subsequently obtain an optimization system specific to that architecture. The features extracted are truly representative of the behavior of the benchmark, and they are generated with an emulator. This approach allows, for instance, accounting for the number of times loops are executed, unlike the other existing works.

3 Software Feature Extraction

Feature extraction can be performed both statically (by checking the object code resulting from compilation) and dynamically (by executing the program and extracting information from its execution). Static program analysis consists of studying the object code written in assembler to extract the number of occurrences of each instruction in the program. However, this approach is ambiguous because it does not allow considering important information about the program execution, such as the number of iterations performed by a loop, the number of function calls, or recursive calls.

Dynamic analysis solves the aforementioned problems of static analysis, since it does allow the information generated during program execution to be extracted. However, the impact of simultaneous measurements of the program execution on its overall performance must be considered.

This study uses the latter approach, using the two tools previously mentioned: `perf` [1] and Intel's SDE [5].

3.1 `perf` Tool

`perf` is a Linux performance analysis tool that provides information about the performance registers located in the Performance Monitoring Unit (PMU Hardware Events) physically inside the CPU. These registers maintain information about hardware events occurring on the CPU and form an excellent basis for specifying intrinsic characteristics of a program on a particular architecture.

On the one hand, it offers a large number of events that can be monitored, although some of them do not work on certain architectures or may suffer from measurement noise. On the other hand, there is no theoretical limit on how many events can be measured simultaneously, this applies without any problem on software events. However, for hardware events there is a limitation due to the number of counters in the CPU. These counters are in charge of monitoring the events occurring at the hardware level and if the number of event types (hardware) to be measured exceeds the number of counters, time division multiplexing is applied.

Time multiplexing is a technique used by `perf` as an attempt to measure all types of events, without being monitored at all times. At the end of the execution, the data is scaled to estimate the number of events that would have been performed based on the time during which that event has been measured and the total execution time. The scaling is calculated as:

$$Total\_Events = Measured\_Events * \frac{t_{execution}}{t_{enabled}} \quad . \tag{1}$$

The scaling percentage returned by `perf` is calculated as:

$$p_{Scaled} = \frac{t_{enabled}}{t_{execution}} \quad , \tag{2}$$

where scaling gives the ratio of the time the event has been measured with respect to the total program execution time.

3.2 SDE Intel Tool

Intel's SDE is a code emulator on 64-bit x86 processors. It offers multiple measurement tools, although the most important one for the present work is the histogram of assembler instructions executed by the processor, either at thread, logical core or global levels. The operation is similar to `perf`: it executes a software binary monitoring which assembler instructions have been executed. However, its main advantage is that because it is a simulator, the computation of the metrics do not have any impact on the performance of the measured programs.

For this study, the total number of x86 assembly instructions executed is taken into account, therefore they are not divided into threads or logical cores.

Intel x86-64 assembler instructions are divided into different categories [3, 4], depending on the scope of operation, type of instruction or intrinsic features of these instructions.

4 Experimentation

This section summarizes the experiments performed for this work. First, Sect. 4.1 presents the setup of the experiments. Then, in Sects. 4.2 and 4.3 the most representative characteristics of the programs are selected for classification, as well as the appropriate number of clusters to be used, respectively. Finally, the main results obtained are discussed in Sect. 4.4.

4.1 Experiment Setup

The tools used for the feature extraction system are Intel SDE version 9.0 and `perf` version 5.13.19. For the compilation of the benchmarks, `clang` version 9.0.1-12 is used, and it is configured so that it applies the lowest level of optimization during the compilation (i.e., using "-O0" compilation flag).

As previously mentioned, the features are dependent on the architecture on which the benchmarks are run. The present study has been performed on an AMD Ryzen 7 3700x processor (x86-64 architecture) with 32 GB of DDR4 RAM and Ubuntu 20.04 LTS as Operating System.

4.2 Features Selection

The features extracted by `perf` and SDE are multiple and of great variety, but not all of them offer relevant information. For example, multiple events of the `perf` list have been detected that either do not provide information due to limitations of the architecture used in the experiments or do not provide relevant information (they are linear combinations of other events).

From `perf`, it has been decided to take only cache-references (sum of `LLC-load`, `LLC-stores` and prefetching of data and instructions, all referred to the `L3 cache`) as a representative feature of the programs. The decision was made thanks to an analysis of the median absolute deviation, which showed how all the benchmarks reported similar values for the rest of the features studied, making them not useful for the purpose of characterizing programs. The other events studied were branch-instructions (jump assembly instructions executed), branch-misses (jump assembly instructions that have failed), `cache-misses` (sum of `LLC-load-misses`, `LLC-stores-misses` and prefetching failure of data and instructions, all referred to the L3 cache), cpu-cycles (total CPU cycles), or instructions (total assembly instructions executed).

As for Intel SDE, it provides information about the types and number of x86 assembly instructions executed by a program. The information returned by SDE uses the Intel x86 manual's own internal classification. This classification has 33 instruction categories, which we propose to group into sets of classes according to their meaning (vector instructions, function calls, binary operators, etc.) to form 17 groups of instruction categories. This decision was motivated after a previous study where a manual and automatic classification of all the x86 instructions that were executed was performed, in addition to an exploratory sensitivity analysis, all of which showed that those instruction categories with similar meaning showed similar values to the resemblant categories. The characteristic corresponding to each group is calculated as:

$$feature_i = \frac{n_i}{t_{instr}} \ , \tag{3}$$

where i is the calculated feature category, n_i is the number of executed assembler instructions of that category, t_{instr} is the total number of executed assembler instructions and $feature_i$ is the value assigned to the feature. Once this value has been calculated for all benchmarks, the data is normalized.

Finally, we obtained 18 potential features (1 from `perf` and 17 from SDE). They adopt the name of the most important representative of each category (e.g., under `SSEx` we have `SSEx`, `AVXx`, `MMX` and `VFMA`) and they are: `SSEx`, `BINARY`, `CALL`, `CMOVBE`, `COND_BR`, `CONVERT`, `DATAXFER`, `LOGICAL`, `MISC`, `POP`, `ROTATE`, `SEMAPHORE`, `BITBYTE`, `SYSCALL`, `BROADCAST`, `XRSTOR`, `NOP` and `cache-references`.

In order to test the strength of the 18 features selected for software classification, two different analyses were performed, the results of which are shown in the Fig. 1:

- **Information gain**: In specific terms, it calculates the entropy reduction produced by the transformation of the data set. In general terms, it measures how much information each feature adds to the data set. This information translates into how much a feature helps to differentiate a benchmark i of a cluster u from another benchmark j of a cluster v. The sklearn library[1] has been used for computing this metric.
  For this study, agglomerative (hierarchical) clustering was used to group the benchmarks into 10 clusters using the Euclidean distance function and ward distance method.
- **Mean absolute deviation**: It calculates the mean absolute error on the mean of each feature (not the mean square error). Therefore, the higher its value, the greater the variability of the data and the more information the feature includes.

For this study, each benchmark was run 30 times with the respective tools (Intel SDE and `perf`) and the median is taken as the representative statistic for each of the characteristics. It is done because of the uncertainty in the measurements, because the number of executed assembly instructions are different in each test. This non determinism is due to cache misses, jump prediction failure, context switches, OS handling, etc., which affect the program execution.

Two thresholds are set for both analyses based on the mean and standard deviation of the values. In the case of information gain (Fig. 1a), the threshold value is 0.3. Therefore, those features whose information gain is below this threshold are considered not to be sufficiently representative. For the mean absolute deviation analysis (Fig. 1b) the threshold is set to 0.125, proceeding as in the previous analysis with those features that do not exceed it.

Thus, the intersection of the categories that do not exceed the above thresholds are classified as not representative. They are: `BROADCAST`, `XRSTOR`, `SYSCALL`, `BITBYTE`, `SEMAPHORE`, `DATAXFER`, `CONVERT` and `Cache-references`. These categories will be unified under the new class called `NOTREPRESENTATIVE`. The only exception is for cache-references, which has a different unit of measurement than the other categories and therefore cannot be directly unified with them. As it has values close to the discard threshold, it is considered itself as one additional category, together with the previously unified and representative categories.

[1] https://scikit-learn.org/.

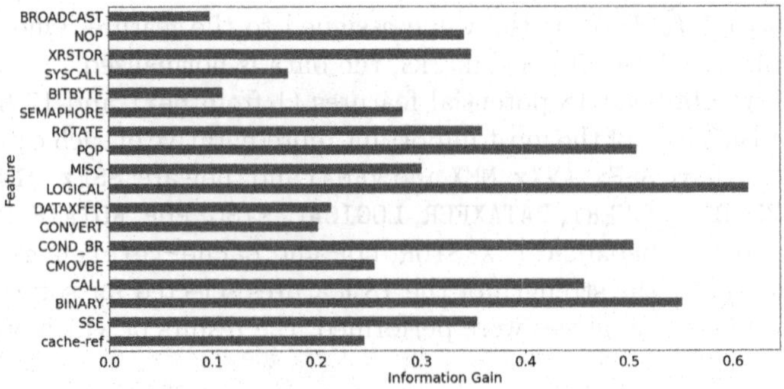

(a) Information gain analysis.

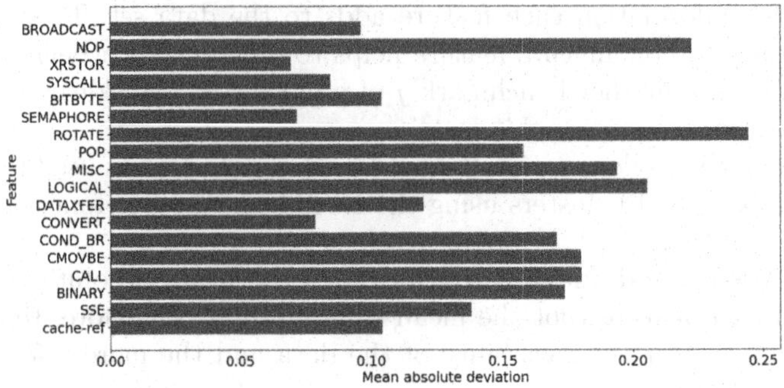

(b) Mean absolute deviation analysis.

Fig. 1. Information gain (a) and mean absolute deviation (b) of the data set formed by the benchmark features.

Finally, the characteristics selected after this study for the final characterization system are listed in Table 1, together with the categories they represent and their description.

4.3 Selecting the Number of Clusters

After identifying the potential features for the program characterization and clustering system, this section presents the final clustering system to be used for software categorization. To do this, it is necessary to determine the appropriate number of clusters. The elbow method [6] is used by applying k-means++ from the sklearn library as a classification technique on the studied data set.

The elbow algorithm receives the data to be clustered and a range with the possible number of clusters to be considered (in this study the range {1,24} is taken). For each possible number of clusters, the k-means++ algorithm is applied on the data, and the degree of distortion is obtained, calculated as the sum of the mean squared errors on the distances of each pattern of a cluster with its representative.

Table 1. Features selected for benchmark characterization.

Feature	Category	Description
SSEx	SSEx, AVXx, AVX2 MMX, VFMA	Vector instructions (SIMD, single instruction multiple data)
BINARY	BINARY, X87_ALU	Arithmetic operations (addition, subtraction, multiplication, division) with binary operators (two registers)
CALL	CALL	Function calls
CMOVBE	CMOVBE, CMOV	Conditional moves
COND_BR	COND_BR, UNCOND_BR	Conditional and unconditional jumps
LOGICAL	LOGICAL, LOGICAL_FP	Logical operations and vector logical operations
MISC	MISC	Miscellaneous (wait, halt...)
POP	POP, PUSH, RET	Stack operations
ROTATE	ROTATE, SHIFT	Rotating or shifting operations
NOP	NOP, WIDENOP	No operation
NOTREPRESENTATIVE	BROADCAST, CONVERT, XRSTOR,	Not representative enough features XSAVE, SYSCALL, SYSTEM, BIT-BYTE, BMI1, SETCC, LODSD, SEMAPHORE
	DATAXFER	Data transfer
Cache-References	Cache-References	Referencias a caché

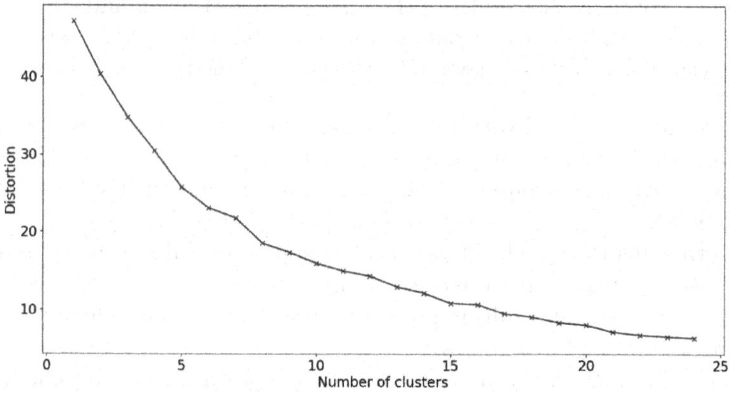

Fig. 2. Elbow method with final features.

The degree of distortion obtained for each cluster is shown in Fig. 2. Based on the theory of the elbow method, the number of clusters corresponding to the inflection point of the graph should be selected. In this case, due to the complexity of the problem there is no sharp inflection point, but rather a gradual reduction in the degree of distortion in the classification. We select 11 as the ideal number of clusters for the considered problem, since between {10,14} some stabilization of the distortion is observed. Other higher values with clearer stabilization can be appreciated, but it is not considered appropriate to select a high number of clusters, since this would imply a situation of over-fitting in

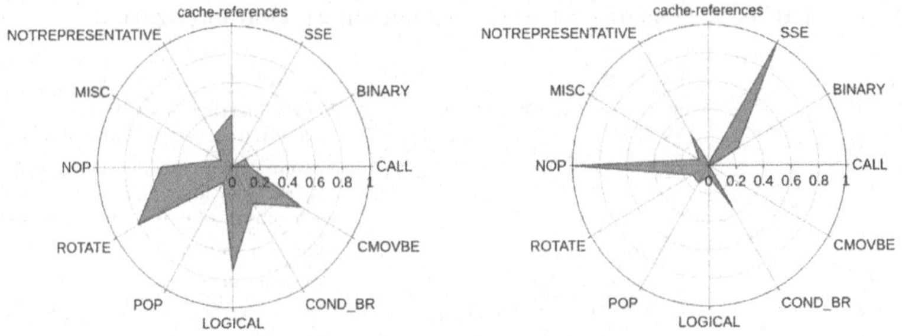

(a) *Benchmark* `bezier01`. (b) *Benchmark* `nbody`.

Fig. 3. Radar graphs of the features of two benchmarks.

which programs with small variations in their characteristics could be placed in different groups.

4.4 Discussion of Results

This section contains the results of the final characterization and clustering system. After the studies carried out in Sects. 4.2 and 4.3, a characterization and clustering model with the following features is obtained:

- Features are extracted with Intel SDE and `perf`, in both cases running each benchmark 10 times using the mean of the runs as a proxy. The selected features have been grouped as shown in Table 1, and their values have been normalized.
- Clustering algorithm: The hierarchical agglomerative algorithm is used, which is suitable for visualizing the order of merging clusters, the distance at which they merge and the resulting groups with a dendrogram. This model allows finding a suitable fit manually.
- Distance method: ward. This method avoids both generating clusters with a single benchmark and the inclusion of most benchmarks on a single main cluster. The ward method works with Euclidean distance.
- Number of clusters: 11. This number is obtained by applying the elbow method (based on k-means++) on the final set of features.

As an example of the proposed software characterization, the characterization of two of the benchmarks is shown in Fig. 3, for which different computational loads are shown. In the case of the `bezier01` benchmark (Fig. 3a) a higher load of operations of the logical type, rotations and conditional movements is shown. However, in the case of the `nbody` benchmark (Fig. 3b) a load mainly of NOP type operations and vector instructions is presented.

As for the results of the proposed clustering model, they are summarized in the dendrograms shown in Fig. 4. It can be seen in Fig. 4a the final grouping of the benchmarks in the 11 selected clusters, while the complete dendrogram of

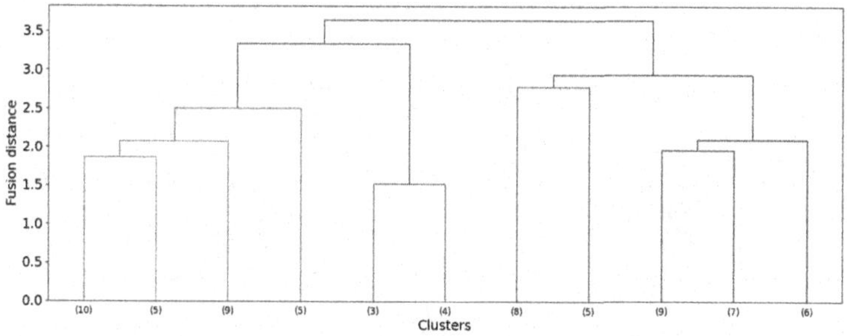

(a) Dendrogram resulting from applying clustering. The leaves show the number of benchmarks in each of the 11 final clusters.

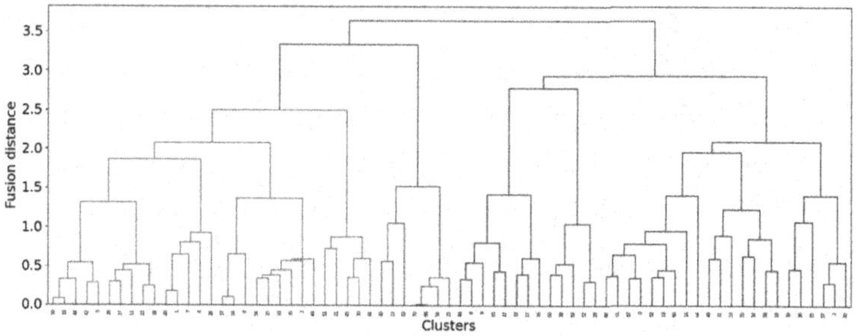

(b) Complete dendrogram of the clustering.

Fig. 4. Dendrograms resulting from the configured clustering algorithm.

the clustering is detailed in Fig. 4b. This way, it is possible to visualize the order in which the benchmarks were merged to end up in the final 11 clusters.

Finally, Fig. 5 and Fig. 6 show the features of all the benchmarks that have been assigned to each of the 11 clusters. It can be seen how each cluster brings together benchmarks with very similar characteristics to each other.

From these graphs it can be deduced that there is a great heterogeneity of existing problem types, and a higher number of clusters could be obtained if necessary. Furthermore, this classification indicates that the initial grouping of the benchmarks hosted in EEMBC corresponds to a large extent with the one obtained. For example, the packages referring to networking (`networking` and `networking_V2`) have been grouped under several clusters with cluster 1 being the main representative. Those referring to image or signal processing (`jpeg` and `mp3`) have also been grouped under the same clusters (groups 7 and 8), thus showing that these types of benchmarks have similar features to those related to matrix processing and yet they are so different from each other as to form distinct clusters.

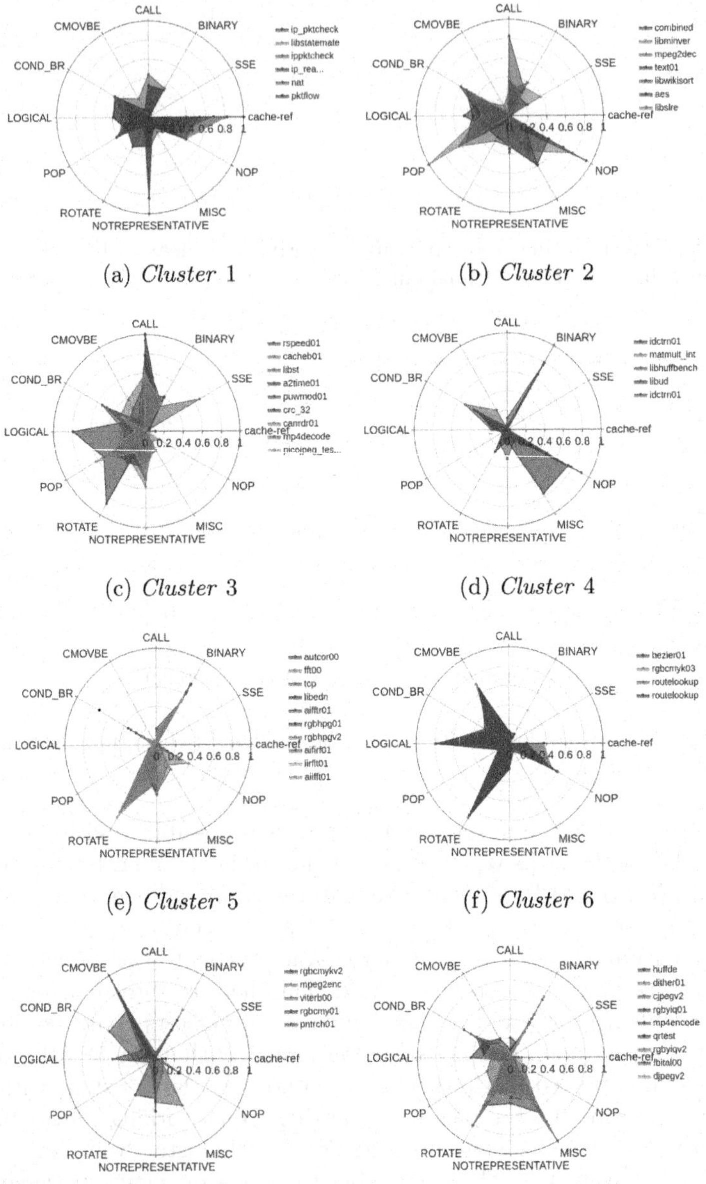

(a) *Cluster* 1 (b) *Cluster* 2

(c) *Cluster* 3 (d) *Cluster* 4

(e) *Cluster* 5 (f) *Cluster* 6

(g) *Cluster* 7 (h) *Cluster* 8

Fig. 5. Clusters resulting from all the benchmarks considered. In each radar graph the benchmarks corresponding to a group are collected.

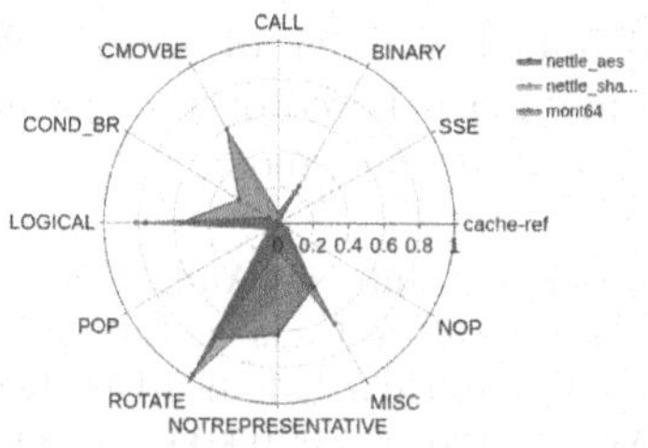

(a) *Cluster* 9

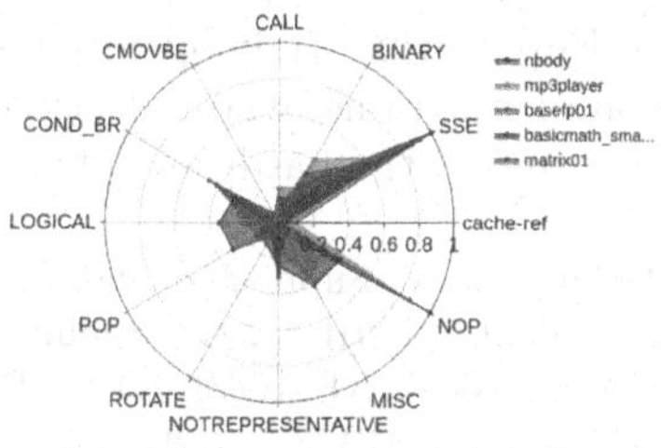

(b) *Cluster* 10

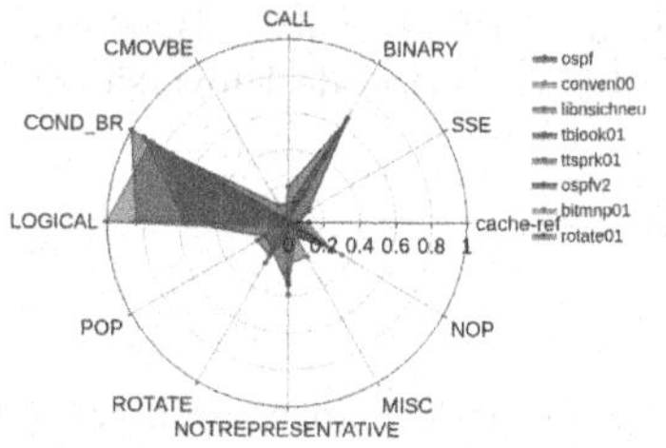

(c) *Cluster* 11

Fig. 6. Clusters resulting from all the benchmarks considered. In each radar graph the benchmarks corresponding to a group are collected.

5 Conclusions and Future Work

In the present study, a complete automatic clustering and characterization system for Intel x86-64 architectures has been designed. Likewise, a series of clusters has been predefined with the EEMBC benchmarks already classified in them. The characterization of the benchmarks and their corresponding grouping in different clusters is a very important process to formally study any system that operates on them. For its achievement, the following has been done:

- Design a complete characterization system based on the intrinsic features of the program on a given architecture, being as deterministic as possible. For this purpose, the main performance tools have been studied, filtering with a sensitivity analysis all the candidate features to be selected as representative.
- To model and configure a clustering algorithm that would be suitable for the problem that concerns us. For this purpose, a study of the possible parameters has been carried out, highlighting among them the analysis with the elbow method, as well as the combination of two sensitivity analyses to ensure the representativeness of the features in the defined model.

As a future work, the main study is to extrapolate feature extraction to ARM architectures. For this purpose, an ARM emulator must be designed, since there is currently no open source emulator that allows us to extract the executed ARM instructions. It is also considered very interesting to work with architecture-independent features, as well as to study other methods of defining the number of clusters. Finally, this study opens a new line of research towards the intelligent compilation of programs, whose decision making is based on the characteristics of the software and the hardware where it will be executed.

Acknowledgements. Supported and financed by Junta de Andalucía and ERDF (project P18-2399, GENIUS), the Spanish Ministerio de Ciencia, Innovación y Universidades and ERDF (NEMOVISION - PID2019-109465RB-I00), FEDER (FEDER-UCA18-108393, OPTIMALE), and MCIN/AEI/10.13039/501100011033 and the European Union "NextGenerationEU"/PRTR (TED2021-131880B-I00, eFracWare). J.C. de la Torre thanks the support of Ministerio de Ciencia, Innovación y Universidades on their FPU program (FPU17/00563). B. Dorronsoro acknowledges "ayuda de recualificación" funding by Ministerio de Universidades and the European Union-NextGenerationEU.

References

1. perf: Linux profiling with performance counters (2022). https://github.com/torvalds/linux/tree/master/tools/perf
2. Conte, T., Hwu, W.M.: Benchmark characterization. Computer **24**(1), 48–56 (1991)
3. Corporation, I.: Intel 64 and IA-32 Architectures Developer's Manual, vol. 2A
4. Corporation, I.: Intel 64 and IA-32 Architectures Developer's Manual, vol. 2B
5. Corporation, I.: Intel software development emulator (2022). https://www.intel.com/content/www/us/en/developer/articles/tool/software-development-emulator.html
6. Cui, M.: Introduction to the k-means clustering algorithm based on the elbow method. Geosci. Remote Sens. **3**, 9–16 (2020)
7. EEMBC: EEMBC Benchmarks, https://www.eembc.org/products/. Accessed 10 May 2022
8. Hoste, K., Eeckhout, L.: Microarchitecture-independent workload characterization. IEEE Micro **27**(3), 63–72 (2007)
9. Joshi, A., Phansalkar, A., Eeckhout, L., John, L.: Measuring benchmark similarity using inherent program characvteristics. IEEE Trans. Comput. **55**, 769–782 (2006)
10. Neath, A.A., Cavanaugh, J.E.: The Bayesian information criterion: background, derivation, and applications. Wiley Interdisc. Rev. Comput. Stat. **4**(2), 199–203 (2012)
11. Poovey, J.A., Conte, T.M., Levy, M., Gal-On, S.: A benchmark characterization of the EEMBC benchmark suite. IEEE Micro **29**(5), 18–29 (2009)

Robot-Assisted Delivery Problems and Their Exact Solutions

Abdullahi Mohammed Jingi[1(✉)] and Xinan Yang[2]

[1] Adamawa State University, P.M.B 25, Mubi, Adamawa State, Nigeria
modijingi@gmail.com
[2] University of Essex, Wivenhoe Park, CO4 3SQ Colchester, Essex, UK
xyangk@essex.ac.uk

Abstract. This study focuses on the use of automated vehicles for last-mile delivery, specifically using local depots and integrated trucks-robots. The researchers developed two MILP models, RADP-1 and RADP-2, with RADP-2 being a modified version of TSP-D by [1]. The models were tested on three different service areas with varying customer densities, and it was found that robots were more efficient in areas with high customer densities. As the problem size increases, RADP-2 becomes more difficult to solve than RADP-1 due to the exponential increase in the number of feasible operations. The computational time of RADP-2 can be improved by removing operations that are unlikely to be part of the optimum solution.

Keywords: Vehicle Routing Problem · Robot · Scheduling · Last-mile delivery

1 Introduction

Recent developments in technology and the growth of e-commerce has led to several researches considering alternative assisted deliveries not only by drones, but also by robots. Some technology and traditional logistics companies like Starship technologies [2] and JD.com [5] have implemented the usage of autonomous robots for small parcel delivery. Most robots are electrically powered and can travel a limited distance owing to its limited battery capacity. Also, robots travel at a very low speed, making their application for long distance delivery problems inefficient. As a result of this, logistics companies are considering integrated truck-robot assisted deliveries to overcome the issue of limited travel distance of robots by transporting them using trucks to dense customer areas for efficient robot usage. Robots though characterized by low travel speed, would provide better alternative solutions, particularly in an urban environment characterized by a high number of stopping points with relatively short distances. Table 1 summarizes the strengths and limitations of both trucks and robots.

In this work, the authors proposed a framework called the "Robot-Assisted Delivery Problem (RADP)". The problem considers the integration of standard

B. Dorronsoro et al. (Eds.): OLA 2023, CCIS 1824, pp. 341–353, 2023.
https://doi.org/10.1007/978-3-031-34020-8_26

Table 1. Comparison of Robot and Truck

Characteristics	Truck	Robot
Load Capacity	large	small
Endurance	Unlimited	short
Speed	high	low
Carbon-Emmision	high	low
Route	Along road network	padestrian walk

truck delivery, robot delivery (robots launched from trucks), and local depots into one model, where customer orders can be served by either the truck, the robot launched from the truck, or a local depot. The truck departs from the distribution center and travels to customer and/or depot locations to serve orders. The truck carries a certain number of robots on board, which can be launched at customer or local depot nodes to fulfill customer demands. The robots are picked up at a different location from the drop-off point to allow for simultaneous service of orders by the two vehicles. The local depot is also equipped with a certain number of localized robots that serve customers covered by the depot. Two different models were developed for the Robot-Assisted Delivery Problem, namely RADP-1 and RADP-2. In RADP-1, each node and each arc are exclusively considered in the modeling process, while RADP-2 uses the concept of operations, where each operation is made up of a combination of arcs and nodes. The two proposed models, RADP-1 and RADP-2, offer different perspectives in solving the problem, and the comparison between the two models provides valuable insights for future research in this area.

Figure 1 illustrates a simple example of the Robot-Assisted Delivery Problem (RADP), which consists of 11 customer nodes and 1 local depot node. Node 1 represents the distribution center where the truck starts and terminates its journey, while node 2 represents a local depot. The truck visits nodes 3, 5, 7, 2, and 4 to serve customer orders. At node 3, two robots are launched from the truck to serve customers 8 and 9 and customers 13 and 12, respectively. The first robot serves customers 8 and 9 and is picked up by the truck at node 4 while the second robot serves customers 13 and 12 and is picked up by the truck at the local depot node 2. Nodes 6, 10, and 11 are within the coverage range of local depot 2 and are therefore served by the local depot's localized robots. This example shows how the RADP framework combines different delivery methods to optimize the delivery system's efficiency.

2 Literature Review

Scheduling procedure proposed in [3] involves a truck-based robot-assisted delivery system that uses autonomous robots to serve deliveries with time windows. The system involves a single truck that loads customer shipments and robots

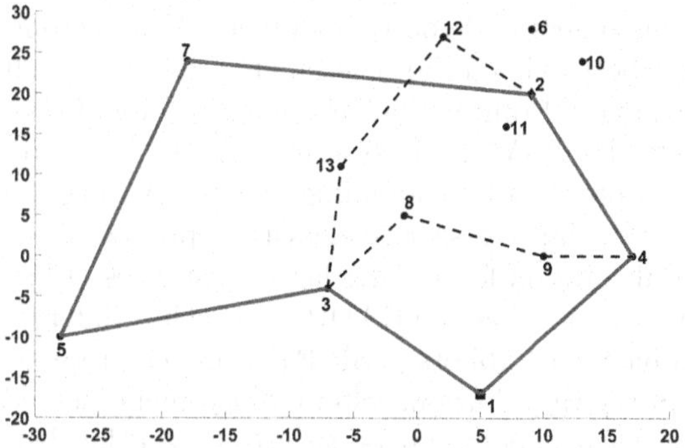

Fig. 1. A simple illustration of RADP

from a central depot and transports them to drop-off points located in the city center. The robots are loaded with the shipments to make deliveries to customers and rejoins the truck. The authors assumed a network with an unlimited number of robots which is unrealistic in practice. In reality, deploying a large fleet of robots and leaving them unused for a significant amount of time would be cost-prohibitive. Therefore, there is a need to develop more realistic scheduling procedures that consider the limited availability of robots and other resources to optimize the delivery system's efficiency while minimizing costs.

A similar idea of carrying robots on a conventional truck and releasing them from the truck is the study presented in [4]. Unlike [3], where robots have a single delivery capacity, it is assumed that a robot can serve up to six customers. A comparison of their approach with a standard delivery truck was made based on the assumption that a truck has the capacity of carrying a maximum of eight robots and a robot can travel up to four miles to make deliveries. The study revealed that autonomous robots may be more efficient when customer densities is high.

The paper by [7] describes a location-routing problem that involves the use of multi-compartment robots and customer time windows. The goal is to determine the optimal locations of robot hubs, the set of tours that the robots will take, and the number of robots needed to minimize the total costs of one working day. The problem is modeled as a mixed-integer programming problem with the objective to minimize the total costs while meeting the delivery requirements and the battery swapping constraints. The results of the study show that the proposed model and algorithm can effectively solve the location-routing problem and provide solutions that are close to the optimal solution. The study also highlights the potential benefits of using multi-compartment robots and battery swapping to improve the efficiency of the delivery process.

[8] presents a two-echelon van-based robot hybrid pickup and deliveries (2E-VRHPD) system. The system consists of vans carrying small robots that travel along tier-1 routes, stopping at parking nodes to drop off and/or pickup robots, and to replenish or swap robot batteries. There are two types of customers in

the system: van customers and robot customers. Van customers can be served by either vans or robots, while robot customers are served exclusively by robots. Robots travel along tier-2 routes that are not accessible by vans to serve robot customers. Overall, the 2E-VRHPD system presents an interesting approach to solving the delivery problem by combining the use of vans and robots in a two-echelon system. The adaptive search algorithm proposed by the authors also provides a practical solution for solving larger instances of the problem.

It is interesting to note that the RADP approach is different from the Travelling Salesman Problem with robots (TSP-R) approach presented in [6] by including local depots in the distribution system. Customers located within a certain distance from the local depots can be served by the local depot. The RADP approach is entirely different from the existing studies in the literature, any customer can be served either by a truck, a robot or a local depot. This allows for greater flexibility and optimization in the distribution system, potentially leading to more efficient and effective delivery of goods.

3 Problem Formulation of RADP Models

This problem involves a combination of conventional truck delivery, robot delivery (carried on trucks), and local depots to optimize the delivery process. The scenario assumes a set of trucks, robots, customers, and local depots. The trucks start from the distribution center, which is located in the countryside and carry robots and customer orders. Since robots have limited battery capacity and travel speed, they are not designed for long distances, and trucks carry them to suitable areas for delivery. However, instead of just serving as mobile depots for robots, trucks can also deliver customer orders as standard delivery vans. The robots' storage is divided into compartments for multiple shipments, and they return to either a customer or local depot pickup point after delivery. The trucks can also service multiple local depots that provide other customer service facilities. The aim of this study is to optimize the delivery schedules of all facilities involved so as to minimize total work span of trucks.

3.1 Mixed Integer Programming Formulation of RADP-1

The important assumptions of the model are summarized below:

1. We assume that there are K trucks and R robots, starting and finishing at the distribution center. We also assume that there are N_D local depots.
2. There are N_C customers to be served, each with known location, order size q_i and service time o_i.
3. Robots carried by trucks are identical with their storage separated into compartments.
4. Each customer is served either by a truck, a localized robot from robot depot or a robot carried on a truck.

5. The drop-off and pick-up locations of robots are different
6. Multiple robots can be launched and retrieved at the same customer node.
7. The robots must be picked up by the truck from which it was dropped.

Variables and Notations

Sets:

$\mathcal{K} := \{1, ..., K\}$ - Set of trucks
$\mathcal{R} := \{1, ..., R\}$ - Set of robots
$\mathcal{N}_C := \{1, ..., N_C\}$ - Set of customers
$\mathcal{N}_D := \{1, ..., N_D\}$ - Set of robot depots
$\mathcal{N}_{DC} := \{1, N_C + N_D + 2\}$ - Warehouse
$\mathcal{N} := \mathcal{N}_C \cup \mathcal{N}_D \cup \mathcal{N}_{DC}$ - All nodes.

Parameters

q_i - order size of node $i \in \mathcal{N}_C$
o_i- service time at node $i \in \mathcal{N}_C$
C_R - capacity of robot
C_T - capacity of truck
v_R - speed of robot
v_T - speed of truck
h - setup time of robot
s - capacity occupation of a robot
d_{ij} - distance between node i and node j $\forall i, j \in \mathcal{N}$.
p_i - maximum capacity that can be served by the local depot.
D_{max} - maximum allowable distance traveled by robot, $r \in R$.

$$\lambda_{ij} = \begin{cases} 1, & \text{if customer } i \text{ is covered by depot } j, \\ 0, & \text{otherwise} \end{cases} \quad i \in \mathcal{N}_C, j \in \mathcal{N}_D.$$

Decision Variables

- $x_{ijk} = \begin{cases} 1, & \text{if truck } k \text{ travels } (i, j), \\ 0, & \text{otherwise} \end{cases} \quad i \neq j \in \mathcal{N}, k \in \mathcal{K}.$
- $y_{ijr} = \begin{cases} 1, & \text{if robot } r \text{ travels } (i, j), \\ 0, & \text{otherwise} \end{cases} \quad i \neq j \in \mathcal{N}_D \cup \mathcal{N}_C, r \in \mathcal{R}.$
- $\delta_{kr} = \begin{cases} 1, & \text{if robot } r \text{ is placed on truck } k, \\ 0, & \text{otherwise} \end{cases} \quad r \in \mathcal{R}, k \in \mathcal{K}.$
- $\gamma_{ikr} = \begin{cases} 1, & \text{if robot } r \text{ is launched from truck } k \text{ at node } i, \\ 0, & \text{otherwise} \end{cases} \quad r \in \mathcal{R}, k \in \mathcal{K}, i \in \mathcal{N}_D \cup \mathcal{N}_C.$
- $\eta_{ikr} = \begin{cases} 1, & \text{if robot } r \text{ is collected by truck } k \text{ at node } i, \\ 0, & \text{otherwise} \end{cases} \quad r \in \mathcal{R}, k \in \mathcal{K}, i \in \mathcal{N}_D \cup \mathcal{N}_C.$
- $z_{ij} = \begin{cases} 1, & \text{if node } i \text{ is served by depot } j, \\ 0, & \text{otherwise} \end{cases} \quad i \in \mathcal{N}_C, j \in \mathcal{N}_D.$
- t_{ik} = the visiting time at node i by truck k, $\forall i \in \mathcal{N}_C \cup \mathcal{N}_D, \forall k \in \mathcal{K}$.
- τ_{ir} = the visiting time at node i by robot r, $\forall i \in N_C \cup N_D, \forall r \in R$.

RADP-1

$$\min \sum_{k} \left(t_{(N_C+N_D+2,k)} - t_{1k} \right) \tag{1}$$

subject to:

$$\sum_{j \in \mathcal{N} \setminus \{1\}} x_{1jk} = 1, \forall k \in \mathcal{K}. \tag{2}$$

$$\sum_{i \in \mathcal{N}_C \cup \mathcal{N}_D} x_{i,N_C+N_D+2,k} = 1, \forall k \in \mathcal{K}. \tag{3}$$

$$\sum_{j \in \mathcal{N}} x_{ijk} = \sum_{j \in \mathcal{N}} x_{jik}, \forall i \in \mathcal{N}_C \cup \mathcal{N}_D, \forall k \in \mathcal{K}. \tag{4}$$

$$\sum_{j \in \mathcal{N}_C \cup \mathcal{N}_D} y_{ijr} + \sum_{k \in \mathcal{K}} \eta_{ikr} = \sum_{j \in \mathcal{N}_C \cup \mathcal{N}_D} y_{jir} + \sum_{k \in \mathcal{K}} \gamma_{ikr}, \forall i \in \mathcal{N}_c \cup \mathcal{N}_D, \forall r \in \mathcal{R}. \tag{5}$$

$$2\gamma_{ikr} \leq \delta_{kr} + \sum_{j \in \mathcal{N}} x_{ijk}, \forall k \in \mathcal{K}, \forall r \in \mathcal{R}, \forall i \in \mathcal{N}_C \cup \mathcal{N}_D. \tag{6}$$

$$2\eta_{ikr} \leq \delta_{kr} + \sum_{j \in \mathcal{N}} x_{ijk}, \forall k \in \mathcal{K}, \forall r \in \mathcal{R}, \forall i \in \mathcal{N}_C \cup \mathcal{N}_D. \tag{7}$$

$$\sum_{k \in \mathcal{K}} \delta_{kr} \leq 1, \forall r \in \mathcal{R}. \tag{8}$$

$$z_{ij} \leq \lambda_{ij}. \forall i \in \mathcal{N}_C, j \in \mathcal{N}_D. \tag{9}$$

$$\sum_{i \in \mathcal{N}} \sum_{k \in \mathcal{K}} x_{ijk} \geq \frac{1}{M} \sum_{i \in \mathcal{N}_C} q_i z_{ij}, \forall j \in \mathcal{N}_D. \tag{10}$$

$$\sum_{j \in \mathcal{N}} \sum_{k \in \mathcal{K}} x_{ijk} + \sum_{j \in \mathcal{N}_C \cup \mathcal{N}_D} \sum_{r \in \mathcal{R}} y_{ijr} + \sum_{j \in \mathcal{N}_D} z_{ij} = 1 + \sum_{r \in \mathcal{R}} \sum_{k \in \mathcal{K}} \gamma_{ikr}, \forall i \in \mathcal{N}_C \tag{11}$$

$$\sum_{i \in \mathcal{N}_C} q_i z_{ij} \leq p_j, \forall j \in \mathcal{N}_D. \tag{12}$$

$$\sum_{i \in \mathcal{N}_C \cup \mathcal{N}_D} \sum_{j \in \mathcal{N}_C \cup \mathcal{N}_D} d_{ij} y_{ijr} \leq D_{max}, \forall r \in \mathcal{R}. \tag{13}$$

$$\sum_{i \in \mathcal{N}_C} q_i \left(\sum_{j \in \mathcal{N}_C \cup \mathcal{N}_D} y_{ijr} - \sum_{k \in \mathcal{K}} \gamma_{ikr} \right) \leq C_R, \forall r \in \mathcal{R}. \tag{14}$$

$$\sum_{i \in \mathcal{N}_C} \sum_{j \in \mathcal{N}} q_i x_{ijk} + \alpha_k + \beta_k + \sum_{r \in \mathcal{R}} s.\delta_{kr} \leq C_T, \forall k \in \mathcal{K}. \tag{15}$$

$$\alpha_k \geq \sum_{i \in \mathcal{N}_c} q_i \left(\sum_{j \in \mathcal{N}_c \cup \mathcal{N}_D} y_{ijr} - \sum_k \gamma_{ikr} \right) - M\left(1 - \delta_{kr}\right), \forall k, \forall r. \quad (16)$$

$$\beta_k \geq \sum_{i \in \mathcal{N}_c} q_i z_{ij} - M \left(1 - \sum_{i \in \mathcal{N}} x_{ijk} \right), \forall k, \forall j. \quad (17)$$

$$t_{jk} \geq t_{ik} + o_i + \frac{d_{ij}}{v_T} + h \sum_{r \in \mathcal{R}} \gamma_{ikr} - M(1 - x_{ijk}), i \neq j \forall i \in \mathcal{N}, \forall k \in \mathcal{K}. \quad (18)$$

$$\tau_{jr} \geq \tau_{ir} + o_i(1 - \gamma_{ikr}) + \frac{d_{ij}}{v_R} - M(1 - y_{ijr}), i \neq j \forall i, j \in \mathcal{N}_c \cup \mathcal{N}_D, \forall k \in \mathcal{K}. \quad (19)$$

$$t_{ik} \geq \tau_{ir} - M(1 - \eta_{ikr}), \forall i \in \mathcal{N}_C \cup \mathcal{N}_D, \forall k \in \mathcal{K}, \forall r \in \mathcal{R}. \quad (20)$$

$$\tau_{ir} \geq t_{ik} + h - M(1 - \gamma_{ikr}), \forall i \in \mathcal{N}_C \cup \mathcal{N}_D, \forall k \in \mathcal{K}, \forall r \in \mathcal{R}. \quad (21)$$

$$x_{ijk} \in \{0, 1\}, i \neq j \in \mathcal{N}, k \in \mathcal{K}. \quad (22)$$

$$y_{ijr}, \delta_{kr}, \gamma_{ikr}, \eta_{ikr} \in \{0, 1\}, i \in \mathcal{N}_C \cup \mathcal{N}_D, k \in \mathcal{K}, r \in \mathcal{R}. \quad (23)$$

$$z_{ij} \in \{0, 1\}, i \in \mathcal{N}_C, j \in \mathcal{N}_D. \quad (24)$$

$$\alpha_k, \beta_k, t_{ik}, \tau_{ir} \geq 0, i \in \mathcal{N}_C \cup \mathcal{N}_D, k \in \mathcal{K}, r \in \mathcal{R}. \quad (25)$$

The objective function (1) minimizes the total working time of all the trucks. Constraints (2), (3) & (4) are standard network flow constraints for truck. Equation (5) is the network flow constraint for robot. Constraints (6) & (7) ensures that robot can be launched from/collected at a node only if it is carried by a truck which visits the node. Constraint (8) guarantees that same robot cannot be carried by more than one truck. Constraint (9) ensures that a customer node is covered by a depot before it can be served by the depot. Constraint (10) ensures that all local depots with positive customer demands are visited by truck. Constraint (11) ensures that all customers are either served by truck, robot launched from truck or by a local depot. Constraint (12) ensures that the total order size served by a local depot does not violate its capacity. Constraint (13) ensures that the total distance travelled by a robot does not violate its maximum travel capacity, while constraint (14) ensures that total capacity of shipments carried on robot does not exceed its carrying capacity. Constraint (15) ensures that total capacity of shipments and capacity occupation of all robots carried on truck does not violate the truck capacity. Constraints (18) & (19) calculate the visiting time at nodes by trucks and robots respectively, while constraints (20) & (21) linked them together. Finally, (22), (23), (24) & (25) defined the values of the decision variables.

3.2 Mixed Integer Programming Formulation of RADP-2

In this section, an IP formulation of the problem is presented using the concept of operation, which is a modification of the Traveling Salesman Problem with Drones (TSP-D) proposed by [1]. An operation o represents a sequence of nodes that can be serviced by a truck with a robot on board or by a truck and robot splitting at the departure node, visiting other nodes in parallel, and rejoining at a pickup node. It is assumed that robots are allowed to visit at most two nodes per launch

An operation is made up of a combination of either truck and robot nodes (if the two modes are used in parallel), or only truck nodes (if they are used jointly). Each Operation is made up of at least two nodes consisting of start and end nodes which are referred drop off and pick up nodes of robots respectively.

Variables and Notations

Sets:

$\mathcal{N} := \mathcal{N}_C \cup \mathcal{N}_D$ - Set of all nodes, $\mathcal{N}_c$: - Set of customer nodes
$\mathcal{N}_D$: - set of depots $\mathcal{O}$: - set of feasible operations
$\mathcal{O}^-(i) \subset \mathcal{O}$:- Set of operations with start node $i \in \mathcal{N}$.
$\mathcal{O}^+(i) \subset \mathcal{O}$:- Set of operations with end node $i \in \mathcal{N}$.
$\mathcal{O}(i) \subset \mathcal{O}$:- Set of all operations that contain node $i \in \mathcal{N}$.

Variables

Decision Variables

- $x_o = \begin{cases} 1, & \text{if operation } o \text{ is chosen} \\ 0, & \text{otherwise} \end{cases} \quad \forall o \in \mathcal{O}.$
- $\beta_i = \begin{cases} 1, & \text{if depot } i \text{ has positive demand} \\ 0, & \text{otherwise} \end{cases} \quad \forall i \in \mathcal{N}_D,$
- v_i : visiting time at node $i \in \mathcal{N}$.
- $y_i = \begin{cases} 1, & \text{if atleast one chosen operation uses node } i \text{ as start node,} \\ 0, & \text{otherwise} \end{cases} \quad \forall i \in \mathcal{N}.$

Parameters

t_o :- time to complete operation $o \in \mathcal{O}$ and list of parameters in Subsect. 3.1

RADP-2

$$\min v_{(nC+nD+2)} \tag{26}$$

subject to:

$$\sum_{o \in \mathcal{O}(i)} x_o + \sum_{j \in \mathcal{D}} \lambda_{ij}\beta_j \geq 1, \forall i \in \mathcal{N}_c \tag{27}$$

$$\sum_{o \in \mathcal{O}^-(i)} x_o = \sum_{o \in \mathcal{O}^+(i)} x_o, \forall i \in \mathcal{N}. \tag{28}$$

$$\sum_{o \in \mathcal{O}^+(i_o)} x_o \geq 1. \tag{29}$$

$$\sum_{o \in \mathcal{O}(i)} x_o \geq \beta_i, \forall i \in \mathcal{N}_D. \tag{30}$$

$$\sum_{o \in \mathcal{O}^-(i)} x_o \leq y_i, \forall i \in \mathcal{N}. \tag{31}$$

$$y_{i_0} = 1. \tag{32}$$

$$v_j \geq v_i + \sum_{o \in (\mathcal{O}^-(i) \cap \mathcal{O}^+(j))} x_o * t_o - M(1 - x_o), \forall i, j \tag{33}$$

$$x_o \in \{0, 1\}, \forall o \in \mathcal{O}. \tag{34}$$

$$y_i \in \{0, 1\}, \forall i \in \mathcal{N}. \tag{35}$$

$$\beta_i \in \{0, 1\}, \forall i \in \mathcal{N}_D, . \tag{36}$$

The objective function (26) minimizes the total time to complete the tour. Constraint (27) ensures that all customer nodes are covered either by an operation or a local depot. Constraint (28) ensures that the chosen operation visits each node once. Constraint(30) ensures that all local depots with positive demands are covered by operations. Constraint (31) ensures that at most one operation with start node i is chosen. Constraint (32) ensures that the tour starts (and end) at the depot and Constraint (33) calculate the visiting time at each node. Finally, constraints (34), (35) and (36) force the variables x_o, y_i, and β_i to be binary.

4 Computational Experiments

The location of all nodes are randomly generated using Matlab, and the road distance between each pair of node calculated. Customer orders are also generated randomly. In all the experiments, we used the Robot set up time and service time of customers as 1 min and 3 min respectively, covering range of local depot to be 0.03 mi, while speed of Truck and robot as 13mph and 4mph. The experiments were performed on different order densities.

The MILP solution approach to the model is coded in MATLAB R2020a and executed on a CPU with Intel(R)Core(TM)i5-7300U processor. The MILP is solved by CPLEX Studio 12.10.0.

4.1 CPLEX MILP Solution of RADP Models

Due to complexity of the RADP models, numerical experiments were performed for small instances. We considered one truck, two robots and one local depot. We test the performance of the models on three different service areas with map range = 0.5, 1.0 and 5.0 mi. Within each service area, we perform five different tests with different order densities. The results of the experiments are presented in Tables 2, 3 and 4 . Table 2 is a result of experiments with map range = 0.5, it represents a solution of an area with high customer density covering around one square miles. Robots and local depots are well utilized in this area with average savings over TSP of up to 25% for both RADP-1 and RADP-2. RADP-1 is solved to optimality for up to 8 nodes within an hour, while RADP-2 is solved to optimality for up to 5 nodes only. The computational time of RADP-1 is worse in Test 1 only and better in Tests 2, 3, 4 and 5 because of the exponential increase in the number of feasible operations in RADP-2 due to increase in the number of nodes.

For more details, a sample solutions of Test 5 in Table 2 are presented graphically in Fig. 2, where sub-figures (a) and (b) are for RADP-1 and RADP-2 respectively. The central depot is node 1 (indicated by the square), the local depot is node 2 (indicated by diamond) and the rest are customer nodes (indicated by dots). With RADP-1, the truck starts from the central depot to launch 2 robots at customer node 9, one of the robots served customer at node 5 while the other served customers at nodes 3 and 7, they both rejoin the truck at customer node 4 for pickup. The truck from customer node 9 visits the local depot node 2, then to customer node 4 and back to the central depot. Customers at nodes 6 and 8 are served from the local depot. Robots and local depots are well utilized because of the higher density of the area. Meanwhile, optimum solution for RADP-2 is not obtained within one hour, the truck instead of visiting the local depot node 2 from customer node 9, it travels to customer node 5 before visiting the local depot, then returns to the central depot passing very close to node 9 that is already visited. Like RADP-1, customers at nodes 6 and 8 are served from the local depot. Table 3 is an example with map range = 1.0 mi. In this example, the customer density is lesser than the area represented in Table 2. In all the Tests, no customer is served by local depot because they are not within the covering area of the local depot due to enlarged service area, however robots are utilized in all test examples. Average savings over TSP for both models are 20%, which is less than the savings obtained in Table 2. Figure 3 is a graphical solution of Test 5 in Table 3 with sub-figures (a) and (b) for RADP-1 and RADP-2 respectively. RADP-1 provides better solution than RADP-2 within the computational time allowed, even though the truck route appears shorter in RADP-2, this could be due to longer waiting time of truck due to slower speed of robot.

Table 4 represents solutions of a less dense area (map range = 5.0 mi), robots and the local depot are not efficiently utilised, this accounts for low average savings over TSP of 7.23%.

Table 2. RADP-1 Vs RADP-2: Map Range = 0.5, $S_T = 13, S_R = 4$

Model	Test	# of customers	# of Local Dep.	CPLEX MILP Solution				Savings over TSP (%)
				# of robots Used	Service Time	Computational Time	Gap (%)	
RADP-1	1	3	1	1	0.1926	0.28	0.00%	9.02
RADP-2	1	3	1	1	0.1926	0.22	0.00%	9.02
RADP-1	2	4	1	2	0.1962	1.27	0.00%	25.65
RADP-2	2	4	1	2	0.1962	1.69	0.00%	25.65
RADP-1	3	5	1	2	0.2433	12.61	0.00%	22.64
RADP-2	3	5	1	2	0.2433	3600	19.90%	22.64
RADP-1	4	6	1	2	0.2448	261.28	0.00%	33.21
RADP-2	4	6	1	2	0.2458	3600	79.07%	32.93
RADP-1	5	7	1	2	0.2450	2748.72	0.00%	41.16
RADP-2	5	7	1	2	0.2625	3600	80.61%	36.96
						Average Savings		**25.89**

Table 3. RADP-1 Vs RADP-2: Map Range = 1.0, $S_T = 13, S_R = 4$

Model	Test	# of customers	# of Local Dep.	CPLEX MILP Solution				Savings over TSP (%)
				# of robots Used	Service Time	Computational Time	Gap (%)	
RADP-1	1	3	1	1	0.2185	0.42	0.00%	2.15
RADP-2	1	3	1	1	0.2185	0.21	0.00%	2.15
RADP-1	2	4	1	2	0.2258	1.69	0.00%	19.73
RADP-2	2	4	1	2	0.2258	1.58	0.00%	19.73
RADP-1	3	5	1	2	0.2560	9.67	0.00%	22.19
RADP-2	3	5	1	2	0.2560	240.04	0.00%	22.19
RADP-1	4	6	1	2	0.2746	454.8	0.00%	28.32
RADP-2	4	6	1	2	0.2990	3600	83.32%	21.95
RADP-1	5	7	1	2	0.2827	3600	11.17%	34.70
RADP-2	5	7	1	2	0.3044	3600	82.98%	29.68
						Average Savings		**20.28**

Table 4. RADP-1 Vs RADP-2: Map Range = 5.0, $S_T = 13, S_R = 4$

Model	Test	# of customers	# of Local Dep.	CPLEX MILP Solution				Savings over TSP (%)
				# of robots Used	Service Time	Computational Time	Gap (%)	
RADP-1	1	3	1	0	0.3166	0.22	0.00%	0.00
RADP-2	1	3	1	0	0.3166	0.31	0.00%	0.00
RADP-1	2	4	1	0	0.4065	2.01	0.00%	0.00
RADP-2	2	4	1	0	0.4065	1.17	0.00%	0.00
RADP-1	3	5	1	2	0.4115	15.55	0.00%	7.51
RADP-2	3	5	1	2	0.4115	53.52	0.00%	7.51
RADP-1	4	6	1	2	0.4368	525.47	0.00%	15.25
RADP-2	4	6	1	2	0.4568	3600	85.91%	11.37
RADP-1	5	7	1	2	0.4442	3600	9.17%	21.28
RADP-2	5	7	1	2	0.5114	3600	88.46%	9.37
						Average Savings		**7.23**

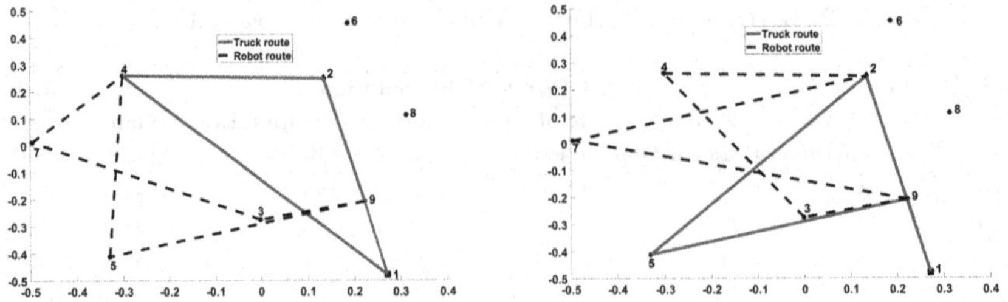

(a) RADP-1 Solution (b) RADP-2 Solution

Fig. 2. RADP Solutions of Test 5 in Table 2

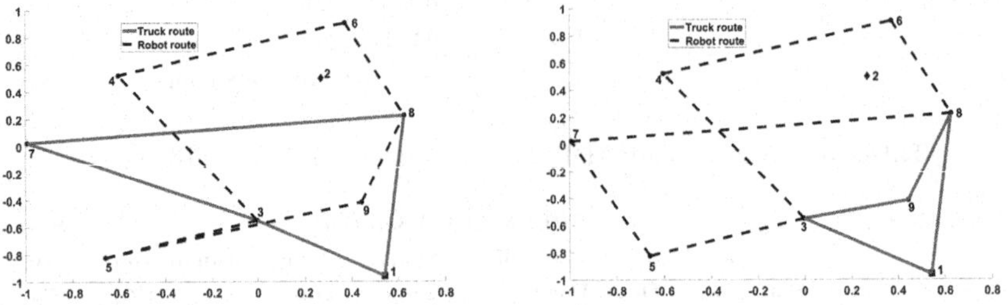

(a) RADP-1 Solution (b) RADP-2 Solution

Fig. 3. RADP Solutions of Test 5 in Table 3

5 Conclusion

From the numerical results presented and discussed above, it can be seen that the two models are consistent since they produced the same result when both of them reached the optimum solution. RADP-1 is easier to solve than RADP-2 in all Tests except Test 1 and sometimes Test 2. Computational difficulty of RADP-2 increases more than that of RADP-1 as the problem size increases because of the exponential growth in the number of feasible operations. The robots being very slow in speed compared to trucks, are more suitable in areas where customer densities is high. Although RADP-1 looks more complex than RADP-2 due to its larger number of variables and constraints, it is still better in terms of computational difficulty. The number of feasible operations used in RADP-2 grows exponentially and the computational time required to solve problems with practical size is impossible. There is potential for RADP-2 to be produced. This can be done by removing operations which are not likely to be part of the optimum solution. It involves ranking the operations based on "time-nodes ratio", that is dividing the total time of performing each operation by the number of nodes covered by the operation and selecting operations with smaller ratios.

References

1. Agatz, N., Bouman, P., Schmidt, M.: Optimization approaches for the traveling salesman problem with drone. Transp. Sci. **52**(4), 965–981 (2018)
2. Andrew, J.H.: Thousands of autonomous delivery robots are about to descend on us college campuses (2019). https://www.theverge.com/2019/8/20/20812184/starship-delivery-robot-expansion-college-campus. Accessed 08 Nov 2022
3. Boysen, N., Schwerdfeger, S., Weidinger, F.: Scheduling last-mile deliveries with truck-based autonomous robots. Eur. J. Oper. Res. **271**(3), 1085–1099 (2018)
4. Jennings, D., Figliozzi, M.: Study of sidewalk autonomous delivery robots and their potential impacts on freight efficiency and travel. Transp. Res. Rec. **2673**(6), 317–326 (2019)
5. Lieping, G.: Delivery robot hit the road in Beijing (2019). https://www.ecns.cn/news/sci-tech/2018-06-20/detail-ifyvmiee7350792.shtml. Accessed 08 Nov 2022
6. Simoni, M.D., Kutanoglu, E., Claudel, C.G.: Optimization and analysis of a robot-assisted last mile delivery system. Transp. Res. Part E: Logistics Transp. Rev. **142**, 102049 (2020)
7. Sonneberg, M.O., Leyerer, M., Kleinschmidt, A., Knigge, F., Breitner, M.H.: Autonomous unmanned ground vehicles for urban logistics: optimization of last mile delivery operations (2019)
8. Yu, S., Puchinger, J., Sun, S.: Van-based robot hybrid pickup and delivery routing problem. Eur. J. Oper. Res. **298**(3), 894–914 (2022)

Modeling and Analysis of Organizational Network Analysis Graphs Based on Employee Data

Abdel-Rahmen Korichi[1,2](✉), Hamamache Kheddouci[1], and Taha Tehseen[2]

[1] Université Claude Bernard Lyon 1 - LIRIS UMR 520, Bâtiment Nautibus 43, bd du 11 novembre 1918, 69622 Villeurbanne, France
[2] Panalyt Pte. Ltd., Singapore, Singapore
abdelrahmen.korichi@gmail.com
https://www.univ-lyon1.fr/, https://www.panalyt.com

Abstract. Business leaders have access to a new set of tools to monitor and improve employees' productivity and wellness as well as to retain talents: People or Human Resource Analytics. However, while People Analytics helps understanding employees at the individual level, it doesn't reflect the social interactions happening in the workplace. For that, you would rather look at Organizational Network Analysis to examine how people communicate across an organization and give a more accurate and realistic understanding of employees' interactions. In this paper, we propose a framework for organizations to supplement their People Analytics with Organizational Network Analysis. We address the different challenges associated to that and provide different case-studies based on real clients' data to demonstrate how Organizational Network Analysis can be used to solve practical business applications.

Keywords: Relational Analytics · Organizational Network Analytics · People Analytics · 4th Industrial Revolution

1 Introduction

The Human Resource (HR) industry is experiencing a revolution led by the profusion of employee data and new possibilities to take data-driven decisions and discover new insights about how organizations function [1]. Business leaders are looking for ways to back up their hypotheses, understand better their employees, manage them better, and improve internal processes. They want to have real-time access to their company's turnover - where it happens and why? They want to know their gender pay gap, see how diverse their organization is, retain their talent, and more. This is called People Analytics (PA) or Human Resource Analytics: the collection and application of talent data to improve critical talent and business outcomes [2]. As a matter of fact, more than 70% of companies now say that they consider PA to be a high priority [3]. PA focuses mostly on employee attribute data, of which there are two kinds:

B. Dorronsoro et al. (Eds.): OLA 2023, CCIS 1824, pp. 354–367, 2023.
https://doi.org/10.1007/978-3-031-34020-8_27

- Static attributes (or traits): facts about individuals that don't change, such as ethnicity, gender, hire date, etc.
- Dynamic attributes (or state): facts about individuals that do change, such as age, company tenure, compensation, etc.

PA is great to check that numbers are in order, benchmark data across different teams, and set Key Performance Indicators (KPIs) for managers. A limitation of PA, as Paul Leonardi and Noshir Contractor argue [4], is that it limits itself to individuals attributes and neglect the interdependence between employees. Indeed, what drives an organization or a team performance is not only the individual attributes, but also the interplay among people and how it evolves over time. Organizational Network Analytics (ONA), also called Relational Analytics, addresses that and captures communication between people to give insights about the nature and the quality of their relationships. It makes it possible to evaluate a collaboration in the context of the surrounding network, which is a critical criterion for success in any collaborative project.

In this paper, we present a framework for not only exploring organizations' key information based on employees' static attributes as we see in standard HR analytics systems, but we will also present how to factor in communication metadata and unlock more insights. At Panalyt [5], we created a Software as a Service (SaaS) tool where we plug in our clients' data and allow them to perform PA as well as ONA using our suite of pre-built dashboards which cover the end-to-end employee lifecycle. Based on our experience building this product and interacting with dozens of clients, we have identified multiple challenges. We will present the main ones in the following sections, and how we overcame them. Later, we will see how the association of PA and ONA can help to deduce transformative insights by performing queries on a partial subgraph. Finally, we will share advanced real case-studies from our client base.

2 Challenges

2.1 Ethical Concerns

The accessibility of employee communication data raises important ethical concerns. Beyond legal considerations, employers should be mindful of the ethical standards that they adhere to while utilizing this information. One study estimated that 81% of people analytics projects are jeopardized by ethics and privacy concerns [7]. At Panalyt, we purposefully refuse to ingest and analyse the content of the communication between employees, due to the sensitivity of the information. According to a recent study, transparency was identified as being one of the most critical considerations for PA projects [8]. The use of aggregated, non-identifiable data is recommended where possible, to demonstrate to employees that the purpose behind PA projects is to capture larger organizational trends.

Organizations should communicate their reasons for pursuing PA projects and the kind of benefits employees should expect from them. We also recommend establishing clear governance around data collection, access and storage, consent, and anonymity.

2.2 Data and Storage

To perform our analysis, we need data of 2 kinds: people data and communication metadata. People data is collected from HR systems, such as human resources information systems (HRIS), payroll tools, absence management tools, performance management tools, and recruitment tools. In many cases, organizations use Software as a service (SaaS) tools to store their data, and most of them would have API connections. In other cases, the data is stored in Microsoft Excel or CSV documents. On the other hand, communication metadata can be retrieved from chat/email tools. Over the last few years, online communication and management tools such as Microsoft Teams/Outlook, Slack, Gmail and others have started providing API access for their customers, making it easier for organizations to collect communication metadata.

Having the experience of working with many different clients, we have a good overview of how people data and communication data is stored, and how they can be retrieved, and used to perform analysis. At Panalyt, after being collected, the data is stored, transformed and queried using SQL, a structured query language that allows fast retrieval and transformation of data.

We recommend to versionize the data, at least monthly, before storing it in one or multiple tables, to make it easier to query snapshots of the data for any particular month and perform join operations across tables. If a record is updated for an employee during a month, we forward propagate the change to the future versionized rows, as we can see in Table 1. We then make query on a partial subgraph to visualize a defined population of the network, where nodes are employees (or their external contact) and where edges represent an information about their interactions. We chose to aggregate the data by month as we realize this granularity is large enough deducing trends, and small enough to notice significant changes quickly in the scale that companies operate in.

Table 1. Employee table

empID	*versionId*	*companyemail*	*department*	*salary*	*JobTitle*	...
0001	30/09/2022	emp1@email.com	Technology	$6,000	Data Engineer	
0001	31/10/2022	emp1@email.com	Technology	$6,000	Data Engineer	
0001	30/11/2022	emp1@email.com	Technology	$6,000	Data Engineer	
0001	31/12/2022	emp1@email.com	Technology	$7,000	Sr Data Engineer	
0002	31/08/2022	emp2@email.com	Marketing	$4000	UI Designer	
0002	30/09/2022	emp2@email.com	Marketing	$4,5000	UI Designer	
...						

Regarding communication data, the main fields that we retrieve are:

- *messageId*: unique identifier of the message
- *sendingFrom*: sender's email

- *sendingTo*: receiver's email
- *sentDatetime*: time when the message was sent
- *threadId*: thread/conversation the message belongs to
- *channelId*: channel where the message has been sent - only for chat tools

We process the data to get one line per sender and receiver pair for each month. We then compute various aggregated metrics, from person 1 to person 2 and person 2 to person 1, including, but not limited to:

- count of messages sent: The volume of messages sent by either or both correspondents.
- average response time: The time taken for a correspondent to reply to the other correspondent.
- response rate: The percentage of messages received by either correspondent that they replied to.
- the reciprocity: the ratio of the volume of communication between a correspondent and another.
- average time first/last message sent: The average time when a correspondent sends his first/last message to another person.
- average time first/last message received: The average time when a correspondent receives his first/last message to another person - emails only.
- channels in common (chat tools only): The number of channels or groups where both correspondents are active (i.e have both send more than X number of messages during that month).

We then create a graph based on the analysis. This could be a direct graph, where arrows point from the sender to the recipient, or an undirected graph. Although some authors propose to use the number of e-mails sent by a node i to a node j divided by the total number of e-mails sent by member i [9], and others suggests to use the geometric mean of sent-received counts, [10], there is no unique way to generate the attributes to connect people in an ONA graph. It really depends on the use case.

At Panalyt, we use simple metrics such as those described above as well as more complex ones that we can't disclose as they are confidential. We have also created a score that includes the information of some of the metrics defined above, especially the count of messages sent, the average response time, the response rate, and the reciprocity. In general, our recommendation is to define a metric based on the use case. For example, if we want to highlight employees and managers who have an asymmetric collaboration, we should query the pairs of employees where the number of direct messages sent one way is at least X times greater than the other way.

For the purpose of this study, we use a metric called *collaboration score* which is defined as follows: For each version (month), we count the number of groups/channels where both employees are active. We define being active by sending at least 5 message in the same group/channel and the same month. For instance, if Mizuki and Hong have both sent more than 5 messages in the channels "Client A" and "Engineers", they would have a collaboration score equal to 2.

2.3 Dashboards

In order to efficiently use network data to understand their teams, managers need access to interactive and easy to use dashboards. Most companies rely on data analysts to get insights related to talent and performance management using BI tools such as Tableau or Power BI. That often creates a bottleneck for multiple reasons:

1. It requires a dedicated data team and decision makers to help them put the data into context, which is very costly.
2. There are not enough data analysts to address all management queries in a timely manner.
3. Employee data is sensitive, and business leaders often hesitate to share this information with regular employees. The BI tools also just do not have a permission system granular enough to give appropriate permissions to users.

Because of the complexity behind obtaining the data, as well as the technical resources and cost needed to process large amounts of sensitive data, it is difficult to build such a solution in-house for most companies. We had cases where clients told us that they liked our product but wanted to build an in-house solution. Many came back more than a year later with no progress. The resources and amount of work required to build such a tool is often underestimated.

We have spent more than 4 years building Panalyt's product at the time of writing this paper. We have a product that efficiently extracts, processes and stores people data and communication data every day and displays interactive visualizations in pre-built dashboards where data from different sources can interact with each other.

We also have a permissions system that allows an admin to give access a user to specific data, aggregated or not, based on the line hierarchy of the user or on defined attributes (e.g. a manager can be limited to see aggregated or individual's values only for the team he manages, or for a specific department/location/etc.)

In our application we use d3.js [6] for our network data visualizations. The library takes in source-target pairs to create nodes and the edges between them. In addition to that, we can also add other attributes like colour to each node based on features like tenure, organization, function, etc.

Here are some of the salient features of the ONA dashboard:

- Colouration: By default, nodes are coloured based on the department of the employee they belong to. In Fig. 1a, the nodes belong to different departments - Sales and Marketing. The nodes in Fig. 1b belong to Engineering.
- Layering: Each node can also be recoloured based on user-selected properties. For instance, in Fig. 1c, the nodes are not coloured based on departments - yellow is applied to all employees that are working part-time and red to all employees with less than one year of tenure. There is one employee that is both part time and has less than one year of tenure. On the other hand, there are three employees that have neither of those properties (in blue).

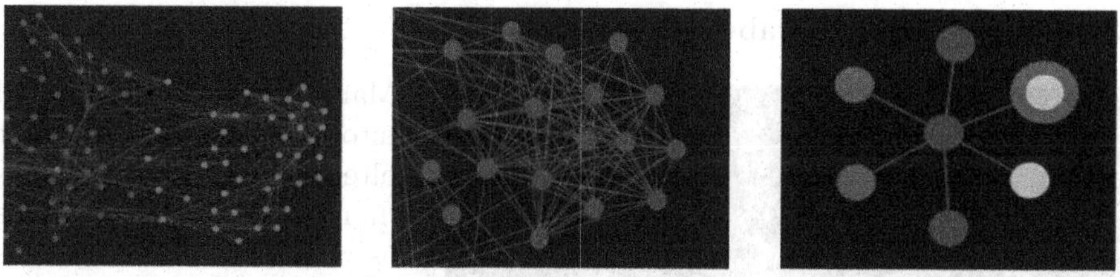

(a) Nodes from two different teams coloured based on their teams.

(b) Nodes from the same team connected by an attribute

(c) Nodes layered with different properties of the data.

Fig. 1. Different types of visualizations from Panalyt platfrom.

- A-B Filtering: The graph can itself be filtered based on employee data attributes. For instance, in Fig. 1a we have filtered the data such that both the senders (A) and receivers (B) are from two departments only.

2.4 Datasets

For the upcoming analyses, we will be using 3 datasets from 3 different clients:

- Client 1: the Slack communication data of 486 internal users of a Japanese company from January 2019 to March 2022. This data includes their public and private channels, but does not include any direct messages. We do not include direct messages as their company's Slack subscription does not allow access to direct messages for extraction due to privacy reasons.
- Client 2: the Slack communication data of 845 internal users of an Australian company from August 2022 to December 2022. Again, the data does not include any direct messages for the same reasons.
- Client 3: we will share a case study with Panasonic Operational Excellence Co., Ltd. that partnered with Panalyt in March 2022 and applied ONA to identify organizational issues. The size of the dataset is confidential.

3 Basic Analysis

We can deduce important information about an organization using its communication data. However, on its own the network data can be imposing and unwieldy. It is when we add attributes about the employees (e.g. their department, age, grade, etc.) and make queries on a partial subgraph that we can get a deeper and more precise understanding of the organization.

3.1 Inter-Team Collaboration

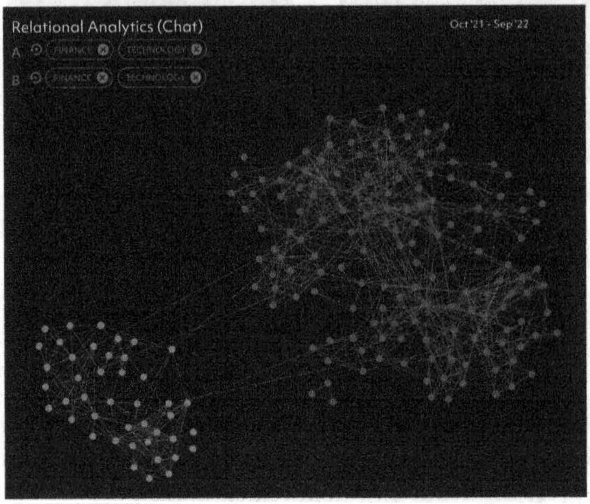

Fig. 2. Collaboration between two teams

Many companies are divided into multiple teams that often work together. Communication within the teams as well as across teams is very important for good and timely results. It is difficult to ascertain the quality of communication between two teams except through the use of surveys or interviews with managers. Not only can they be difficult to arrange, it is not feasible to have every member of the team take a survey and companies often have to defer to managers for their perspective on the communication level between teams. Even if managers are unbiased in their responses, surveys can be prone to human errors of judgement as we will see later in a case-study. A faster, more accurate, and reproducible way to judge communication health between two teams is through the use of their ONA data.

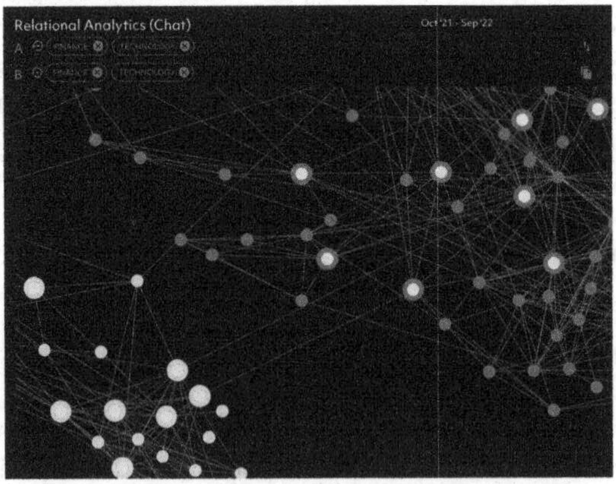

Fig. 3. Inter-Team Collaboration with Manager layer

In Fig. 2 (client 2), we look at ONA data for the Finance and Technology teams (by applying the A-B filters to those teams only) and we see that while there is an abundance of connections within each team, the flow of communication between the two teams is limited to only 5 channels, i.e., there are 5 pairs of employees that are responsible for the communication between the two teams. This can be crucial information for the organization. If the two teams are supposed to be mostly independent of each other, then this communication model will work fine. However, in a technology company, where the product is the technology itself, the communication between the two teams is rather slim. It would make sense to open up more channels of communication to prevent a backlog of information. Moreover, if we use the layering function to add a layer for managers(yellow), we uncover more detail on the health of communication between the two teams. We see in Fig. 3 that even among the five channels, only one of them is between managers. This means that there are only two decision makers between the two teams that are connected. This can be an additional concern.

3.2 Better Onboarding of New Employees

Many companies spend a lot of resources on getting the best talents. The cost of hiring does not end with a signed contract - rather, it continues for the next few months with companies spending money and months on onboarding the new hires. At this stage, it is important to make sure that the new hires are sufficiently connected to their team members to have a smooth onboarding process. They need people to teach them the team's business practices and the tools they use. In addition, they also need to be able to get along with their teammates and get to know their working styles.

Managers can make use of ONA data to pinpoint new hires that might not be getting an adequate onboarding experience or who seem to be left out. They can do that by looking at their communication within the team, and see if they are connected to their manager and team members. As an example, we can use A-B filters to restrict Client 2's data to the Finance team and use layering to pinpoint the employees with less than six months of tenure. The resultant graph is displayed in Fig. 4. We can see that among the five new hires for the Finance team, three seem to be adequately integrated. They have multiple connections/collaborations with team members. However, we also see two new hires that are only in communication with one other team mate. This can be worth looking into as these two hires might not be getting adequate resources to transition into their roles and this can prolong the time required to get them fully on board.

3.3 Gender Diversity in Teams

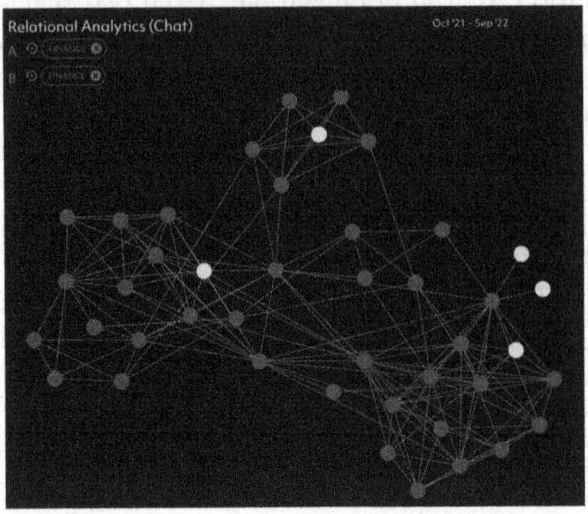

Fig. 4. New Hires Communications

Companies strive for an inclusive work culture where employees from various genders and demographics are not only represented but also included readily in conversations. Foma [11] details the benefits of workplace diversity such as plurality of ideas owing to diverse experiences and the likelier retention of employees. But they also note that there can be significant communication gaps between the different groups. We see that companies already spend time and effort on streamlining their hiring practices to make them more inclusive. For instance, there has been a sustained effort, spanning many decades, to include more women in the workplace, especially in the higher positions and male-dominated sectors like technology and finance.

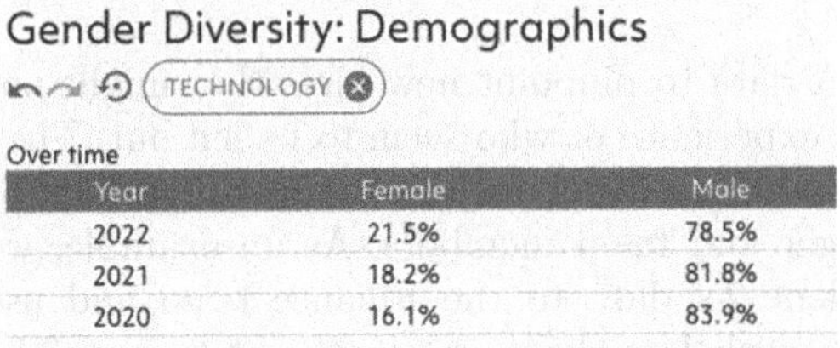

Year	Female	Male
2022	21.5%	78.5%
2021	18.2%	81.8%
2020	16.1%	83.9%

Fig. 5. Gender Diversity chart

Looking at client 2, we focused on looking at the gender diversity in the Technology department. From our Panalyt dashboard, we can see that for a particular department the female headcount has been steadily rising by more than 13% per year to reach 21.5% of women from January 2020 to November 2022 (Fig. 5). However, when we look at the communication graph (Fig. 6) for the same department, we found that many females (in red) cluster together, and there are entire subsections of the graph where there are no collaborations with females - like in the bottom right, with just male employees communicating with each other. Gaps like this in communication between demographics can impede the company from fully reaping the benefits of diversity. Our clients also benefit from similar use cases by studying how populations with different attributes (ages, tenure, etc.) work together.

4 Advanced Case Studies

4.1 Attrition vs Amplitude of Communication

For client 1, to understand how to help them promote a healthier company culture and retain talents, we studied their historical communication data to under-

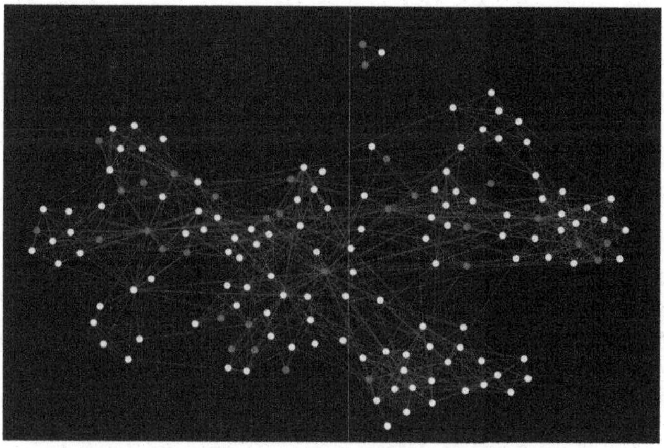

Fig. 6. Network Graph layered by gender

stand the impact of the amplitude of communication in attrition. We started by looking at the count of messages sent by active employees by month (Fig. 7a). We then averaged the results for all active employees by month. The average time of the first message sent was roughly between 10am and 11am, while the average time of the last message sent was between 6pm and 7pm. After that, we focused our attention on the employees who quit (37 terminated employees with at least 12 months tenure) to see if we could find any pattern in their amplitude of work in the last 15 months before they left. We uncovered from Fig. 7b and Fig. 7c that on average before employees quit, they start communicating earlier, and send their last message later, until they reach a point around 5 months before termination where the trend starts to reverse. In Fig. 7d, we looked at the total amplitude of communication (average time of last message sent - average time of first message sent). As expected, we saw a similar pattern: the amplitude of communication keeps increasing until it reaches a pic and start decreasing again. Finally, in Fig. 7e, we wanted to make sure that this trend was observable in comparison to their peers. We benchmarked the average first/last time of messages sent by terminated employees, as well as the amplitude, against the averages for the active population (380+ active employees) for that month for each month leading up to their eventual termination date.

We found again a pattern of employees that start communicating earlier and finish later compared to the average employee, especially between 5 to 7 months before their termination where it peaks to 40 min more on average compared to the active employees.

Our conclusions, based on those findings, was that the amplitude of communication is positively correlated to the overwork of employees and factors greatly in the attrition. Interestingly, employees gradually disengage from 5 to 6 months before they quit. This study is a good example that shows that it is possible to uncover insightful information based on aggregated, non-identifiable data.

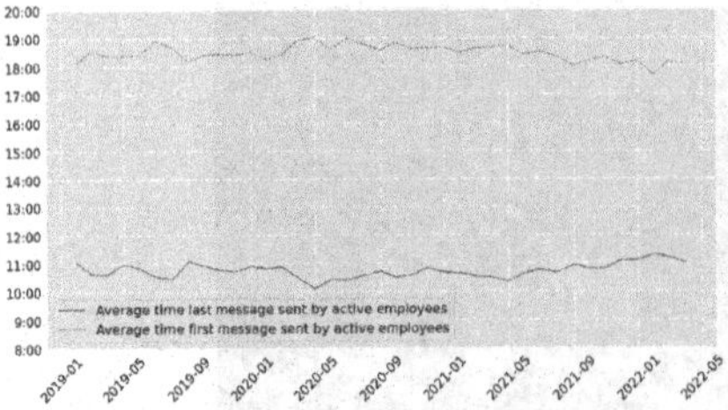

(a) Average time of the day for the first and last message sent by each employee across the entire population for each month.

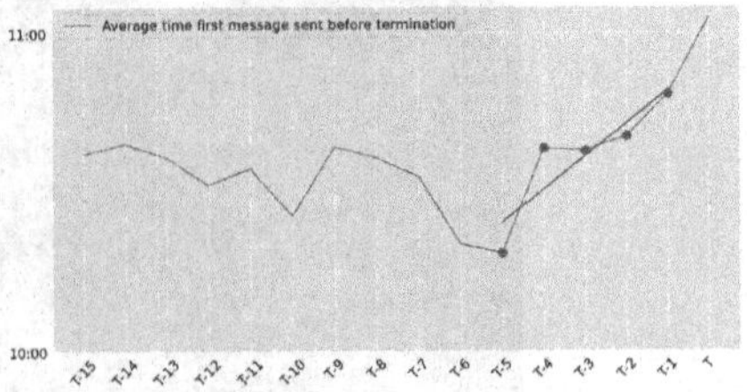

(b) Average time of the day for the first message sent by each terminated employee for each month leading up to their eventual termination date.

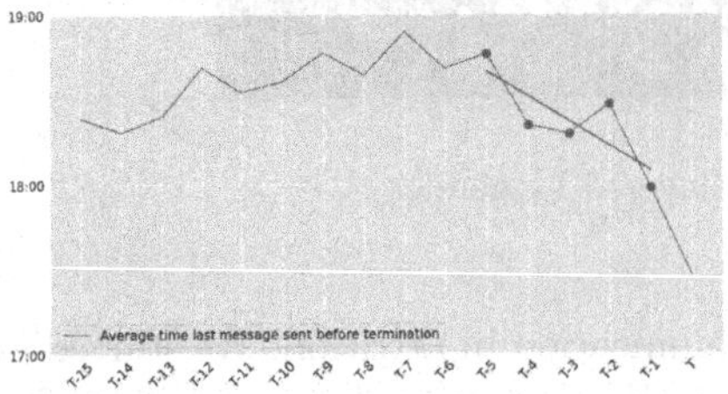

(c) Average time of the day for the last message sent by each terminated employee for each month leading up to their eventual termination date.

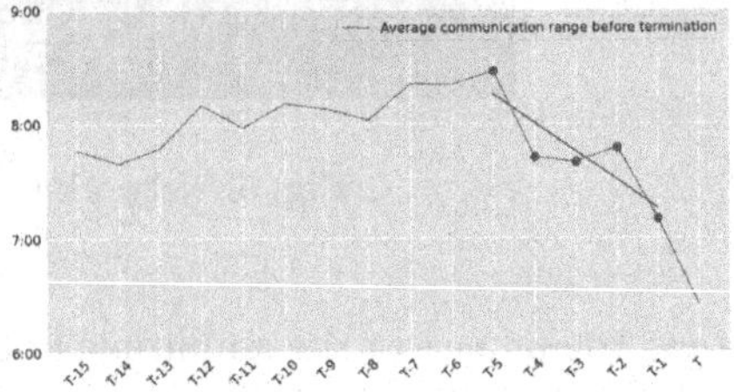

(d) Average total hours of communication for each terminated employee for each month leading up to their eventual termination date.

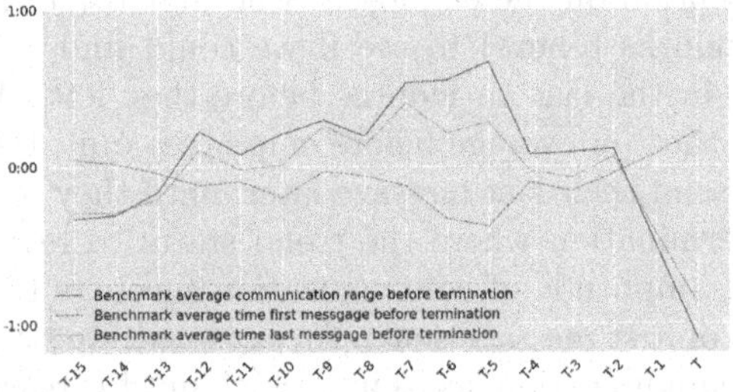

(e) Average time of the day for first and last message as well as total hours of communication for terminated employee benchmarked against the averages for the active population for that month for each month leading up to their eventual termination date.

Fig. 7. Analysis of communication times

Companies can then use it to create new policies and limit overwork by discouraging communication after a certain hour when possible, and to identify groups, departments or talents where employees tend to overwork and disincentivize it.

4.2 Improving Manager Effectiveness and Innovation

Our second case study has been done with Panasonic Operational Excellence Co., Ltd., that partnered with Panalyt in March 2022 and applied ONA to identify organizational issues that could not be discovered through conventional methods such as surveys. The detailed case-study is accessible online [12]. Following a merger and a group restructuring, their goal was to further strengthen collaboration between different departments that used to historically work independently. Initially, the team had introduced a pulse survey to understand and improve the collaboration between employees with a relatively short tenure and their managers, as well as their motivation. They quickly found that pulse surveys alone were not effective in identifying from where the communication issues were coming from, presumably because the survey answers are subjective and that bias could not be completely eliminated.

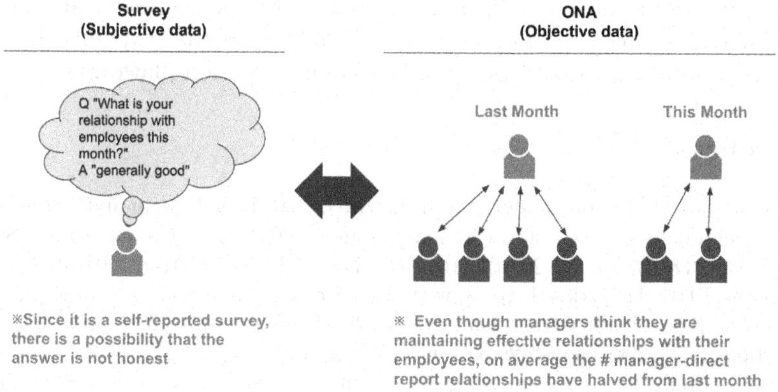

Fig. 8. Surveys vs ONA

Some of their initial assumptions were that "an employee with low motivation in a pulse survey may not be able to interact with other employees and may be isolated." and that "employees who interact with many other employees have high motivation." After analyzing their ONA data with Panalyt, they observed that some employees were "isolated but maintain high motivation", or had "low motivation despite being in the center of the organization and interacting with many other employees including those from other departments". The initial assumptions were therefore proven to be false. Another assumption was that the higher the collaboration between an employee and his manager, the more likely the employee would meet expectations of their roles and responsibilities. However, they realized that managers were only able to have a grasp on their relationships with their direct reports; but they were unaware of the work their team members were doing in collaboration with members from other departments - making it impossible to notice signs of collaboration overload that can lead to burnout unless the employees speak out themselves (Fig. 8). This was especially true at the time of the imposed working from home period during the COVID-19 pandemic.

5 Discussions and Conclusion

People analytics has introduced new ways to make evidence-based decisions to improve organizations' processes. But most companies have been focused on studying the attributes of their employees, forgetting the reality of interdependence. As collaboration platforms such as Slack, Teams and others become prominent and are increasingly used in virtual teams, organizations can now better understand what drives group or organizational performance. This is confirmed by Keith G. Provan when he says: "only by examining the whole network can we understand such issues as how networks evolve, how they are governed, and, ultimately, how collective outcomes might be generated" [13].

In future work, we would like to develop ways to take communication metrics from different sources and formulate a combined score that explains the overall communication behaviour across multiple communication platforms.

References

1. DiClaudio, M.: People analytics and the rise of HR: how data, analytics and emerging technology can transform human resources (HR) into a profit center. Strateg. HR Rev. **18**(2), 42–46 (2019). https://doi.org/10.1108/SHR-11-2018-0096
2. Gartner: HR Analytics. https://www.gartner.com/en/human-resources/glossary/hr-analytics#:~:text=HR%20analytics%20(also%20known%20as,and%20promote%20positive%20employe%20experience)
3. People analytics: Recalculating the route. https://www2.deloitte.com/us/en/insights/focus/human-capital-trends/2017/people-analytics-in-hr.html. Accessed 3 Dec 2022
4. Better People Analytics. https://hbr.org/2018/11/better-people-analytics. Accessed 3 Dec 2022
5. Panalyt. https://panalyt.com/. Accessed 4 Dec 2022
6. D3.js. https://d3-graph-gallery.com/network.html. Accessed 4 Dec 2022
7. Data ethics: 6 steps for ethically sound people analytics, Visier. https://www.visier.com/blog/six-steps-ethically-sound-people-analytics/
8. Tursunbayeva, A., Pagliari, C., Di Lauro, S., Antonelli, G.: The ethics of people analytics: risks, opportunities and recommendations. Pers. Rev. **51**(3), 900–921 (2021). https://doi.org/10.1108/PR-12-2019-0680
9. Michalski, R., Palus, S., Kazienko, P.: Matching organizational structure and social network extracted from email communication. In: Abramowicz, W. (ed.) BIS 2011. LNBIP, vol. 87, pp. 197–206. Springer, Heidelberg (2011). https://doi.org/10.1007/978-3-642-21863-7_17
10. De Choudhury, M., Mason, W.A., Hofman, J.M., Watts, D.J.: Inferring relevant social networks from interpersonal communication. In: Proceedings of the 19th International Conference on World Wide Web (WWW 2010), pp. 301–310. Association for Computing Machinery, New York, NY, USA (2010). https://doi.org/10.1145/1772690.1772722

11. Foma, E.: Impact of workplace diversity. Rev. Integr. Bus. Econ. Res. **3**(1), 382 (2014)
12. How Panasonic is Improving Manager Effectiveness and Innovation with Organizational Network Analysis. https://panalyt.com/case/panasonic-ona/#panalyt
13. Provan, K.G., Fish, A., Sydow, J.: Interorganizational networks at the network level: a review of the empirical literature on whole networks. J. Manag. **33**(3), 479–516 (2007)

Time Series Forecasting for Parking Occupancy: Case Study of Malaga and Birmingham Cities

José Ángel Morell[1,2(✉)], Zakaria Abdelmoiz Dahi[1,2], Francisco Chicano[1,2], Gabriel Luque[1,2], and Enrique Alba[1,2]

[1] ITIS Software, University of Malaga, Malaga, Spain
{jamorell,zadahi,chicano,gabriel,eat}@lcc.uma.es
[2] Department of Fundamental Computer Science and Its Applications, Faculty of NTIC, University of Constantine 2, El Khroub, Algeria
zakaria@univ-constantine2.dz

Abstract. The smart city concept refers principally to employing technology to deal with different problems surrounding the city and the citizens. Urban mobility is one of the most challenging aspects considering the logistical complexity as well as the ecological relapses. More specifically, parking is a daily tedious task that citizens confront especially considering the large number of vehicles compared to the limited parking, the rush hours peaks, etc. Forecasting parking occupancy might allow citizens to plan their parking better and therefore enhance their mobility. Time-series forecasting methods have proved their efficiency for such tasks, and this work goes in the same line by exploring how to provide more accurate parking occupancy forecasting. Concretely, its contributions stand in a complete pipeline, including (I) the automatic extra-transform-load data module and (II) the time-series forecasting methods themselves, where four have been studied: one additive regression model (Prophet), the Seasonal Auto-Regressive Integrated Moving Average (SARIMAX), and two deep learning models, the Long Short Term memory neural networks (LSTM), and Neural Prophet. Experiments have been performed on data of 3 and 28 parking from the city of Malaga (Spain) and Birmingham (England) using data recorded through 6 months (June-November 2022) and two and a half months (October-December, 2016), for Malaga and Birmingham, respectively. The results showed that Prophet provided very competitive results compared to the literature.

Keywords: Deep Learning · Time Series Forecasting · Smart Cities

1 Introduction

The world's technological and societal evolution has substantially increased cities' size and needs. Statistics have shown that 56% of the world's popula-

B. Dorronsoro et al. (Eds.): OLA 2023, CCIS 1824, pp. 368–379, 2023.
https://doi.org/10.1007/978-3-031-34020-8_28

tion (around 4.4. billion) are living in the cities today and 8.8 billion by 2050[1]). This increases the complexity of city-life, whether from the citizen's or the city manager's point of view. This makes maintaining the same quality of life harder, especially since the logistics requirements are way more difficult to handle. Considering these facts, the smart city concept appeared to encompass the use of cutting-edge technologies to enhance the city-life at several levels (e.g. education, mobility, etc. [5]). The citizens' mobility is increasingly attracting the interest and efforts of academia and industry. The complexity of mobility emerges considering factors such as the number of vehicles, their type, the restricted transportation paths, the economic and ecological relapses, etc. Previous works have investigated how to enhance mobility from a moving-vehicle point of view (e.g. by computing the shortest path, optimising the traffic lights [6], etc.). However, enhancing the mobility of vehicles does not only depend on moving vehicles but also on stationary ones. Actually, parking [2] turns out to be quite problematic considering factors such as parking vs vehicles imbalance ([2]). In this case, the parking quality depends on how two questions are answered: (Q_1) *When* and (Q_2) *Where to Park?*.

To answer the above-stated questions, the idea consists in giving the citizens a step ahead to plan their parking better. Technically, this will be done by forecasting parking occupancy at a given moment. Bearing in mind this, times-series techniques have already demonstrated their efficiency for forecasting parking tasks [4,10]. Thus, this work goes along this research axis by building a complete parking-occupancy forecasting pipeline composed of (I) the Extraction, Transformation and Loading (ETL) as well as (II) the forecasting module based on times-series forecasting methods including one additive regression model (Prophet) [12], the Seasonal Auto-Regressive Integrated Moving Average (SARIMAX) [3], and two deep learning models, the Long Short Term Memory neural networks (LSTM) [7], and Neural Prophet [13]. The study has been done using real-life data of 3 parking in the city of Malaga (Spain) as well as 28 parking in the city of Birmingham (UK) using data recorded throughout 6 months (from 01/06/2022 to 30/11/2022) and two and a half months (from 04/10/2016 to 19/12/2016) for the city of Malaga and Birmingham [10], respectively.

The remainder of the paper is structured as follows. In Sect. 2, both time-series and parking forecasting concepts are depicted. Then, in Sect. 3, the proposed approach is introduced, while Sect. 4 presents the experimental analysis as well as their discussion. Finally, Sect. 5 concludes the paper.

2 Basic Background

This section gives an overview of times-series forecasting and parking occupancy.

[1] World Cities' Statistics: https://ourworldindata.org/urbanization.

[2] Vehicles Statistics: https://en.wikipedia.org/wiki/List_of_countries_by_vehicles_per_capita.

2.1 Techniques for Time Series Forecasting

Let $Y = \{y_1, \ldots, y_n\}$ denote times series data. Forecasting consists in estimating the next values y_{n+h} of Y, where the variable h denotes the forecasting horizon. Two families of forecasting methods exist: parametric (e.g. statistical methods) and non-parametric (e.g. machine learning) ones [9]. Unlike machine learning models, statistical methods require prior knowledge of the data distribution to build the predictive models, which makes the model depend on some hyper-parameters needed to optimise the prediction's results. Depending on the data characteristics, a given forecasting method might be more suitable than others. As an instance of these characteristics, one can cite the stationary nature of data (i.e. mean and variance constant over time), including the presence/absence of trends (i.e. a constant inc/decreasing behaviour) and seasonality (i.e. cyclic patterns). In this work, both statistical models such as SARIMAX and Prophet as well as machine learning models like LSTM and Neural Prophet are being researched to faithfully represent the plethora of state-of-the-art time-series prediction techniques. Section 3 will give more in-depth of the studied techniques.

2.2 Parking Occupancy Forecasting

Some previous works have studied some parking-related aspects, such as car parking occupancy detection [2], but still, parking occupancy forecasting has been of keen interest in both academia and industry [4,10,11]. In [4,10], the authors studied the parking data of the city of Birmingham (UK), while in [11], the cities of Birmingham, Glasgow, Norfolk, and Nottingham (UK) were researched. In both [10,11], the forecasting has been made using several techniques such as polynomial fitting, Fourier series, k-means clustering, and time series predictors, while in [4], deep learning techniques have been used. Therefore, this work takes a further step towards more advanced techniques (e.g. LSTM, Prophet, Neural Prophet and SARIMAX) to attain more accurate parking occupancy forecasting. Also, this work goes beyond current literature in terms of the used data by studying more fresh and never-used-before data representing 6 months of parking records in the city of Malaga (Spain). These data represent different time laps and complexities since they include parking behaviours in seasonal periods (e.g. summer and winter), and different societal influences (e.g. beginning of the academic year, end of summer and beginning of winter holidays, the end of the natural year, etc.) which provide more diversity in the study.

3 The Proposed Approach

This section presents the contributions of this work including: (I) the ETL module as well as (II) the time-series forecasting module (see Fig. 1).

3.1 The Extract-Transform-Load Module

This section presents the ETL components where the data being extracted is the one of Malaga, while those of Birmingham were proposed in [4,10]. The complete source-code of the devised approach is available in [1].

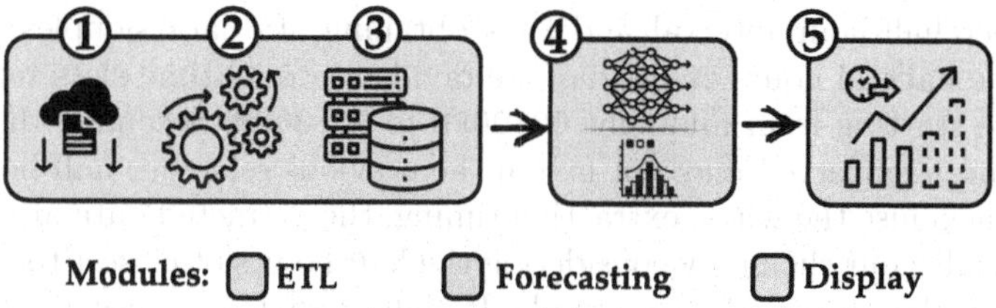

Fig. 1. The workflow of the proposed forecasting pipeline

Extraction. The extraction module uses the API provided by Malaga city hall ([3]) to extract, every 15 min, real-life parking occupancy data of 13 public parking in the city of Malaga with are updated every 60 s: Salitre, Cervantes, El Palo, Av. de Andalucía, Camas, Cruz De Humilladero, Alcazaba, San Juan De La Cruz, Pz. de la Marina, Tejón y Rodriguez, El Carmen, Mármoles, El Limonar and Bailén (see Fig. 2). The pseudo-code of the extraction module devised in this work is made publicly available for further use by the community at [1]. It is also worth stating that some historical data of the same type that are made publicly available at ([4]) have been considered as well in this work.

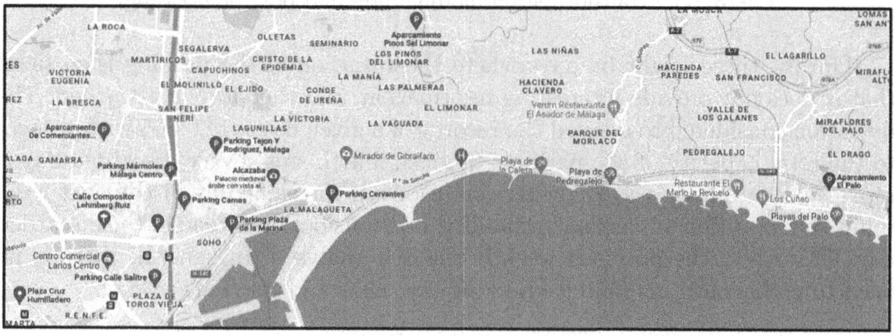

Fig. 2. Geographic localisation of the studied parking

Transformation. The transformation module starts by converting, to CSV files, the raw data downloaded by the extraction module, where the header (i.e. the name of the columns) is similar to the one used in other open access repositories. This aims to ease the generalisation of the ETL and comparison against previous works [4,10].

The transformation includes data pre- and post-processing (see Fig. 3). The pre-processing stands in dealing with missing or incoherent information (e.g. HTTPS requests' failure returning incomplete/empty data). First, the data is

[3] Malaga City Hall API: https://datosabiertos.malaga.eu/dataset/ocupacion-aparcamientos-publicos-municipales.

[4] Malaga parking history data: https://github.com/javisenberg/malaga-parking-data.

divided per half-hour interval and thus obtaining 48 data samples/day. Concretely, the natural hours of the day are divided into 48 time slots representing half-hours, starting from midnight `00:00PM` to `23:30PM`. Secondly, the time the data has been extracted may not match the previous reference half-hour slicing. So, to homogenise the whole extraction timing, the extracted data are fitted into reference half-hour slicing by considering the half-hour slot closest to the extraction time of the processed data sample. If multiple data samples are assigned to the same half-hour slot, one is chosen randomly. Some intervals may not have assigned data samples at all. These orphans' half-hour slots will be assigned data sample closer to them in terms of "extraction timing".

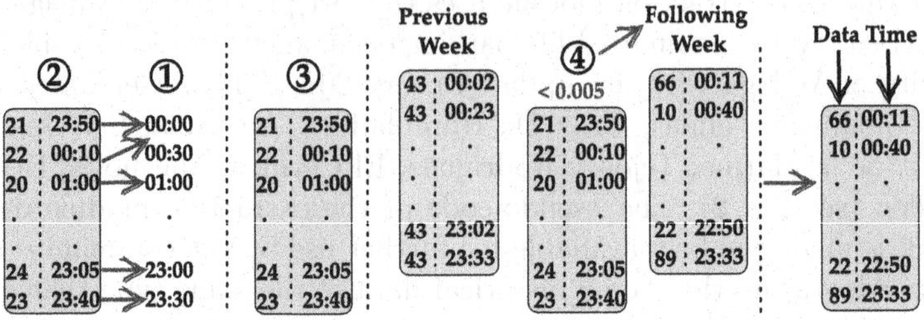

Fig. 3. The transformation module: pre- and post-preocessing

Thirdly, there might be days where the variation of the parking is so small that it does not provide interesting information, and therefore is discarded. The deviation considered to discard the records of a given day is of 0.005 with respect to the total capacity of the parking. So, as a fourth step, every discarded day is replaced by applying the following conditions: if the same day of the next week fulfils the condition of parking variability, it is considered instead of the discarded day. Otherwise, the previous week is the one analysed. In general, there is no need to go beyond this double-checking (i.e. next and previous weeks).

Loading. Once complete and homogenised data are obtained, the loading module stores the test and training data in separate files according to whether it is dedicated to testing using one to four last weeks of the whole data samples. The training and test data are stored in `CSV` files with the following appellation to ease their further use: `"train/test_parking_name_weeks_number"`. Also, for further use and comparison with this work, Malaga's parking occupancy data are constantly being extracted and stored every 24 h at [1].

3.2 The Times-Series Forecasting Methods

This section explains the time-series forecasting methods studied in this work.

The SARIMAX Model. SARIMAX model [3] is an extension of the SARIMA model, which by its turn is an extension of the ARIMA model. It combines three parts that capture the stationary, seasonal and exogenous variables (i.e.

external data). Concretely, it is defined by Eq. (1) that includes Auto-Regressive (AR(p)), Integration (I) and Moving Average (MA(q)) terms, where p, d and q represents the nonseasonal variables, P, D, Q, S for the seasonal part and r for the exogenous variable. More specifically, the variable p is the number of previous data points used to predict the current data point, d is the number of times a differencing transformation is applied to the times series until it becomes stationary before applying the AR-MA(p,q), and q is the number of past forecast errors used to predict the current data point. The same applies to the parameters P, D, and Q for seasonality if present. Next, S represents the duration of the repetitive pattern. The parameters θ, ϕ, α, β, and γ represent the parameters of the model, y_t and ϵ_t represent the time series value as well as the error at an instant t. The variables y, y', and y'' represent the original time series data, the transformed one after applying the differencing using non-seasonal and seasonal difference operator degree d and D, respectively (see Eqs. (2) and (3)).

$$y_t = c + \sum_{i=1}^{p} \theta_i y'_{t-i} + \sum_{i=1}^{q} \phi_i \epsilon_{t-i} + \sum_{i=1}^{r} \alpha_i x_{(i,t)} + \sum_{i=1}^{P} \beta_i y''_{t-iS} + \sum_{i=1}^{Q} \gamma_i \epsilon_{t-iS} + \epsilon_t \quad (1)$$

$$y'_t = \Delta(\Delta^{d-1} y_t), \Delta y_t = y_t - y_{t-1} \ (2) \qquad y''_t = \Delta(\Delta^{D-1} y_t), \Delta y_t = y_t - y_{t-1} \quad (3)$$

The Prophet Model. The Prophet model [12], developed by Facebook, is an additive regression model which extends the basic AR models. It specialises in handling time series with cyclical patterns (daily, weekly and yearly), trends and holidays effects as well. In addition to using lagged values of the target variable, the model applies the Fourier series to the input variable as an additional form of feature engineering. That allows further model refinement and decomposition of the results for better interpretation. Prophet can also use exogenous variables such as holidays and automatically detect trend change points. It has four main components, including (I) a piecewise linear or logistic growth curve trend to automatically detect changes in trends by selecting change points from the data, (II) a yearly seasonal component modeled using Fourier series, (III) a weekly seasonal component using dummy variables, and (IV) a user-provided list of important holidays.

The Neural Prophet. The Neural Prophet model [13] is based on Prophet but with an embedded neural network which allows handling time series with more complex patterns. Neural networks, in contrast to autoregressive models, have the advantage of being non-parametric, which implies that they do not make strong assumptions about the shape of the mapping function. So, they can learn any nonlinear function to approximate any continuous function from the training data. However, their disadvantages are that they: (I) require a large volume of data, (II) the fitting of hyperparameters is not as straightforward as with parametric models, and (III) are less interpretable than parametric models. Neural Prophet tries to maintain the advantages of Prophet, such as good performance, interpretability and ease of configuration while improving its accuracy and scalability. Technically it replaces the Stan backend for a PyTorch one and uses Auto-Regressive Network (AR-Net), a one-layer Neural Network trained to mimic the AR process in a time-series signal.

Stacked Long Short-Term Memory Model. The LSTM model [7] is a type of Recurrent Neural Network (RNN) designed to remember information over extended periods and resist vanishing and exploding gradients significantly affecting RNNs. Unlike the RNNs that have one interacting layer in their repeating module, LSTMs have four (see Equations (4)–(9)), where f_t, i_t, o_t and $\widetilde{c}_t$ are the activation vectors of forget, input, output, and cell input gates. c_t, h_t and x_t are the cell, hidden and input state vectors. Input and recurrent connections weights are stored in matrices of the form W_q and U_q. The σ_h and σ_c variables denote the hyperbolic tangent and sigmoid functions, respectively [8].

$$f_t = \sigma_g(W_f x_t + U_f h_{t-1} + b_f) \quad (4) \qquad i_t = \sigma_g(W_i x_t + U_i h_{t-1} + b_i) \quad (5)$$

$$o_t = \sigma_g(W_o x_t + U_o h_{t-1} + b_o) \quad (6) \qquad \widetilde{c}_t = \sigma_h(W_c x_t + U_c h_{t-1} + b_c) \quad (7)$$

$$c_t = f_t \circ c_{t-1} + i_t \circ \widetilde{c}_t \quad (8) \qquad h_t = o_t \circ \sigma_h(c_t) \quad (9)$$

4 The Experimental Study, Results and Analysis

This section presents the experiments used to assess the proposal, including the experimental settings, the benchmarks, the obtained results and their analysis.

4.1 Benchmarks, Experiments and Techniques Settings

The LSTM model is composed of three layers with 14880, 29040 and 61 parameters and training via Adam optimiser over 5000 epochs using a batch size of 32 and a sliding window size of 48 (48 input data equals one day). In the case of neural/Prophet, the variability interval of the forecasting (i.e. confidence interval) has been set to 0.80, the seasonality has been set to 1 day, and experiments were performed by considering and neglecting the holidays. Regarding the SARIMAX, the $(p, d, q)(P, D, Q)$ values were selected by `auto_arima` ([5]) for each case, and we set a seasonality S equal to 48 for all cases. The experiments were designed to assess the forecasting in different time laps by considering testing and training using `(I)` single days as well as `(II)` natural weeks. Since parking tasks have short-term purposes where the citizens do not make real use of forecasting beyond 7 days, the test was done on the last week of the data samples recording. That means that in the case of training using a single day, the testing has been

[5] Auto ARIMA: https://alkaline-ml.com/pmdarima/modules/generated/pmdarima.arima.auto_arima.

done using the same day of the week (e.g. Thursday) of the last week of data samples, while when training using the entire week, the test is done using the last complete natural week. The Mean Absolute Error (MAE) (see Eq. (10)) has been used as a metric during the training and testing steps.

$$\texttt{MAE} = \frac{1}{N}\sum_{i=1}^{N}|\hat{y}_i - y_i| \tag{10}$$

As a benchmark, two datasets have been considered: one of Malaga and one of Birmingham cities. For the data of the city of Malaga, an explanation has already been provided in Sect. 3.1. Although 13 parking were extracted, only three had changing parking occupancy rates: `Salitre`, `San Juan de La Cruz`, `Cruz de Humilladero`. The remaining 10 parking data occupancy is constant for some reason (e.g. sensors failure), so they are neglected in this work. In the case of Birmingham, the data being used is of 28 parking previously studied in [10].

4.2 Obtained Results and Discussion

Tables 1–4 show the parking occupancy forecasting for the city of Malaga and Birmingham, respectively. In Tables 1–4, the metrics `Median`, `Mean`, `Best`, `Worst` and `STD` represent the median, mean, minimum, maximum and the standard deviation of the `MAE` found for each technique across all the studied parking. The best results are highlighted in bold based on the `Median` metric.

Parking Occupancy in the City of Malaga. Tables 1 and 2 show the results obtained by the studied techniques when addressing the parking data of the city of Malaga. Table 1 shows that for the case of the parking "Cruz de Humilladero" and "Salitre", the best predictor was found to be Neural Prophet, while for the case of the parking "San Juan de la Cruz", Prophet was found to be competitive. It is to be noted that overall Prophet and Neural Prophet were found to be relatively better than SARIMAX and LSTM when addressing data without holidays. So, a decision was made to discard executing SARIMAX and LSTM when considering holidays. If we consider the mean and median of the solutions obtained, the predictors that achieve the best results are Neural Prophet and Prophet when separate training is carried out for both each day of the week as well as the entire week (see Table 2).

Since predictors such as Prophet provide a range of predictions (i.e. confidence interval), Table 3 indicates how many times the predictor has attained a prediction range that includes the real forecasting. It is to be noted that the confidence interval is wider when considering the whole week (around 0.2), while when using single days it is narrower (around 0.1). This explains why better hit rates are obtained in the first case and not the second one (see Fig. 7). When using Prophet, the advantage over other techniques is that the forecasting can be found with a hit rate of 80%. Also, one can note that there is no improvement when using "holidays". A hypothesis for enhancing this aspect is the use of more training data. This being said, for these data the Prophet model seems to be the most appropriate.

Figures 4 and 5 present the concatenation of the results of applying the Prophet for predicting the parking occupancy using all days of the week and per day, respectively. Figures 6–8 represent the parking occupancy prediction using Prophet for the "Salitre", "San Juan de la Cruz" and "Cruz de Humilladero" parking during Sundays, Tuesdays and Wednesdays, respectively. The results show that Prophet obtains forecasting intervals of confidence that include, in general, the real parking occupancy value.

Table 1. MAE results of Prophet, Neural Prophet and LSTM in the case study of Malaga

Parking	Technique	Without Holidays		With Holidays	
		Per Week	Per day	Per Week	Per day
Cruz de Humilladero	Prophet	0.071	0.088	0.074	0.089
	Neural Prophet	**0.070**	**0.078**	**0.070**	**0.078**
	LSTM	0.109	0.107	-	-
	SARIMAX	0.072	0.120	-	-
Salitre	Prophet	0.112	0.078	**0.109**	0.077
	Neural Prophet	**0.110**	**0.073**	0.110	**0.073**
	LSTM	0.215	0.082	-	-
	SARIMAX	0.124	0.109	-	-
San Juan de la Cruz	Prophet	**0.081**	0.074	**0.081**	**0.070**
	Neural Prophet	0.084	0.074	0.083	0.074
	LSTM	0.096	**0.059**	-	-
	SARIMAX	**0.081**	0.061	-	-

Table 2. Summary of MAE results in the case study of Malaga

Type	Per	Technique	Best	Worst	Mean	Median	STD
Without Holidays	Week	Prophet	0.071	0.112	0.085	**0.081**	0.017
		Neural Prophet	0.070	0.110	0.085	0.084	0.017
		LSTM	0.096	0.215	0.140	0.109	0.031
		SARIMAX	0.072	0.124	0.092	**0.081**	0.023
	Day	Prophet	0.074	0.088	0.080	0.078	0.006
		Neural Prophet	0.073	0.078	0.075	**0.074**	0.002
		LSTM	0.059	0.107	0.083	0.082	0.020
		SARIMAX	0.061	0.120	0.097	0.109	0.026
With Holidays	Week	Prophet	0.074	0.109	0.086	**0.081**	0.015
		Neural Prophet	0.070	0.110	0.085	0.083	0.017
	Day	Prophet	0.070	0.089	0.078	0.077	0.008
		Neural Prophet	0.073	0.078	0.075	**0.074**	0.002

Table 3. Summary of hit rate results in the case study of Malaga

Type	Per	Technique	Best	Worst	Mean	Median	STD
Without Holidays	Week	Prophet	0.878	0.735	0.801	**0.804**	0.058
		Neural Prophet	0.554	0.390	0.480	0.533	0.073
	Day	Prophet	0.756	0.649	0.684	**0.658**	0.048
		Neural Prophet	0.479	0.310	0.381	0.390	0.069
With Holidays	Week	Prophet	0.875	0.774	0.809	**0.786**	0.045
		Neural Prophet	0.554	0.390	0.480	0.533	0.073
	Day	Prophet	0.768	0.649	0.690	**0.664**	0.053
		Neural Prophet	0.485	0.310	0.389	0.411	0.072

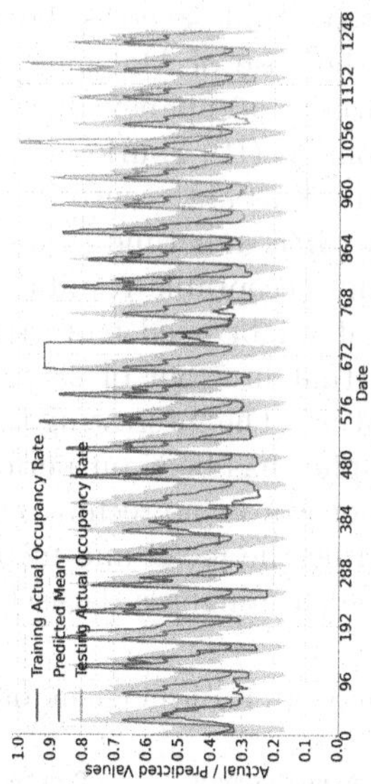

Fig. 4. Training one model on Salitre using all days of the week.

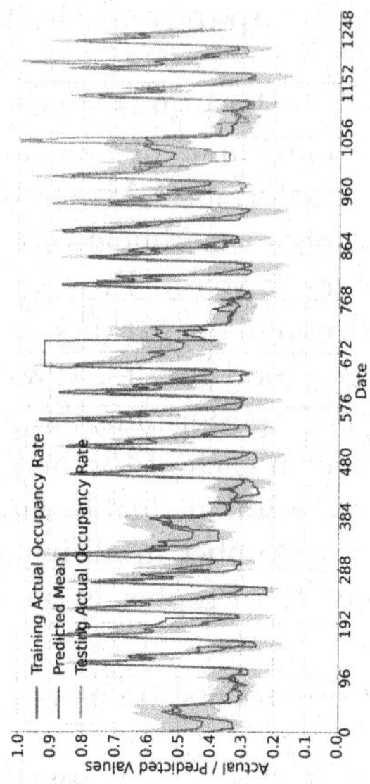

Fig. 5. Training one model on Salitre per day of the week.

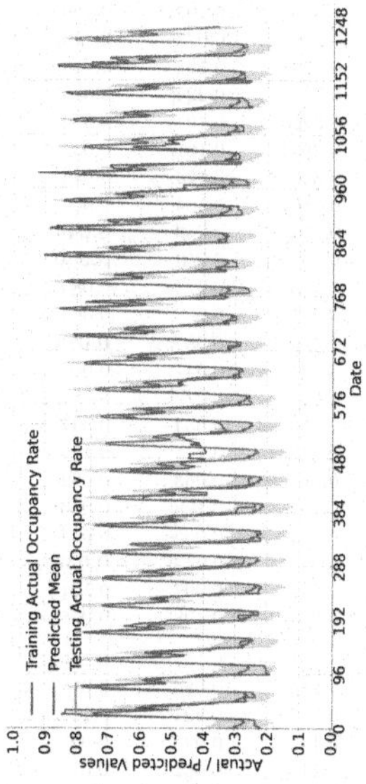

Fig. 6. Training one model on Salitre only considering Sundays.

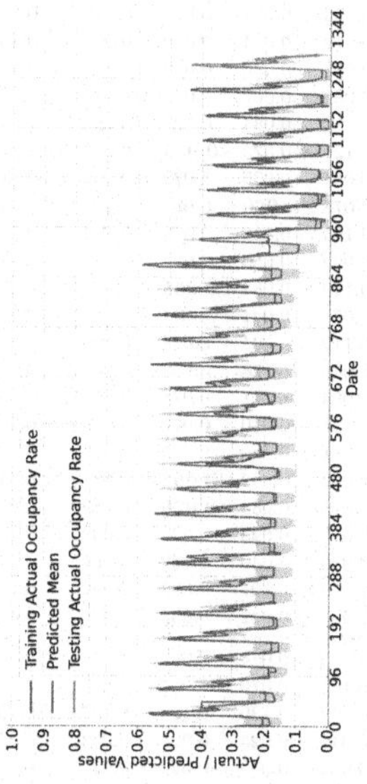

Fig. 7. Training one model on San Juan de la Cruz only considering Tuesdays.

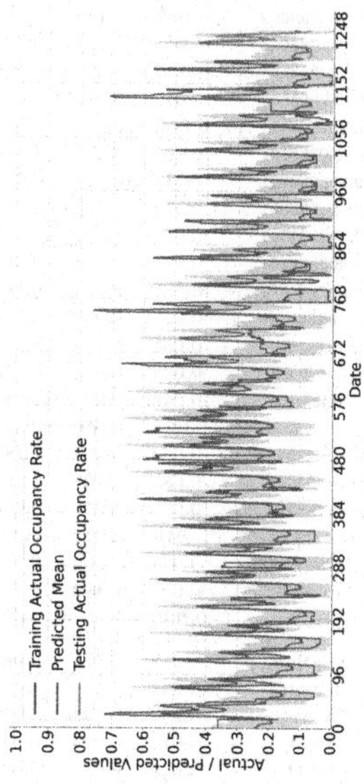

Fig. 8. Training one model on Cruz de Humilladero only considering Wednesdays.

Parking Occupancy in the City of Birmingham. Table 4 shows the results obtained by the best of the predictors found in the case of Malaga city and compared to the results given in [4] using the parking occupancy data of the city of Birmingham. The authors in [4] used polynomials (P), Fourier series (F), k-means clustering (KM), polynomials fitted to the k-means' centroids (KP), shift and phase modifications to KP polynomials (SP), and time series (TS). In the case of Neural Prophet, it could not get any acceptable results on any of the Birmingham parking, which indicates that it might need more training data to get good results. However, it can be seen that in 7 out of 28 parking, Prophet is outperforming the state-of-the-art literature. The interesting thing to note is that in each of the instances where the Prophet has been outperformed, a different technique has been able to do so. In other words, no technique could outperform Prophet in all instances. Now, considering the mean and median of the results, the Prophet is the best predictor on the Birmingham dataset.

Table 4. MAE results: state-of-the-art predictors vs Prophet on Birmingham dataset

Parking	State-of-the-art [4]							Neural Prophet		Prophet	
	P	F	KM	KP	SP	TS	RNN	Per Week	Per Day	Per Week	Per Day
BHMBCCMKT01	**0.041**	0.053	0.087	0.086	0.059	0.067	0.063	0.377	0.276	0.076	0.054
BHMBCCPST01	0.076	**0.072**	0.148	0.149	0.083	0.111	0.137	0.447	0.420	0.114	0.107
BHMBCCSNH01	0.132	0.141	0.150	0.148	0.139	**0.069**	0.117	0.559	0.336	0.100	0.081
BHMBCCTHL01	0.122	0.142	0.134	0.131	0.123	0.080	0.103	0.355	0.435	**0.058**	0.069
BHMBRCBRG01	0.101	0.148	0.148	0.149	0.133	0.095	0.123	0.501	0.679	0.101	**0.085**
BHMBRCBRG02	**0.087**	0.116	0.122	0.122	0.097	0.088	0.112	0.389	0.430	0.107	0.096
BHMBRCBRG03	0.068	0.085	0.113	0.112	0.074	0.059	0.076	0.432	0.200	0.073	**0.058**
BHMEURBRD01	**0.044**	0.057	0.087	0.085	0.036	0.042	0.077	0.438	0.424	0.194	0.066
BHMEURBRD02	0.072	0.078	0.064	0.063	0.067	0.068	**0.062**	0.722	0.442	0.224	0.089
BHMMBMMBX01	**0.063**	0.067	0.074	0.072	0.084	0.129	0.084	0.338	0.322	0.101	0.099
BHMNCPHST01	0.060	0.079	0.130	0.127	0.073	**0.034**	0.050	0.212	0.363	0.101	0.037
BHMNCPLDH01	**0.030**	0.034	0.087	0.084	0.036	0.072	0.072	0.311	0.329	0.124	0.054
BHMNCPNHS01	0.072	0.084	0.082	0.078	**0.060**	0.082	0.102	0.340	0.340	0.184	0.064
BHMNCPNST01	0.085	0.083	0.150	0.154	0.117	0.074	0.124	0.237	0.274	0.063	**0.044**
BHMNCPPLS01	0.078	0.088	0.067	0.066	0.080	0.058	0.076	0.176	0.209	0.059	**0.050**
BHMNCPRAN01	0.083	0.094	0.143	0.140	0.084	**0.055**	0.089	0.470	0.486	0.221	0.089
BroadStreet	0.047	0.057	0.073	0.071	0.041	**0.034**	0.064	0.540	0.395	0.203	0.064
BullRing	0.088	0.119	0.113	0.112	0.101	0.074	0.100	0.529	0.409	0.085	**0.071**
NIACarParks	0.033	0.033	0.048	0.049	**0.028**	0.054	0.033	0.087	0.121	0.047	0.052
NIASouth	0.040	0.036	0.064	0.064	**0.031**	0.078	0.053	0.158	0.183	0.077	0.081
Others-CCCPS105a	**0.032**	0.050	0.119	0.121	0.050	0.072	0.065	0.207	0.263	0.054	0.047
Others-CCCPS119a	0.090	0.092	**0.081**	**0.081**	0.089	0.095	0.091	0.135	0.248	0.111	0.116
Others-CCCPS133	0.083	0.108	0.093	0.092	0.087	**0.061**	0.091	0.343	0.288	0.063	0.069
Others-CCCPS135a	0.057	0.075	0.078	0.076	0.058	**0.029**	0.049	0.277	0.345	0.129	0.043
Others-CCCPS202	**0.016**	0.024	0.074	0.075	0.025	0.023	0.033	0.332	0.138	0.055	0.023
Others-CCCPS8	0.038	0.055	0.081	0.079	0.040	0.047	0.061	0.353	0.238	0.043	**0.035**
Others-CCCPS98	0.092	**0.089**	0.177	0.179	0.101	0.097	0.092	0.273	0.254	0.097	0.101
Shopping	**0.035**	0.054	0.065	0.066	0.032	0.032	0.037	0.494	0.353	0.108	0.036
Median	0.070	0.078	0.087	0.086	0.074	0.069	0.077	0.348	0.333	0.101	**0.065**
Mean	0.067	0.079	0.102	0.101	0.073	0.067	0.079	0.286	0.285	0.086	0.058
Worst	0.132	0.148	0.177	0.179	0.139	0.129	0.137	0.722	0.679	0.224	0.116
Best	0.016	0.024	0.048	0.049	0.025	0.023	0.033	0.087	0.121	0.043	0.023
STD	0.029	0.033	0.035	0.035	0.033	0.026	0.028	0.144	0.116	0.052	0.024

5 Conclusions and Perspectives

This paper prototypes a parking occupancy forecasting pipeline including (I) the ETL module that automatically extracts and processes real-time parking data and (II) the forecasting module composed of four time-series techniques: LSTM, Prophet, Neural Prophet and SARIMAX. The experiments have been made using real-life data of parking occupancy rate of 3 parking in the city of Malaga (Spain) during 6 months, while the second represents 28 parking spaces in the city of Birmingham (UK) for two and a half months. The obtained results have shown that the Prophet technique is the one achieving the best forecasting in both benchmarks. In future works, it is planned to expand the current study to other advanced time-series forecasting techniques and extend the ETL module to include other cities with more complex parking behaviours, city size, etc.

References

1. The proposed parking occupancy forecasting prototype. https://github.com/NEO-Research-Group/Parking-Ocuppancy-Forcasting.git. Accessed 27 Jan 2023
2. Amato, G., Carrara, F., Falchi, F., Gennaro, C., Meghini, C., Vairo, C.: Deep learning for decentralized parking lot occupancy detection. Expert Syst. Appl. **72**, 327–334 (2017)
3. Banaś, J., Utnik-Banaś, K.: Evaluating a seasonal autoregressive moving average model with an exogenous variable for short-term timber price forecasting. Forest Policy Econ. **131**, 102564 (2021)
4. Camero, A., Toutouh, J., Stolfi, D.H., Alba, E.: Evolutionary deep learning for car park occupancy prediction in smart cities. In: Battiti, R., Brunato, M., Kotsireas, I., Pardalos, P.M. (eds.) LION 12 2018. LNCS, vol. 11353, pp. 386–401. Springer, Cham (2019). https://doi.org/10.1007/978-3-030-05348-2_32
5. Camero, A., Alba, E.: Smart city and information technology: a review. Cities **93**, 84–94 (2019)
6. Cintrano, C., Ferrer, J., López-Ibáñez, M., Alba, E.: Hybridization of evolutionary operators with elitist iterated racing for the simulation optimization of traffic lights programs. Evolut. Comput. 1–21 (2022)
7. Hochreiter, S., Schmidhuber, J.: Long short-term memory. Neural Comput. **9**(8), 1735–1780 (1997)
8. Nath, P., Saha, P., Middya, A.I., Roy, S.: Long-term time-series pollution forecast using statistical and deep learning methods. Neural Comput. Appl. **33**(19), 12551–12570 (2021). https://doi.org/10.1007/s00521-021-05901-2
9. Parmezan, A.R.S., Souza, V.M., Batista, G.E.: Evaluation of statistical and machine learning models for time series prediction: Identifying the state-of-the-art and the best conditions for the use of each model. Inf. Sci. **484**, 302–337 (2019)
10. Stolfi, D.H., Alba, E., Yao, X.: Predicting car park occupancy rates in smart cities. In: Alba, E., Chicano, F., Luque, G. (eds.) Smart Cities. Smart-CT 2017. LNCS, vol. 10268, pp. 107–117. Springer, Cham (2017). https://doi.org/10.1007/978-3-319-59513-9_11
11. Stolfi, D.H., Alba, E., Yao, X.: Can i park in the city center? predicting car park occupancy rates in smart cities. J. Urban Technol. **27**(4), 27–41 (2020)
12. Taylor, S.J., Letham, B.: Forecasting at scale. Am. Stat. **72**(1), 37–45 (2018)
13. Triebe, O., Hewamalage, H., Pilyugina, P., Laptev, N., Bergmeir, C., Rajagopal, R.: Neuralprophet: explainable forecasting at scale (2021)

E-scooters Routes Potential: Open Data Analysis in Current Infrastructure. Malaga Case

Diego Daniel Pedroza-Perez(✉), Jamal Toutouh, and Gabriel Luque

ITIS Software, Universidad de Málaga, Málaga, Spain
{pedroza,jamal,gluque}@uma.es

Abstract. There is an increasing interest in alternative vehicle mobility, such as electric scooters (e-scooters). E-scooters are getting attention not only for their environmental impact but also because they are easy to ride on. However, could our current infrastructure support e-scooter trips? Could e-scooter offer a better way to move on our present streets and roads? Using current open data such as the census segments of Malaga city (Spain), the list of non-tertiary institutions, and a map of OpenStreetMap, this work explores the potential of e-scooter in a growing European city by simulating routes trips in Origin-Destination Matrix. The main results are 1) in the current streets, a person can reduce at least 40% of the time using an e-scooter to commute compared to walking; 2) the road infrastructure offered for e-scooters downtown is adequate, especially for routes of more than one kilomete; however this infrastructure needs to be improved in the outskirts of the city.

1 Introduction

Modern city planning must consider urban mobility because it impacts citizens' everyday lives and a city's overall sustainability and livability [17]. As well as the advent of new technology and services like ride-sharing and e-scooters, there has been a shift in recent years toward more environmentally friendly and active means of transportation, including walking, cycling, and public transportation.

For this reason, the study of mobility has increased in recent years, especially those related to the public system, such as buses and subways. For example, Toutouh et al. [19] used open data to analyze the bus service in the city of Melilla and assumed an older pedestrian target. In the work of Toto et al. [18], they used an analysis of smart card data to predict crowd subway to offer recommendations about the route to take. The same type of data was studied by Fabbiani et al. to optimize bus schedules, and bus-stop location [6]. In our work, we use an alternative vehicle, electric scooters, to characterize its current mobility infrastructure in a growing European city. This type of work is helpful to realize the potential of alternative mobility in today's infrastructure.

In this context, e-scooters are electric-powered vehicles that, as an inexpensive, practical, and greener alternative to cars and other forms of public transit,

B. Dorronsoro et al. (Eds.): OLA 2023, CCIS 1824, pp. 380–392, 2023.
https://doi.org/10.1007/978-3-031-34020-8_29

have grown in popularity in cities worldwide [9,11]. According to the study by Wanganoo *et al.* [20], e-scooters have the potential to reduce carbon emissions and traffic congestion in urban areas.

However, introducing e-scooters has also brought some difficulties, such as issues with regulation, accessibility, and safety. For instance, a research study discovered that between 2014 and 2018, e-scooter injuries climbed by 222% in the United States, with head injuries accounting for most cases [7]. E-scooters have the potential to significantly influence how urban transportation will develop in the future, despite these obstacles.

We can find a few studies about electric e-scooters mobility in the literature. For example, Hardt et al. [9] studied the usage of e-scooters in urban environments and found that people used e-scooters mainly to commute. Moreover, Almannaa et al. [2] analyzed the trips in e-scooter and e-bikes in Austin, Texas, in the USA. They realized that people riding an e-scooter go at 2.78 m per second and supported the hypothesis that people used e-scooters to commute. However, these works are quite limited, and more research is required to fully comprehend how e-scooters affect urban mobility, particularly traffic congestion, road infrastructure impact, air quality, public health, and social equity.

Many cities are modifying (albeit slightly) their road infrastructure to make it suitable for e-scooters. They are using cycle lanes or creating speed-limited lanes to facilitate e-scooter traffic. This study will analyze whether these changes make it possible to offer quality routes for this transport mode. In our work, we assumed that people over 16 years old, especially teenagers, would use e-scooters to commute to educative institutions in a real scenario: Malaga, a medium-sized city. Also, we computed our simulations in the current infrastructure mobility model. The final objective is to analyze if there is a significant difference between making the trip by walking.

The rest of this paper is organized as follows. Section 2 presents some crucial details about mobility in Malaga, the use case in this work. The methodology applied in this study and the results are discussed in Sects. 3 and 4, respectively. Finally, in Sect. 5, we summarize our main achievements and give some hints about the following analyses that can extend this study.

2 Malaga City Use Case

Malaga is located in the region of Andalusia in the South of Spain. It is a top-10-highest populated city in the country. Its density is comparable with cities such as Alicante in Spain and Austin (Texas). Malaga city has an area of 394.98 km^2. In 2022, Malaga had a population of 577 405 people. About 85% of its population is over 15 years old; more precisely, 38% are from 15 to 44 years old, where females and males share a similar percentage of the population [1]. It is administratively divided into 11 districts and 441 census segments. The districts near the coast are the most populated ones (66% of the population live near the coast) [4]. As shown in Fig. 1a, the highest density of schools is found in these districts.

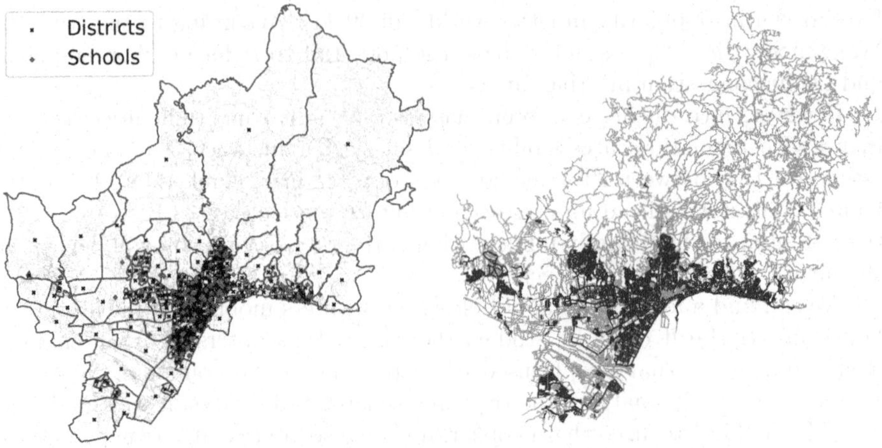

(a) Districts centroids (black) and schools (red) (b) E-scooter (blue) and walk (yellow) streets.

Fig. 1. Maps of Malaga city with the information used in this study.

In Malaga, the Official State Gazette n° 297 of November 11, 2020, and the Official Bulletin of the province of Malaga, n° 252/2021, published from January 11 to 19, 2021, set the regulations regarding Personal Mobility Vehicles (PMV) [14,16] administrative regulations include electric scooters. The main regulations established that the minimum age to ride an e-scooter is 16 years old. In turn, they fixed the maximum speed between 6 km per hour (km/h) to 25 km/h. Moreover, they restricted the type of roads and streets not permitted to ride on with an e-scooter. The e-scooters cannot be used on footways, pedestrians-only roads, primary and intercity highways, cars-only streets, and tunnels [14]. Figure 1b shows in blue the streets allowed for riding e-scooters. On the contrary, they let the use of e-scooters in cycleways, residential roads, and highways with maximum speeds of 20 km/h or 30 km/h.

3 Materials and Methods

This section presents the primary artifacts employed in this research to analyze the road infrastructure quality offered to e-scooter users in Malaga. We focused on a significant user segment of this type of transport: young people over 16 years old who still go to educational institutions [5]. In Malaga, these users use e-scooters daily to go to their centers of study. Thus, we evaluated the routes users can take from their homes (i.e., the centroids of the Malaga census segments) to the schools. The open data, methods, and metrics used in this analysis are presented below.

3.1 Open Data Used

In this research, we employed real-world data to characterize the quality of road infrastructure for e-scooter mobility. We use mainly two types of available data sources: official open data and collaboratively open data.

The official open data portal from the region of Andalusia (i.e., *Junta of Andalusia*) provided data about educational institutions [10]. This data was evaluated and processed to get the location of high schools with 16-year-old or older students. These institutions are the destination of the evaluated routes in the analysis. Figure 1a illustrates the location of the schools in the map of Malaga. Most of the information about the city of Malaga used, such as the census segments, was gathered from the Malaga Open Data portal [13]. The census segments centroids are the origin of the trips studied here. These two open data portals were developed as part of the transparent governance initiatives that allow citizens to know more about their institutions.

This study required spatial information about the road mobility infrastructure. This information includes several aspects of the roads, such as the type of road, the number, and type of lanes, etc. In order to get this spatial information, we downloaded and processed data from the collaborative initiative project OpenStreetMap (OSM) [8]. Figure 1b shows the road data used to perform the analysis.

3.2 Methods and Tools for the Evaluation

The Origin-Destination matrix (OD matrix) is widely used in mobility studies to enhance the analysis of routes. It is formed by two sets of points, origin and destination, and allows it to represent a sample of local mobility. The OD matrix generates structured traffic flow data that helps to characterize the mobility and can be used as a stepstone for other analysis [15].

In our study, the origins were the centroids of the census segments and the destinations were educational institutions. On the one hand, we used schools and educational institutions as destinations because the age to ride an e-scooter is over 16. In the case of Malaga, the type of non-tertiary institutions for people of that age include high schools and adult education institutions.

On the other hand, we used census segments as origin points in our OD matrix. We used the data from Malaga Open Data portal to find the geospatial information of each district and segment. Then, we calculated every census segment's more geometric representative point and transformed them to represent them in our map. Figure 1a shows the map with both markers, red for schools and black for census segments centroids.

The experiments were carried out using a graph representing the city's different types of roads/streets. This graph was constructed by using OSM graph of Malaga as a base. The OSM graphs consist of nodes, edges, and a dictionary of attributes related to the streets and points they represent. The length, the street type, the maximum speed allowed, and if it has a cycleway were some of the most valuable characteristics used.

In our analysis, we did two types of simulations: a) on foot-only (`walking`), which represents the trips in which the student only walks from the origin to the school, and b) using an e-scooter (`e-scooter`), which represents the users that use the e-scooter if it is allowed on a given road/street. Thus, for each pair origin-destination, there are two different types of trips, i.e., `walking` and `e-scooter`, respectively.

This study considered 44 541 routes (441 census segments $\times$ 101 schools). The e-scoter mobility was simulated by finding the shortest route from each pair origin-destination of our OD matrix. The routing paths were computed using the Dijkstra algorithm, which uses the time between two nodes as a weight metric to optimize the routes. In order to compute the time weight of the edges between two nodes, we relied on the length of the edge and assumed linear straight-line motion, i.e., people and e-scooters had a constant speed. We did not consider gender in our study for the reason that, as reported by Campisi et al. [3], less than 20% of women use e-scooter. However, we used gender-neutral speed metrics. For `walking` trips, we used the speed specified by Knoblauch et al. of 1.25 m per second (m/s) for young people [12]. For `e-scooter` simulation, we considered 2.78 m/s speed following the study of Almannaa et al. in Austin city [2].

The software required for the computations has been developed using `Python`. Specific software libraries were applied to implement the code for the data analysis: Numpy, scikit-learn, and Pandas. Geopandas and Arcpy were used for the GIS computations. Finally, OSMnx and NetworkX software libraries were used to computed the routes and paths.

3.3 Metrics Evaluated

This paper aims to use open data to study the quality of road infrastructure to be used by e-scooter users, emphasizing the service provided to young users that use this transportation mean to go to school every day. The metrics evaluated here are associated with the trip time, path length, and users' route decision-making. These metrics evaluated are presented below:

1. *Total duration trip time*: It measures the travel time in minutes when the user travels from the census segments to the schools. It allows understanding the traffic flow between every two points in the OD matrix in each type of simulation (`e-scooter` and `walking`).
2. Walking distance length: It evaluates the total distance a user walked in the trip considering `e-scooter` and `walking`.
3. *Trip time reduction*: Given a route between a census segment s and a high school h, it takes into account the difference relative of time that a non e-scooter user walked that route ($t_{ow}^{(s,h)}$) and with the time took to the e-scooter rider ($t_{sw}^{(s,h)}$). It is computed according to Eq. 1.
4. *Percentage of mobility mode switching*: To explore the quality of the road infrastructure, we calculate the number of times a user had to switch the mobility mode when changing the street/road during the trip, i.e., how many

times it changed from riding to walking or to walking to riding. It was measured in terms of the ratio (percentage) between the number of times the user changed the mobility mode c_m and the number of times the user changed the street/road c_r. It is computed according to Eq. 2

$$\frac{t_{ow}^{(s,h)} - t_{sw}^{(s,h)}}{t_{ow}^{(s,h)}} \% \tag{1}$$

$$1 - \frac{c_m}{c_r} \% \tag{2}$$

4 Empirical Analysis

This section presents the experimental analysis results to evaluate e-scooter mobility in Malaga. A preliminary analysis is performed on the whole city by considering all the trips between any census segment centroid and any educational institution. Then, the road e-scooters' quality for a selected route type is studied.

4.1 Preliminary Analysis

The `e-scooter` and `walking` trips between all the origins and destinations are computed. The main idea is to provide a global overview of the degree of connectivity among the census segments and high schools using e-scooters. Thus, the route length and time are considered to compare the `e-scooter` and `walking` trips.

Table 1 summarizes the trip time and length results. As the distributions of the results are non-normally distributed according to the Shapiro-Wilk statistical test, this table presents the median, interquartile range (IQR), and maximum values for each distribution. Figures 2 and 3 show the distribution of the trip time and length results as probability density functions, respectively. Finally, Fig. 4 illustrates the distribution of travel time reduction that occurs when riding an e-scooter compared to walking. This way of graphically representing the results has been used because 44 541 different routes have been analyzed, and the probability distributions encapsulate well the result distributions when there are so many of them.

Table 1. Trip time and length distance results for `e-scooter` and `walking` routes.

Trip type	Time (in minutes)			Length (in meters)		
	median	IQR	maximum	median	IQR	maximum
`e-scooter`	35.30	35.95	473.21	5 353.18	5 676.95	44 262.19
`walking`	62.97	69.32	562.79	4 723.08	5 199.57	42 209.56

The results in Table 1 show how using e-scooters considerably reduces travel time. The median and IQR values for the `e-scooter` routes are almost half those of the `walking` routes. The Kruskal-Wallis H-test statistical test confirmed that `e-scooter` trip times are significantly shorter than `walking` times (p-value $<$ 0.001).

The longest routes require 473.21 and 562.79 min for `e-scooter` and `walking` trips, respectively. These routes occur from the most remote census segments that do not have nearby high schools and are far from roads suitable for scooter use.

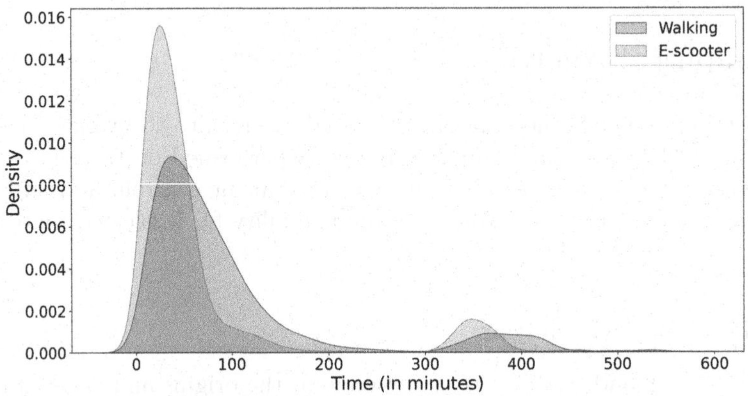

Fig. 2. Trip times probability density distribution for `e-scooter` and `walking` trips.

Figure 2 shows that the probability density distributions representing the trip time results for `e-scooter` and `walking` are bimodal (i.e., two modes group different types of results). The first mode with the highest frequency of results (on the left side of the figure) includes the trip times for routes involving downtown districts. In these central districts, there are a significant number of (nearby) high schools, and more roads suitable for scooter use are installed. Therefore, this mode includes the shortest trips. The second mode, with a much lower frequency, includes the longest trips and involves the districts on the outskirts of the city that are less well-connected and have fewer educational centers nearby.

Focusing on the routes distance length, Table 1 shows that the scooter routes are longer than the `walking` ones. The median distance of the `e-scooter` trips is 5 353.18 m, and the median length of the `walking` ones is 4 723.08 m. This distance difference shows that pedestrians can find shorter alternatives to walking. But as mentioned above, pedestrians require more travel time. Figure 3 confirms that the `walking` routes lengths are shorter than the `e-scooter` ones. Besides, the distribution of route length results is bimodal, where the first mode (with the highest frequency) represents the shorter routes and the second includes the very long-distance trips.

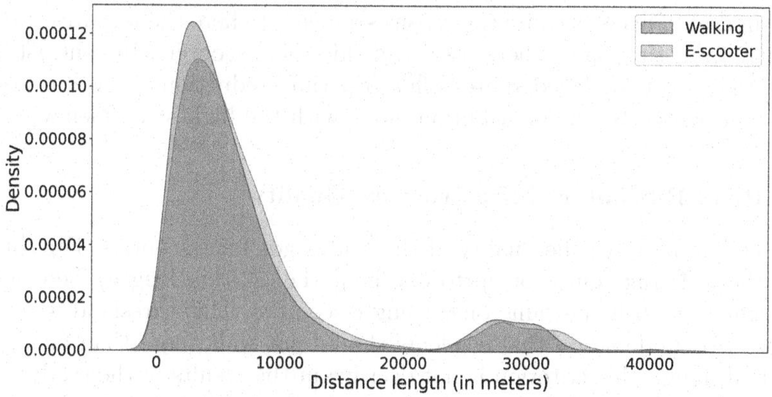

Fig. 3. Route lengths probability density distribution for `e-scooter` and `walking` trips.

The distance length results show that the trips that required maximum times (of more than seven hours) represent routes of more than 40 km. This shows that census segments from the city's outskirts are unconnected from some high schools (i.e., the students cannot walk or ride from their homes to given educational centers).

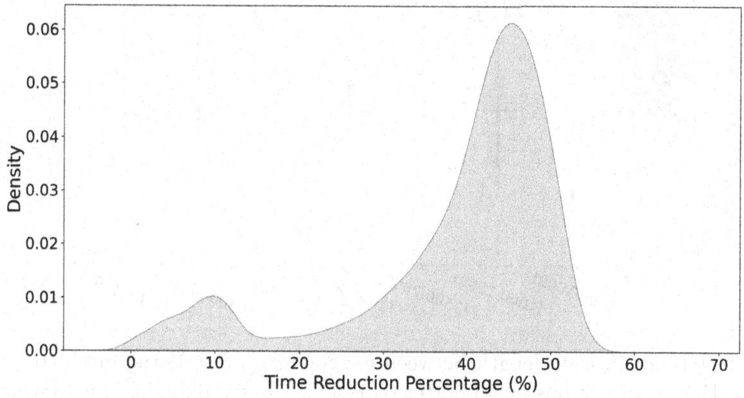

Fig. 4. Time reduction percentage probability density distribution.

As is shown in Table 1 and Fig. 2 e-scooter riders require shorter times to get to their destination. We studied the improvement in terms of time when moving using an e-scooter as the trip time reduction. Again, the probability function representing the travel time reduction results is bimodal (see Fig. 4). However, the shape of this probability density function is symmetric to that shown in Fig. 1b (travel time results). That means that the most distant routes improve travel times the least (represented by the distribution mode on the left of the figure). This is because the census segments (where these routes start) are far away from the roads where users can ride the e-scooters. In contrast, Fig. 4 shows that most analyzed routes offer trip time reductions between 40% and 50% (represented by the distribution mode with the highest frequency).

4.2 Road E-Scooters Infrastructure Quality

This section analyzes the quality of the routes got for e-scooters. For this purpose, the `walking` routes of up to one hour (i.e., 4.5 km length) were selected and compared with the same ones using e-scooters. This threshold was chosen because it is understood that students should not walk more than one hour to school, although this criterion is not decisive in the results of the study.

The metrics evaluated were: the reduction in walking distance when using the e-scooter versus only walking (see Fig. 5), the reduction in travel times (see Fig. 6), and the number of times the travel mode had to be changed (from walking to riding or from riding to walking) during the route (see Fig. 7). In order to study these three metrics, the trips have been grouped into 500-meter length walking routes. Thus, we analyzed nine classes of routes defined by c, where each class is defined as $c \geq \textit{route length} > 500 \times (c + 1)$.

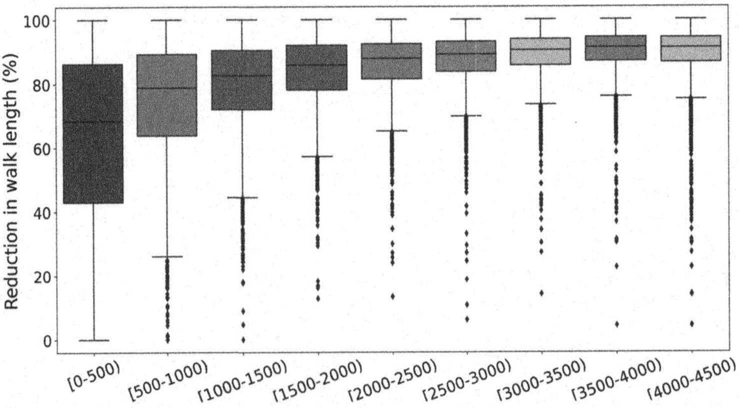

Fig. 5. Difference in walk length between `e-scooter` and `walking` grouped by c class of 500 m. Higher reductions mean better trips, i.e., lower walking distances when riding e-scooters.

The boxplots in Fig. 5 show the distribution of the percentage reduction in walking distance relative to the one walked when using an e-scooter. In general, reductions are obtained for all classes of routes studied. These reductions go up to 100% (i.e., the person can ride the entire path on an e-scooter).

The shortest routes (from 0 to 500 m) offer the slightest improvement in walking distance (see Fig. 5). This is because the educational centers are too close, so there are few alternatives to riding the e-scooter. As the route length between the census segments and the high schools increases, the benefit of using the e-scooter is more significant. These results illustrate that there is adequate infrastructure to use e-scooters.

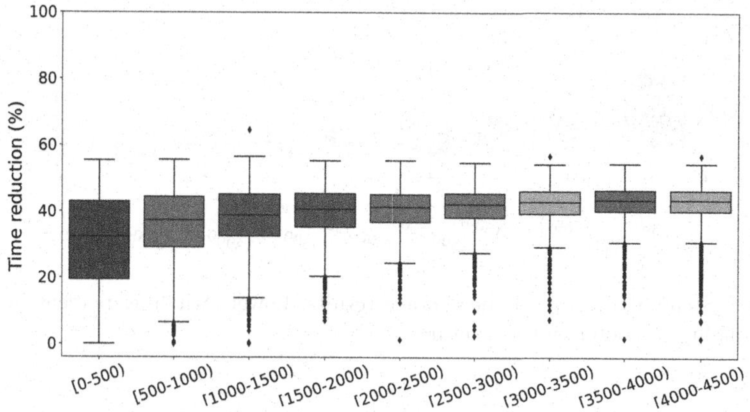

Fig. 6. Time reduction percentage per group of 500 m. Higher reduction times mean better trips, i.e., lower walking distances when riding e-scooters.

Something similar occurs with the improvements in travel times depicted in Fig. 6. For shorter routes, the reduction in travel time by employing an e-scooter is less than for longer routes. In the worst case (routes from 0 to 500 m), the median value is 33% of the reduction in trip times. At the same time, Fig. 6 shows that 40% of the travel time is saved in general. There are even cases where this travel time is reduced by more than 60% (e.g., trips ranging from 1000 to 1500 m).

Finally, a metric has been studied that reflects the quality of the infrastructure provided to ride e-scooters in terms of the number of times the user has to change travel mode. This metric considers the number of times the e-scooter users have to switch between riding to walking or walking to riding when changing streets/roads while moving to the destination. This metric is a ratio of the number of mode changes over the total number of street/road changes. It is understood that a lower ratio of changes is indicative that the roads that e-scooters can use are well-designed because the rider does not have to switch modes continuously.

Figure 7 illustrates the results on the switching mode percentage. The shortest routes are those with the least competitive results. This is mainly because, even though the number of times the mode is switched is not very high, as the number of possible changes is also small, the resultant rate is higher than for the other routes studied. Overall, these results show that mode switching is required less than 10% of the time when the e-scooter user changes from the street or road. Thus, the e-scooter infrastructure design allows users to use comfortable routes (i.e., not requiring multiple mode switching).

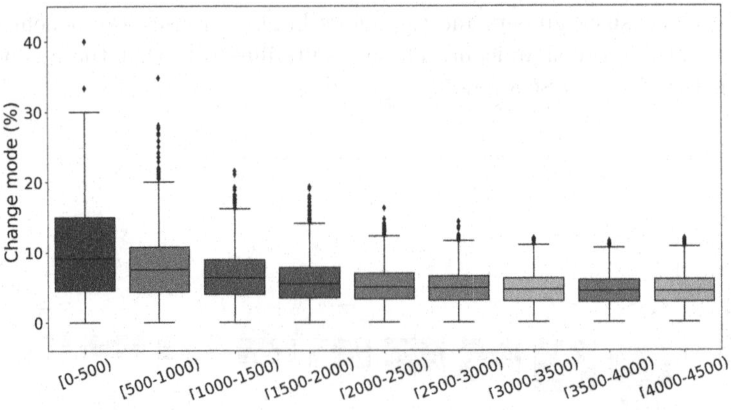

Fig. 7. Switching travel mode percentage results. Lower switching percentages mean better quality e-scooters infrastructure.

In general, the analysis of the results shows that for the trips evaluated here (those of less than 4.5 km), the road infrastructure offered for e-scooters is adequate, especially for routes of more than one kilometer. This outcome is positive because high school students can walk such paths in less than 15 min.

5 Conclusions and Future Work

This paper has studied how infrastructure affects the quality of routes using e-scooters. The study was carried out on the trips that young people have to take to attend their educational centers in Malaga, a medium-sized city.

In our analysis, we observed two very different behaviors detected in this city. On the one hand, we have peripheral districts without educational institutions or large segments of roads suitable for scooters where this type of individual vehicle only contributes a little.

On the other hand, we have another widely populated area (downtown districts) with many schools and road infrastructure suitable for scooters. This second scenario shows the advantages of this vehicle, allowing a significant reduction in time (about 40%) and walking time (more than 80% when the route exceeds half a kilometer).

All these observations have shown that an infrastructure suitable for e-scooters can enable this alternative transport mode in medium-sized cities. Adopting this mode of transport could reduce car use, which would significantly impact emissions, congestion, and even road safety (fewer cars in educational centers at school start and finish times).

In future work, we plan to extend this study by including other types of public transport, such as buses, to study longer routes that can cover the entire city. We are also interested in using the system developed to apply different optimization problems allowing the e-scooter users to select routes by optimizing other metrics, such as mode switching, battery consumption, safety, etc. Moreover, we are also considering applying more demographic characteristics such as power energy, gender, purchasing power, and geographic aspects such as terrain and topological elevation. Finally, we will apply the same analysis in other cities with different road systems to study their impact on this type of vehicle.

Acknowledgements. This research is partially funded by the Universidad de Málaga; under grant PID 2020-116727RB-I00 (HUmove) funded by MCIN/AEI/10.13039/501100011033; under grant number PRE2021-100645 by MCIN/AEI/10.13039/501100011033 and by the FSE+; and TAILOR ICT-48 Network (No 952215) funded by EU Horizon 2020 research and innovation programme.

References

1. Sistema de informacin multiterritorial de andaluca: Instituto de estadística y cartografía de andalucía. https://www.juntadeandalucia.es/institutodeestadisticaycartografia/sima/index2.htm
2. Almannaa, M.H., Ashqar, H.I., Elhenawy, M., Masoud, M., Rakotonirainy, A., Rakha, H.: A comparative analysis of e-scooter and e-bike usage patterns: Findings from the city of Austin, tx. Int. J. Sustain. Transp. **15**(7), 571–579 (2021). https://doi.org/10.1080/15568318.2020.1833117
3. Campisi, T., Skoufas, A., Kaltsidis, A., Basbas, S.: Gender equality and e-scooters: mind the gap! A statistical analysis of the sicily region, Italy. Soc. Sci. **10**(10) (2021). https://doi.org/10.3390/socsci10100403, https://www.mdpi.com/2076-0760/10/10/403
4. CEMI - Centro Municipal de Informática: Estadísticas/informes. https://gestrisam.malaga.eu/estadisticas-informes/#estadisticas-de-poblacion
5. Department for Transport: Perceptions of current and future e-scooter use in the UK, January 2021. https://assets.publishing.service.gov.uk/government/uploads/system/uploads/attachment_data/file/1024151/perceptions-of-current-and-future-e-scooter-use-in-the-uk-summary-report.pdf
6. Fabbiani, E., Nesmachnow, S., Toutouh, J., Tchernykh, A., Avetisyan, A., Radchenko, G.: Analysis of mobility patterns for public transportation and bus stops relocation. Program. Comput. Softw. **44**(6), 508–525 (2018)
7. Farley, K.X., et al.: Estimated incidence of electric scooter injuries in the us from 2014 to 2019. JAMA Netw. Open **3**(8), e2014500 (2020)
8. Haklay, M., Weber, P.: Openstreetmap: user-generated street maps. IEEE Pervasive Comput. **7**(4), 12–18 (2008). https://doi.org/10.1109/MPRV.2008.80
9. Hardt, C., Bogenberger, K.: Usage of e-scooters in urban environments. Transp. Res. Procedia **37**, 155–162 (2019)
10. Junta Andalusia Open Data Portal: Directorio de centros docentes no universitarios de andalucía - portal de datos abiertos. https://www.juntadeandalucia.es/datosabiertos/portal/dataset/directorio-de-centros-docentes-de-andalucia
11. Kazemzadeh, K., Sprei, F.: Towards an electric scooter level of service: a review and framework. Travel Behav. Soc. **29**, 149–164 (2022)
12. Knoblauch, R.L., Pietrucha, M.T., Nitzburg, M.: Field studies of pedestrian walking speed and start-up time. Transp. Res. Rec. **1538**(1), 27–38 (1996). https://doi.org/10.1177/0361198196153800104

13. Malaga Open Data Portal: Sistema de información cartográfica - sección censal. https://datosabiertos.malaga.eu/dataset/sistema-de-informacion-cartografica-seccion-censal
14. Málaga: Boletín oficial de la provincia de málaga. edicto 252/2021 - published from january 11 to 19, 2021, January 2021. http://www.bopmalaga.es/cve.php?cve=20210119-00252-2021
15. Moreira-Matias, L., Gama, J., Ferreira, M., Mendes-Moreira, J., Damas, L.: Time-evolving OD matrix estimation using high-speed GPS data streams. Expert Syst. Appl. **44**, 275–288 (2016). https://doi.org/10.1016/j.eswa.2015.08.048
16. Official State Gazette Agency: Official state gazette n° 297 of november 11, 2020 (boe), November 2020. https://www.boe.es/eli/es/rd/2020/11/10/970
17. Olayode, I., Tartibu, L., Okwu, M., Uchechi, U.: Intelligent transportation systems, un-signalized road intersections and traffic congestion in johannesburg: a systematic review. Procedia CIRP **91**, 844–850 (2020)
18. Toto, E., et al.: PULSE: a real time system for crowd flow prediction at metropolitan subway stations. In: Berendt, B., et al. (eds.) ECML PKDD 2016. LNCS (LNAI), vol. 9853, pp. 112–128. Springer, Cham (2016). https://doi.org/10.1007/978-3-319-46131-1_19
19. Toutouh, J., Lebrusán, I., Cintrano, C.: Using open data to analyze public bus service from an age perspective: melilla case. In: Nesmachnow, S., Hernandez Callejo, L. (eds.) Smart Cities. ICSC-Cities 2021. CCIS, vol. 1555, pp. 223–239. Springer, Cham (2022). https://doi.org/10.1007/978-3-030-96753-6_16
20. Wanganoo, L., Shukla, V., Mohan, V.:Intelligent micro-mobility e-scooter: revolutionizing urban transport. In: Agrawal, P., Madaan, V., Sharma, A., Sharma, D., Agarwal, A., Kautish, S. (eds.) Trust-Based Communication Systems for Internet of Things Applications, chap. 11, pp. 267–290. Wiley (2022). https://doi.org/10.1002/9781119896746.ch11. https://onlinelibrary.wiley.com/doi/abs/10.1002/9781119896746.ch11. ISBN 9781119896746

Automatic Generation of Subtitles for Videos of the Government of La Rioja

Mirari San Martín(✉), Jónathan Heras, and Gadea Mata

University of La Rioja, Logroño, Spain
{miren.san-martin,jonathan.heras,gadea.mata}@unirioja.es

Abstract. Nowadays, public institutions usually provide videos that contain important information in their webpages. However, people suffering from hearing impairment have difficulties accessing content provided by that mean, and the manual transcription of those videos is a time-consuming task. This problem can be faced by means of Automatic Speech Recognition (ASR) systems. In this work, we have evaluated the performance of several ASR systems when applied to videos from the Government of La Rioja, Spain. Our study shows that the Whisper medium model provides the best trade-off between accuracy and speed. Using this model, we have generated the transcription of all the videos from the YouTube channel of the Government of La Rioja. In addition, we have created a tool to facilitate this task for other YouTube Spanish channels. Hence, this can be seen as a step towards improving the accessibility of the information and contents produced by Spanish public administrations.

Keywords: Accessibility · Automatic Speech Recognition · Subtitles

1 Introduction

The Spanish law of 2002, 11 July [4], says that all the information provided by electronic means in web sites of public institutions must be accessible to people with disabilities. Among those sources of information, videos are an important information carrier that presents visual and audio content in live form, and play a key role in people's daily life due to the advances to capture and stream them. However, people suffering from hearing impairment have great difficulty in comprehending video content as auditory information is lost or incomplete in their hearing. This issue can be faced by transcribing the audio of videos and synchronously showing such a transcription when the videos are played [8].

Manual video transcription is a time-consuming task; therefore, computational methods that can automatically generate subtitles can considerably reduce

This work was partially supported by Ministerio de Ciencia e Innovación [PID2020-115225RB-I00 / AEI / 10.13039/501100011033], OTRI OTCA221110 and, by a regional project of the Government of La Rioja.

B. Dorronsoro et al. (Eds.): OLA 2023, CCIS 1824, pp. 393–402, 2023.
https://doi.org/10.1007/978-3-031-34020-8_30

that burden. Automatic Speech Recognition (ASR) systems help us achieve this goal. Such systems allow a computer to take an audio file and convert it into text [16]. As in many other fields, deep learning techniques have also revolutionized ASR systems reaching a performance close to humans [18]. Unfortunately, most ASR systems are focused on the English language, and do not work with other languages. In addition, ASR systems available in other languages than English are usually tested with short audios of good quality.

In this work, we have conducted a thorough study of existing Spanish ASR systems when applied to the videos of the YouTube channel of the Government of La Rioja, Spain. In particular, there are three contributions in this work:

1. We evaluate state-of-the-art Spanish open ASR models when applied to the transcription of videos from different quality and length.
2. We provide the transcriptions of all the videos of the YouTube channel of the Government of La Rioja.
3. Finally, we have created a tool to automatically generate the transcriptions of all the videos from a Spanish YouTube channel.

All the code associated with this project is available at https://github.com/mirenmirari/TFM.

2 Materials and Methods

In this section, we present the studied ASR models, the dataset of videos of La Rioja council, and the evaluation metric employed to measure the performance of the studied ASR models.

2.1 Spanish ASR Models

We start by providing a brief introduction to 6 open deep learning ASR models that can be directly apply to transcribe Spanish audio files. The selected ASR models have been chosen for this work mainly because they have been trained with datasets in Spanish, have an open-source licence, and are easily accessible through the HuggingFace platform [10]. Other models have been seen to be competitive (for example, Deep Speech [6] or Listen, Attend and Spell (LAS) [3] but they only work right for English videos, or have a privative licence. A complete description of the techniques employed by these models is out of the scope of this paper, and we refer the interested reader to [12] for an introduction to the topic; in the rest of this section, we just provide a brief overview of the selected models.

Conformer-CTC and *Conformer-Transducer* [5] are two variants of Conformer, an architecture that combines convolutions with self-attention to produce an ASR system. Conformer-CTC is a non-autoregressive model (that is, it transcribes speech without using any historical context or previous transcriptions as input), whereas Conformer-Transducer is an autoregressive model (that is, it transcribes speech based on previous transcriptions or historical context). For the Spanish version of these models, they were trained using the nemo NVIDIA library [9] and using the LibriSpeech [17] and Common Voice [1] datasets. *PocketSphinx* [11] is an open-source speech recognition software library that is designed to be lightweight and run on resource-constrained devices, such as mobile phones or low-power embedded systems. PocketSphinx uses a Hidden Markov Model (HMM) approach to speech recognition, and supports many languages including, among others, English, Chinese, French, Spanish, German, or Russian. PocketSphinx is the only analyzed systems that is not provided by Hugging Face, but it is available as a Python package.

STT Quartznet [13] is a convolutional model based on Jasper [15]. In our study, we used a fine-tuned version of this model with around 944 h of Spanish data gathered or developed by the CIEMPIESS-UNAM Project [7].

Wav2Vec [2] is a self-supervised framework for learning speech representations by training a convolutional model to predict the next frame of an audio waveform, given the previous frames as input. Such representations can then be used for other tasks such as speech recognition or music classification by applying fine-tuning — a technique that transfers knowledge from one model, usually trained on a large dataset, to another model, where the data is scarce, by making small changes to the original model [19]. In the case of the Spanish ASR system based on Wav2Vec, the model was trained using the Common Voice dataset [1].

Whisper [18] is a transformer based model trained on 680,000 h of multilingual and multitask supervised data collected from the web. The Whisper model process audio by splitting it into 30 s chunks that are converted into a log-Mel spectrograms. Whisper is provided in 6 sizes: tiny, base, small, medium, large, and large-v2 — the difference between the large and large-v2 versions is that the large-v2 was trained longer than the large version. The 6 versions of Whisper can be directly applied for transcribing audios in multiple languages including Spanish. Moreover, we have analysed whether Whisper models (in particular the small and medium versions) fine-tuned on Spanish datasets (namely, the Common Voice dataset [1]) improve with respect to their original counterparts.

It must be taken into account that a limitation of this type of models is the duration of the audio files that can be processed. Most models allow audio files that last at most 120 s. However, in this work, we broke them into 60 s chunks in order to generalize the process to more restrictive models. This fragmentation of the videos implies that a post-processing step to reconstruct the transcription is necessary. To this end, an overlap time between fragment and fragment of 8 s is considered and will serve as a reference to join the transcriptions of all the fragments.

Once that we have presented the studied ASR models, we introduce the dataset that has been used to evaluate those models in a real scenario.

2.2 Dataset

The YouTube channel of the Government of La Rioja[1] contains 3246 videos[2], approximately 60% of the videos include subtitles that were automatically generated by YouTube, but 1256 do not. The videos represent a diverse range of subjects, including news, documentaries, and education. In spite of the existence of videos with subtitles, all the videos of the channel have been downloaded, using the YouTube API, since YouTube transcriptions cannot be edited.

The videos are in MP4 format and have a resolution of 1080p, and have a duration that ranges from less than a minute (19% of the videos), to more than an hour (1.6% of the videos); but most of the videos last between 1 and 10 min (60% of the videos). In order to transcribe the videos the audio was extracted to acc format using the ffmpeg library. From the 3246 audio files, 74 of them were randomly picked to evaluate the ASR models. These 74 were manually transcribed, and such a transcription was stored in txt files.

2.3 Evaluation

In order to evaluate the aforementioned ASR models on our dataset of videos, we have used the *Word Error Rate (WER)* [20], a common metric to measure the performance of ASR systems. WER is defined from the Levenshtein distance [14] and works at the word level. Given a reference sentence and an automatically generated sentence, WER is computed using the following formula:

$$WER = \frac{S + D + I}{N}$$

where S is the number of substitutions, D is the number of deletions, I is the number of insertions, and N is the number of words in the reference; therefore, the lower the WER value, the better. Moreover, we can consider a normalized version of WER, from now on WER_N, that uses the same formula than WER, but the reference sentence and an automatically generated sentence are normalized by removing punctuation marks and lower casing the sentences.

3 Results and Discussion

In this section, we analyse the performance of the ASR models in our dataset. Table 1 presents the WER and WER_N of the studied ASR models. From those results we can draw several conclusions. First of all, PocketSphinx is far from the state-of-the-art models since it is the only model with a WER and a WER_N over 1 — the rest of the models obtained a WER value below 0.4, and a WER_N value below 0.25. On the contrary, all the Whisper models achieved a better WER value than the rest of the models, showing the outstanding results produced by these models, even when using the smallest version of this model. Note that the

[1] https://www.youtube.com/@GobiernoDeLaRiojaES.
[2] On July 28th, 2022.

WER metric evaluates unnormalized text; so, this indicates that Whisper models are more adjusted than the rest of the models to transcriptions given by humans. If the text is normalized, the smallest versions of Whisper (tiny and base) are overcame by the Conformer Transducer, but the bigger version still produce the best results. For the NVIDIA models, Conformer Transducer achieved the best results followed by STT and Conformer CTC.

Table 1. Mean (std) results obtained by the different ASR models. In bold, the best results.

Model	WER	WER_N
Conformer-CTC Large	0.393 (0.158)	0.242 (0.154)
Conformer-Transducer Large	0.229 (0.068)	0.068 (0.046)
PocketSphinx	1.720 (0.374)	1.709 (0.373)
STT Quartznet	0.317 (0.093)	0.176 (0.093)
Wav2Vec	0.312 (0.098)	0.169 (0.090)
Whisper tiny	0.216 (0.086)	0.129 (0.073)
Whisper base	0.169 (0.073)	0.083 (0.053)
Whisper small	0.132 (0.064)	0.053 (0.041)
Whisper small fine-tuned	0.872 (0.443)	0.847 (0.452)
Whisper medium	0.112 (0.060)	**0.042 (0.037)**
Whisper medium fine-tuned	0.224 (0.177)	0.140 (0.179)
Whisper large	0.111 (0.063)	**0.042 (0.036)**
Whisper large-v2	**0.108 (0.062)**	0.044 (0.046)

If we compare the different versions of Whisper, the performance of the models increases with the size. Namely, the best results are obtained with the medium, large, and large-v2 versions of the Whisper model that achieved a similar WER value (0.112, 0.111, and 0.108) and obtained a similar WER_N, with a value of 0.042 in the case of medium and large versions, and 0.044 in the case of the large-v2 version. In contrast with the success of fine-tuning in other areas like Computer Vision or Natural Language Processing [21], the fine-tuned versions of the Whisper models obtain worse parts than their original counterparts, the Whisper small fine-tuned obtained a WER of 0.872 (the original version achieved 0.132), and the medium fine-tuned version obtained a WER of 0.2241 (the original counterpart achieved 0.112). This might happen because the fine-tuned versions of Whisper are over-tuned on the Common Voice dataset (the medium model obtained state-of-the-art results) and they are no able to generalize to other kinds of audios. On the contrary, the original versions of Whisper were trained on a diverse multi-lingual dataset, which make them robust to different conditions. Further research is necessary to confirm this hypothesis using other datasets.

We have also studied the inference times of each ASR models for a video that lasts 7 min. To this aim, we have used a GPU NVIDIA GeForce RTX 3080, and the results are presented in Table 2 — note that the fine-tuned versions of Whisper and their original counterparts are the same model with different weights, and the same happens with the large and large-v2 versions; therefore, only one of the versions of those models has been included. The worst models are PocketSphinx and Wav2Vec that took more than 5 min to process the video. On the contrary, the fastest models are the NVIDIA models that can process the video in less than 12 s — this shows the optimizations that have been introduced in the nemo library to use GPUs. In the case of the Whisper models, the bigger the model, the slower. Namely, the tiny version took approximately 30 s to process the video, but the large model took almost 3 min. If we compare the medium and large versions of Whisper, that were the models with a smaller WER; the medium version took one minute less than the large model. Hence, we can conclude that the Whisper medium model provides the best trade-off between accuracy and speed, but if speed is more important, it is better to use the Conformer-Transducer model.

Table 2. Inference times of the ASR models. In bold, the best result.

Model	Time (secs)
Conformer-CTC Large	11.047
Conformer-Transducer Large	8.625
PocketSphinx	522.13
STT Quartznet	**7.729**
Wav2Vec	306.54
Whisper tiny	33.30
Whisper base	37.29
Whisper small	60.83
Whisper medium	106.67
Whisper large	167.29

For our project, and since we did not have a time constraint, we have employed the Whisper medium model to generate the transcription of the 3246 videos from the YouTube channel of the Government of La Rioja[3]. It is worth mentioning that Whisper models not only generate subtitles, but also phrase-level timestamps that indicate when the text must be shown. Hence, the subtitles can be easily uploaded to YouTube or shown in video players, see Fig. 1.

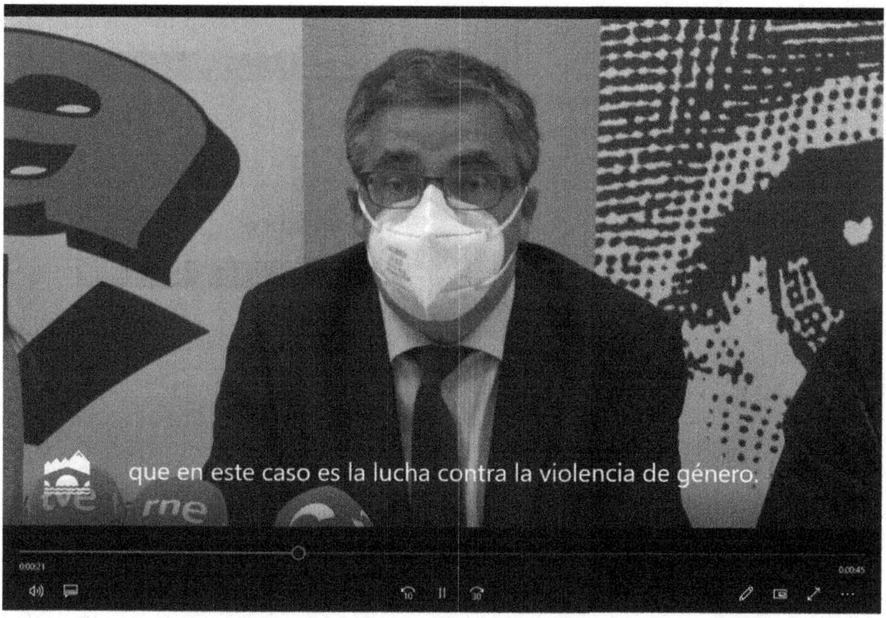

Fig. 1. Video player showing the transcription generated by the Whisper model.

Finally, to facilitate the automatic transcription for other videos, we have created a HuggingFace space[4], see Fig. 2. This space is a freely available web application where any user can provide a URL from a Spanish YouTube video and obtain automatically the transcription of the video.

[3] The transcriptions of the videos are available at https://github.com/mirenmirari/subtitulos_canalgobierno.

[4] Available at https://huggingface.co/spaces/mirari/Whisper-Youtube.

Fig. 2. HuggingFace space to automatically obtain transcription of YouTube videos.

4 Conclusions and Further Work

In this work, we have analyzed several ASR systems to automatically transcribe the videos of the official webpage of the Government of La Rioja. Our study has shown that the Whisper medium model produces the best transcriptions in a reasonable time, obtaining better results than the large version of this model. In addition, we have shown that, for our particular dataset, fine-tuning Whisper models on a dataset of Spanish audios produces worse models than their original counterparts.

The final result of this project is a webpage that contains the transcriptions of all the videos available in the YouTube channel of the Government of La Rioja; and the project can be applied to any Spanish YouTube channel. Therefore, this is a step towards improving the accessibility of the information and contents produced by Spanish public administrations.

As further work, several tasks remain to improve the accessibility of video transcriptions. First of all, none of the analyzed models can produce transcriptions on real time, this is an important for facilitating the access to stream events. Moreover, the models do not incorporate information about who is speaking; so, dialogues might be difficult to follow. Finally, it would be interesting to produce simplified transcriptions or summaries of videos, and apply the ASR technology to other contents like podcasts.

References

1. Ardila, R., et al.: Common voice: a massively-multilingual speech corpus. arXiv preprint arXiv:1912.06670 (2019)
2. Baevski, A., Zhou, Y., Mohamed, A., Auli, M.: wav2vec 2.0: a framework for self-supervised learning of speech representations. Adv. Neural Inf. Process. Syst. **33**, 12449–12460 (2020)
3. Chan, W., Jaitly, N., Le, Q.V., Vinyals, O.: Listen, attend and spell. arXiv preprint arXiv:1508.01211 (2015)
4. de España, C.G.: Ley 34/2002, de 11 de julio, de servicios de la sociedad de la información y de comercio electrónico. No 166 12 (2002)
5. Gulati, A., et al.: Conformer: convolution-augmented transformer for speech recognition. arXiv preprint arXiv:2005.08100 (2020)
6. Hannun, A., et al.: Deep speech: scaling up end-to-end speech recognition. arXiv preprint arXiv:1412.5567 (2014)
7. Hernandez Mena, C.D.: Acoustic model in spanish: stt_es_quartznet15x5_ft_ep53_944h. (2022). https://huggingface.co/carlosdanielhernandezmena/stt_es_quartznet15x5_ft_ep53_944h
8. Hong, R., et al.: Video accessibility enhancement for hearing-impaired users. ACM Trans. Multimed. Comput. Commun. Appl. (TOMM) **7**(1), 1–19 (2011)
9. Hrinchuk, O., et al.: Nvidia nemo offline speech translation systems for IWSLT 2022. In: Proceedings of the 19th International Conference on Spoken Language Translation (IWSLT 2022), pp. 225–231 (2022)
10. Hugging Face: Hugging Face Hub (2022). https://huggingface.co/docs/hub/index
11. Huggins-Daines, D., Kumar, M., Chan, A., Black, A.W., Ravishankar, M., Rudnicky, A.I.: Pocketsphinx: a free, real-time continuous speech recognition system for hand-held devices. In: 2006 IEEE International Conference on Acoustics Speech and Signal Processing Proceedings, vol. 1, p. I. IEEE (2006)
12. Jurafsky, D., Martin, J.H.: Speech and language processing (3rd draft ed.), 2019 (2022)
13. Kriman, S., et al.: QuartzNet: deep automatic speech recognition with 1d time-channel separable convolutions. In: ICASSP 2020–2020 IEEE International Conference on Acoustics, Speech and Signal Processing (ICASSP), pp. 6124–6128. IEEE (2020)
14. Levenshtein, V.I., et al.: Binary codes capable of correcting deletions, insertions, and reversals. In: Soviet physics doklady, vol. 10, pp. 707–710. Soviet Union (1966)
15. Li, J., et al.: Jasper: an end-to-end convolutional neural acoustic model. arXiv preprint arXiv:1904.03288 (2019)
16. Malik, M., Malik, M.K., Mehmood, K., Makhdoom, I.: Automatic speech recognition: a survey. Multimed. Tools Appl. **80**, 9411–9457 (2021)
17. Pratap, V., Xu, Q., Sriram, A., Synnaeve, G., Collobert, R.: MLS: a large-scale multilingual dataset for speech research. arXiv preprint arXiv:2012.03411 (2020)
18. Radford, A., Kim, J.W., Xu, T., Brockman, G., McLeavey, C., Sutskever, I.: Robust speech recognition via large-scale weak supervision. arXiv preprint arXiv:2212.04356 (2022)

19. Sharif Razavian, A., Azizpour, H., Sullivan, J., Carlsson, S.: CNN features off-the-shelf: an astounding baseline for recognition. In: Proceedings of the IEEE Conference on Computer Vision and Pattern Recognition Workshops, pp. 806–813 (2014)
20. Woodard, J., Nelson, J.: An information theoretic measure of speech recognition performance. In: Workshop on Standardisation for Speech I/O Technology, Naval Air Development Center, Warminster, PA (1982)
21. Zhuang, F., et al.: A comprehensive survey on transfer learning. Proc. IEEE **109**(1), 43–76 (2020)

Estimation of the Distribution of Body Mass Index (BMI) with Sparse and Low-Quality Data. The Case of the Chilean Adult Population

Fernanda Suazo-Morales[1,2] and Óscar C. Vásquez[1,2(✉)]

[1] Faculty of Engineering, Program for the Development of Sustainable Production Systems (PDSPS), University of Santiago of Chile (USACH), Santiago, Chile
oscar.vasquez@usach.cl

[2] Faculty of Engineering, Industrial Engineering Department, University of Santiago of Chile (USACH), Santiago, Chile

Abstract. Obesity is a non-communicable disease that has a major impact on people's health, increasing the risk of other chronic diseases such as diabetes, hypertension, and cardiovascular problems. Usually, the nutritional status of the population is determined by the body mass index (BMI) applied on a population sample via a national health survey (NHS), whose results are extrapolated. Except for highlighted cases such as the United States of America, these NHSs are infrequently carried out with different sampling methodologies. The outcomes are sparse and low-quality data, which complicate the estimation and forecasting of the population's BMI distribution. In this work, this problem is addressed by considering the case of Chile, one of the countries with the highest prevalence of obesity, with an NHS every 7 years. Our approach proposes a maximum entropy optimization model to estimate the probability transition between different nutritional states, considering age and sex, which is based on the analogy with the determination of the origin-destination trip matrix used in the transport setting. The obtained results show that for the year 2024, there will be an increase of 798,898 (35%) and 758,124 (30%) men and women respectively, with overweight and obesity.

Keywords: Sparse Data · Obesity · Non-linear Programming · Transition Probabilities

1 Introduction

Nutritional status can be evaluated using the body mass index (BMI), which is calculated by the ratio of weight in kilograms to the square of height in meters (kg/m^2). This indicator defines the categories in which the individual is classified according to the value of their BMI: (a) Underweight, BMI $<$ 18.5; (b) Normal weight, 18.8 $\leq$ BMI $\leq$ 24.9; (c) Overweight, 25.0 $\leq$ BMI $\leq$ 30.0; (d) Obesity, 30.0 $<$ BMI [3].

B. Dorronsoro et al. (Eds.): OLA 2023, CCIS 1824, pp. 403–413, 2023.
https://doi.org/10.1007/978-3-031-34020-8_31

It is alarming that obesity has become a constantly increasing public health problem worldwide, considered a global pandemic. According to the World Health Organization (WHO), in 2016, more than 1.9 billion adults worldwide were overweight, of which more than 650 million were obese. Additionally, WHO reports that obesity has tripled in many countries since 1975 [12]. A study published in The Lancet in 2016 [4] suggests that if current trends in obesity prevalence continue, it would not be possible to achieve the global non-communicable diseases goal, which aims to maintain 2010 levels by 2025. In that study, it is estimated that if current trends continue, the global prevalence of obesity will reach 18% in men and exceed 21% in women.

Obesity is a growing public health problem in the United States of America, with a significantly higher prevalence compared to previous decades. According to the Centers for Disease Control and Prevention (CDC), in 2019, the prevalence reached 41.9%, showing an increase of 11.4%, while the prevalence of severe obesity reached 9.2% according to the National Health and Nutrition Examination Survey (NHANES) data in 2021. In addition, obesity has been observed to be more common in adults over 40 years of age and is more prevalent among certain populations, such as African Americans, Latinos, and people with low incomes [6].

In recent decades, a significant increase in the prevalence of obesity worldwide has been observed. According to the Aprovian study in 2016 [1], an increase of 27.5% in adults and 47% in children has been recorded. Additionally, the author established that an increase of 5 units in the Body Mass Index (BMI) above $25\,\text{kg/m}^2$ is associated with an increase in overall mortality of 29%, in vascular mortality of 41%, and in diabetes-related mortality of 210%.

Obesity also has a significant economic impact. In fact, a 77% increase in drug costs and a 36% increase in annual medical costs are due to comorbidities associated with obesity, such as diabetes and hypertension [13]. Additionally, Kelly et al. [9] suggest that a higher level of BMI may have negative effects on economic growth. For Latin American countries, Elgart et al. [7] note that BMI shows a high incidence in the expenditure of drugs for the treatment of type 2 diabetes and cardiovascular diseases in Latin American countries. In particular, Cuadrado [5] determines an average of 2.29% of total annual health expenditure is attributed to obesity and related conditions for the Chilean case.

1.1 Related Work

The estimation of obesity prevalence and transition rates in developing countries, such as Chile, is crucial for understanding and addressing the public health phenomenon, which often faces challenges due to the sparse and low-quality data. However, one way to address this limitation is through the use of mathematical and statistical models.

In recent studies [10,17], it has been seen an increase in the use of systems dynamics (SDM) and agent-based models (ABM) in the field of health, the first of which is particularly useful in non-communicable diseases such as obesity. Differential equations models are seen that consider social and non-social parameters for measurement, with data from the United States and the United Kingdom [15]. In Avalos et al. [2] the same line of research is studied, proposing a non-linear programming model to characterize the population in a disaggregated way, assuming that transition probabilities depend on gender and age. Another study, published in 2022, uses heteroscedastic longitudinal mixed models to adapt to data scarcity and predict trends of overweight and growth delay in the European region [14]. It is important to note that the use of these models not only allows adapting to data scarcity but also allows for a more accurate characterization of the population, which is essential for implementing effective prevention and treatment strategies. Additionally, the use of mathematical and statistical models also allows for simulating future scenarios and thus taking preventive measures and preparing for changes in obesity prevalence.

1.2 Our Contribution

This research focuses on addressing the problem of estimating the nutritional status of the population with sparse and low-quality data. Chile has been chosen as a case study, as according to the State of Food Security and Nutrition in the World 2021 [16], a prevalence of 28% of obesity in adults over 18 years old has been identified in this country, which is above the average for Latin America and the Caribbean, which was 24.2%. Additionally, Chile has a shortage of data in its health surveys, as only three surveys have been conducted between 2003 and 2017. Therefore, for this research, the national health surveys (NHS) and data from the National Institute of Statistics (NIS) will be used to understand the population projections. The model used is based on the maximum entropy transport model [11], as this model determines transition rates assuming high levels of misinformation or uncertainty. In this model, only data by age range and Body Mass Index (BMI) will be taken into account, making a differentiation between men and women. Finally, an estimation for the year 2024 of the levels of obesity in Chile will be given.

2 Material and Methods

2.1 Statement of the Problem

The issue at hand focuses on estimating BMI of the adult Chilean population using spaced and lower quality data. To tackle this problem, a method composed of several phases is employed. Firstly, a maximum entropy model is used to determine the existing transition rates between the years 2003 and 2010, as well as between 2010 and 2017. Then, these rates are projected into the future starting from the year 2017, normalizing the population for each age range, this is due to the expansion factor of the health surveys consider the population projections made by the National Institute of Statistics. Finally, the simple average between both projected rates for the years 2017–2024 is calculated and the estimated population is obtained.

2.2 Mathematical Model

Initially, the population data is distributed among different age ranges i and BMI j, denoted as $x_{i,j}$ for each combination of origin. Since we work with populations, and in order to maintain the equality constraints present in the model, it can also be formulated under the approach of probability distributions. The same applies for the destination data. Therefore, we consider the variable $O_{i,j} \in [0,1]$ and $D_{i,j} \in [0,1]$.

$$O_{ij} = \frac{x_{ij}}{\sum\sum x_{ij}} \qquad\qquad D_{ij} = \frac{d_{i'j'}}{\sum\sum d_{i'j'}}$$

The mathematical model used is based on a maximum entropy model in which the transition rates, $t_{ij,i'j'}$, are considered, with O_{ij} as the source data where i is the age range, j is the BMI, and $D_{i'j'}$ as the destination data where i' is the age range and j' is the BMI.

$$\max \; - \sum_{i\in\mathcal{I}}\sum_{j\in\mathcal{J}}\sum_{i'\in\mathcal{I}}\sum_{j'\in\mathcal{J}} t_{iji'j'} \ln t_{iji'j'} \tag{1}$$

$$\sum_{i'\in\mathcal{I}}\sum_{j'\in\mathcal{J}} t_{iji'j'} = O_{ij} \qquad \forall i \in \mathcal{I}; \forall j \in \mathcal{J} \qquad (\tau_{ij}) \tag{2}$$

$$\sum_{i\in\mathcal{I}}\sum_{j\in\mathcal{J}} t_{iji'j'} = D_{i'j'} \qquad \forall i' \in \mathcal{I}; \forall j' \in \mathcal{J} \qquad (\gamma_{i'j'}) \tag{3}$$

$$\sum_{i\in\mathcal{I}}\sum_{j\in\mathcal{J}}\sum_{i'\in\mathcal{I}}\sum_{j'\in\mathcal{J}} t_{iji'j'} C_{iji'j'} = \mathcal{C}' \qquad (\beta) \tag{4}$$

$$t_{iji'j'} > 0 \tag{5}$$

Expression (1) defines the objective function. Constraints (2) and (3) are used to ensure conservation of the input and output population. Constraint (4) states a value for the sum over the proportional cost $C_{i,j,i'.j'}$ associated with the transition rate $t_{i,j,i'.j'}$. Additionally, expression (5) defines the nature of the variables. It should be noted that the values of O_{ij} and $D_{i'j'}$ are previously known. In order to provide flexibility to the formulation, intervals were chosen to be used for the source and destination data, which allows the model to maintain feasibility.

$$\max - \sum_{i \in \mathcal{I}} \sum_{j \in \mathcal{J}} \sum_{i' \in \mathcal{I}} \sum_{j' \in \mathcal{J}} t_{iji'j'} \ln t_{iji'j'} \tag{6}$$

$$\sum_{i' \in \mathcal{I}} \sum_{j' \in \mathcal{J}} t_{iji'j'} = 0.95 \cdot \rho_{ij} \cdot O_{ij} + 1.05 \cdot (1 - \rho_{ij}) \cdot O_{ij} \qquad \forall i \in \mathcal{I}; \forall j \in \mathcal{J} \tag{7}$$

$$\sum_{i \in \mathcal{I}} \sum_{j \in \mathcal{J}} t_{iji'j'} = 0.95 \cdot \alpha_{i'j'} \cdot D_{i'j'} + 1.05 \cdot (1 - \alpha_{i'j'}) \cdot D_{i'j'} \quad \forall i' \in \mathcal{I}; \forall j' \in \mathcal{J} \tag{8}$$

$$\sum_{i \in \mathcal{I}} \sum_{j \in \mathcal{J}} \sum_{i' \in \mathcal{I}} \sum_{j' \in \mathcal{J}} t_{iji'j'} = 1 \tag{9}$$

$$t_{iji'j'} > 0 \tag{10}$$

Expression (6) is used as the objective function, aimed at maximizing entropy. Additionally, constraints (7) and (8) are employed to ensure the conservation of the input and output population. Since only one data point is available for each O_{ij} and $D_{i'j'}$, a range between 95% and 105% is established, and the variables ρ_{ij} and $\alpha_{i'j'}$ are incorporated, respectively, to provide slack.

Furthermore, constraint (9) is included to limit the sum of t_{ijij} to one. It is worth noting that there are constraints that restrict population mobility based on their age range. This is because people cannot decrease their age, and according to the data used in the research, they can only stay in the same age range or move to the next one. Finally, reference is made to the nature of the variables in Eq. (10).

2.3 Case Study

This case study focuses on the analysis of health data of the Chilean population, taking the country itself as a reference, which is in a process of development. To do this, two main sources of information have been used: health surveys conducted every 7 years and population projections developed by the National Institute of Statistics (NIS).

The National Health Survey of Chile is a valuable resource for analyzing the health status of the Chilean population. One of the indicators evaluated in the survey is the BMI, which is used to determine if an individual is in a healthy weight, overweight or obese. The data collected is divided into 7 age intervals, each one considering people from 10 years, starting at 15 years and ending with people of 75 years or more. In addition, the data is also segmented into 20 different BMI categories.

However, because the survey has only been carried out in the years 2003, 2009–2010, and 2016–2017, the data is scarce for long-term research. Despite this, it is expected that the information collected will be valuable for understanding the evolution of health in Chile and taking measures to improve it.

3 Results

Graphs are presented showing the proportions obtained for the year 2024, projecting the transition rates for the years 2003–2010 and the transition rates for the years 2010–2017, normalizing the proportions for each age range, for both men and women (see Fig. 1 and Fig. 2, respectively). Subsequently, a simple average of these data was obtained to obtain the population estimate for men for 2024 (see Table 2) and the population estimate for women for 2024 (see Table 3).

From the obtained graphs, it is notable that for both men and women and for different age ranges, the proportions are higher among those with a BMI of 23–24, which corresponds to a classification of normal weight, and a BMI of 25–30, which corresponds to individuals with overweight. Additionally, it can be seen that individuals in the youngest age range, 15–24 years old, have the highest proportions of obesity.

The National Institute of Statistics (NIS) projects that by 2024, the population of men over 15 years of age will reach 8.027.132 people, and the population of women over 15 years of age will reach 8.361.220 people [8] as shown in Table 1.

Table 1. Population projection 2024 INE

Age range	15–24	25–34	35–44	45–54	55–64	65–75	75+
Men	1.297.516	1.629.350	1.506.681	1.279.988	1.107.086	757.796	448.715
Women	1.255.894	1.584.519	1.475.987	1.298.457	1.183.687	881.158	681.518

Table 2. Male estimates 2024

Estimated ratios								Estimated population of men							
IMC	Age range							IMC	Age range						
	15–24	25–34	35–44	45–54	55–64	65–75	75+		15–24	25–34	35–44	45–54	55–64	65–75	75+
36	0,0202	0,0011	0,0002	0,0002	0,0001	0,0006	0,0024	36	26.154	1.827	327	246	136	470	1.084
35	0,0241	0,0030	0,0008	0,0007	0,0005	0,0017	0,0048	35	31.257	4.833	1.243	922	535	1.321	2.168
34	0,0405	0,0080	0,0029	0,0025	0,0017	0,0050	0,0110	34	52.613	13.026	4.341	3.176	1.930	3.772	4.957
33	0,0599	0,0178	0,0080	0,0068	0,0051	0,0117	0,0215	33	77.665	28.956	12.104	8.760	5.594	8.902	9.658
32	0,0788	0,0333	0,0183	0,0156	0,0121	0,0234	0,0365	32	102.211	54.266	27.643	19.905	13.362	17.751	16.360
31	0,0938	0,0491	0,0351	0,0298	0,0243	0,0402	0,0546	31	121.658	79.951	52.907	38.163	26.913	30.496	24.506
30	0,1022	0,0700	0,0575	0,0492	0,0421	0,0606	0,0734	30	132.618	114.113	86.621	63.017	46.583	45.913	32.922
29	0,1431	0,1292	0,1096	0,1057	0,0953	0,1186	0,1286	29	185.704	210.440	165.156	135.337	105.513	89.869	57.724
28	0,0574	0,0629	0,0643	0,0573	0,0536	0,0604	0,0610	28	74.455	102.517	96.947	73.322	59.393	45.757	27.366
27	0,0866	0,1084	0,1062	0,1065	0,1027	0,1076	0,1038	27	112.358	176.634	159.953	136.373	113.685	81.506	46.559
26	0,0732	0,1056	0,1094	0,1128	0,1122	0,1088	0,1003	26	94.936	171.996	164.905	144.381	124.210	82.459	45.023
25	0,0591	0,0958	0,1047	0,1100	0,1124	0,1022	0,0912	25	76.688	156.028	157.797	140.777	124.405	77.468	40.920
24	0,0610	0,1097	0,1237	0,1287	0,1419	0,1216	0,1057	24	79.086	178.736	186.330	164.746	157.138	92.159	47.446
23	0,0194	0,0392	0,0479	0,0487	0,0519	0,0428	0,0367	23	25.196	63.854	72.143	62.340	57.513	32.410	16.465
22	0,0251	0,0501	0,0634	0,0675	0,0727	0,0586	0,0500	22	32.587	81.676	95.549	86.414	80.512	44.371	22.443
21	0,0178	0,0382	0,0484	0,0512	0,0556	0,0439	0,0375	21	23.144	62.273	72.969	65.475	61.541	33.259	16.817
20	0,0124	0,0276	0,0345	0,0369	0,0403	0,0315	0,0270	20	16.075	45.046	51.918	47.284	44.650	23.894	12.138
19	0,0086	0,0191	0,0236	0,0255	0,0279	0,0218	0,0189	19	11.212	31.122	35.565	32.700	30.900	16.525	8.469
18	0,0062	0,0125	0,0156	0,0170	0,0185	0,0146	0,0128	18	8.107	20.291	23.564	21.747	20.487	11.047	5.729
17	0,0106	0,0195	0,0257	0,0273	0,0290	0,0243	0,0222	17	13.792	31.765	38.698	34.900	32.086	18.448	9.960

Table 3. Female estimates 2024

Estimated ratios								Estimated population of women							
IMC	Age range							IMC	Age range						
	15–24	25–34	35–44	45–54	55–64	65–75	75+		15–24	25–34	35–44	45–54	55–64	65–75	75+
36	0,0235	0,0058	0,0012	0,0004	0,0002	0,0003	0,0048	36	29.544	9.125	1.744	455	284	223	3.295
35	0,0262	0,0093	0,0026	0,0010	0,0008	0,0008	0,0070	35	32.886	14.727	3.837	1.255	917	723	4.753
34	0,0426	0,0186	0,0064	0,0028	0,0024	0,0026	0,0136	34	53.489	29.518	9.488	3.609	2.882	2.278	9.237
33	0,0613	0,0303	0,0136	0,0067	0,0063	0,0067	0,0233	33	76.930	47.988	20.103	8.739	7.433	5.876	15.902
32	0,0792	0,0449	0,0252	0,0140	0,0136	0,0144	0,0361	32	99.405	71.067	37.176	18.175	16.121	12.723	24.578
31	0,0931	0,0617	0,0412	0,0254	0,0254	0,0268	0,0506	31	116.951	97.739	60.872	33.022	30.050	23.632	34.508
30	0,1009	0,0772	0,0577	0,0409	0,0414	0,0435	0,0652	30	126.718	122.269	85.193	53.151	49.013	38.338	44.463
29	0,1411	0,1264	0,1062	0,0879	0,0893	0,0930	0,1117	29	177.168	200.335	156.707	114.126	105.683	81.958	76.127
28	0,0567	0,0574	0,0600	0,0484	0,0491	0,0507	0,0528	28	71.151	91.001	88.596	62.878	58.092	44.695	35.993
27	0,0857	0,0953	0,0978	0,0924	0,0932	0,0956	0,0907	27	107.665	151.009	144.321	119.938	110.268	84.214	61.832
26	0,0727	0,0902	0,1007	0,1015	0,1019	0,1035	0,0898	26	91.328	142.991	148.698	131.801	120.642	91.242	61.221
25	0,0590	0,0812	0,0968	0,1007	0,1035	0,1041	0,0846	25	74.039	128.666	142.806	130.777	122.504	91.707	57.662
24	0,0609	0,0943	0,1184	0,1256	0,1349	0,1340	0,1038	24	76.535	149.479	174.827	163.092	159.684	118.049	70.771
23	0,0194	0,0331	0,0472	0,0527	0,0513	0,0504	0,0382	23	24.377	52.488	69.700	68.392	60.670	44.372	26.059
22	0,0250	0,0459	0,0623	0,0769	0,0744	0,0725	0,0549	22	31.446	72.710	91.883	99.878	88.080	63.888	37.449
21	0,0177	0,0356	0,0506	0,0624	0,0600	0,0579	0,0443	21	22.193	56.449	74.654	81.048	71.008	50.982	30.225
20	0,0121	0,0274	0,0371	0,0484	0,0463	0,0442	0,0348	20	15.255	43.411	54.831	62.832	54.752	38.929	23.687
19	0,0082	0,0211	0,0262	0,0360	0,0343	0,0324	0,0265	19	10.240	33.387	38.701	46.730	40.548	28.566	18.079
18	0,0057	0,0152	0,0179	0,0258	0,0245	0,0229	0,0198	18	7.161	24.011	26.349	33.466	28.949	20.221	13.474
17	0,0091	0,0291	0,0308	0,0501	0,0474	0,0437	0,0472	17	11.411	46.148	45.503	65.093	56.107	38.544	32.201

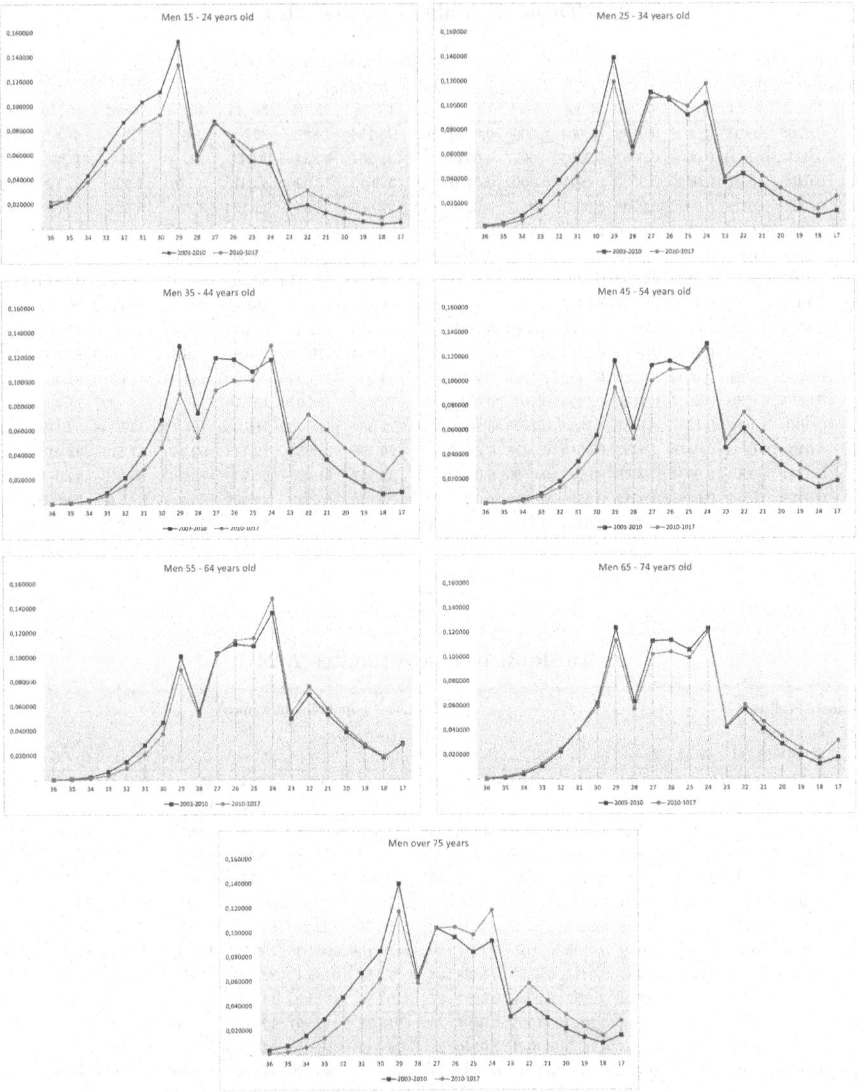

Fig. 1. Distribution graphs for 2024 by age range in males

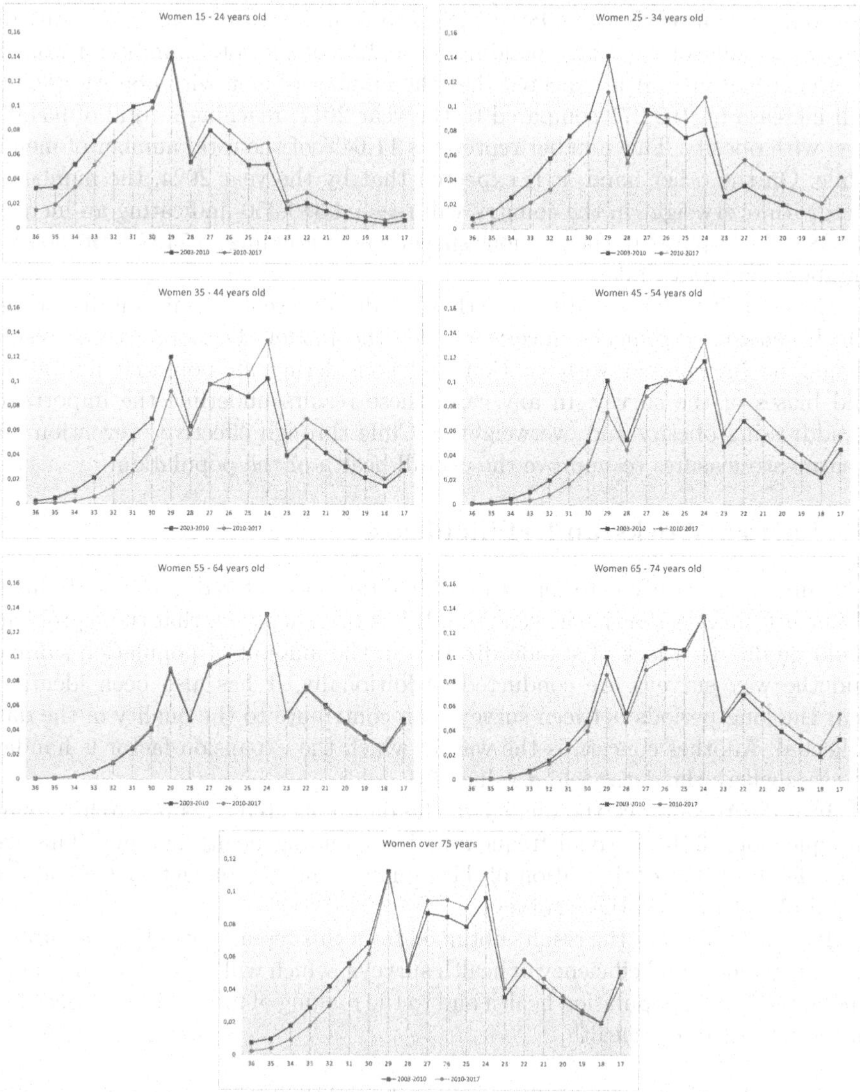

Fig. 2. Distribution graphs for 2024 by age range in females

4 Discussion and Conclusions

According to recent data provided by health surveys in Chile, it is expected that the number of women with obesity will increase by 295,277 compared to the year 2017, resulting in a total of 1,073,261 women with obesity in the country. This number represents 12.84% of the total number of women in Chile. Additionally, it is expected that by the year 2024, the number of women with overweight in the country will reach 4,189,988, which means an increase of 462,847 compared to the previous survey, corresponding to 50.22% of the total number of women.

Regarding men, it is expected that the number of men with obesity in Chile will increase by 160,271 compared to the year 2017, reaching a total of 934,071 men with obesity. This number represents 11.64% of the total number of men in Chile. On the other hand, it is expected that by the year 2024, the number of men with overweight in the country will reach 4,380,350, indicating an increase of 638,627 compared to the previous survey, corresponding to 54.57% of the total number of men in Chile.

However, it is important to note that the data presents significant disparities, which generate significant uncertainty for the model. Therefore, these results should be interpreted with caution and considering all potential limitations and biases of the survey. In any case, these results underline the importance of addressing obesity and overweight in Chile through effective prevention and treatment measures to improve the overall health of the population.

5 Future Works and Difficulties

The aim of this study is to report on difficulties encountered in the work due to the health surveys that have been used. It has been observed that these problems could be due to a lack of standardization in the amount of population samples and the way surveys are conducted. Additionally, it has also been identified that the long periods between surveys can contribute to the quality of the data collected. Another concern is the way in which the expansion factor is handled in cases where the data yield a value of 0.

In order to improve the quality of the data collected, this research presents the question of the optimal frequency of conducting health surveys. This will allow for both the optimization of data quality and the budgetary expenditure required to carry out these surveys.

It is expected that the results obtained from this research will help to improve the effectiveness and efficiency of health surveys, which will contribute to a better understanding of population health and to the making of more informed decisions in the field of public health.

Acknowledgements. The authors are grateful for partial support from ANID, FONDECYT No1211640.

References

1. Apovian, C.M.: Obesity: definition, comorbidities, causes, and burden. Am. J. Manag. Care **22**, S176–S185 (2016). https://www.ajmc.com/view/obesity-definition-comorbidities-causes-burden
2. Ávalos, D., et al.: Mathematical model for estimating nutritional status of the population with poor data quality in developing countries: the case of Chile. In: Parlier, G., Liberatore, F., Demange, M. (eds.) Proceedings of the 10th International Conference on Operations Research and Enterprise Systems, ICORES 2021, pp. 408–415. SciTePress (2021)

3. CDC, HW: Assessing your weight (2021). Accessed 26 Jan 2023
4. NRF, Collaboration et al.: Trends in adult body-mass index in 200 countries from 1975 to 2014: a pooled analysis of 1698 population-based measurement studies with 19 · 2 million participants. Lancet **387**(10026), 1377–1396 (2016)
5. Cuadrado, C.: The health and economic burden of obesity in Chile–an epidemiological and economic simulation model. Value Health **19**(7), A584 (2016)
6. Centers for Disease Control and Prevention et al.: Adult obesity facts. Centers for disease control and prevention (2021). https://www.cdc.gov/obesity/data/adult.html. Accessed 26 Jan 2023
7. Elgart, J.F., Prestes, M., Gonzalez, L., Rucci, E., Gagliardino, J.J., QUALIDIAB Net Study Group: Relation between cost of drug treatment and body mass index in people with type 2 diabetes in Latin America. PLoS ONE **12**(12), e0189755 (2017)
8. Instituto Nacional de Estadísticas: Proyecciones de población. https://www.ine.gob.cl/estadisticas/sociales/demografia-y-vitales/proyecciones-de-poblacion. Accessed 26 Jan 2023
9. Kelly, I.R., Doytch, N., Dave, D.: How does body mass index affect economic growth? A comparative analysis of countries by levels of economic development. Econ. Hum. Biol. **34**, 58–73 (2019)
10. Liu, S., Xue, H., Li, Y., Xu, J., Wang, Y.: Investigating the diffusion of agent-based modelling and system dynamics modelling in population health and healthcare research. Syst. Res. Behav. Sci. **35**(2), 203–215 (2018)
11. López-Ospina, H., Cortés, C.E., Pérez, J., Peña, R., Figueroa-García, J.C., Urrutia-Mosquera, J.: A maximum entropy optimization model for origin-destination trip matrix estimation with fuzzy entropic parameters. Transportmetrica A: Transp. Sci. **18**(3), 963–1000 (2022)
12. World Health Organization: Obesidad y sobrepeso (2021). https://www.who.int/news-room/fact-sheets/detail/obesity-and-overweight. Accessed 26 Jan 2023
13. Reátegui, R., Ratté, S., Bautista-Valarezo, E., Duque, V.: Cluster analysis of obesity disease based on comorbidities extracted from clinical notes. J. Med. Syst. **43**, 1–9 (2019)
14. Saraswati, C.M., et al.: Estimating childhood stunting and overweight trends in the European region from sparse longitudinal data. J. Nutr. **152**(7), 1773–1782 (2022)
15. Thomas, D.M., et al.: Dynamic model predicting overweight, obesity, and extreme obesity prevalence trends. Obesity **22**(2), 590–597 (2014)
16. UNICEF, et al.: The state of food security and nutrition in the world 2021 (2021)
17. Xue, H., Slivka, L., Igusa, T., Huang, T., Wang, Y.: Applications of systems modelling in obesity research. Obes. Rev. **19**(9), 1293–1308 (2018)

A New Automated Customer Prioritization Method

Amira Ben Hadid[1,2](✉), Hamamache Kheddouci[1], and Seddik Hadjadj[2]

[1] Université de Lyon - LIRIS UMR 5205, Bâtiment Nautibus 43, bd du 11 novembre 1918, 69622 Villeurbanne, France
amira.ben-hadid@etu.univ-lyon1.fr, hamamache.kheddouci@univ-lyon1.fr
[2] Béton Direct, Lyon, France
s.hadjadj@betondirect.fr

Abstract. One of the most important business decisions is how to acquire customers. Indeed, whatever the size of the company, it is faced with the problem of limited resources. Therefore, it is necessary to know which customers are worthy of marketing activities and efforts and which are not. Knowing the journeys and characteristics of customers that lead to successful sales would allow businesses to optimize their spending by targeting those likely to make purchases. Customer prioritization involves assigning a score (i.e., a buying probability) to each possible lead generated for the business. An accurate scoring process can help marketing and sales teams prioritize and respond appropriately to selected leads in an optimal time frame, increasing their propensity to become customers. The purpose of this article is to develop a new automated method to prioritize customers. A complete process for implementing a prioritization solution is described. We present experiments that show positive results using a real-world dataset.

Keywords: Customer Prioritization · Lead Scoring · Data Cleaning · Data Quality

1 Introduction

Customer prioritization has become a common business practice and an important area of marketing research. Companies often set priorities and assign resources based on priorities when resources are limited. Indeed, client prioritization is defined as the degree to which customers are treated differently and preferentially in terms of marketing instruments according to their importance to the business [1]. On one hand, several studies have focused on the impact of prioritization on desirable and revenue-generating customer behaviors such as retention, loyalty, and positive word of mouth. The proponents of customer prioritization argue that firms use marketing tools differently for various levels of their customers. They affirm that such differentiation in marketing efforts allows companies to achieve high profits. Actually, firms can achieve greater effectiveness and efficiency by focusing these efforts on priority customers [3]. On the

B. Dorronsoro et al. (Eds.): OLA 2023, CCIS 1824, pp. 414–425, 2023.
https://doi.org/10.1007/978-3-031-34020-8_32

other hand, opponents of customer prioritization argue that customer prioritization is not always profitable. They assume that client prioritization can leave demoted clients dissatisfied [2]. As a result, these dissatisfied customers could fail or spread negative word of mouth, leading to lower sales and long-term profits.

Against this background, this article shows the positive impact of customer prioritization on the company's sales and work organisation.

In our work, we collaborate with a company specialized in the sale of ready-mix concrete. It acts as an intermediary between concrete factories and consumers, its work as follows:

1. The customer request an estimate on-line via a dedicated platform.
2. A sales advisor calls the customer back to validate his order.
3. If it works for both, the customer can pay for their order and the advisor will schedule the delivery with the plant concerned.

The second step of the work process is essential to the way the company operates. In this step, an estimate needs to be validated before the client pays and the order is placed. Indeed, considering the sensitivity of the product, the sales advisor should contact the customer to verify the order and delivery details. This verification stage guarantees a correct and compliant delivery to satisfy the client.

The company receives hundreds of requests every day which sales advisors must process it and call back customers to convert it into purchases. As the number of sales advisors is limited, they are not always able to process all requests on time. For this reason, we have decided to prioritize client requests. The idea is to assign a score (i.e., a buying probability) to each request and the higher the score the higher the priority is given to process the command. In this way, the sales advisors will not spend time randomly contacting all prospects and will only focus their efforts on the most likely converts. In this article, we describe the entire process followed to develop the scoring algorithm Fig. 1.

The rest of this article is organized into four sections. In the following Sect. 2, the data cleaning process is described. In Sect. 3, the prioritization algorithm is presented. Section 4 details the experimental setup and discusses the obtained results and Sect. 5 concludes the paper.

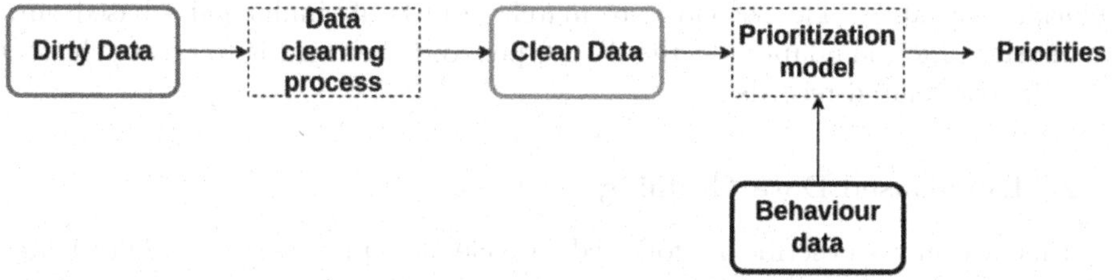

Fig. 1. General architecture of our proposed system.

2 Data Cleaning

To better understand market demands and extract features to be used to prioritize customers, data analysis is required. Data analysis is defined as a process of cleaning, transforming, and modeling data to discover useful information for business decision-making. It is important to know that high-quality decision depends not only on the appropriate decision-making methods but also on high-quality data. Actually, over the past several years, industry and academia have shown an increased interest in data quality and data cleaning issues [4]. One of the most anticipated data quality challenges, which becomes especially critical when data comes from multiple or unique real-world data sources, is duplication or non-uniqueness. Indeed, duplication is one of the main causes of poor data quality in databases. To resolve this issue, the process of detecting and correcting duplicate records is required.

In this section, we describe the data cleaning process.

2.1 Related Works

Data quality represents a great challenge in real-world datasets. Data quality and Database researchers have discussed various types of data anomalies types. Recent research suggests workflows and methodologies for cleaning and repairing data sets using different approaches, such as: statistical analysis [4], integrity constraints [6], neuronal networks and machine learning models [7]. In addition, a data cleaning process does not only require sophisticated methods, but it also requires contextual and integrative expertise that generates rules based on the nature of the data [8]. Experts can help answer the three key questions posed during data processing to find errors in the data: what to detect, how to detect, and where to detect. Within this context, there has been much discussion and research on the detection of duplicates in the data. It is important that the procedure be efficient and precise in order to produce high quality data [15].

Probably, the most popular one is the data similarity-based approach [9]. It identifies the similarity of the data to determine whether the data objects in question are two unique objects or duplicates. It is a matter of finding every possible duplication and eliminating them. The duplication detection is generally done in 3 steps. The first step is to standardize and to index records using blocking variables. The second is to match records of similar pairs based on a similarity function adapted to the data type. And the third is to create clusters of coherent related records.

2.2 Rule-Based Data Cleaning

In this section, we describe our followed data cleaning process. Figure 2 illustrates the general architecture of the system.

The process starts with the detection and correction of null and empty values then the elimination of outliers. Next we proceed to the detection and elimination of duplicates by following these steps:

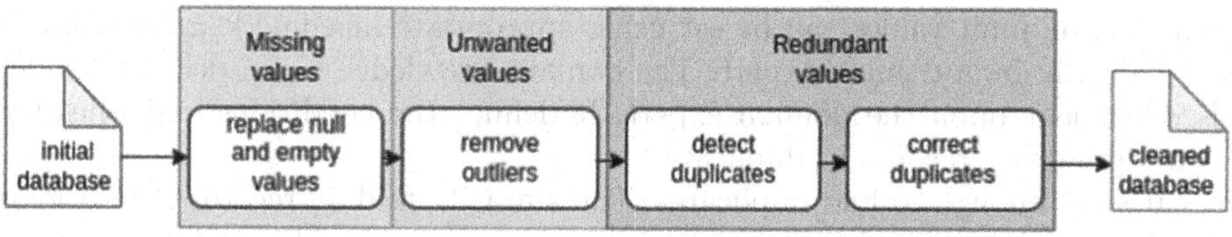

Fig. 2. Data cleaning process.

Prepare the Data. To start, data is brought in a standardized form to facilitate the identification of duplicates. This standardization may vary considerably depending on the field concerned. As an example, when data is strings all uppercase letters are often converted into lowercase letters or when we deal with dates it should be brought in a uniform format.

Search Area Definition. This step focuses on the definition of the search area. To detect duplicate records in a database D, the maximum number of possible comparisons is $|D|(|D| - 1)/2$ because each D record must be compared with all other records. The task of defining research areas is to reduce the number of comparisons and minimize the resources to be spent. Two ways to accomplish this are the Sorted Neighbourhood Method (SNM) and Blocking. Sorted Neighbourhood is a method where rows are sorted according to a suitable key. Next, a fixed-size window is moved across the rows and only the rows in the window are considered for comparison. The selection of the key is important in this method, it is necessary to ensure that the duplicates are reconciled in the order of the key. About Blocking, it is a method whereby the search area is divided into blocks where the detection of duplicates is carried out inside those blocks. Splitting into blocks may be done in various ways. The principle is to generate or use a block key, where all tuples with the same block key are grouped together in the same block for processing. In the literature, there are several Blocking-based and Sorting-based indexing techniques that allow reducing the possibly very large number of record pairs that need to be compared in the case of large datasets [10].

Matching and Clustering Duplicates. To compare two record pairs E1 and E2, similarity functions S = sim (E1, E2) are used. According to the type of attribute value, the similarity function can vary. For example, for dates, it could be the number of seconds between, or for numbers the difference could be an appropriate option. Using the similarity vector, a decision model is used to determine whether the tuples compared are duplicates or not. There are two most common ways to design a decision model: probability-based decision and domain knowledge-based decision. The probability-based decision determines two conditional probabilities: the probability that the two compared pairs are duplicates and the probability that they are not. The resulting probabilities are then compared to boundary values to determine whether or not the compared pairs match. The limit values can be set using appropriate machine learning models or manually by a domain expert. The domain knowledge-based decision model is dependent upon the domain expert. It defines the conditions and rules for considering two records as duplicates.

It's not enough to find duplicates. The aim is to find all records that belong to the same real-world entity so duplicates must be combined into clusters. Generally, this clustering is carried out using certain forms of transitivity [13]. Indeed, if E1 is a duplicate of E2, and E2 is a duplicate of E3, it is true that E1 is a duplicate of E3 and the three records belong to the same cluster. Finally, the recognized clusters are properly merged or the decision on the remaining records and those to be removed is made.

3 Automated Customer Prioritization

The use of data to resolve business problems and support business decisions has become standard practice today. A Harvard Business Review article by Andrew McAfee and Erik Brynjolfsson claims that *"Data-driven decisions tend to be better decisions. Leaders will either embrace this fact or be replaced by others who do"* [11]. One of the most important business decisions involves the acquisition of customers. Indeed, no matter how large the enterprise it is, it is confronted with problem of limited resources, which is in principle the central economic problem. Therefore, there is a need to know what clients are worth of marketing activities and efforts and what are not.

In this section, we provide a literature review of lead scoring. Then, we present our automated customer prioritization method.

3.1 Related Works

Lead scoring is used to help decision-makers to identify which leads to target. The idea is to assign scores to all leads based on how their features fit the pre-established profile of a converted client. In this way, the salesforce professionals will focus their efforts on prospects more likely to convert.

In traditional or manual lead scoring, a marketing expert or a senior sales executive assigns points to the customer's actions based on how important those actions are to the business. Figure 3 presents an example of traditional lead scoring. However, there are several problems related to it. Indeed, there is no statistical support for manual lead scoring. In addition, due to human nature, the lead scoring officer's decisions can be biased or based on prior prejudices. Also, it is very long to always adjust the scores manually and the time used could be spent more efficiently somewhere else.

On the other hand, smart or predictive lead scoring is based on a mathematical and statistical approach called propensity modeling [12]. It aims at predicting the chances that a visitor will perform certain actions (reservation, registration, purchase, etc.). Data Mining and Machine Learning algorithms are capable of automatically detecting useful patterns for lead scoring from historical

Fig. 3. Traditional lead scoring.

sales data. K. Prasad et al. [17] suggested a comparative analysis between Logistic Regression and SVM algorithms in the construction of propensity predictive models and evaluated their performance. R. Nygård et al. [18] proposed a supervised learning approach to lead scoring based on algorithms such as Decision Trees to predict the purchase probability. The authors discovered that the best performing algorithm is the Random Forest model. Y. Zhang et al. [19] used the machine leaning to identify the most valuable prospects. The authors compared the predictive capacity of Logistic Regression and Random Forest. The results showed that Logistic Regression model outperformed the other one.

3.2 Proposed Customer Prioritization Approach

The lead scoring allows companies to optimize their spending by targeting people who are likely to convert and it is an important and effective practice for optimizing conversion rates. A conversion rate registers the percentage of users that have performed a desired action. It's calculated by taking the total number of users who 'convert' (for example, by placing an order), dividing it by the overall size of the audience, and converting it to a percentage. For example, let's say that our company records 3000 requests for estimates this month. Out of those 3000 estimates, a total of 400 estimates are paid (which is the conversion event). Therefore, the conversion rate of the company for this month can be calculated as follows: 400/3000 = 0.133 or 13.3%. The main objective of any commercial enterprise is to maximize this conversion rate.

A more thorough exploitation of the conversion rate is used. Indeed, companies look for the elements (product, service, etc.) that transform the most to highlight them and improve those that do not transform effectively. If we want to use the conversion rate for lead scoring, we can say that the higher the conversion rates of the order details and customer characteristics, the higher the customer score. For instance, a customer who requests a product with a 30% transformation rate will have a higher score than one who requests a product with a 5% transformation rate.

If we switch from the business language to a formal one, we could present a customer conversion rate-based score as follows:

$$Score_i = \sum_{j=1}^{n} w_j * x_{i,j} \tag{1}$$

where:

- $Score_i$ the score of the ith client;
- w_j the weight of the jth characteristic;
- $x_{i,j}$ the conversion rate of the jth characteristic of the ith client;
- n number of characteristics of a client taken into account for scoring.

In general, the set of features chosen reflects experts' understanding of the situation and preferences of managers. Characteristics can be very varied depending on the context, e.g:

- Features about customer profile (number of previous purchases, professional or individual, etc.);
- Features about customer's behaviour (number of pages visited, time spent to complete the estimate form, etc.);
- Features about the order (the type of the requested product, the quantity, etc.)

For weights, it will be more efficient to calculate them automatically. The weight of a feature will be its importance in a classification model. Feature Importance refers to techniques that calculate a score for all the input features for a given model. The feature scores simply represent their "importance". A higher score means that the specific feature will have a larger effect on the model that is being used to predict a certain variable.

To the end, all clients are sorted by $Score_i$ from maximum to minimum. The highest rated client has the highest priority.

4 Experiments

In this section, we introduce our test dataset and discuss the experimental results.

4.1 Dataset

For our experiments, we use the dataset of the company. It contains a sample of estimate records (123 287 records). For the data cleaning step, the dataset is reduced to 10 attributes including estimate information (id, creation date, requested quantity, etc.) and customer information (id, number of previous orders, acquisition source, etc.). Then, for the prioritization model, data on user behaviour on the site is added (time spent on the site, pages per visit, etc.).

In the initial dataset exploration, redundant information was found. The disturbing cause is the creation of various variants or versions of an estimate to place a single order since customers and advisors can make more than one to correct some details or make another proposal. This practice results in an imbalance in the database in terms of paid and unpaid estimates, as only one of the estimates created, will be paid and the others will not, even if they are for the same project. For this reason, the objective of the data cleaning is to pool estimates of the same order and derive a representative one.

4.2 Results and Discussion

For the data cleaning, we followed the steps mentioned in Sect. 2. After detecting and correcting the empty values and outliers, we resolved the duplication problem as follows:

Prepare the Data. We have changed client names to their unique identifiers, we have also put customer addresses into a standard format and substituted the required product types with their unique technical references.

Search Area Definition. We have been inspired by the two indexing methods cited previously in Sect. 2.2. First, tuples are grouped together in the same block. The blocking-key used is the customer's unique identifier. This step reduces significantly the number of comparisons to be made. Actually, it ensures that the same person's estimates will be compared to each other as there is no need to compare the estimates of two different clients. Then, records from each block are sorted based on their creation date. This step will also reduce the number of comparisons needed because estimates related to the same order are normally created one by one.

Matching and Clustering Duplicates. Using the transitivity principle to group duplicates is not effective in our case. In fact, we can find two estimates that are detected as similar but do not belong to the same project or order in reality. A client can make two separate similar orders which, if we limit ourselves to transitivity, will be detected as a single project while they are two. In this way, the conversion rate used to analyze the data will be biased. To this end, we have added additional rules when forming clusters. These rules are created by domain experts. In addition to the result of the decision model to determine whether two records are duplicated or not, external rules are used to decide whether they belong to the same group or not.

Through our data cleaning method, the Imbalanced Ratio [16] increases from **21,38** to **8,93** and the number of samples in the majority class (unpaid estimates) is decreased by about 42%.

Figure 4 shows the duplicate detection result using the transitivity principle (1) and our modified method (2) for the formation of clusters.

We consider E1, E2, E3, E4, and E5 are 5 estimates from the same customer and with the same characteristics. The domain knowledge-based decision model used to determine whether the tuples compared are duplicates or not provides the following results: E1 is similar to E2, E2 is similar to E3, E3 is similar to E4 and E4 is similar to E5. The next step is the clustering of duplicates. The result of the transitivity-based model is shown in (1) in Fig. 4. All estimates are put in the same group as by transitivity they are all similar. This clustering is not effective in this example case. Actually, E3 is a paid estimate and it was paid before E4 and E5 were created. In this case, E1, E2, and E3 belong to the same cluster (order or project) and E4 and E5 belong to another one as shown in (2) in Fig. 4. The two main expert rules used in this duplication clustering are according to the state of the estimate when comparing (paid or unpaid) and the creation and payment dates of the estimates. Using this method, a more correct

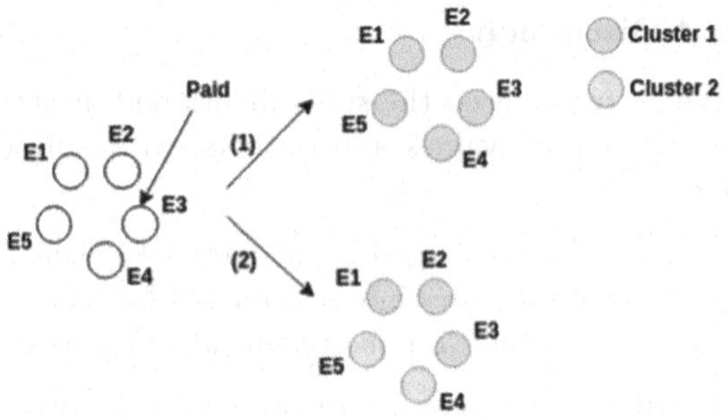

Fig. 4. Data duplication detection results.

conversion rate is obtained. The results of method (1) show that the customer has transformed 100% (1 cluster = 1 project and 1 paid estimate in a project = 1 paid project) whereas in reality, it is 50% (1 project paid out of 2).

We apply 10-fold cross-validation to provide a robust estimate of the performance of our model which is trained using Random Forest, Decision Tree, and Logistic Regression.

Table 1. Models results before and after data cleaning.

Methods	Before cleaning				After cleaning			
	Accuracy	Precision	Recall	AUC	Accuracy	Precision	Recall	AUC
Random Forest	0,92	0,05	0,02	0,50	0,90	**0,57**	**0,25**	**0,62**
Decision Tree	0,93	0,07	0,05	0,51	0,87	**0,36**	**0,32**	**0,63**
Logistic Regression	0,95	0,26	0,00	0,50	0,91	**0,76**	**0,21**	**0,60**

Apart from the accuracy, all the other metrics are higher for the cleaned dataset. The models trained before the data cleaning process fails to identify the paid estimates compared to the models after the data cleaning Table 1. Furthermore, our method allows more effective data analysis and more real profitability value because it removes the noise from the dataset. Due to the duplication detection and clustering, we have obtained a new useful feature *number of estimates per project* (number of the samples in the cluster) for prioritization. Actually, the analysis of this new feature shows that the more a person requests estimates, the more likely he will place an order.

To evaluate our customer prioritization model and after the data cleaning and data preparation step, we started by calculating the weights of the used features. The Random Forest algorithm is chosen for this task due to its optimal results in feature selection found in the literature [20]. Figure 5 shows the importance of certain features of our database. According to the Eq. (1), the resulting weightings are used with the corresponding feature conversion rates to

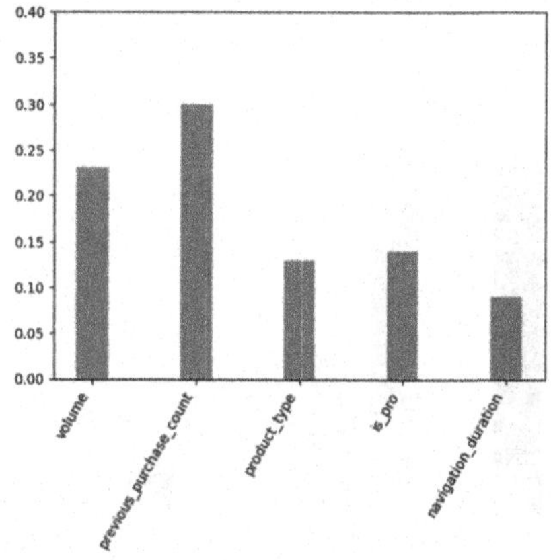

Fig. 5. Feature importance scores.

calculate the client's final score. The obtained scores are then normalized using the following formula:

$$normalizedScore_i = \frac{Score_i - Score_{min}}{Score_{max} - Score_{min}} \quad (2)$$

where:

- $Score_i$ the score of the ith client;
- $Score_{min}$ the lowest score;
- $Score_{max}$ the highest score;

In order to evaluate our model, we create intervals of 0.1 between 0 and 1 where if $normalizedScore_i \in [0, 0.1[$ the ith client takes the priority 10 and if $normalizedScore_i \in]0.9, 1]$ the ith client takes the priority 1. A good prioritization model will contain the maximum number of estimates paid in the top priorities. it will be able to detect customers who are more likely to convert and assign them a high priority.

We tested the algorithm on a database that contains 1000 estimates from different clients which 300 are paid. We compare our method to the Random Forest (RF) and Logistic Regression (LR) algorithms used in the literature for lead scoring. As shown in Fig. 6 our algorithm classifies over 70% of paid estimates compared to RF algorithm which classifies about 36% and to the LR algorithm which classifies about 37% Fig. 7 in the top 3 priorities. Furthermore, our algorithm classifies very few paid estimates in the latest priorities, unlike LR and RF. The classification of estimates that are likely to be paid in low priorities results in lost opportunities. As a result, our customer prioritization algorithm provides a better understanding of which estimates are more likely to be paid. The complete model is implemented in the back office of the company and showed good results on this year's conversion rate.

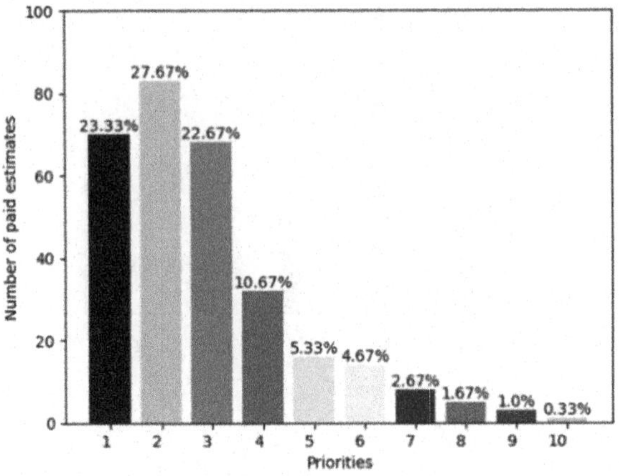

Fig. 6. Number of estimates paid by priority using our method.

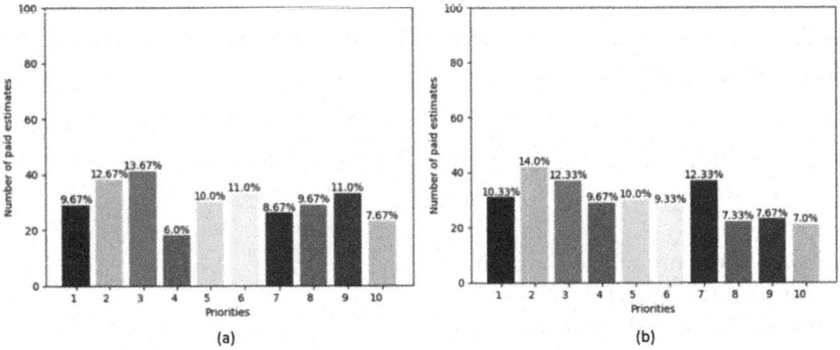

Fig. 7. Number of estimates paid by priority using (a) RF and (b) LR

5 Conclusion

Prioritizing customers is a hot topic for businesses looking for new opportunities. In this paper, we have presented a complete process for customer prioritization using a real-world dataset. For the data cleaning process, we used a rules-based method for detecting and eliminating duplicates. It has contributed to reducing the imbalance in the dataset and improving the quality of the data for analysis. The algorithm is also used for the automatic matching of new recorded data. In the second section, a new lead scoring method has been introduced. It uses the conversion rate and the importance of the features to calculate the customer's scores. The calculated score helps marketing and sales teams focus these efforts on priority customers to achieve high profits. Our solution is currently being used by the company and work is underway to improve it.

References

1. Homburg, C., Droll, M., Totzek, D.: Customer prioritization: does it pay off, and how should it be implemented? J. Mark. **72**, 110–130 (2008)
2. Gerstner, E., Libai, B.: Invited commentary-why does poor service prevail? Mark. Sci. **25**, 601–603 (2006)
3. Rust, R., Lemon, K., Zeithaml, V.: Return on marketing: using customer equity to focus marketing strategy. J. Mark. **68**, 109–127 (2004)
4. Chu, X., Ilyas, I., Krishnan, S., Wang, J.: Data cleaning: overview and emerging challenges. In: Proceedings of the 2016 International Conference on Management of Data, pp. 2201–2206 (2016)
5. Khayyat, Z., et al.: BigDansing: a system for big data cleansing. In: Proceedings of the 2015 ACM SIGMOD International Conference on Management of Data, pp. 1215–1230 (2015)
6. Hu, K., Li, L., Hu, C., Xie, J., Lu, Z.: A dynamic path data cleaning algorithm based on constraints for RFID data cleaning. In: 2014 11th International Conference on Fuzzy Systems and Knowledge Discovery (FSKD), pp. 537–541 (2014)
7. Ge, C., Gao, Y., Miao, X., Yao, B., Wang, H.: A hybrid data cleaning framework using Markov logic networks. IEEE Trans. Knowl. Data Eng. **34**, 2048–2062 (2020)
8. Alipour-Langouri, M., Zheng, Z., Chiang, F., Golab, L., Szlichta, J.: Contextual data cleaning. In: 2018 IEEE 34th International Conference on Data Engineering Workshops (ICDEW), pp. 21–24 (2018)
9. Christen, P.: A survey of indexing techniques for scalable record linkage and deduplication. IEEE Trans. Knowl. Data Eng. **24**, 1537–1555 (2011)
10. Ramadan, B., et al.: Indexing techniques for real-time entity resolution. The Australian National University (2016)
11. McAfee, A., Brynjolfsson, E., Davenport, T., Patil, D., Barton, D.: Big data: the management revolution. Harv. Bus. Rev. **90**, 60–68 (2012)
12. Artun, O., Levin, D.: Predictive Marketing: Easy Ways Every Marketer Can Use Customer Analytics and Big Data. Wiley, Hoboken (2015)
13. Naumann, F., Herschel, M.: An introduction to duplicate detection. Synthesis Lect. Data Manage. **2**, 1–87 (2010)
14. Béton Direct. https://www.betondirect.fr/
15. Elmagarmid, A., Ipeirotis, P., Verykios, V.: Duplicate record detection: a survey. IEEE Trans. Knowl. Data Eng. **19**, 1–16 (2006)
16. Fern'andez, A., Garc'ia, S., Galar, M., Prati, R., Krawczyk, B., Herrera, F.: Learning from Imbalanced Data Sets. Springer, Heidelberg (2018). https://doi.org/10.1007/978-3-319-98074-4
17. Prasad, K., Anjaneyulu, G.: A comparative analysis of support vector machines & logistic regression for propensity based response modeling. Int. J. Bus. Anal. Intell. **3**, 7 (2015)
18. Nygård, R., Mezei, J.: Automating lead scoring with machine learning: an experimental study (2020)
19. Zhang, Y.: Prediction of customer propensity based on machine learning. In: 2021 Asia-Pacific Conference on Communications Technology and Computer Science (ACCTCS), pp. 5–9 (2021)
20. Speiser, J., Miller, M., Tooze, J., Ip, E.: A comparison of random forest variable selection methods for classification prediction modeling. Expert Syst. Appl. **134**, 93–101 (2019)

References

1. Homburg, C., Droll, M., Totzek, D.: Customer prioritization: does it pay off, and how should it be implemented? J. Mark. 72, 110–130 (2008)
2. [illegible] J. Mark. [illegible] (2000)
3. [illegible] J. Mark. 68, [illegible] (2004)
4. [illegible] In: [illegible] 2010 International [illegible] Conference on [illegible] pp. [illegible] (2010)
5. [illegible] In: Proceedings of the [illegible] International Conference on Management of Data [illegible] (2014)
6. [illegible] In: 2014 [illegible] International [illegible] pp. [illegible]
7. [illegible]
8. [illegible] Conference [illegible] International Conference on Data Engineering Workshops [illegible]
9. [illegible]
10. [illegible]
11. [illegible] (2012)
12. [illegible]
13. [illegible] (2010)
14. [illegible] https://www.[illegible]
15. [illegible] IEEE Trans. [illegible] (2007)
16. Fernandez, A., [illegible] Springer, Heidelberg (2018). [illegible]
17. [illegible] modeling. [illegible] (2019)
18. [illegible] machine learning [illegible] (2020)
19. [illegible] machine learning. In: [illegible] International Conference on Communications, Technology and [illegible] (2017)
20. [illegible] random forest [illegible] prediction modeling. [illegible] Appl. [illegible] (2019)

Author Index

B. Dorronsoro et al. (Eds.): OLA 2023, CCIS 1824, pp. 427–428, 2023.
https://doi.org/10.1007/978-3-031-34020-8

Printed in the United States
by Baker & Taylor Publisher Services